Advanced Methodologies and Technologies in Engineering and Environmental Science

Mehdi Khosrow-Pour, D.B.A.
Information Resources Management Association, USA

A volume in the Advances in Environmental Engineering and Green Technologies (AEEGT) Book Series

IGI Global
DISSEMINATOR OF KNOWLEDGE

Published in the United States of America by
IGI Global
Engineering Science Reference (an imprint of IGI Global)
701 E. Chocolate Avenue
Hershey PA, USA 17033
Tel: 717-533-8845
Fax: 717-533-8661
E-mail: cust@igi-global.com
Web site: http://www.igi-global.com

Library of Congress Cataloging-in-Publication Data

Names: Khosrow-Pour, Mehdi, 1951- editor.
Title: Advanced methodologies and technologies in engineering and environmental science / Mehdi Khosrow-Pour, D.B.A., editor.
Description: Hershey, PA : Engineering Science Reference, [2018] | Includes bibliographical references.
Identifiers: LCCN 2018024577| ISBN 9781522573593 (h/c) | ISBN 9781522573609 (eISBN)
Subjects: LCSH: Engineering. | Environmental sciences.
Classification: LCC TA7 .A26 2018 | DDC 620--dc23 LC record available at https://lccn.loc.gov/2018024577

This book is published in the IGI Global book series Advances in Environmental Engineering and Green Technologies (AEEGT) (ISSN: 2326-9162; eISSN: 2326-9170)

British Cataloguing in Publication Data
A Cataloguing in Publication record for this book is available from the British Library.

Advances in Environmental Engineering and Green Technologies (AEEGT) Book Series

Sang-Bing Tsai
University of Electronic Science and Technology of China Zhongshan Institute, China
Ming-Lang Tseng
Lunghwa University of Science and Technology, Taiwan
Yuchi Wang
University of Electronic Science and Technology of China Zhongshan Institute, China

ISSN:2326-9162
EISSN:2326-9170

Mission

Growing awareness and an increased focus on environmental issues such as climate change, energy use, and loss of non-renewable resources have brought about a greater need for research that provides potential solutions to these problems. Research in environmental science and engineering continues to play a vital role in uncovering new opportunities for a "green" future.

The **Advances in Environmental Engineering and Green Technologies (AEEGT)** book series is a mouthpiece for research in all aspects of environmental science, earth science, and green initiatives. This series supports the ongoing research in this field through publishing books that discuss topics within environmental engineering or that deal with the interdisciplinary field of green technologies.

Coverage

- Alternative Power Sources
- Water Supply and Treatment
- Biofilters and Biofiltration
- Contaminated Site Remediation
- Pollution Management
- Electric Vehicles
- Green Transportation
- Renewable energy
- Policies Involving Green Technologies and Environmental Engineering
- Sustainable Communities

IGI Global is currently accepting manuscripts for publication within this series. To submit a proposal for a volume in this series, please contact our Acquisition Editors at Acquisitions@igi-global.com or visit: http://www.igi-global.com/publish/.

The Advances in Environmental Engineering and Green Technologies (AEEGT) Book Series (ISSN 2326-9162) is published by IGI Global, 701 E. Chocolate Avenue, Hershey, PA 17033-1240, USA, www.igi-global.com. This series is composed of titles available for purchase individually; each title is edited to be contextually exclusive from any other title within the series. For pricing and ordering information please visit http://www.igi-global.com/book-series/advances-environmental-engineering-green-technologies/73679. Postmaster: Send all address changes to above address.

Titles in this Series

For a list of additional titles in this series, please visit: www.igi-global.com/book-series

Biostimulation Remediation Technologies for Groundwater Contaminants
Ashok K. Rathoure (Independent Researcher, India)
Engineering Science Reference • copyright 2018 • 355pp • H/C (ISBN: 9781522541622) • US $215.00 (our price)

Innovations and Trends in Environmental and Agricultural Informatics
Petraq Papajorgji (University of New York Tirana, Albania) and Francois Pinet (Irstea Centre de Clermont-Ferrand, France)
Engineering Science Reference • copyright 2018 • 317pp • H/C (ISBN: 9781522559788) • US $205.00 (our price)

Effective Solutions to Pollution Mitigation for Public Welfare
Ahmet Ozan Gezerman (Yildiz Technical University, Turkey) Burcu Didem Corbacioglu (Yildiz Technical University, Turkey) and Bhola R. Gurjar (Indian Institute of Technology Roorkee, India)
Engineering Science Reference • copyright 2018 • 274pp • H/C (ISBN: 9781522533795) • US $180.00 (our price)

Handbook of Research on Microbial Tools for Environmental Waste Management
Vinay Mohan Pathak (Gurukul Kangri University, India) and Navneet (Gurukul Kangri University, India)
Engineering Science Reference • copyright 2018 • 496pp • H/C (ISBN: 9781522535409) • US $285.00 (our price)

Food Systems Sustainability and Environmental Policies in Modern Economies
Abiodun Elijah Obayelu (Federal University of Agriculture – Abeokuta (FUNAAB), Nigeria)
Engineering Science Reference • copyright 2018 • 371pp • H/C (ISBN: 9781522536314) • US $195.00 (our price)

Microbial Biotechnology in Environmental Monitoring and Cleanup
Pankaj (G. B. Pant University of Agriculture and Technology, India) and Anita Sharma (G. B. Pant University of Agriculture and Technology, India)
Engineering Science Reference • copyright 2018 • 427pp • H/C (ISBN: 9781522531265) • US $235.00 (our price)

Handbook of Research on Renewable Energy and Electric Resources for Sustainable Rural Development
Valeriy Kharchenko (Federal Scientific Agroengineering Center VIM, Russia) and Pandian Vasant (Universiti Teknologi PETRONAS, Malaysia)
Engineering Science Reference • copyright 2018 • 672pp • H/C (ISBN: 9781522538677) • US $345.00 (our price)

Green Production Strategies for Sustainability
Sang-Binge Tsai (University of Electronic Science and Technology of China (Zhongshan Institute), China & Civil Aviation University of China, China) Bin Liu (Shanghai Maritime University, China) and Yongian Li (Nankai University, China)
Engineering Science Reference • copyright 2018 • 325pp • H/C (ISBN: 9781522535379) • US $235.00 (our price)

IGI Global
DISSEMINATOR OF KNOWLEDGE

701 East Chocolate Avenue, Hershey, PA 17033, USA
Tel: 717-533-8845 x100 • Fax: 717-533-8661
E-Mail: cust@igi-global.com • www.igi-global.com

List of Contributors

Table of Contents

Section 3
Computer Engineering

Section 4
Electrical Engineering

Section 5
Environmental Science and Agriculture

Section 8
Optical Engineering

Preface

Advanced Methodologies and Technologies in Engineering and Environmental Science is a single-volume major reference work comprised of 29 unique chapters focusing on some of the most advanced and innovative technologies and methodologies within various branches of engineering and environmental science. The concepts presented within these chapters are sure to support the successful strategic management of operations and new research initiatives.

Aiming to serve the needs of libraries, corporations, research organizations, independent researchers, practitioners, scientists, policymakers, and instructors who are seeking advanced cross-disciplinary knowledge on the conceptual and technical aspects, this publication offers a wide breadth of coverage, with topics ranging from the engineering of prosthetics to green IT.

Attention is paid to emerging subject matter including but not limited to environmental sustainability, building information modeling, 3D printing, electrical conductivity, and spatial analysis. The chapters within this publication will provide readers with the tools necessary for further research and discovery in their own respective industries and/or fields. Additionally, this publication could be extremely beneficial for use in coursework by instructors of various engineering and environmental science programs.

Advanced Methodologies and Technologies in Engineering and Environmental Science is organized into eight sections: Biomedical Engineering, Civil Engineering, Computer Engineering, Electrical Engineering, Environmental Science and Agriculture, Geographic Information Systems, Industrial Engineering and Informatics, and Optical Engineering.

The following paragraphs provide a summary of what to expect from this invaluable reference source:

Section 1, "Biomedical Engineering," is comprised of three chapters and opens this extensive reference source by highlighting the latest technological advancements in the creation of biomedical tools and materials. The first chapter in this section, "General Perspectives on Electromyography Signal Features and Classifiers Used for Control of Human Arm Prosthetics," authored by Mr. Faruk Ortes, from Istanbul University, Turkey, Ms. Derya Karabulut, from Halic University, Turkey, and Prof. Yunus Ziya Arslan, from Istanbul University, Turkey, focuses on the use of electromyography signal features and classifiers to revolutionize the design of human arm prosthetics with the hope that there can be an improvement to the locomotor tasks performed by the human body during daily activity.

The second chapter in this section, "The Principle and Process of Digital Fabrication of Biomedical Objects," authored by Profs. S. H. Choi, H. H. Cheung, and W. K. Zhu, all from The University of Hong Kong, Hong Kong, discusses the fabrication of biomedical objects of multiple using MMLM technology. It also addresses the limitations of MMLM and the development of multi-material virtual prototyping and manufacturing (MMVPM) technology for digital fabrication of complex biomedical objects in a convenient and cost-effective manner.

The third and final chapter in this section, "Reverse Engineering in Rehabilitation," authored by Prof. Emilia Mikołajewska, from Nicolaus Copernicus University, Poland, Prof. Marek Macko, from Kazimierz Weilki University, Poland, Mr. Zbigniew Szczepański, from Kazimierz Wielki University, Poland, and Prof. Dariusz Mikołajewski, from Kazimierz Wielki University, Poland, investigates the extent to which the available opportunities of the application of reverse engineering in rehabilitation are being exploited, including concepts, studies, and observations. In sum, through all these perspectives on reverse engineering, digital fabrication, and artificial neural networks, this section demonstrates general perspectives on electromyography signal features and classifiers used for control of human arm prosthetics and other biomedical innovations.

Section 2, "Civil Engineering," is comprised of two chapters and explores emerging innovations on the use of digital technologies in architectural design and engineering. The first chapter in this section, "Digital Animation for Representing Architectural Design," authored by Prof. Roberta Spallone, from Politecnico di Torino, Italy, highlights how the ability to access the fourth dimension through the construction of a sequence of images constitutes a specific prerogative of digital representation that goes beyond the static constraint imposed by conventional methods of representation.

The second and final chapter in this section, "Literature Review of Augmented Reality Application in the Architecture, Engineering, and Construction Industry With Relation to Building Information," authored by Prof. Aydin Tabrizi and Prof. Paola Sanguinetti, from University of Kansas, USA, explores the current trends in the development of augmented reality applications and the application of augmented reality technologies in the architecture, engineering, and construction industries.

Section 3, "Computer Engineering," is comprised of two chapters. The first chapter in this section, "Architecture of an Open-Source Real-Time Distributed Cyber Physical System," authored by Prof. Stefano Scanzio, from CNR-IEIIT, Italy, presents a completely open-source architecture of a cyber-physical system based on the Linux operating system. The proposed architecture exploits the RTAI or XENOMAI hard real-time schedulers to guarantee the required degree of determinism of nodes.

The second and final chapter of this section, "Consistency Is Not Enough in Byzantine Fault Tolerance," authored by Prof. Wenbing Zhao, from Cleveland State University, USA, presents a novel integrity-preserving replica coordination algorithm for Byzantine fault tolerant systems. The central idea behind this CD-BFT algorithm is that all random numbers to be used by the replicas are collectively determined based on the contributions made by a quorum of replicas, at least f+1 of which are not faulty.

Section 4, "Electrical Engineering," is comprised of one chapter titled "Mechanisms of Electrical Conductivity in Carbon Nanotubes and Graphene" and authored by Prof. Rafael Vargas-Bernal, from Instituto Tecnológico Superior de Irapuato, Mexico. It presents coverage on the mechanisms of electrical conductivity in carbon nanotubes and graphene. Also highlighted are other electrical variables and their role in the electrical conductivity of these materials.

Section 5, "Environmental Science and Agriculture," is comprised of 10 chapters on emerging technologies and applications for maintaining environmental functions and safeguarding the planet. The first chapter in this section, "Carbon Capture From Natural Gas via Polymeric Membranes," authored by Profs. Nayef Mohamed Ghasem, Nihmiya Abdul Rahim, and Mohamed Al-Marzouqi, all from UAE University, UAE, illustrates the potential for an energy-efficient and effective separation of CO2/CH4 gas mixture via lean solvent and regenerating of the rich solvent through absorption/stripping mechanism taking place in a hollow fiber GLMC process.

The second chapter in the section, "Enhancing the Resiliency of Smart Grid Monitoring and Control," authored by Prof. Wenbing Zhao, from Cleveland State University, USA, presents the justification and a feasibility study of applying the Byzantine fault tolerance (BFT) technology to electric power grid health monitoring.

The third chapter in the section, "E-Waste, Chemical Toxicity, and Legislation in India," authored by Prof. Prashant Mehta, from National Law University Jodhpur, India, provides a comprehensive overview of India's current e-waste scenario, analyzes hazardous metals and considers environmental and health risks posed by them, examines the existing legal framework and strategic interventions, and explores the immediate technical solutions to manage and minimize its impact on all.

The fourth chapter in the section, "Green IT and the Struggle for a Widespread Adoption," authored by Prof. Edward T. Chen, from University of Massachusetts – Lowell, USA, describes the struggle of adoption and provides basic concepts and sustainable solutions of Green IT for businesses and individuals.

The fifth chapter in the section, "Identification of Green Procurement Drivers and Their Interrelationship Using Fuzzy TISM and MICMAC Analysis," authored by Dr. Surajit Bag, from Tega Industries South Africa Pty Ltd., South Africa, identifies the leading green procurement drivers and the interrelationships using fuzzy total interpretive structural modeling and MICMAC approach.

The sixth chapter in the section, "Load Flow Analysis in Smart Grids," authored by Prof. Osman Hasan, National University of Sciences and Technology, Pakistan, Prof. Awais Mahmood, from National University of Sciences and Technology, Pakistan, and Prof. Syed Rafay Hasan, from Tennessee Technological University, USA, describes smart grids in terms of their basic components and then categorizes the factors that affect the loads in smart grids. This is followed by a comprehensive survey of various existing load flow analysis techniques (i.e., numerical, computational intelligence, and probabilistic).

The seventh chapter in the section, "Methodology of Climate Change Impact Assessment on Forests," authored by Prof. Mostafa Jafari, from the Regional Institute of Forest and Rangelands (RIFR), Iran, explores climate change, climate variability, and global warming and its effects on natural resources, plants, animals, and human life.

The eighth chapter in the section, "Model for Assessment of Environmental Responsibility in Healthcare Organizations," authored by Prof. María Carmen Carnero, from University of Castilla-La Mancha, Spain and University of Lisbon, Portugal, sets out a multicriteria assessment system constructed by extension to a fuzzy environment of the Technique for Order Preference by Similarity to Ideal Situation (TOPSIS) to assess the environmental responsibility of a healthcare organization. This model allows joint evaluation of a significant number of decision criteria.

The ninth chapter in the section, "Potential Benefits and Current Limits in the Development of Demand Response," authored by Prof. Clementina Bruno, from University of Piemonte Orientale, Italy, explores the potential benefits of demand response (DR) and illustrates a set of challenges to related demand response development, while also discussing the regulatory, technical, and socio-economic challenges related to DR.

The tenth and final chapter in the section, "Waste Gas End-of-Pipe Treatment Techniques in Italian IPPC Chemical Plants," authored by Profs. Gaetano Battistella, Giuseppe Di Marco, Carlo Carlucci, Raffaella Manuzzi, Federica Bonaiuti, and Celine Ndong, from ISPRA, Italy, includes the latest results of a screening of Italian IPPC plants. It highlights the operating conditions of abatement devices and possible existing improvements for several compounds removal. The abatement techniques analyzed operate mainly on VOC content reduction or on inorganic compounds abatement.

Section 6, "Geographic Information Systems," is comprised of eight chapters on the application of geographic information systems and remote sensing across various disciplines. The first chapter in the section, "Application of Geospatial Mashups in Web GIS for Tourism Development," authored by Prof. Somnath Chaudhuri, from Maldives National University, Maldives and Prof. Nilanjan Ray, from Adamas University, India, examines current development in Web GIS with the implementation of geospatial mashup technologies, such as Google Map, in the context of map mashups and presents a classification of map mashups and their application in tourism management and promotion.

The second chapter in the section, "Archaeological GIS for Land Use in South Etruria Urban Revolution in IX-VIII Centuries B.C.," authored by Prof. Giuliano Pelfer, from University of Florence, Italy, focuses on the use of GIS for archaeological predictive modeling of ancient land use by critically applying this new technology and exploring its theoretical and analytical implications.

The third chapter in this section, "Exploring Tourism Cluster in the Peripheral Mountain Area Based on GIS Mapping," authored by Prof. Ya-Hui Hsueh, from National Taichung University of Education, Taiwan, Prof. Huey-Wen Chuang, from National Taichung University of Education, Taiwan, and Prof. Wan-Chiang Hsieh, from National Taichung Girls' Senior High School, Taiwan, locates a set of points of tourist spots distributed on a peripheral mountain area by GIS mapping, illustrates that accessibility and neighbor-to-community are the influencing factors of tourism cluster in a peripheral mountain area, and further analyzes the benefits of tourism cluster for establishing cooperation network by tourism associations.

The fourth chapter in this section, "Geographic Information System (GIS) Modeling Analysis and the Effects of Spatial Distribution and Environmental Factors on Breast Cancer Incidence," authored by Akram Gasmelseed, from University of Science and Technology, Sudan and Prof. Ali H. Alharbi, from Qassim University, Saudi Arabia, contains an analysis of the breast cancer distribution in the United States by comparing the spatial distribution of breast cancer cases against physical environmental factors using a geographic information system (GIS).

The fifth chapter in this section, "Geographic Information Systems," authored by Profs. Paula Remoaldo, Vitor P. Ribeiro, Hélder Silva Lopes, and Sara Catarina Gomes Silva, from University of Minho, Portugal, underwrites the concept of geographic information systems, distinguishes the diversity of applications from the past until the present, and identifies new paths to accommodate recent scientific approaches with an extensive range of application possibilities.

The sixth chapter in this section, "Geospatial Influence in Science Mapping," authored by Prof. Carlos Granell-Canut, from Universitat Jaume I of Castellón, Spain and Prof. Estefanía Aguilar-Moreno, from Universitat Jaume I of Castellón, Spain, aims at drawing attention to the possibilities that geospatial technologies can bring to science mapping. It distinguishes the notion of mapping between the geospatial information science (GIScience) and librarianship and information science (LIS).

The seventh chapter in the section, "Parallel Development of Three Major Space Technology Systems and Human Side of Information Reference Services as an Essential Complementary Method," authored by Prof. Joyce Gosata Maphanyane, from University of Botswana, Botswana, elaborates upon GIS/remote sensing. It compares and contrasts the four globally used GIS systems: GOES, the Geosynchronous Orbiting Environmental Satellite; LANDSAT; SPOT, Satellite Pour l'Observation de la Terre; and WorldView.

The eighth and final chapter of the section, "Use of GIS and Remote Sensing for Landslide Susceptibility Mapping," authored by Prof. Arzu Erener, from Kocaeli University, Turkey, Prof. Gulcan Sarp, from Suleyman Demirel University, Turkey, and Prof. Sebnem Duzgun, from Middle East Technical University, Turkey, focuses on the use of geographic information systems and remote sensing data for landslide susceptibility mapping. Five factors including normalized difference vegetation index (NDVI) and topographic wetness index (TWI), slope, lineament density, and distance to roads were used for the grid-based approach for landslide susceptibility mappings.

Section 7, "Industrial Engineering and Informatics," is comprised of two chapters. The first chapter in this section, "Cuckoo Search Algorithm for Solving Real Industrial Multi-Objective Scheduling Problems," authored by Prof. Mariappan Kadarkarainadar Marichelvam, from Mepco Schlenk Enginering College, India and Prof. Mariappan Geetha, from Kamaraj College of Engineering and Technology, India, explores the meta-heuristic algorithms inspired by the social behavior of cuckoos. The cuckoo search algorithm is presented to solve the hybrid flow shop scheduling problems. To illustrate the proposed algorithm, data sets from a steel furniture manufacturing company are used.

The second and final chapter in this section, "The Trends and Challenges of 3D Printing," authored by Profs. Edna Ho Chu Fang and Sameer Kumar, from University of Malaya, Malaysia, examines the trends and challenges of 3D printing in various industries including footwear, jewelry, architecture, engineering and construction, aerospace, dental and medical industries, education, consumer products, automotive, and industrial design.

Section 8, "Optical Engineering," contains one chapter titled "Visible Light Communication Numerous Applications," authored by Prof. Ala' Fathi Khalifeh, from German Jordan University, Jordan and Profs. Hasan Farahneh, Christopher Mekhiel, and Xavier Fernando, from Ryerson University, Canada. It presents coverage on emerging innovations in using the visible light spectrum for data communication. This inclusive information assists in understanding the potential for the adoption of this technology in indoor and outdoor applications.

The comprehensive coverage this publication offers is sure to contribute to an enhanced understanding of all topics, research, and discoveries pertaining to engineering and environmental science. Furthermore, the contributions included in this publication will be instrumental to the expansion of knowledge offerings in this area. This publication will inspire its readers to further contribute to recent discoveries and will surely progress future innovations.

Mehdi Khosrow-Pour, D.B.A.
Information Resources Management Association, USA

Section 1
Biomedical Engineering

Chapter 1

General Perspectives on Electromyography Signal Features and Classifiers Used for Control of Human Arm Prosthetics

Faruk Ortes
Istanbul University, Turkey

Derya Karabulut
Halic University, Turkey

Yunus Ziya Arslan
Istanbul University, Turkey

ABSTRACT

Physically handicapped people encounter various kinds of obstacles and difficulties in their daily lives due to the restricted ability of motion. Assistive technologies represent a crucial challenge of scientific studies to overcome such an issue of reducing quality of life. Assistive devices such as wheelchairs, orthoses, and prostheses are designed and built to contribute rehabilitation progress and to regain lost functions. Although human body parts have intricate forms and functions, artificial devices and components integrating to the body are anticipated to compensate the fundamental functions related to user's demands. Upper- or lower-arm amputations also result in severe cosmetic matters. However, what is more important and obtrusive is the loss of primary functions including manipulating and grasping the objects besides the locomotor tasks which are performed by the human body during daily activity.

DOI: 10.4018/978-1-5225-7359-3.ch001

BACKGROUND

Development of human arm prosthetics, which are improved to regain lost functions of amputated limbs, encounters critical and challenging problems to carry out various dexterous tasks. To date, many of revolutionizing design of human arm prosthetics including Boston Arm (Mann & Reimers, 1970), Deka Arm (Resnik, 2010), Otto Bock trans carpal hand (Otto Bock Health Care, Minneapolis, MN), and Shanghai Kesheng Hands (Shanghai Kesheng Prosthese Corporation Ltd.) have been developed. Intuitive and precise control of such prostheses is still one of the main interests of scientific studies. The main deduction from researches could be stated as control of the prosthetics is a particular concern of understanding the nature of the electrical activations of muscles. Imitation of the fundamental patterns of human arm motion depends highly upon the transformation of the neuromuscular activities of residual limbs to a specific control signal for controlling the artificial arm. In this respect, myoelectric signals provide a base of intuitive control, unlike the conventional or direct control. The dexterous control of such myoelectric-based prostheses requires a clear extraction of features from recorded surface electromyography (SEMG) signals and pattern recognition to discriminate the motion and force intentions of the prosthetics users. The progress of feature extraction from SEMG signals has an extensive coverage of myoelectric controlled prostheses studies due to the features in both time and frequency domains have the great potential on representing clear and meaningful information of EMG signals. Additionally, the feature classifiers have been given a special scientific interest by researchers. Selection and developing of the case-specific classifiers, which are desired to have the optimal performance to specify motion classes, still continue to be the main goal of current studies. Although, various types of classifiers such as linear discriminant analysis (LDA), support vector machine (SVM), artificial neural networks (ANN) and fuzzy logic (FL) techniques have been utilized to classify human arm motion patterns, merits, shortcomings and pitfalls of the classifiers are still required to be discussed extensively.

FUNDAMENTAL ASPECTS OF EMG

EMG is the electrical activity of skeletal muscles (Basmajian & Deluca, 1985). It represents the summation of the muscle action potentials which cause the contraction of muscle fibers. Recorded EMG data by means of electrodes are amplified and filtered to eliminate the motion artifacts, as well as the environment and device related noises. Rejection of ambient influences on natural muscle activation improves the accuracy and usability of EMG signals. One of the most widely usage of EMG signals is to control the myoelectric-based prosthetics which are used by amputated people. Control scheme for EMG-driven human arm prosthetics includes a sequential series of signal processing (Figure 1).

A condensed and clear control signal is needed to control the EMG-based prosthetics. In order to reduce calculation and to provide stability of signal, EMG data are scanned by sliding segmented windows (Figure 2). Because the raw (amplified+ pre-processed) EMG signal contains a huge burden of data, this signal is needed to be represented in a concise, but accurate ways. Widely used time domain features extracted from signals includes mean absolute value (MAV), root mean square (RMS), Willison amplitude (WAMP), waveform length (WL), variance of EMG (VAR), simple square integral (SSI), zero crossing (ZC) and integrated EMG (IEMG) (Phinyomark et al., 2013). In frequency domain, mean frequency, median frequency, peak frequency, mean power, total power, and spectral power features are commonly preferred (Phinyomark et al., 2013).

Figure 1. Control scheme of multifunctional human arm prosthetics

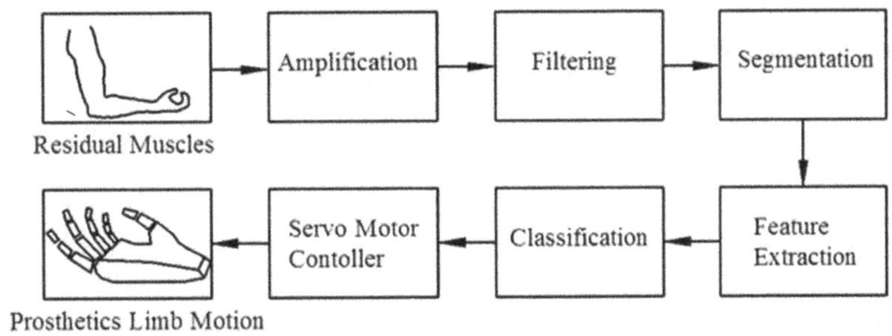

Figure 2. The basic representation of sliding windows

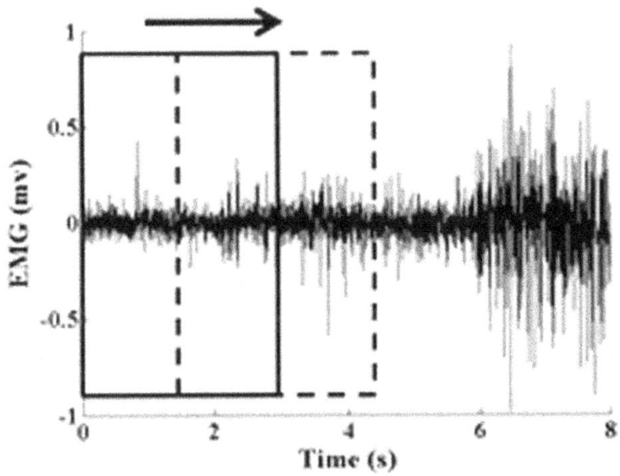

EMG SIGNAL FEATURES

Obtained EMG signals during contraction of a muscle or muscle groups are needed to be quantified in order to relate these signals with some certain sets of movement types (Zecca, Micera, Carrozza, & Dario, 2002). Mathematical expression of EMG signals could be defined using feature extraction approach. An EMG signal could be expressed in two domains including time and frequency domains.

Time Domain Features

Features expressed in time domain are useful for pattern recognition process due to no transformation process is required. Easy and fast calculation of features provides to reduce delay which is a critical concern in control of human arm prosthetics. A wide range of time domain features have been proposed by researchers for the purpose of movement or force classification process (Phinyomark, Phukpattaranont, & Limsakul, 2012). While x_k is the *k*th EMG sample and *N* is the number of samples in each segment, the most widely used time domain features are given as follows.

Mean Absolute Value

Mean absolute value (MAV) of an EMG signal is the average of absolute value of sequential signal amplitudes. MAV is one of the mostly used features and defined as,

$$MAV = \frac{1}{N}\sum_{k=1}^{N}\left|x_k\right| \tag{1.1}$$

Root Mean Square

Root mean square (RMS) feature represents a calculation of amplitude modulated Gaussian random process relating to constant force and non-fatiguing contraction. The mathematical expression of the RMS is given as,

$$RMS = \sqrt{\frac{1}{N}\sum_{k=1}^{N}x_k^2} \tag{1.2}$$

Willison Amplitude

Willison amplitude (WAMP) feature is the number of times the EMG signal amplitude exceeds a pre-defined threshold. WAMP is an indicator of motor unit action potentials (MUAP) and contraction force in muscles and can be expressed mathematically as,

$$WAMP = \sum_{k=1}^{N}\left[f\left(\left|x_k - x_{k+1}\right|\right)\right] \tag{1.3}$$

$$f(x) = \begin{cases} 1, & if \;\; x \geq threshold \\ 0, & otherwise \end{cases}$$

Waveform Length (WL)

Waveform length (WL) of EMG signal is the cumulative length of the waveform over the time segment. WL feature can be calculated as,

$$WL = \sum_{k=1}^{N-1}\left|x_{k+1} - x_k\right| \tag{1.4}$$

Variance of EMG (VAR)

Variance of EMG (VAR) implies the second-order moment of EMG signal and is a measure of power. VAR feature can be defined as follows,

$$VAR = \frac{1}{N-1}\sum_{k=1}^{N} x_k^2 \tag{1.5}$$

Simple Square Integral (SSI)

Simple square integral (SSI) of an EMG signal represents the summation of square values of EMG signal amplitude over time segment. SSI can be expressed as,

$$SSI = \sum_{k=1}^{N} x_k^2 \tag{1.6}$$

Zero Crossing (ZC)

Zero crossing (ZC) feature measures how many times the amplitude of EMG signal crosses zero level. Threshold value is assigned to prevent voltage fluctuations or noises effects. ZC feature calculation could be defined as,

$$ZC = \sum_{k=1}^{N-1}\left[\operatorname{sgn}(x_k \times x_{k+1}) \cap \left|x_k - x_{k+1}\right| \geq threshold\right] \tag{1.7}$$

$$\operatorname{sgn}(x) = \begin{cases} 1, & if \;\; x \geq threshold \\ 0, & otherwise \end{cases}$$

Integrated EMG (IEMG)

Integrated EMG refers to the summation of absolute values of EMG amplitude for each time segment. IEMG feature is also used for clinical applications and could be expressed as,

$$IEMG = \sum_{k=1}^{N}\left|x_k\right| \tag{1.8}$$

Frequency Domain Features

Investigation of EMG signal characteristics in frequency (or spectral) domain is mainly carried out to analyze both the fatigue phenomenon in muscles and the motor unit recruitment (Kallenberg, Schulte, Disselhorst-Klug, & Hermensa, 2007). Various types of features have been proposed to handle EMG signal behavior in the frequency domain. While f_j, P_j and, M represent a frequency value at a frequency bin j, the EMG power spectrum at a frequency bin j and the length of frequency bin, respectively, some of frequency domain features are given as follows.

Mean Frequency (MNF)

Mean frequency (MNF) is basically the calculation of average frequency dividing the sum of product of EMG power spectrum and frequency by total sum of the spectrum intensity. The mathematical expression is given as,

$$MNF = \frac{\sum_{j=1}^{M} f_j P_j}{\sum_{j=1}^{M} P_j} \quad (1.9)$$

Median Frequency (MDF)

Median frequency (MDF) of an EMG signal is the frequency at which the spectrum is partitioned into two equal amplitude. The MDF feature can be calculated as,

$$\sum_{j=1}^{MDF} P_j = \frac{1}{2} \sum_{j=1}^{M} P_j \quad (1.10)$$

Peak Frequency (PKF)

The frequency containing the maximum power is called peak frequency (PKF). The PKF can be calculated as follows,

$$PKF = \max(P_j) \quad (1.11)$$

Mean Power (MNP)

Average of power spectrum of EMG signals is used to determine the characteristics of signal. The feature could be expressed mathematically as,

$$MNP = \frac{1}{M}\sum_{j=1}^{M} P_j \tag{1.12}$$

Total Power (TTP)

The sum of the power spectrum of EMG signal reveals another feature, namely, the total power (TTP) and it is obtained as,

$$TTP = \sum_{j=1}^{M} P_j \tag{1.13}$$

Spectral Moments (SM)

Spectral moments (SM) is another important approach for feature extraction. Although higher order spectral moment could be calculated, the mathematical expressions of the first (SM_1) and the second (SM_2) order moments are given as follows, respectively.

$$SM_1 = \sum_{j=1}^{M} P_j f_j \tag{1.14}$$

$$SM_2 = \sum_{k=1}^{M} P_j f_j^2 \tag{1.15}$$

EMG signals can also be characterized in joint time-frequency domain (von Tscharner, 2000). In order to observe more accurate description of the signal in physical manner, EMG signals could be transformed to the area at which both frequency and time domain features exist. However, this transformation requires heavy computational costs and likely causes delay in controlling assistive devices. Main features of time-frequency domain are Wavelet Transform (WT), Wavelet Packet Transform (WPT) and Short-time Fourier Transform (STFT).

The above mentioned signal features are needed to be classified to specify the intended motion and force production. To achieve this task, various types of classifiers such as artificial neural networks (ANN) (Arslan, Adli, Akan & Baslo, 2010), fuzzy logic (FL) (Chan, Yang, Lam, Zhang, & Parker, 2000), support vector machines (SVM) (Oskoei & Hu, 2008) and linear discriminant analysis (LDA) (Lorrain, Jiang, & Farina, 2011) are widely employed in literature.

FEATURE CLASSIFICATION AND PATTERN RECOGNITION

Extracted time or frequency domain features are required to be classified to determine the motion or applied force patterns (Oskoei & Hu, 2007). Characterization capability of prosthetic hands is firmly related to the classification performance of the selected classifier due to the classification accuracy reflects the fundamental neuromuscular activity of human muscles. The main consideration of pattern recognition progress of myoelectric signal is that each force or motion class is described by the corresponding muscle activation which is represented by a set of extracted features (Farina et al., 2014). The chosen classifier discriminates separate tasks using trial and test approaches, so that a relation between muscle activation, features and real-world tasks could be built. Thus, selection of appropriate classifier for pattern recognition process is a key issue which is expected to identify accurate patterns and to perform fast sufficiently. A great amount of literature exists to propose an optimal performance of classifiers and thereby selecting the most suitable one (Lorrain et al., 2011). The next section involves fundamental structures and applications of the widely used classifiers including ANN, FL, SVM and LD for the purpose of using in control of myoelectric based prosthetics.

Artificial Neural Networks (ANN)

ANN is an artificial intelligence method inspired by the biological structure of human brain and generally referred to as "neural networks" (Haykin, 1999). In human brain, the neural networks is the central of the decision making process. The receptors receive stimuli from the external environment and convert to electrical impulses in order to transmit them to neural nets. Then, neural nets perceive the information and make decision. Finally, the decision is transmitted the effectors to convert the impulses to response as outputs. Setting a linear or nonlinear relation between inputs and outputs, biological and artificial neural networks makes a specific decision.

Surface EMG feature classification using ANN is a popular subject among scientific studies related to control of human arm prosthetics. The basic structure of ANN which is used in pattern recognition process of EMG signals is shown in Figure 3.

The structure, namely the multilayer perceptron (MLP), which consists of a set of one input layer, one output layer, and a number of hidden layers is one of the most simple and commonly used type of ANN. Typical structure of an ANN, which is used for classification of EMG signal features, includes input, hidden, and output layers, so that the features could be related to different force or position classes. Extracted EMG features are fed to ANN as a set of inputs and classified into different force or motion classes as the set of outputs. Each connection between neurons in neighboring layers such as input/hidden layers and hidden/output layers has a weighting factor (w). Moreover, hidden and output

Figure 3. Schematic representation of an artificial neural network

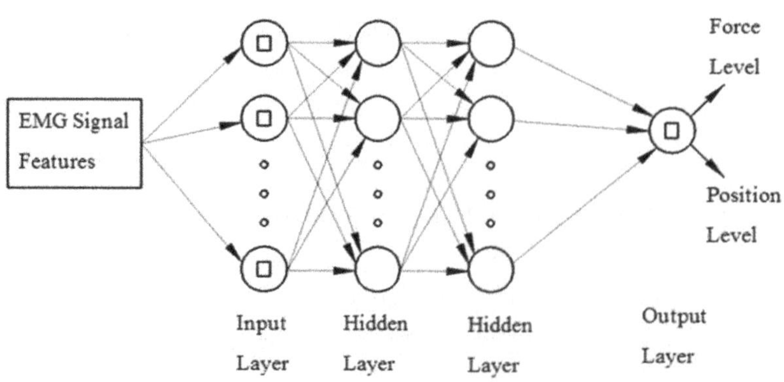

layer neurons implement a transfer function to make a mathematical relation between inputs and outputs. The transfer function $f(x)$ of input arrays x, which sets a relation between input and output data arrays, could be selected according to characteristics of the problem. For instance, a logistic sigmoid transfer function is given as,

$$f(x) = \frac{1}{1 + e^{-\beta x}} \tag{2.1}$$

where

$$x = \sum_{n=1}^{k} w_n a_n \tag{2.2}$$

means the total input of neuron, where w_n, β and a_n are the weight, coefficient and input of ith element, respectively. Although many transfer functions are available, the most widely used transfer functions in pattern recognition of EMG signals are logistic sigmoid and hyperbolic tangent sigmoid transfer functions. The number of layers and neurons are adjustable based on obtained results. Using huge numbers of training data and neurons could lead to overfitting and make a complex networks structure which has to carry out more tasks and likely produces delays. To overcome such issue, some dropout techniques could be operated and train-test proportion of data is proposed to be adjusted.

Neural networks have been utilized to obtain the closest values of output for targeting the real world results by changing the weights (training stage). Adaptation of weights is implemented according to the desired results which is called supervised learning. In the cases of supervised learning of EMG signals, desired results can be position, hand/muscle force, joint torque or motion trajectory. ANN have been operated as a classifier to predict arm and joint trajectories (Cheron, Draye & Bourgeios, 1996), to estimate hand and wrist motion trajectories in the control of a virtual hand (Sebelius et al., 2005), to classify types of limbs motion (Hudgins et al., 1993), to recognize motion patterns based on signal time scale features (Zhao et al., 2006) and to predict the kinematics of shoulder and elbow (Luh, Chang, Cheng, Lai, & Kuo, 1999).

Fuzzy Logic (FL) System

FL systems are beneficial in signal processing and classification, especially for biomedical signals which are not always repeatable, and may even be conflicting (Zadeh, 1973). One of the most useful properties of fuzzy logic systems is that discrepancies in the data can be tolerated. Moreover, it is possible to detect the patterns in data which are not easily identified by other methods using trainable fuzzy systems. Thus, the experience of medical experts or clinicians could be integrated and benefitted. It is possible to incorporate this incomplete but precious knowledge into the fuzzy logic system, due to the system's reasoning style, which is similar to that of a human being. This is a substantial advantage over the artificial neural network (ANN). Fuzzy logic systems better reflect the human decision-making ability than the ANN (Chan et al., 2000). The fundamental of a fuzzy system is the fuzzy inferring engine. Fuzzy production rules are identified according to the available knowledge or well-classified examples (Wang, 1994).

In the fuzzy method, none of operations are random. Information involving a certain amount of suspense is expressed as reliable as possible, without the deformation of forcing it into a "crisp" mold, and it is then handled in a suitable manner. Figure 4 shows the schematic representation of a fuzzy logic system with a fuzzifier and a defuzzifier phases. Fuzzy Logic Systems architecture has three main (Figure 4). These are

1. Fuzzification Module (transforms the system inputs into fuzzy sets),
2. Fuzzy Inference Engine (simulates the human decision making process by making fuzzy inference on the inputs),
3. Defuzzification Module (transforms the fuzzy sets into output parameters).

The inference engine maps each rule's fuzzy input sets into each rule's fuzzy output set. Rules have a critical influence on the performance of a FL system. The rules operate only when the inputs are applied to them.

In order to predict the rules for a fuzzy logic system, dataset is trained. To begin, a certain number of input-output training pairs are selected. The next step is to convert the training dataset into a set of fuzzy rules (IF-THEN, IF-THEN-ELSE, etc.).

The fuzzy rules are mapping from the inputs to the outputs and this mapping can be denoted quantitatively. This kind of FL system is very common and widely used in many engineering applications, such as in fuzzy logic controllers and signal processors. It is also known as fuzzy system, fuzzy controller, fuzzy model or fuzzy expert system.

In recent years, FL systems have been used for decision making process in biomechanical science (Reaz, Hussain, & Mohd-Yasin, 2006). FLS system has been performed to control the elbow and shoulder joint angles of the exoskeletons to design a controller of multifunction prosthetics (Kiguchi, Tanaka, Watanabe, & Fukuda, 2003).

Support Vector Machines

Support vector machines (SVM) is a modern and sophisticated machine learning method (Cortes & Vapnik, 1995). Since EMG-based classification process for prosthetic control problems requires high accuracy and short duration of time to obtain outputs, SVM has become a prevalent and widely used classifier (Lorrain et al., 2011). Although the main notion of the classification progress is to assign the inputs to predefined groups or categories, SVM basically separates the classes operating an optimal hyperplane. In order to discriminate the data among a vast number of classes, a combination of multiple SVM is used. SVM classification process, briefly, is described as follows (Leon, Gutierrez, Leija & Munoz, 2011),

Figure 4. Schematic representation of a fuzzy logic system

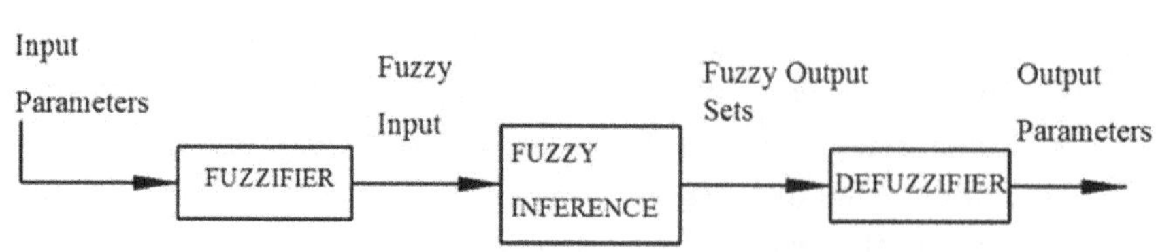

Let x_i and y_i are inputs and outputs, respectively, for $x_i \in R^i$ and $y_i \in \{-1,1\}^l$. The hyperplane, which divides them into two previously determined groups, is defined as,

$$w^T\phi(x) + b = 0 \tag{2.3}$$

where *w* and *b* are weight and bias parameters of hyperplane. Additionally, ϕ is a mapping function which transforms x_i vector into higher dimensional space.

For a classification case, a various number of hyperplane could separate data into two classes. However, there must be only hyperplane, which satisfies maximum margin between the classes, is defined as,

$$\min[\frac{1}{2}\|w\|^2 + C\sum_{i=1}^{m}\xi_i] \tag{2.4}$$

subject to $y_i(w\phi(x_i) + b) \geq 1 - \xi_i$ and $\xi_i \geq 0$ where ξ_i is the slack variables that related to error between training data. In order to obtain optimal hyperplane with limited error equation, equation (2.4) is solved, While α_i and $k(x_i, x_j)$ are Lagrange multipliers and Kernel function, respectively, the equation is reduced as follows,

$$\max[\sum_{i=1}^{\infty}\alpha_i - \frac{1}{2}\sum_{i,j=1}^{m}\alpha_i\alpha_j y_i y_j k(x_i, x_j)] \tag{2.5}$$

The equation of optimal hyperplane is expressed as,

$$w = \sum_{i=1}^{m} y_i\alpha_i\phi(x_i, x_j)$$

which satisfies

$$\sum_{i=1}^{m}\alpha_i y_i = 0 \text{ and } 0 \leq \alpha_i \leq C \tag{2.6}$$

The inputs x_i which satisfy $\alpha_i \neq 0$ are called support vectors. The maximization process to build decision function of the classifier is related to choose suitable kernel function which is generally selected based on inputs type and structure. The most common used kernel functions are linear, polynomial, sigmoid and radial basis functions. The major components of SVM is shown in Figure 5.

Lucas et al. (2008) implemented the SVM method as a supervised classification of multi-channel surface electromyographic signals with the aim of controlling myoelectric prostheses. They concluded that the SVM classification rule can be effectively implemented with fast algorithms (after training) for real-time applications.

Figure 5. The basic representation of support vector machine. Optimal hyperplane is represented with solid line which divides optimal margin between two classes. Square and triangle on dashed lines are support vectors.

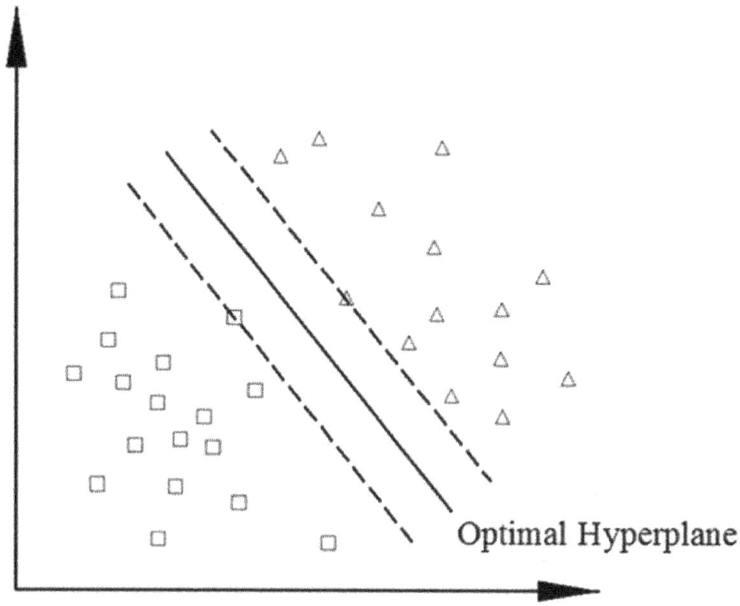

Linear Discriminant Analysis (LDA)

Linear discriminant analysis (LDA) has become a prominent classifier with the intent of grouping very complex EMG data arrays. This section summarizes, briefly, that how LDA method works. The method is based on deriving the combination of parameters that optimally discriminates the priori defined groups (Cao & Sanders, 1996). It is assumed that the vector of features is given as $X = [x_1, x_1, ..., x_m]$. The mean values of *X* for the *i*th class are expressed as $\mu_i = [\mu_{i1}, \mu_{i2}, ..., \mu_{im}]$. The main procedure of LDA method is to maximize the following function which is known as linear discrimination or gate function (Kim, Choi, Moon, & Mun, 2011),

$$f_i(x) = x^T S^{-1} \mu_i - \frac{1}{2} \mu_i^T S^{-1} \mu_i + \log(\pi_i) \tag{2.7}$$

where *S* is the pooled covariance matrix of input data and π_i is a prior probability of inputs coming from class *i*. Using the combinations of the equation, misclassification could be minimized by obtaining higher likelihood index for each defined class.

LDA method has been applied to identify EMG signals to discriminate the patterns of EMG linear envelope of healthy subjects and patients with anterior cruciate ligament injury (Alkan & Gunay, 2012) and to classify the features to enhance the controllability of a powered prosthetics (Hargrove, Scheme, Englehart, & Hudgins, 2010).

It is necessary to make an extensive evaluation of the performance of these listed classifiers in pattern recognition process besides shortcomings and merits of them. Arslan et al. (2010) employed ANN to predict externally applied forces to human hands using EMG signal features. The study, which was aimed to estimate the forces accurately, showed that the classifier predicted the targeted force values in a range of 0.34 and 0.05, and of 0.24 and 0.09 root mean square difference (RMSD) for isometric and anisometric contraction experiments, respectively. In this study it was clearly stated that ANN method could built a successful non-linear relation between force and EMG signal features. However, the authors highlighted that it is not possible to propose a standard ANN design for training the EMG signal efficiently. Even tough ANN is a powerful classifier, the absence of a conceptualized and standardized training method represents an important disadvantage.

Chan et al. (2000) performed a fuzzy logic based classification procedure to control prosthetics. They also provided a comparison between ANN and fuzzy systems in pattern recognition process. In the study, it was shown that 8% and 11.3% of error rates were obtained by Fuzzy and ANN classifiers in pattern recognition, respectively. Some advantages of fuzzy systems were listed as *i)* slightly higher recognition rate than obtained by ANN, *ii)* insensitivity to overtraining, and *iii)* consistent outputs demonstrating higher reliability. The main drawback of the method was noted that requiring more human intervention at initialization stage in order to get the minimum inter-class cross-over. Hence it was stated that the procedure is not automatic to the same extent as ANN.

LDA is becoming a prominent tool for pattern recognition in EMG studies. Chu et al. (2007) conducted a study which includes an EMG feature discrimination process. After a real-time pattern-recognition progress, it was shown that the proposed method achieves the recognition accuracy rate of 97.4%. Phinyomark et al. (2013) also reported that LDA shows a better performance in the classification of fluctuating EMG signals compared to several classifiers such as quadratic discriminant analysis (QDA), random forests (RFs) and k-nearest neighbor (KNN).

The performance of SVM as a machine learning method is needed to be assessed. Subasi (2012) compared the performance of a group of classifiers and reported that classical SVM method provided 96.75% accuracy, while the kNN and the radial basis function (RBF) classifiers achieved the process with 95.17% and 94.08% accuracy, respectively. In the study, it was also noted that SVM performance could be enhanced with some modifications.

FUTURE RESEARCH DIRECTIONS

Though rapid improvements of assistive technologies and specifically artificial human hand prosthetics have been observed, challenging problems are still remaining to be solved in terms of EMG signal pre- and post-processing operations with the intention of providing dexterous control of prosthetics.

Feature Selection

Features extracted from EMG signals for both in time and frequency domains are used at present. However, new features may be proposed for better representation of EMG signals. Furthermore, effect of sliding windows for calculation of features should be investigated in detail. Used features presently are calculated by means of differentiation or summation of neighboring EMG signal amplitudes. New approaches such as measuring energy consumption for each EMG signal maybe employed for time domain in future studies, as well.

Classification Methods

The classifier performance is investigated by researchers extensively. Selection of training and test data for discrimination process has a great influence on classification accuracy and operation duration. Selection of the optimum cross-validated EMG signal arrays should be a purpose of next studies. Additionally, using the combinations of time and frequency domain features for training and test may provide higher accuracy of classification. Furthermore, semi-supervised learning, which is one of the fundamental aspects of classification methodology for cases that it is hard to obtain sufficient training data, should be studied in more detail.

CONCLUSION

The chapter provides a general overview on the EMG signal features and their classification methodologies which are critical issues for controlling of human arm prosthetics. EMG-driven human arm prosthetics are highly sensitive to the scientific and technological advances. Through the last decades, many of EMG signal features calculation and discrimination methods have been proposed and applied to prostheses. Precise and intuitive control of prosthetics depends mainly upon the type of extracted feature and classification techniques. Making a significant difference or advancing the dexterity in the control of prosthetic devices depend on achieving the optimum signal feature and classifier architecture. Needless to say that in addition to the control structure, mechanical structure of the prosthetics also plays a major role in the completion of complex motor tasks which deserves to be extensively dealt with in a separate report.

REFERENCES

Alkan, A., & Gunay, M. (2012). Identification of EMG signals using discriminant analysis and svm classifier. *Expert Systems with Applications*, *39*(1), 44–47. doi:10.1016/j.eswa.2011.06.043

Arslan, Y. Z., Adli, M. A., Akan, A., & Baslo, M. B. (2010). Prediction of externally applied forces to human hands using frequency content of surface EMG signals. *Computer Methods and Programs in Biomedicine*, *98*(1), 36–44. doi:10.1016/j.cmpb.2009.08.005 PMID:19762107

Basmajian, J. V., & DeLuca, C. J. (1985). *Muscles alive: Their functions revealed by electromyography*. Baltimore, MD: The Williams & Wilkins Company.

Cao, J., & Sanders, D. B. (1996). Multivariate discriminant analysis of the electromyographic interference pattern: Statistical approach to discrimination among controls, myopathies and neuropathies. *Medical & Biological Engineering & Computing*, *34*(5), 369–374. doi:10.1007/BF02520008 PMID:8945863

Chan, F. H. Y., Yang, Y. S., Lam, F. K., Zhang, Y. T., & Parker, P. A. (2000). Fuzzy EMG classification for prosthesis control. *IEEE Transactions on Rehabilitation Engineering*, *8*(3), 305–311. doi:10.1109/86.867872 PMID:11001510

Cheron, G., Draye, J. P., Bourgeios, M., & Libert, G. (1996). A dynamic neural network identification of electromyography and trajectory relationship during complex movements. *IEEE Transactions on Bio-Medical Engineering*, *43*(5), 552–558. doi:10.1109/10.488803 PMID:8849468

Chu, J. U., Moon, I., Lee, Y. J., Kim, S. K., & Mun, M. S. (2007). A supervised feature-projection-based real-time EMG pattern recognition for multifunction myoelectric hand control. *IEEE/ASME Transactions on Mechatronics*, *12*(3), 282–290. doi:10.1109/TMECH.2007.897262

Cortes, C., & Vapnik, V. (1995). Support-vector networks. *Machine Learning*, *20*(3), 273–297. doi:10.1007/BF00994018

Farina, D., Jiang, N., Rehbaum, H., Holobar, A., Graimann, B., Dietl, H., & Aszmann, O. (2014). The extraction of neural information from the surface EMG for the control of upper limb prostheses: Emerging avenues and challenges. *IEEE Transactions on Neural Systems and Rehabilitation Engineering*, *22*(4), 797–809. doi:10.1109/TNSRE.2014.2305111 PMID:24760934

Hargrove, L. J., Scheme, E. J., Englehart, K. B., & Hudgins, B. S. (2010). Multiple binary classifications via linear discriminant analysis for improved controllability of a powered prosthesis. *IEEE Transactions on Neural Systems and Rehabilitation Engineering*, *18*(1), 49–57. doi:10.1109/TNSRE.2009.2039590 PMID:20071277

Haykin, S. (1999). *Neural Networks: A Comprehensive Foundation*. Englewood Cliffs, NJ: Prentice-Hall.

Hudgins, B., Parker, P., & Scott, R. N. (1993). A new strategy for multifunction myoelectric control. *IEEE Transactions on Bio-Medical Engineering*, *40*(1), 82–94. doi:10.1109/10.204774 PMID:8468080

Kallenberg, L. A. C., Schulte, E., Disselhorst-Klug, C., & Hermensa, H. J. (2007). Myoelectric manifestations of fatigue at low contraction levels in subjects with and without chronic pain. *Journal of Electromyography and Kinesiology*, *17*(3), 264–274. doi:10.1016/j.jelekin.2006.04.004 PMID:16790358

Kiguchi, K., Tanaka, T., Watanabe, K., & Fukuda, T. (2003). Exoskeleton for human upper-limb motion support, *IEEE International Conference on Robotics and Automation*, 2206 – 2211.

Kim, K. S., Choi, H. H., Moon, C. S., & Mun, C. W. (2011). Comparison of k-nearest neighbor, quadratic discriminant and linear discriminant analysis in classification of electromyogram signals based on the wrist-motion directions. *Current Applied Physics*, *11*(3), 740–745. doi:10.1016/j.cap.2010.11.051

Leon, M., Gutierrez, J. M., Leija, L., & Munoz, R. (2011). EMG pattern recognition using support vector machines classifier for myoelectric control purposes. *Pan American Health Care Exchanges (PAHCE)*, 175–178.

Lorrain, T., Jiang, N., & Farina, D. (2011). Influence of the training set on the accuracy of surface EMG classification in dynamic contractions for the control of multifunction prostheses. *Journal of Neuroengineering and Rehabilitation*, *8*(1), 25. doi:10.1186/1743-0003-8-25 PMID:21554700

Lucas, M. F., Gaufriau, A., Pascual, S., Doncarli, C., & Farina, D. (2008). Multi-channel surface EMG classification using support vector machines and signal-based wavelet optimization. *Biomedical Signal Processing and Control*, *3*(2), 169–174. doi:10.1016/j.bspc.2007.09.002

Luh, J. J., Chang, G. C., Cheng, C. K., Lai, J. S., & Kuo, T. S. (1999). Isokinetic elbow joint torques estimation from surface EMG and joint kinematic data: Using an artificial neural network model. *Journal of Electromyography and Kinesiology*, *9*(3), 173–183. doi:10.1016/S1050-6411(98)00030-3 PMID:10328412

Mann, R. W., & Reimers, S. D. (1970). Kinesthetic sensing for the EMG controlled Boston Arm. *IEEE Transactions on Man-Machine Systems*, *11*(1), 110–115. doi:10.1109/TMMS.1970.299971

Oskoei, M. A., & Hu, H. (2007). Myoelectric control systems-a survey. *Biomedical Signal Processing and Control*, *2*(4), 275–294. doi:10.1016/j.bspc.2007.07.009

Oskoei, M. A., & Hu, H. (2008). Support vector machine-based classification scheme for myoelectric control applied to upper limb. *IEEE Transactions on Bio-Medical Engineering*, *55*(8), 1956–1965. doi:10.1109/TBME.2008.919734 PMID:18632358

Otto Bock Gmb. (2012). Retrieved from http://www.ottobock.com/

Phinyomark, A., Phukpattaranont, P., & Limsakul, C. (2012). Feature reduction and selection for EMG signal classification. *Expert Systems with Applications*, *39*(8), 7420–7431. doi:10.1016/j.eswa.2012.01.102

Phinyomark, A., Quaine, F., Charbonnier, S., Serviere, C., Tarpin-Bernard, F., & Laurillau, Y. (2013). EMG feature evaluation for improving myoelectric pattern recognition robustness. *Expert Systems with Applications*, *40*(12), 4832–4840. doi:10.1016/j.eswa.2013.02.023

Reaz, M. B. I., Hussain, M. S., & Mohd-Yasin, F. (2006). Techniques of EMG signal analysis: Detection, processing, classification and applications. *Biological Procedures Online*, *8*(1), 11–35. doi:10.1251/bpo115 PMID:19565309

Resnik, L. (2010). Research update: VA study to optimize DEKA arm. *Journal of Rehabilitation Research and Development*, *47*(3), ix–x. doi:10.1682/JRRD.2010.03.0034 PMID:20665342

Sebelius, F., Eriksson, L., Holmberg, H., Levinsson, A., Lundborg, G., Danielsen, N., ... Montelius, L. (2005). Classification of motor commands using a modified self-organising feature map. *Medical Engineering & Physics*, *27*(5), 403–413. doi:10.1016/j.medengphy.2004.09.008 PMID:15863349

Shanghai Kesheng Prosthese Corporation Ltd. (2012). Retrieved from http://www.keshen.com/index-en.asp

Subasi, A. (2012). Classification of EMG signals using combined features and soft computing techniques. *Applied Soft Computing*, *12*(8), 2188–2198. doi:10.1016/j.asoc.2012.03.035

von Tscharner, V. (2000). Intensity analysis in time-frequency space of surface myoelectric signals by wavelets of specified resolution. *Journal of Electromyography and Kinesiology*, *10*(6), 433–445. doi:10.1016/S1050-6411(00)00030-4 PMID:11102846

Wang, L. X. (1994). Adaptive fuzzy systems and control. *Design and Stability Analysis,* 29–31.

Wehner, M. (2012). Man to machine, applications in electromyography. In M. Schwartz (Ed.), *EMG Methods for Evaluating Muscle and Nerve Function* (pp. 427–454). InTech. doi:10.5772/26495

Zadeh, L. A. (1973). Outline of a new approach to the analysis of complex systems and decision process. *IEEE Transactions on Systems, Man, and Cybernetics*, *SMC-3*(1), 28–44. doi:10.1109/TSMC.1973.5408575

Zecca, M., Micera, S., Carrozza, M. C., & Dario, P. (2002). Control of multifunctional prosthetic hands by processing the electromyographic signal. *Critical Reviews in Biomedical Engineering*, *30*(4-6), 459–485. doi:10.1615/CritRevBiomedEng.v30.i456.80 PMID:12739757

Zhao, J., Xie, Z., Jiang, L., Cai, H., Lio, H., & Hirzinger, G. (2006). EMG control for a five-fingered interactuated prosthetic hand based on wavelet transform and sample entropy. *Proceedings of the IEEE/RSJ International Conference on Intelligent Robots and Systems.*

KEY TERMS AND DEFINITIONS

Assistive Technology: A branch of technology is used to regain the lost functions of human body parts.

Feature Classification: A pattern recognition technique that is used to categorize a huge number of data into different classes.

Feature Extraction: A method to obtain meaningful and clear data of a signal.

Human Arm Prostheses: Assistive devices which enable to perform lost functions of human arm due to upper or lower arm amputations.

Pattern Recognition: A machine learning process which identifies the pattern of physical systems using data belong to investigated systems.

Rehabilitation: A series of therapy to make injured or amputated people regained lost skills or functions.

Surface Electromyography: A type of electromyography signal recording method carrying out by means of adhering electrodes to skin surface.

This research was previously published in the Encyclopedia of Information Science and Technology, Fourth Edition edited by Mehdi Khosrow-Pour, pages 492-504, copyright year 2018 by Information Science Reference (an imprint of IGI Global).

Chapter 2
The Principle and Process of Digital Fabrication of Biomedical Objects

S. H. Choi
The University of Hong Kong, Hong Kong

H. H. Cheung
The University of Hong Kong, Hong Kong

W. K. Zhu
The University of Hong Kong, Hong Kong

ABSTRACT

Biomedical objects are used as prostheses to repair damaged bone structures and missing body parts, as well as to study complex human organs and plan surgical procedures. They are, however, not economical to make by traditional manufacturing processes. Researchers have therefore explored the multi-material layered manufacturing (MMLM) technology to fabricate biomedical objects from CAD models. Yet, current MMLM systems remain experimental with limited practicality; they are slow, expensive, and can only handle small, simple objects. To address these limitations, this chapter presents the multi-material virtual prototyping (MMVP) technology for digital fabrication of complex biomedical objects cost-effectively. MMVP integrates MMLM with virtual reality to fabricate biomedical objects for stereoscopic visualization and analyses to serve biomedical engineering purposes. This chapter describes the principle of MMVP and the processes of digital fabrication of biomedical objects. Case studies are presented to demonstrate these processes and their applications in biomedical engineering.

INTRODUCTION

Mounting pressure of market globalization and intensifying competition has ferociously been driving the manufacturing industry to survive on incessant reductions in cost and lead-time. However, conventional manufacturing methods can no longer satisfy increasingly diverse customer demands, tight cost control, and complex new products.

DOI: 10.4018/978-1-5225-7359-3.ch002

Against this background, much research efforts have been devoted to developing various technologies to help the manufacturing industry, and layered manufacturing (LM, or now often called 3D printing) and virtual reality (VR) simulation have been among the most significant technologies developed over the past couple of decades.

Despite recent proliferation of LM (3D printing) for free-form fabrication, most of the current systems can only fabricate objects of a single material (Wohlers Report, 2013). There are imminent demands for multi-material layered manufacturing (MMLM) processes to fabricate advanced products and biomedical objects comprising of multiple materials. A few experimental MMLM machines have been developed based on the conventional single material LM systems for relatively simple objects (Bellini, 2002; Wang & Shaw, 2006; Wachsmuth, 2008; Li et al, 2009; Åklint et al, 2013). However, their fabrication speed is unsatisfactory for most complex, large products or medical objects for emergency cases.

More recently, many researchers have worked on virtual prototyping and virtual manufacturing (VPM) (Bracht and Masurat, 2006; Wang and Li, 2006), which is regarded as one of the most important technological advancements for product design and development. VPM has been successfully used in shipbuilding and car industries (Kim et al, 2002; Wöhlke and Schiller, 2005). It uses simulation techniques to analyze and improve a product design and validate the fabrication processes and production schedules.

Through simulations in a VR environment, key factors such as the product shape, manufacturability, and durability that may affect the profitability of manufactured products are optimized. VPM enhances profitability by reducing production cost and material usage, etc. Moreover, it reduces time and tooling cost by eliminating the need for multiple physical prototypes. This allows the users to review and validate a product design to "get it right the first time" for delivery of quality products to market on time and within budget.

This chapter describes the principle of virtual prototyping and virtual manufacturing, with a focus on the processes of modeling and subsequent digital fabrication of multi-material biomedical objects. Case studies of modeling and digital fabrication of biomedical objects using a multi-material virtual prototyping and manufacturing (MMVPM) system will be presented to demonstrate its principle and possible applications in biomedical engineering.

BACKGROUND

There has been a huge surge in demand for biomedical objects in recent years for various medical and dental purposes (Khan & Dickens, 2014; Lee et al., 2001; Maji et al., 2014; Pinnock et al., 2016; Ripley et al., 2016; Sanghera et al., 2001; Winder et al., 1999).

Biomedical objects have been traditionally used as prostheses to repair damaged bone structures or to replace missing body parts (D'Urso et al., 2000; Eufinger et al., 1995; Sannomiya et al., 2008). They are now commonly used by medical students, surgeons, and dentists to help study the intricate anatomical details of human organs and bone structures, as well as to facilitate planning of implantations and surgical procedures (Singare et al., 2009). For example, artificial hip joints, and bone and jaw structures are often used in hospitals to assist complex medical operations. In addition, they are used as specimens for experiments in pharmaceutical manufacturing enterprises.

Depending on the required properties and applications, biomedical objects can be made of either homogeneous (single) material, or heterogeneous (discrete multiple) materials, or functionally graded materials (FGM) (Pompe et al., 2003; Sun et al., 2005). Watari et al. (2004) described the fabrication of an FGM dental implant by powder metallurgy. Experimental results showed that the implant could achieve better mechanical properties and biocompatibility, and that it could control the tissue response through the gradient function of FGM.

But over the years, biomedical objects have been getting more complex, both geometrically and structurally, with more intricate internal details and delicate material variations. As such, most biomedical objects are not economical, and very often not possible, to make by the traditional manufacturing processes.

To fabricate biomedical objects of multiple materials, some researchers have explored the MMLM technology, which is a layer-by-layer additive process that fabricates a heterogeneous object of a number of different materials, which can be discrete with distinctive boundary interfaces, or functionally graded with composition gradients changing gradually from one to another. This process requires a computer-aided design (CAD) model with sufficient material information (Gu and Li, 2002; Gupta et al., 2015; Jafari et al. 2000; Sun et al., 2005).

MMLM, however, remains experimental, and its practical application for fabricating biomedical objects is limited. Indeed, most current MMLM systems can only handle relatively small, simple objects of few materials. More importantly, they are slow and expensive to operate. To address the limitations of MMLM, research efforts have recently focused on developing multi-material virtual prototyping and manufacturing (MMVPM) technology for digital fabrication of complex biomedical objects in a convenient and cost-effective manner.

MMVPM provides a digital platform that integrates virtual reality (VR) simulation technique with MMLM processes. It is an effective tool that can digitally fabricate complex biomedical objects for use in lieu of physical ones. The user can model a complex biomedical object and digitally fabricate its prototypes. Subsequently, the resulting digital biomedical prototypes can be visualized and analyzed in a VR environment, as if the user is manipulating physical objects, for some medical and dental purposes.

The following sections describe the workflow of the MMVPM system in detail. Case studies of modeling and subsequent digital fabrication of biomedical objects using MMVPM would be presented to demonstrate the principle and process of digital fabrication of complex objects for possible applications in biomedical engineering.

THE MMVPM SYSTEM FOR DIGITAL FABRICATION OF BIOMEDICAL OBJECTS

The MMVPM system consists of a suite of software packages for design and visualization of discrete and functionally graded multi-material biomedical objects and simulation of MMLM process in a VR environment. These packages include a color modeler for coloring monochrome STL models, a slicing algorithm, a topological hierarchy-sorting algorithm for grouping random slice contours, a topological hierarchy-based toolpath planning algorithm for generation of sequential and concurrent multi-toolpaths, and a virtual prototyping and virtual manufacturing (VPM) module for digital fabrication of the multi-material biomedical objects (Choi and Cheung, 2005, 2007, 2009; Choi and Kwok, 2004).

Figure 1 summarizes the three major processes for digital fabrication of biomedical objects, namely:

1. Data acquisition for model generation of biomedical objects;
2. Model processing, contour slicing, and toolpath planning for digital fabrication;
3. Digital fabrication, visualization and analysis of biomedical objects.

Figure 1. The main processes of digital fabrication of biomedical objects

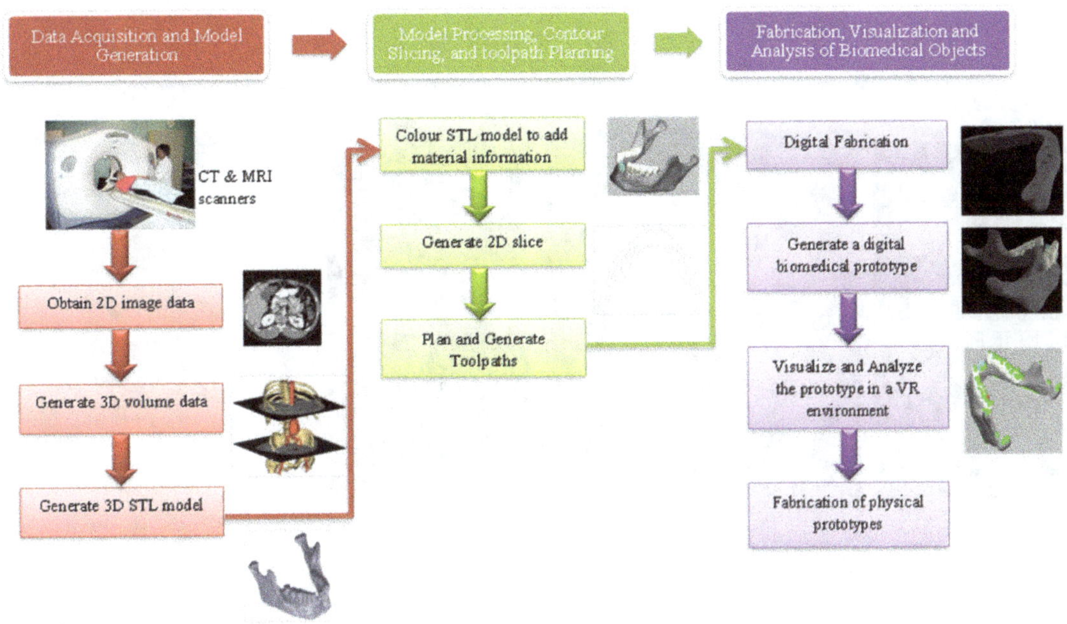

Data Acquisition for Model Generation of Biomedical Objects

To acquire data for generation of a biomedical model, a target anatomical structure is firstly scanned by a medical imaging device, like CT, μ-CT, and MRI, to generate 2D image slices. The resulting 2D slice images in grayscale are then imported into a 3D construction software, such as Materialise Mimics and Able Software Corp 3D-DOCTOR, to extract characteristic data points and subsequently generate the required biomedical model, normally in STL format, which is the de-facto industry standard for layered manufacturing (Sun et al., 2005; Wang et al., 2010). Figure 2 shows two screen captures of the 3D-DOCTOR generating a 3D pelvis model and a 3D human skull model from respective 2D slice images (Able Software Corp, 2016).

Model Processing, Contour Slicing, and Toolpath Planning for Digital Fabrication

The resulting biomedical STL model is then processed in the MMVPM system to add material information, generate slice contours, and plan toolpaths for subsequent digital fabrication and visualization of the resulting biomedical object, as follows:

1. An in-house package is used to paint the biomedical STL model (which is originally monochrome of a single material), with each color representing a specific material;
2. The color STL model is sliced into a number of layers of a predefined thickness. The resulting slice contours and material information are stored in a modified Common Layer Interface (CLI) file;
3. The slice contours are sorted with a contour sorting algorithm to establish explicit topological hierarchy;

Figure 2. Generation of a 3D pelvis model and a 3D human skull model from 2D image slices by 3D-DOCTOR
Source: Able Software Corp, 2011

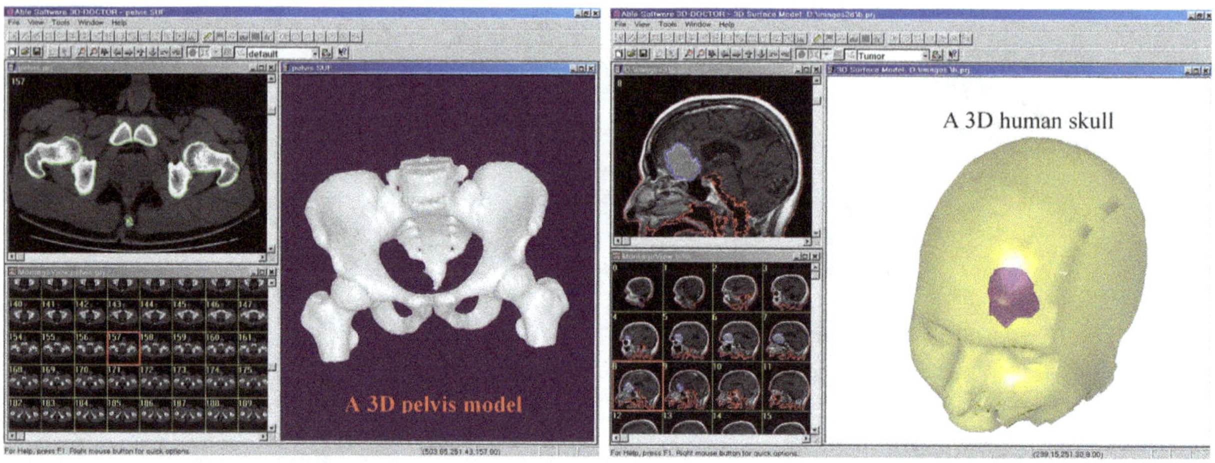

4. Based on the hierarchy information, a multi-toolpath planning algorithm is used to plan and generate toolpaths for concurrent deposition of materials by hatching the slice contours with a predefined hatch space. The hatch vectors are stored in the modified CLI file for fabrication of biomedical object and build-time estimation.

Digital Fabrication, Visualization, and Analysis of Biomedical Objects

The toolpaths together with the hatch vectors are subsequently input to a VP module to perform digital fabrication of the biomedical object. The resulting digital biomedical object can be visualized and analyzed, as if manipulating a physical one, to review and improve the design of the biomedical objects conveniently. Although the digital biomedical object can be displayed on a general computer monitor or projection screen, it is preferable to visualize and analyze it in a VR environment to take advantage of the stereoscopic views and immersive feelings.

The following section uses a human mandible to be inserted with surgiguides to demonstrate how the MMVPM system can model and fabricate multi-material objects for biomedical applications, such as surgical training and planning, patient's education, and implantations.

Figure 3 shows a multi-material assembly of a human mandible inserted with two surgiguides. To ensure the required implant precision, virtual models of the human mandible will be useful to simulate and study possible treatment plans. The MMVPM system can be used to assist surgeons to study a treatment plan through stereoscopic visualization and analysis of a human mandible prototype with surgiguides, as shown in Figure 4.

The surgiguides facilitate positioning the implant accurately. Based on the treatment plan resulted from the simulation, surgeons can improve the design of the surgiguides to fit the patient's bone structure. Figure 5 shows the complete process of digital fabrication of the mandible prototype, while its quality can be visualized accordingly in Figure 6.

Figure 3. A color STL model of a human mandible with two surgiguides

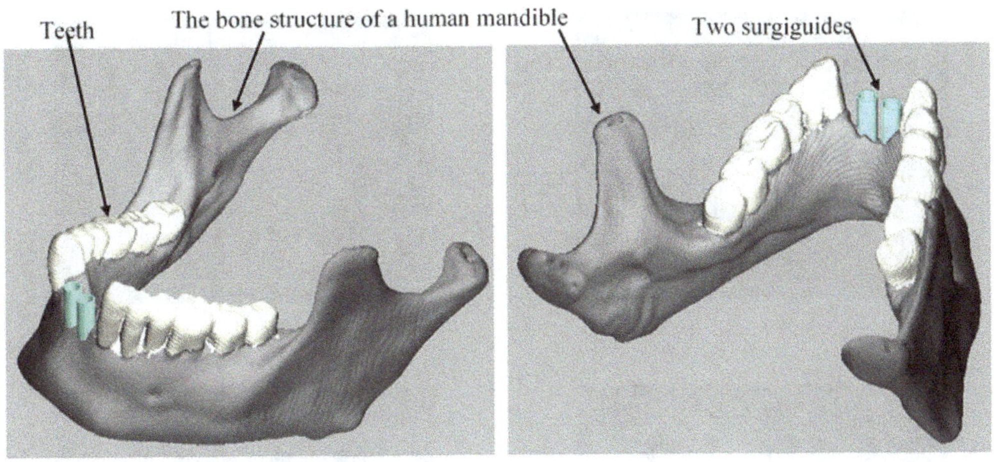

Figure 4. Stereoscopic visualization of a human mandible prototype with surgiguides

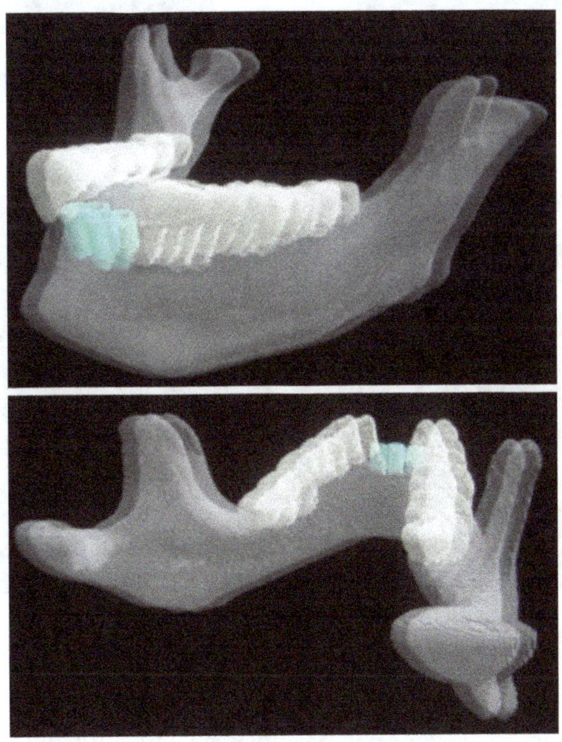

It can be seen in Figure 6 that areas of dimensional deviations scattered mainly around the teeth and the bottom of the mandible bone, while the surgiguides are relatively smooth. This is considered satisfactory as we are more concerned with the accuracy of the surgiguides.

Nevertheless, the simulation process can be iterated with a new combination of process parameters to improve the accuracy of the surgiguides, if necessary. Therefore, based on stereoscopic visualization

Figure 5. Digital fabrication of the human mandible prototype with surgiguides

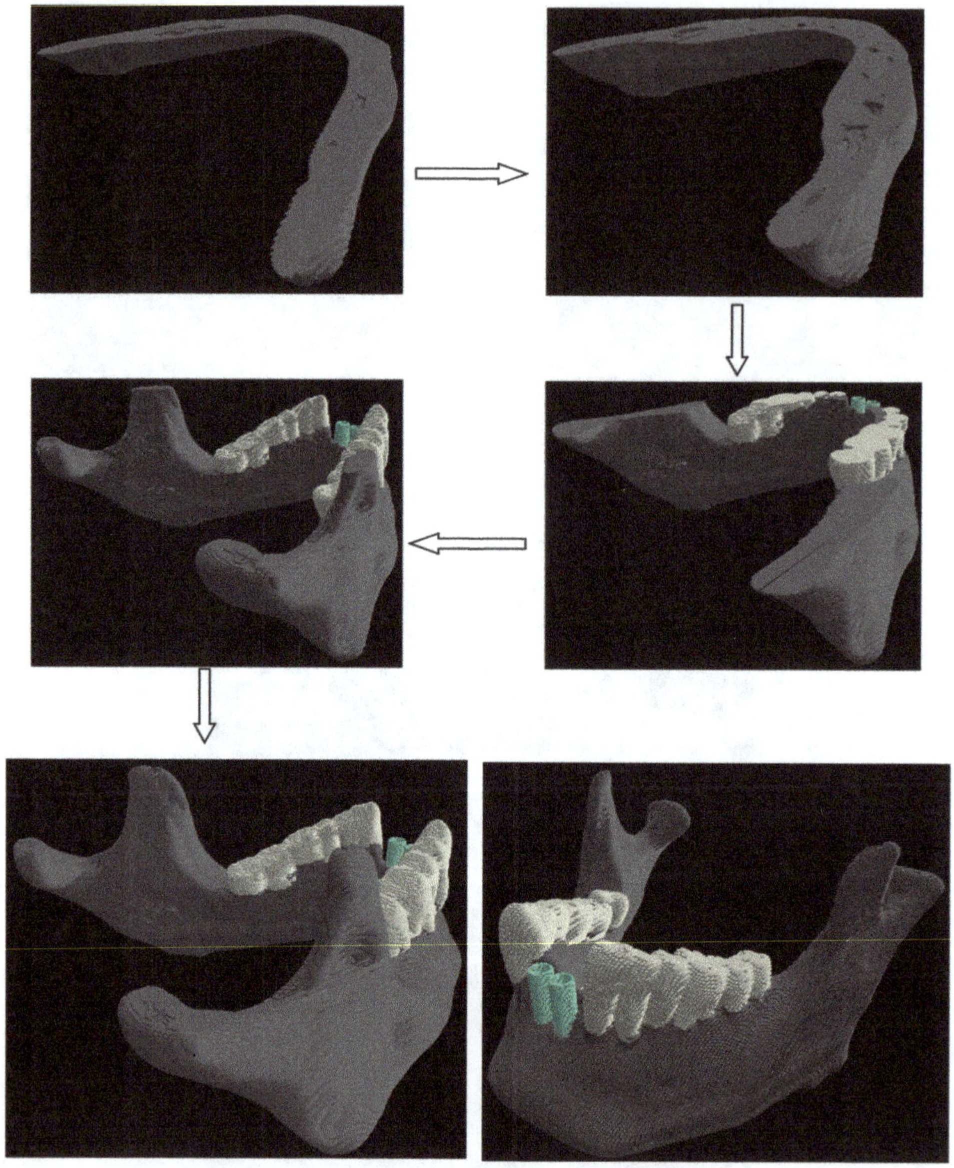

Front View Back View

and deviation analysis, satisfactory designs can be enhanced and validated for subsequent fabrication of dental implants in a convenient and cost-effective way.

To achieve best surgical results possible, an implant should preferably be made of functionally graded materials with properties that would mimic the biological and mechanical characteristics of human organs or tissues (Krishna et al., 2008; Lin et al., 2009; Watari et al., 2004). The MMVPM system can design and digitally fabricate FGM objects to assist such purposes.

For example, it would be desirable to have dental implants made of functionally graded materials, such as titanium (Ti) and hydroxyapatite (HAP), to satisfy both mechanical and bio-compatible properties. Figure 7 shows the resulting FGM dental implant, with material variation represented by blending of red (100% HAP) and green (100% Ti) colors along the Z-axis. Figure 8 shows the resulting FGM teeth.

CASE STUDIES

To further demonstrate possible use of the MMVPM system for modeling and digital fabrication of biomedical objects to facilitate complex surgical procedures, a human ear model and a human thorax model are illustrated in this section.

Figure 6. Areas of the human mandible prototype with dimensional deviations beyond design limits

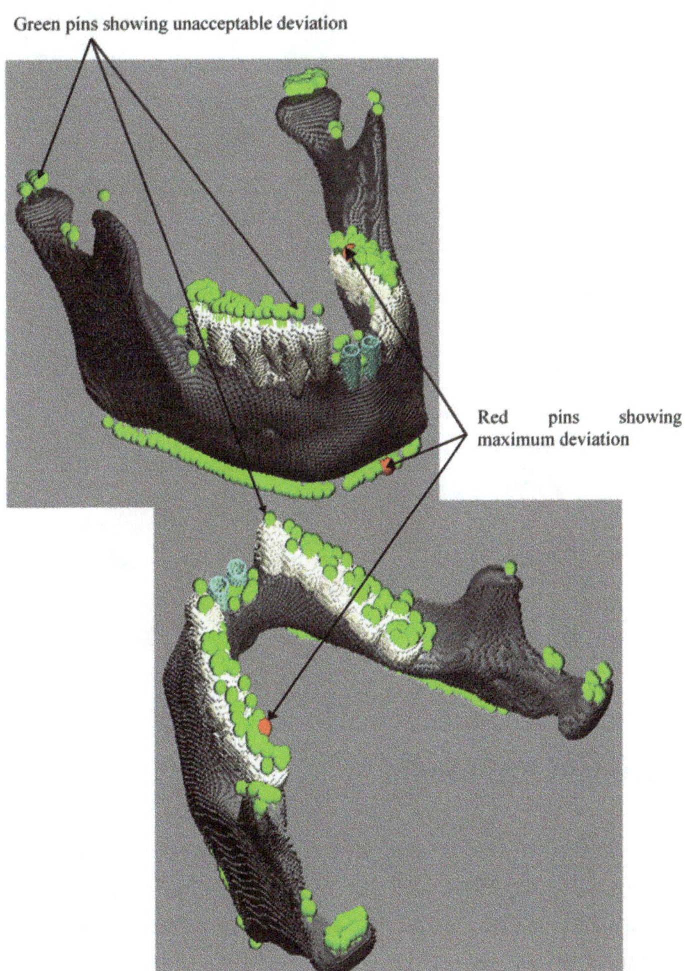

Figure 7. A functionally graded dental implant generated by the MMVPM system

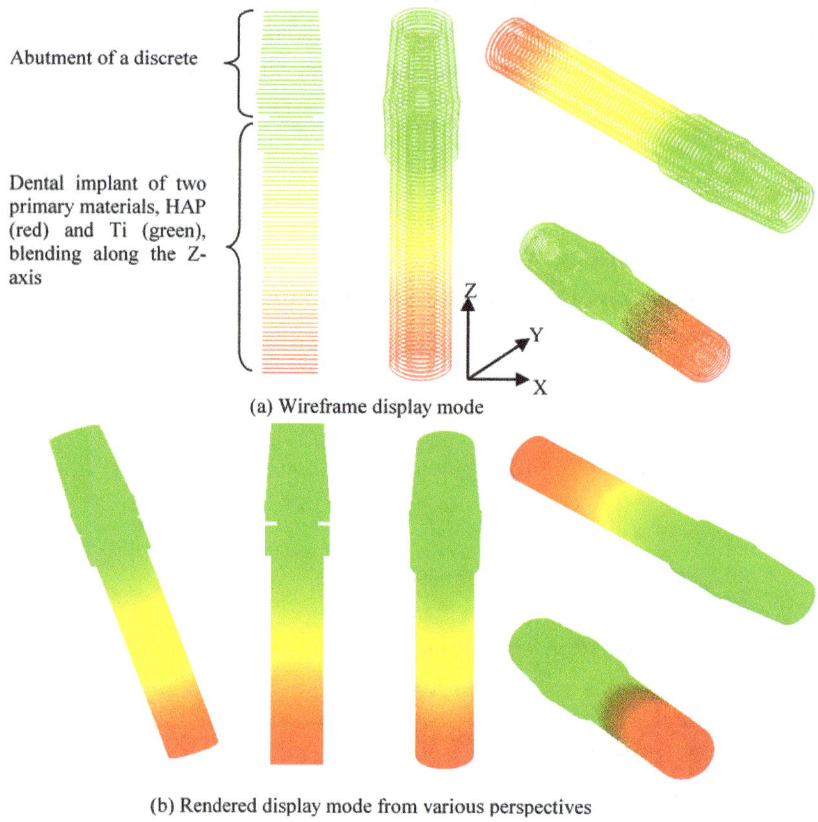

Figure 8. FGM teeth

Case Study One: A Human Ear Model

Hearing loss is one of the common diseases of human ears. It may be caused by genetic problems, harmful noises from working environments, and ageing, etc. (McCullagh, 2011). Some possible clinical approaches include cochlear implant and tympanum (eardrum) repair (Huber & Kipman, 2011). For this purpose, a digitally fabricated human ear model would be particularly cost-effective and convenient to help surgeons diagnose the disease and plan the possible surgical operations. Therefore, to make one, a monochrome human ear STL model is firstly painted with an in-house package to highlight various parts of the model, as shown in Figure 9.

Figure 9. A color STL model of an anatomical human ear

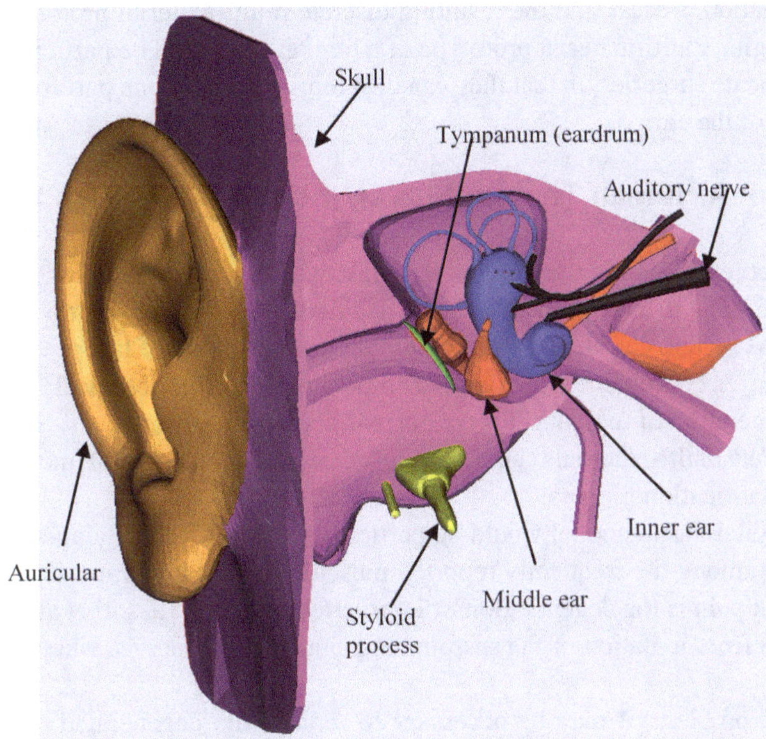

Based on the geometric and material information provided by the color STL model, the ear model is sliced into a number of layers with a predefined layer thickness. Figure 10 shows the slice contours of a particular layer to be made of four materials colored in orange, pink, blue, and grey. The contours of each layer are then sorted to obtain explicit topological hierarchy information for planning and generation of toolpaths for concurrent deposition of materials. Subsequently, a human ear prototype with

Figure 10. A slice of the ear model consisting of four materials

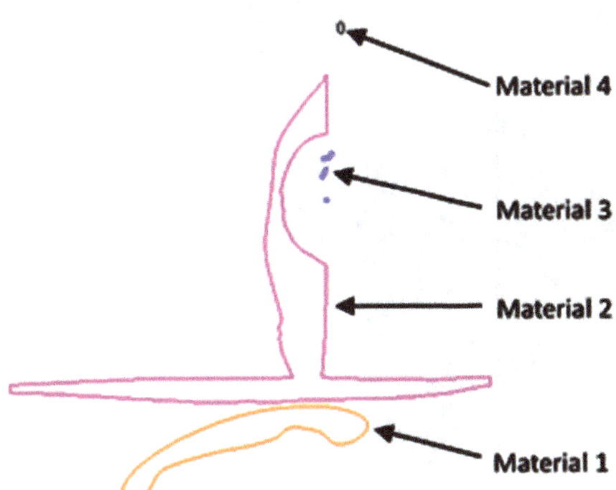

discrete multi-materials can be digitally fabricated using the VPM module. Figure 11 shows some stages of the digital fabrication process and the resulting discrete multi-material prototype of the human ear model. Indeed, a digital multi-material prototype of a human ear would be particularly useful for study and planning of delicate surgeries, in that they can differentiate clearly one part from another, or tissues from blood vessels of the ear.

Case Study Two: A Human Thorax With an Intervertebral Disc Spacer

Figure 12 shows a color STL model of a human thorax, which consists mainly of three parts: the ribs, the sternum, and the spine. Based on the geometric and material information provided by the color STL model, the MMVPM system can slice the thorax model into a number of layers, one of which is shown in Figure 13. The layer contours are then sorted to obtain explicit topological hierarchy information for planning and generation of toolpaths for concurrent deposition of materials. Subsequently, a thorax prototype with discrete multi-materials can be digitally fabricated using the VP module. Figure 14 shows parts of the digital fabrication process.

The resulting digital thorax model would be particularly useful for studying related symptoms like back pain, which is among the frequently reported musculoskeletal problems. The most common factor causing low back pain is the degeneration of intervertebral discs. Taksali et al. (2004) reported that back pain can result from irritation of the surrounding sinuvertebral nerves when the nucleus extrudes into annular tears.

To alleviate this problem, it may be necessary to replace the degenerated disc with an artificial intervertebral disc spacer to remove the primary pain generator while preserving functional motion. As shown in Figure 15, an artificial intervertebral disc spacer has to be inserted between two adjacent

Figure 11. Digital fabrication of a prototype of the human ear model

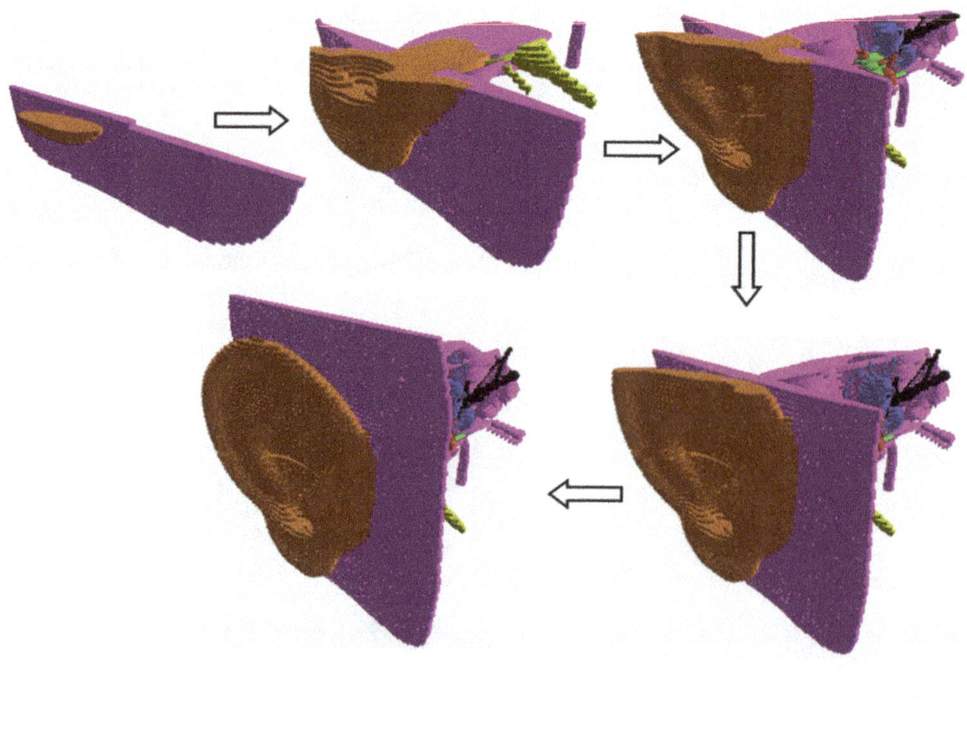

Figure 12. A color STL model of a human thorax

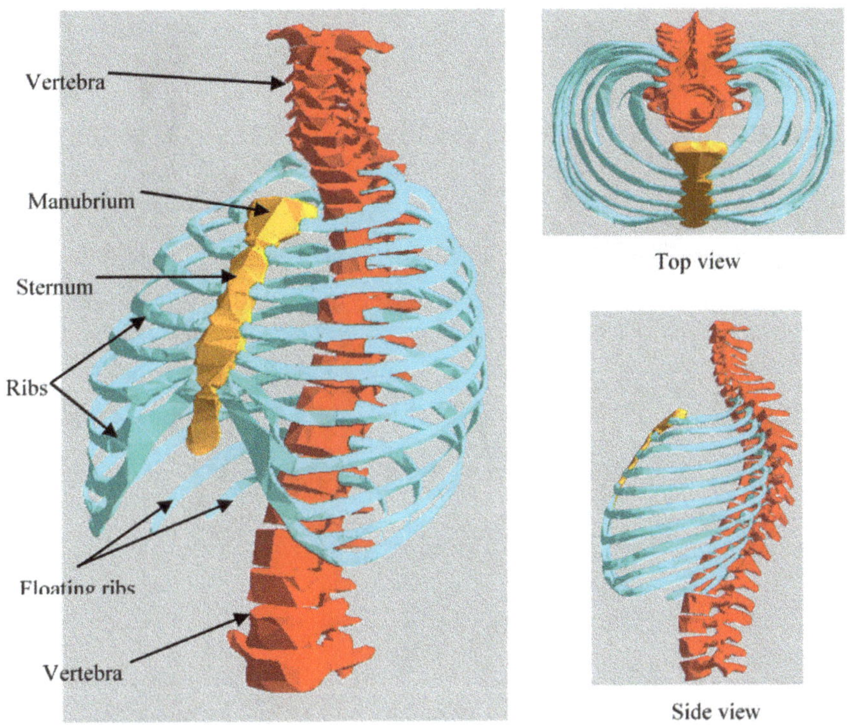

Figure 13. A slice of the thorax model consisting three materials

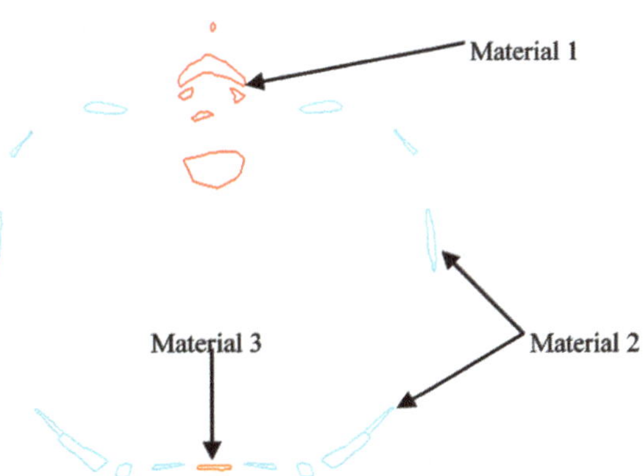

vertebrae. For this purpose, it would be medically preferable to make such a disc spacer with materials having functional and biocompatible properties similar to a normal disc.

The artificial intervertebral disc is to be composed of three components, namely the nucleus pulpousus, the annulus fibrosus, and the vertebral endplates, as shown in Figure 16. In comparison with the material property of the annulus fibrosus, the nucleus pulpousus is softer while the endplates are harder.

Parsons et al. (1992) invented a molding method to fabricate a functional and biocompatible intervertebral disc spacer. The disc spacer consisted of:

Figure 14. Digital fabrication of a prototype of the human thorax model

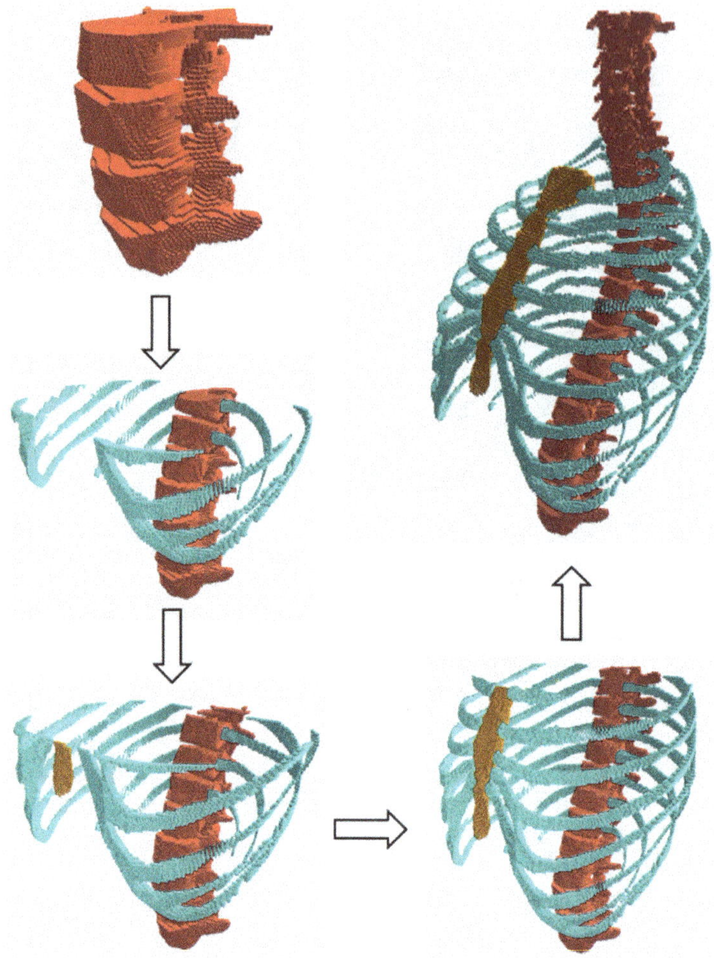

Figure 15. An artificial intervertebral disc spacer inserted between two adjacent vertebrae

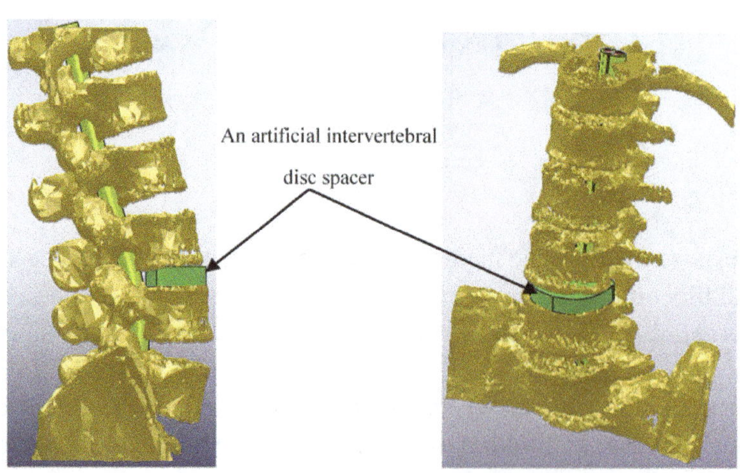

Figure 16. The structure of an intervertebral disc

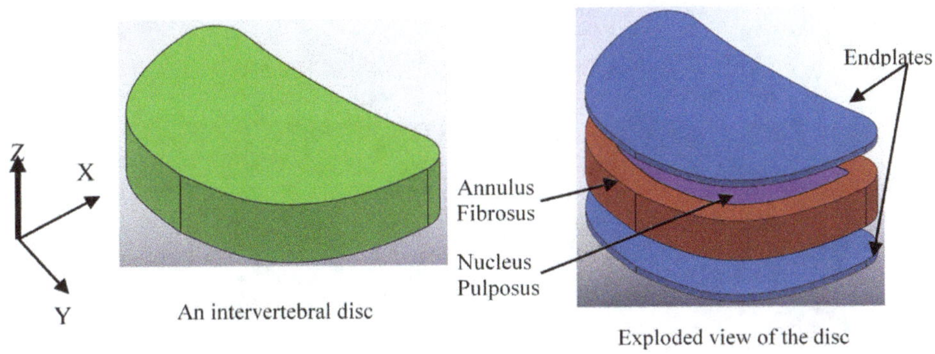

1. A central core of a biocompatible elastomer with a hardness in the range of 20-70 shore-A, shaped and sized to approximate the nucleus pulposus of a natural intervertebral disc, and
2. An outer ring of elastomeric material with a hardness in range of 40–80 shore-A, surrounding the central core to approximate the size and shape of a natural annulus fibrous.

The endplates comprised a biocompatible elastomer with a hardness in range of 90-100 shore-A, and affixed, one to each end, to the outer ring / central core. The relative size of the central core and the thickness of the outer ring in the radial direction, and the selection of material hardness, could be varied in order to better match the mechanical properties of the composite prosthesis with that of the normal disc. The disc spacer could thus achieve the desired properties by varying the hardness of the elastomeric material along the circumferential and the Z-axis directions of the spacer, respectively, by changing the composition of the elastomeric material.

It is therefore desirable to design and fabricate an FGM intervertebral disc spacer with a gradual change of compositions of the elastomeric material to achieve the desired hardness properties in order to mimic the normal disc in an intervertebral disc prosthesis.

The MMVPM system can also be a practical tool to model an FGM intervertebral disc spacer with a functional gradation in hardness property, and then build a digital prototype to facilitate visualization and subsequent fabrication.

To represent the material variations of the intervertebral disc spacer, it is firstly sliced into a set of homogeneous layer contour-based model, and explicit topological hierarchy information is then built for each layer. Secondly, feature layers are selected for assigning primary materials and material control functions for calculation of property values of material composition. Thirdly, each layer is discretized into sub-regions of constant material composition.

Figure 17 shows the resulting FGM intervertebral disc spacer with 3D variations of material composition. Along the Z-axis, it has a material variation represented by blending of blue (100 shore-A elastomer) and red (60 shore-A elastomer) colors; at the same time, there is material variation of each layer in the X-Y plane.

Now, the resulting contour-based FGM intervertebral disc spacer model has both geometric and material information to be processed for visualization of internal material variation of each layer, multi-toolpath planning, and subsequent digital fabrication. Figure 18 shows various stages of the digital fabrication process of an FGM intervertebral disc spacer prototype. The MMVPM system allows adjustment of the

Figure 17. An FGM intervertebral disc spacer with material variations along the Z-axis and in the X-Y plane

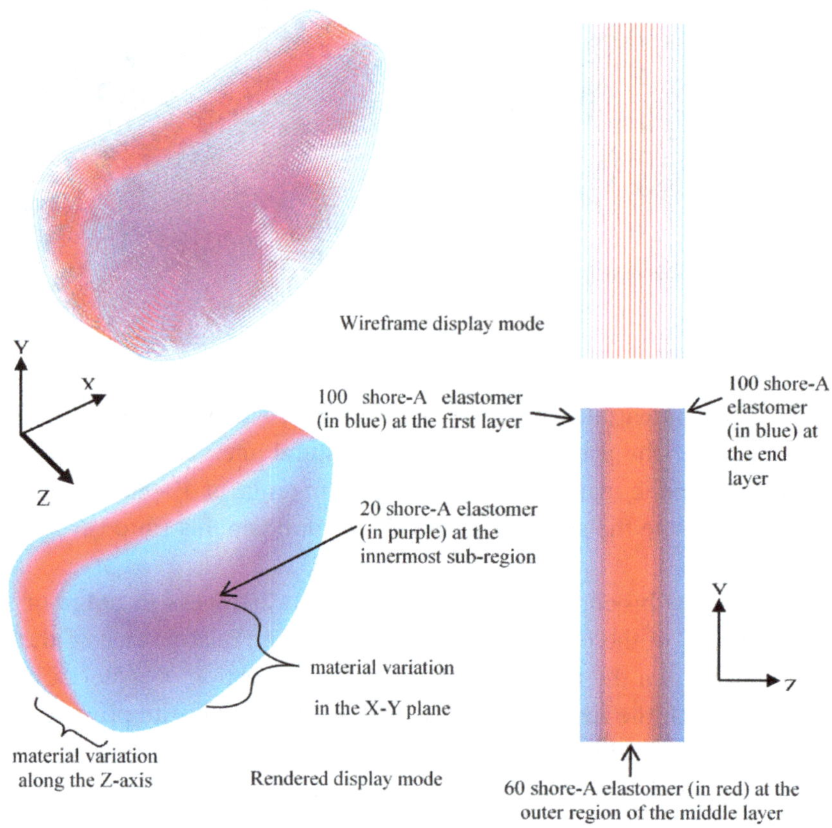

resolution of material composition to suit practical visualization and fabrication requirements, simply by changing the discretization of layer contours, which is the number of layers and the number of sub-regions. Therefore, it is a practical tool for modeling and digital fabrication of biomedical objects with FGM and discrete materials.

FUTURE RESEARCH DIRECTIONS

Although the MMVPM system presented above provides a convenient digital platform that integrates VR simulation technique with MMLM processes for virtual prototyping and virtual manufacturing of complex multi-material objects for advanced product development and biomedical applications, it is limited by its local connectivity and there is a lack of cloud computing capability for remote processing.

Therefore, a worthwhile future development of the MMVPM system is to enhance the algorithms of its main modules for integration with cloud-based computing technology to facilitate remote image processing and visualization of digital fabrication processes. As such, users who are geographically separated can participate in the discussion and sharing of digital fabrication of complex multi-material objects simultaneously for different applications.

Figure 18. Digital fabrication of an FGM intervertebral disc spacer prototype

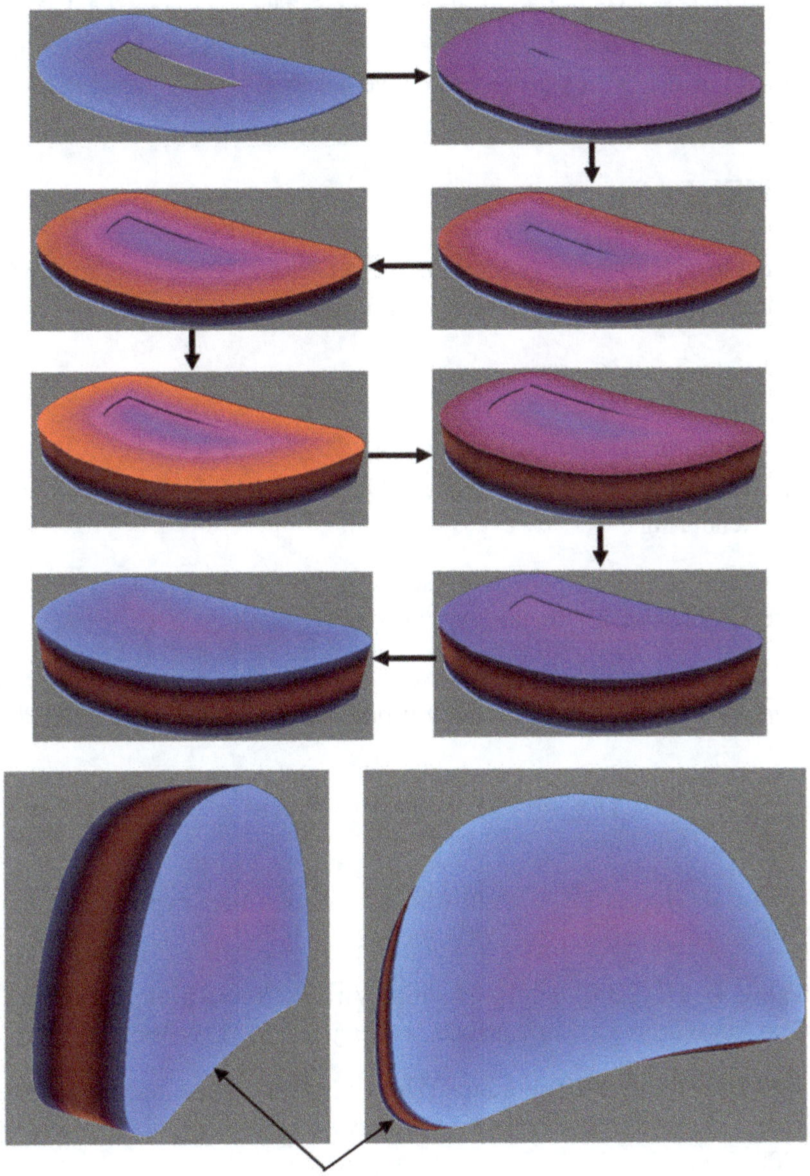

Completed FGM prototype from two perspectives

CONCLUSION

This chapter described the workflow of a multi-material virtual prototyping and manufacturing (MMVPM) system, which is a digital platform that integrates VR simulation technique with MMLM processes. Case studies of modeling and digital fabrication of biomedical objects using the MMVPM system were presented to demonstrate its principle and possible applications in biomedical engineering. The case studies showed that the MMVPM system is a convenient and cost-effective tool that can digitally fabricate complex biomedical objects for use in lieu of physical ones. The user can model a complex biomedical object and perform fabrication simulation. Subsequently, the resulting digital biomedical object can be visualized and analyzed in a VR environment conveniently for some medical and dental purposes.

ACKNOWLEDGMENT

The authors would like to acknowledge the Committee on Research and Conference Grants (CRCG) of the University of Hong Kong for partial financial support for this project.

REFERENCES

Able Software Corp. 3D-DOCTOR. (2016). Retrieved June 28, 2016, from http://ablesw.com/3d-doctor/images.html#SurfaceRendering

Åklint, T., Carlström, E., Johander, P., & Stiernstedt, J. (2013, March 7). Layered manufacturing of free-form multi-material micro-components. *Patent, WO2013030064*, A1.

Bellini, A. (2002). *Fused deposition of ceramics: a comprehensive experimental, analytical and computational study of material behavior, fabrication process and equipment design* (Ph.D thesis). Drexel University.

Bracht, U., & Masurat, T. (2006). The Digital Factory between vision and reality. *Computers in Industry*, *56*(4), 325–333. doi:10.1016/j.compind.2005.01.008

Choi, S. H., & Cheung, H. H. (2005). A Multi-Material Virtual Prototyping System. *Computer Aided Design*, *37*(1), 123–136. doi:10.1016/j.cad.2004.06.002

Choi, S. H., & Cheung, H. H. (2007). Multi-Material Virtual Prototyping for Rapid Product Development and Biomedical Engineering. *Computers in Industry*, *53*(5), 557–566.

Choi, S. H., & Cheung, H. H. (2009). A Topological Hierarchy-Based Approach to Layered Manufacturing of Functionally Graded Multi-Material Objects. *Computers in Industry*, *60*(5), 349–363. doi:10.1016/j.compind.2009.01.008

Choi, S. H., & Kwok, K. T. (2004). A Topological Hierarchy-Sorting Algorithm for Slice Contours. *Rapid Prototyping Journal*, *10*(2), 98–113. doi:10.1108/13552540410526971

DUrso, P. S., Earwaker, W. J., Barker, T. M., Redmond, M. J., Thompson, R. G., & Effeney, D. J. (2000). Custom Cranioplasty Using Stereolithography and Acrylic. *British Journal of Plastic Surgery*, *53*(3), 200–204. doi:10.1054/bjps.1999.3268 PMID:10738323

Eufinger, H., Wehmöller, M., Harders, A., & Heuser, L. (1995). Prefabricated Prostheses for the Reconstruction of Skull Defects. *Internal Journal of Oral and Maxillofacial Surgery*, *24*(1), 104-110.

Gu, P., & Li, L. (2002). Fabrication of Biomedical Prototypes with Locally Controlled Properties Using FDM. *CIRP Annals – Manufacturing Technology*, *51*(1), 181-184.

Gupta, V., & Tandon, P. (2015). Heterogeneous Object Modelling with Material Convolution Surfaces. *Computer Aided Design*, *62*, 236–247. doi:10.1016/j.cad.2014.12.005

Huber, M., & Kipman, U. (2011). The Mental Health of Deaf Adolescents with Cochlear Implants Compared to Their Hearing Peers. *International Journal of Audiology*, *53*(3), 146–154. doi:10.3109/14992027.2010.533704 PMID:21309643

Jafari, M. A., Han, W., Mohammadi, F., Safari, A., Danforth, S. C., & Langrana, N. (2000). A Novel System for Fused Deposition of Advanced Multiple Ceramics. *Rapid Prototyping Journal*, *6*(3), 161–175. doi:10.1108/13552540010337047

Khan, M., & Dickens, P. (2014). Selective Laser Melting (SLM) of Pure Gold for Manufacturing Dental Crowns. *Rapid Prototyping Journal*, *20*(6), 471–479. doi:10.1108/RPJ-03-2013-0034

Kim, H., Lee, J. K., Park, J. H., Park, B. J., & Jang, D. S. (2002). Applying digital manufacturing technology to ship production and the maritime environment. *Integrated Manufacturing Systems*, *13*(5), 295–305. doi:10.1108/09576060210429748

Krishna, B. V., Xue, W. C., Bose, S., & Bandyopadhyay, A. (2008). Functionally Graded Co-Cr-Mo Coating on Ti-6Al-4V Alloy Structures. *Acta Biomaterialia*, *4*(3), 697–706. doi:10.1016/j.actbio.2007.10.005 PMID:18054298

Lee, S. J., Jung, I. Y., Lee, C. Y., Choi, S. Y., & Kum, K. Y. (2001). Clinical Application of Computer-Aided Rapid Prototyping for Tooth Transplantation. *Dental Traumatology. Munksgaard*, *17*(3), 114–119. doi:10.1034/j.1600-9657.2001.017003114.x PMID:11499760

Li, L., Saedan, M., Feng, W., Fuh, J., Wong, Y., Loh, H., ... Lu, L. (2009). Development of a multi-nozzle drop-on-demand system for multi-material dispensing. *Journal of Materials Processing Technology*, *209*(9), 4444–4448. doi:10.1016/j.jmatprotec.2008.10.040

Lin, D., Li, Q., Li, W., Zhou, W. L., & Swain, M. V. (2009). Design Optimization of Functionally Graded Dental Implant for Bone Remodeling. *Composites. Part B, Engineering*, *40*(7), 668–675. doi:10.1016/j.compositesb.2009.04.015

Maji, P. K., Banerjee, A. F., Banerjee, P. S., & Karmakar, S. (2014). Additive Manufacturing in Prosthesis Development – A Case Study. *Rapid Prototyping Journal*, *20*(6), 480–489. doi:10.1108/RPJ-07-2012-0066

McCullagh, M. C. (2011). Effects of a Low Intensity Intervention to Increase Hearing Protector Use among Noise-exposed Workers. *American Journal of Industrial Medicine*, *54*(3), 210–215. doi:10.1002/ajim.20884 PMID:20721900

Parsons, J. R., Le, C. K., Langrana, N. A., Clemow, A. J., & Chen, E. H. (1992). *Functional and Biocompatible Intervertebral Disc Spacer Containing Elastomeric Material of Varying Hardness.* United States Patent.

Pinnock, C. B., Meier, E. M., Joshi, N. J., Wu, B., & Lam, M. T. (2016). Customizable Engineered Blood Vessels using 3D Printed Inserts. *Methods (San Diego, Calif.)*, *99*, 20–27. doi:10.1016/j.ymeth.2015.12.015 PMID:26732049

Pompe, W., Worch, H., Epple, M., Friess, W., Gelinsky, M., Greil, P., ... Schulte, K. (2003). Functionally Graded Materials for Biomedical Applications. *Materials Science and Engineering A*, *362*(1-2), 40–60. doi:10.1016/S0921-5093(03)00580-X

Report, W. (2013). *State of the industry, Annual worldwide progress report, Wohlers Associates Inc.* United States of America.

Ripley, B., Kelil, T., Cheezum, M. K., Goncalves, A., Carli, M. F. D., Rybicki, F. J., ... Blankstein, R. (2016). 3D Printing based on Cardiac CT Assists Anatomic Visualization prior to Transcatheter Aortic Valve Replacement. *Journal of Cardiovascular Computed Tomography*, *10*(1), 28–36. doi:10.1016/j.jcct.2015.12.004 PMID:26732862

Sanghera, B., Naique, S., Papaharilaou, Y., & Amis, A. (2001). Preliminary Study of Rapid Prototype Medical Models. *Rapid Prototyping Journal*, *7*(5), 275–284. doi:10.1108/13552540110410486

Sannomiya, E. K., Brito, A. A., & Agelieri, F. (2008). Surgical Planning for Resection of an Ameloblastoma and Reconstruction of the Mandible Using a Selective Laser Sintering 3D Biomodel. *Oral Surgery, Oral Medicine, Oral Pathology, Oral Radiology, and Endodontics*, *106*(1), 36–40. doi:10.1016/j.tripleo.2008.01.014 PMID:18585610

Singare, S., Lian, Q., Wang, W. P., Wang, F., Liu, Y. X., Li, D. C., & Lu, B. H. (2009). Rapid Prototyping Assisted Surgery Planning and Custom Implant Design. *Rapid Prototyping Journal*, *15*(1), 19–23. doi:10.1108/13552540910925027

Sun, W., Starly, B., Nam, J., & Darling, A. (2005). Bio-CAD Modeling and its Applications in Computer-Aided Tissue Engineering. *Computer Aided Design*, *37*(11), 1097–1114. doi:10.1016/j.cad.2005.02.002

Taksali, S. M. D., Jonathan, N., Grauer, M. D., Alexander, R., & Vaccaro, M. D. (2004). Material Considerations for Intervertebral Disc Replacement Implants. *The Spine Journal*, *4*(11), 231–238. doi:10.1016/j.spinee.2004.07.012 PMID:15541671

Wachsmuth, J. (2008). *Multiple Independent Extrusion Heads for Fused Deposition Modeling. Master of Science*. Virginia Polytechnic Institute and State University.

Wang, C. S., Wang, W. H. A., & Lin, M. C. (2010). STL Rapid Prototyping Bio-CAD Model for CT Medical Image Segmentation. *Computers in Industry*, *61*(3), 187–197. doi:10.1016/j.compind.2009.09.005

Wang, J. W., & Shaw, L. L. (2006). Fabrication of functionally graded materials via inkjet color printing. *Journal of the American Ceramic Society*, *89*(10), 3285–3289. doi:10.1111/j.1551-2916.2006.01206.x

Wang, Q. J., & Li, J. R. (2006). Interactive visualization of complex dynamic virtual environments for industrial assemblies. *Computers in Industry*, *57*(4), 366–377. doi:10.1016/j.compind.2005.11.002

Watari, F., Yokoyama, A., Omori, M., Hirai, T., Kondo, H., Uo, M., & Kawasaki, T. (2004). Biocompatibility of Materials and Development to Functionally Graded Implant for Bio-medical Application. *Composites Science and Technology*, *64*(6), 893–908. doi:10.1016/j.compscitech.2003.09.005

Winder, J., Cooke, R. S., Gray, J., Fannin, T., & Fegan, T. (1999). Medical Rapid Prototyping and 3D CT in the Manufacture of Custom Made Cranial Titanium Plates. *Journal of Medical Engineering & Technology*, *23*(1), 26–28. doi:10.1080/030919099294401 PMID:10202700

Wöhlke, G., & Schiller, E. (2005). Digital Planning Validation in automotive industry. *Computers in Industry*, *56*(4), 393–405. doi:10.1016/j.compind.2005.01.010

KEY TERMS AND DEFINITIONS

Biomedical Object: An object made of bio-compatible materials for prosthetic or bioengineering applications.

Digital Fabrication: Graphical simulation of a fabrication process to make an object in digital form.

Multi-Material Layered Manufacturing: An additive manufacturing process for fabrication of a multi-material object layer by layer.

Multi-Material Object: An object made of a number of heterogeneous (discrete) materials or functionally graded materials.

Multi-Material Virtual Prototyping and Virtual Manufacturing: Graphical simulation of a multi-material layered manufacturing process to make a multi-material object in digital form.

This research was previously published in the Encyclopedia of Information Science and Technology, Fourth Edition edited by Mehdi Khosrow-Pour, pages 505-520, copyright year 2018 by Information Science Reference (an imprint of IGI Global).

Chapter 3
Reverse Engineering in Rehabilitation

Emilia Mikołajewska
Nicolaus Copernicus University, Poland

Marek Macko
Kazimierz Weilki University, Poland

Zbigniew Szczepański
Kazimierz Wielki University, Poland

Dariusz Mikołajewski
Kazimierz Wielki University, Poland

ABSTRACT

Reverse engineering is relatively novel technology, which may revolutionize clinical practice in rehabilitation. This technology may constitute a next step toward patient-tailored therapy, providing customized medical products and increasing effectivity and accessibility of rehabilitation procedures and decreasing cost of manufacturing and time of delivery. Such opportunities need separate research, assessment of associated threats, and dedicated solutions. This chapter aims at investigating the extent to which the available opportunities in the area of application of reverse engineering in rehabilitation are being exploited, including concepts, studies, and observations.

INTRODUCTION

Reverse engineering is relatively novel technology, which may revolutionize clinical practice in rehabilitation. This technology may constitute next step toward patient-tailored therapy, providing customized medical products increasing effectivity and accessibility of rehabilitation procedures and decreasing cost of manufacturing and time of delivery. Such opportunities need separate research, assessment of associated threats, and dedicated solutions.

DOI: 10.4018/978-1-5225-7359-3.ch003

In this chapter authors investigate the extent to which the available opportunities in the area of application of reverse engineering in rehabilitation are being exploited, including own concepts, studies and observations.

BACKGROUND

The additive manufacturing (AM), called also 3D printing or stereolitography is relatively novel technology developing since 1980s. It constitutes iterative technology based on construction the real objects layer by layer, translating this way digital file (digitized object) into a solid object. Features of such object depend on technology and material used to print, but number of both of them rapidly increases, providing important alternative for traditional manufactiring techniques. Moreover some objects have unique features (e.g. shapes) not comparable with products of traditional manufacturing.

Recent editorial article by Maruthappu & Keogh paid particular attention to potiential of 3D printing applications to transform healthcare technologies and organization. Authors divided possible healthcare applications of additive manufacturing into three main groups:

- Internet as decentralised store of blueprints (drugs, equipment, devices, and even body parts) for Early patients-tailored interventions, much quicker and cheaper than traditional delivery solutions,
- Patient–tailored therapy based on medical imaging combined with 3D printing,
- Engineering of 3D printed tissues (Maruthappu & Keogh, 2014; Murphy & Atala, 2014; Seol et al., 2014).

Further implementation of reverse engineering needs additional interdiciplinary research (including randomized controlled trials on patients where available), dedicated methodology, careful assessment of opportunities and threats as far as dedicated solutions.

Reverse Engineering as a Complex Process

Reverse engineering is regarded as quick and cost-effective method of creating functional or nonfunctional copies of existing objects. Process of reverse engineering for rehabilitation purposes is unified to several subsequent stages covered by semi-automated process:

- Digital acquisition of the 3D geometric data: directly from the patient or based on his/her medical records (e.g. using computed tomography – CT or magnetic resonance imaging – MRI),
- Modification/adaptation procedures,
- Creation of 3D model or final product on 3D printer and control of its feasibility: material features, shape, dimensions, patient comfort, etc.

Reverse Engineering for Rehabilitation Support Purposes

Rehabilitation aims at restoration of patient's functions to the maximum possible degree. Scientists and clinicians are aware that in such person the full capacity available in healthy people can not be achievable despite efforts of patient, therapists and caregivers. Increasing number of severely ill, disabled and elderly people makes this task even more difficult. Thus there is still need for novel solutions increasing effectivity of the current rehabilitation procedures.

Rehabilitation should be common, early, comprehensive, and continuous. Reverse engineering products are beneficial for a wide variety of applications in the area of rehabilitation providing:

- Short track between measurements in particular patient and final ready-to-use patient-tailored product (assistive technology device, drug, artificial organ),
- The same quality, geometric accuracy and shape reconfiguration possibilities as original product (e.g. in the case of replacement),
- Features modification possibility (according to the current patient's need) by therapists/manufacturer,
- Low price thanks to cheaper commercial technologies of 3D printing, lack of transport and storage costs,
- Cost-effective production of customized products.

Reverse engineering can provide cells based bones or soft tissues for modelling, testing and therapy purposes thanks to the use of bio-ink (composition of cells and hydrogel materials). Moreover assistive technology products can be cheaper and patient-tailored. Use of such solutions can provide higher therapy efficacy, life quality in patients with severe disorders, usually associated with long-term disability. The huge breakthrough in neurology, neurosurgery and neurorehabilitation can cause novel nerve repair technique: microstructured scaffolds to promote nerves regeneration (Chang et al., 2008; Zhu et al., 2008).

SOLUTIONS AND RECOMMENDATIONS

There is need for further development of interdisciplinary collaboration to identify possible threats and limitations emerging from novel applications of the reverse engineering in everyday clinical practice. Current limitations in the area of reverse engineering cover following topics:

- Ethical issues concerning use of living tissue to print artificial organs, and possible tissue modifications,
- Legal issues including copyright laws to printed copies, patents and original-based hybrid solutions,
- Lack of commonly approved standards of quality,
- Lack of quality and development of internet-based blueprints supervision,
- Lack of control on printed medical devices, especially in poor and developing countries,
- More complex emerging issues concerning eugenics and military applications (Maruthappu & Keogh, 2014).

Novel imaging techniques, printing techniques and software encouraging reverse engineering will constitute challenge for medical education system, due to required specialistic knowledge and experience. Graduate and postgraduate education of medical staff should incorporate novel standards of interdisciplinary co-operation, within patient-tailored therapy. Local scientific and clinical authorities should incorporate modified evidence-based clinical guidelines, paying particular attention to possible secondary changes in patients.

Important limitation constitutes variability of materials and associated techniques of 3D printing. Current materials for 3D printers are limited to plastic, metal, ceramics and living cells. Material engineering faces to biomaterials with features desired for surgical handling, encouraging tissue reintegration, anti-allergic, water-resistant, non-fragile or even biodegradable materials. Natural materials (wood, leather) may be hard to replace. Also combining various materials (e.g. metal and plastic) within one printable product still constitutes a challenge. Increasing use of E-health technologies can change face of the healthcare in the future.

Traditional manufacturing technologies (e.g. welded) in selected cases can provide better solutions than 3D printed. Thus key element of further development in the ares of reverse engineering is integration of 3D printed objects with existing approaches.

Despite aforementioned problems many successful products have resulted from this approach:

- Surgical management and strategy of rehabilitation of acetabular fracture showed accurate and safer surgical management, shorter postoperative recovery time and significantly decreases costs (Deng et al., 2014, Klein et al., 2013),
- Full-arch prostheses of edentulous mandibles (De Santis et al., 2013),
- Customised foot orthoses (Telfer et al., 2013),
- Customized Gensingen braces (Bibb & Brown, 2000; Weiss, 2010),
- Dental technologies (Shahmiri at al., 2014; Yuan et al., 2013; Giordano et al., 2012; Leijnse & Spoor, 2012),
- Urological practice (Youssef et al, 2015),
- Various other easy to use and effective assistive technology solutions (Watanabe et. al., 2015).

Rapid development of hardware and software allows for expectations concerning novel technologies in the area of 2D to 3D images conversion as far as rapid reconstruction of the anatomical parts based on CT and MRI. There is need for effective and accurate, but relatively simple, quick and cheap solutions dedicated for medical applications. Errorless (even semi-automatic)) reproducing of natural affected anatomy can significantly decrease time of recovery in patients after severe injuries.

Development of stem cell based 3D printing can make another breakthrough not only in everyday clinical practice (e.g. reduce mortality in burns or vascular diseases) but also in medical ethics and law regulations concerning transplantation procedures. Lack of physical person (donor) and possibility to print artificial organ as copy of original organ can positively affect public awareness and public opinion playing an important role in increasing organ donation. Clinicians should be aware that it may take a lot of years to print some complex organs as heart and avoid complications.

Another challenge remains integration with other future therapies like nanomedicine, nanorobotics and bioMEMS. There is need for closest co-operation of whole interdisciplinary research teams.

FUTURE RESEARCH DIRECTIONS

Important directions for further research are multifunctional designs that combine geometric and material complexity, such as self-folding 3D printed structures, actuated in a heating environment (Deng & Chen, 2015; Gao et al., 2013; Ishida et al., 2014). Further research on bio-inspired design can help understand the mechanisms underlying many associated cognitive processes (Fu et al. 2014), as well as results of the Human Brain Project (Rose, 2014; Markram, 2012).

Current evidences concerning use of reverse engineering in rehabilitation are regarded as insufficient. There is strong need for further research exploring novel area of applications, e.g. semi-natural bones and tissues, associated with bone kinematics and dynamics simulation (Fang et al., 2015; Wei & Dong, 2011; He et al., 2014). Our own research concern three main topics:

- Use of various technologies and materials to print 3D patient-tailored assistive technology (Lei et al., 2014),
- Research on mechanical properties of printed objects,
- Easy acquiring and modification of existing patients' electronic health records, including medical imaging, as an useful source of templates for reverse engineering.

Further challenge constitutes development of family of novel reverse engineering tools (including user-friendly software) for clinical staff - allowing for easy adjustment of the digital models/patterns to the needs of the particular patient (Macko et al., 2016). Industrial 3D printers and associated technologies may reach extremely high resolution, but everyday use of 3D biomaterial printers and 3D desktop printers is limited. Moreover cost-effect factors associated with aforementioned solutions are hard to compare. Use of bio-ink has still many limitations: limited number of bio-plotting materials, limited resolution and strength, complex technology, timing of gelation time, degradation kinetics, and influence of byproducts (even biodegradable). Simultaneous printing using multiple biofactors and materials is hard to achieve. But the room temperature of processing and homogenous distribution of cells makes bio-printing the most useful and promising technology do far.

Even attempts of low-cost fabrication the soft prostheses with the help of a desktop 3D printer were recently reported (He et al., 2014), but every novel medical technology needs careful studies to prove its safety for patients and therapists, and avoid possible complications and secondary changes (even as long-term effects). On-demand production of drugs must provide patient safety (e.g. dose precision, drug features and effects) and quality similar to the drugs manufactured using traditional technologies (Weismann et al. 2015).

Quick development of 3D printing and reverse engineering creates novel solutions and challenges every day: 95.8% of plastic surgeons want CT- or MRI-derived models for their patients (especially as low-cost solutions for facial transplantation and reconstruction purposes). There is common belief that use of 3D printed models and reverse engineering should be incorporated into graduate and postgraduate medical education. This way preoperative planning, developing intraoperative guidance tools, producing patient-specific prosthetics, and teaching patients and surgeons in everyday clinical practice may be easier, and post-operative comfort and esthetics can be optimized (Chia & Wu, 2015; AlAli et al., 2015; Chae et al., 2015; Jones et al., 2015, Chi & Kim, 2015; Wu & Hsu, 2015; Gerstle et al., 2014).

CONCLUSION

Proposed approach – reverse engineering - enables us to design and manufacture relatively complex objects from its original structures or digitized images. It poses intriguing research questions to its further applications in rehabilitation and possible novel research. It can make key contribution to safe human health and improve the quality of life of patients with functional deficits. Such powerful solution needs for careful further research on larger samples, follow-up studies, and more advanced analytical measurements and technical assistance.

REFERENCES

AlAli, A. B., Griffin, M. F., & Butler, P. E. (2015). Three-Dimensional Printing Surgical Applications. *Eplasty*, *15*, e37. PMID:26301002

Bibb, R., & Brown, R. (2000). The application of computer aided product development techniques in medical modelling topic: Rehabilitation and prostheses. *Biomedical Sciences Instrumentation*, *36*, 319–324. PMID:10834252

Chae, M. P., Rozen, W. M., McMenamin, P. G., Findlay, M. W., Spychal, R. T., & Hunter-Smith, D. J. (2015). Emerging Applications of Bedside 3D Printing in Plastic Surgery. *Frontiers in Surgery*, *2*, 25. doi:10.3389/fsurg.2015.00025 PMID:26137465

Chang, W. C., Kliot, M., & Sretavan, D. W. (2008). Microtechnology and nanotechnology in nerve repair. *Neurological Research*, *10*(10), 1053–1062. doi:10.1179/174313208X362532 PMID:19079980

Chia, H. N., & Wu, B. M. (2015). Recent advances in 3D printing of biomaterials. *Journal of Biological Engineering*, *9*(1), 4. doi:10.118613036-015-0001-4 PMID:25866560

Choi, J. W., & Kim, N. (2015). Clinical application of three-dimensional printing technology in craniofacial plastic surgery. *Archives of Plastic Surgery*, *3*(3), 267–277. doi:10.5999/aps.2015.42.3.267 PMID:26015880

De Santis, R., Gloria, A., Russo, T., DAmora, U., Varriale, A., Veltri, M., ... Ambrosio, L. (2014). Reverse engineering of mandible and prosthetic framework: Effect of titanium implants in conjunction with titanium milled full arch bridge prostheses on the biomechanics of the mandible. *Journal of Biomechanics*, *16*, 3825–3829. doi:10.1016/j.jbiomech.2014.10.020 PMID:25468297

Deng, A., Xiong, R., He, W., Wei, D., & Zeng, C. (2014). Postoperative rehabilitation strategy for acetabular fracture: Application of 3D printing technique. *Nan Fang Yi Ke Da Xue Xue Bao*, *4*, 591–593. PMID:24752117

Deng, D., & Chen, Y. (2015). Origami-based self-folding structure design and fabrication using projection based stereolithography. *Journal of Mechanical Design*, *2*(2), 021701–021712. doi:10.1115/1.4029066

Fang, A., Zheng, M., & Fan, D. (2015). Application status of rapid prototyping technology in artificial bone based on reverse engineering. *Sheng Wu Yi Xue Gong Cheng Xue Za Zhi*, *1*, 225–228. PMID:25997297

Fu, K., Moreno, D., Yang, M., & Wood, K. L. (2014). Bio-inspired design: An overview investigating open questions from the broader field of design-by-analogy. *Journal of Mechanical Design*, *11*(11), 111102. doi:10.1115/1.4028289

Gao, W., Ramani, K., Cipra, R. J., & Siegmund, T. (2013). Kinetogami: A reconfigurable, combinatorial, and printable sheet folding. *Journal of Mechanical Design*, *11*(11), 111009. doi:10.1115/1.4025506

Gerstle, T. L., Ibrahim, A. M., Kim, P. S., Lee, B. T., & Lin, S. J. (2014). A plastic surgery application in evolution: Three-dimensional printing. *Plastic and Reconstructive Surgery*, *2*(2), 446–451. doi:10.1097/01.prs.0000436844.92623.d3 PMID:24469175

Giordano, M., Ausiello, P., & Martorelli, M. (2012). Accuracy evaluation of surgical guides in implant dentistry by non-contact reverse engineering techniques. *Dental Materials*, *9*(9), e178–e185. doi:10.1016/j.dental.2012.06.006 PMID:22796037

He, Y., Xue, G. H., & Fu, J. Z. (2015). Fabrication of low cost soft tissue prostheses with the desktop 3D printer. *Scientific Reports*, *4*, 6973. doi:10.1038rep06973 PMID:25427880

Ishida, S., Nojima, T., & Hagiwara, I. (2014). Mathematical approach to model foldable conical structures using conformal mapping. *Journal of Mechanical Design*, *136*(9), 091007. doi:10.1115/1.4027848

Jones, D. B., Sung, R., Weinberg, C., Korelitz, T., & Andrews, R. (2016). Three-Dimensional Modeling May Improve Surgical Education and Clinical Practice. *Surgical Innovation*, *2*(2), 189–195. doi:10.1177/1553350615607641 PMID:26423911

Klein, G. T., Lu, Y., & Wang, M. Y. (2013). 3D printing and neurosurgery - ready for prime time? *World Neurosurgery*, *3-4*(3-4), 233–235. doi:10.1016/j.wneu.2013.07.009 PMID:23871811

Lei, S., Frank, M. C., Anderson, D. D., & Brown, T. D. (2014). A Method to Represent Heterogeneous Materials for Rapid Prototyping: The Matryoshka Approach. *Rapid Prototyping Journal*, *5*, 390–402. doi:10.1108/RPJ-10-2012-0095 PMID:26120277

Leijnse, J. N., & Spoor, C. W. (2012). Reverse engineering finger extensor apparatus morphology from measured coupled interphalangeal joint angle trajectories - a generic 2D kinematic model. *Journal of Biomechanics*, *3*(3), 569–578. doi:10.1016/j.jbiomech.2011.11.002 PMID:22134183

Macko, M., Mikołajewska, E., Szczepański, Z., Augustyńska, B., & Mikołajewski, D. (2016). Repository of images for reverse engineering and medical simulation purposes. *Medical and Biological Sciences*, *3*, 23–29.

Markram, H. (2012). The human brain project. *Scientific American*, *6*(6), 50–55. doi:10.1038cientificamerican0612-50 PMID:22649994

Maruthappu, M., & Keogh, B. (2014). How might 3D printing affect clinical practice? *BMJ (Clinical Research Ed.)*, *349*, g7709. PMID:25550064

Murphy, S. V., & Atala, A. (2014). 3D bioprinting of tissues and organs. *Nature Biotechnology*, *32*(8), 773–785. doi:10.1038/nbt.2958 PMID:25093879

Rose, N. (2014). The Human Brain Project: Social and ethical challenges. *Neuron*, *82*(6), 1212–1215. doi:10.1016/j.neuron.2014.06.001 PMID:24945767

Seol, Y. J., Kang, H. W., Lee, S., Atala, A., & Yoo, J. J. (2014). Bioprinting technology and its applications. *European Journal of Cardio-Thoracic Surgery*, *46*(3), 342–348. doi:10.1093/ejcts/ezu148 PMID:25061217

Shahmiri, R., Das, R., Aarts, J. M., & Bennani, V. (2014). Finite element analysis of an implant-assisted removable partial denture during bilateral loading: Occlusal rests position. *The Journal of Prosthetic Dentistry*, *5*(5), 1126–1133. doi:10.1016/j.prosdent.2014.04.023 PMID:24951387

Telfer, S., Abbott, M., Steultjens, M., Rafferty, D., & Woodburn, J. (2013). Dose-response effects of customised foot orthoses on lower limb muscle activity and plantar pressures in pronated foot type. *Gait & Posture*, *3*(3), 443–449. doi:10.1016/j.gaitpost.2013.01.012 PMID:23391752

Watanabe, T., Hatakeyama, T., & Tomiita, M. (2015). Improving Assistive Technology Service by Using 3D Printing: Three Case Studies. *Studies in Health Technology and Informatics*, *217*, 1047–1052. PMID:26294609

Wei, X., & Dong, F. (2011). Development of computer aided forming techniques in manufacturing scaffolds for bone tissue engineering. *Zhongguo Xiu Fu Chong Jian Wai Ke Za Zhi*, *12*, 1508–1512. PMID:22242356

Weisman, J. A., Nicholson, J. C., Tappa, K., Jammalamadaka, U., Wilson, C. G., & Mills, D. K. (2015). Antibiotic and chemotherapeutic enhanced three-dimensional printer filaments and constructs for biomedica l applications. *International Journal of Nanomedicine*, *10*, 357–370. PMID:25624758

Weiss, H. R. (2010). Brace technology thematic series - the Gensingen brace in the treatment of scoliosis. *Scoliosis*, *5*(1), 22. doi:10.1186/1748-7161-5-22 PMID:20942970

Wu, G. H., & Hsu, S. H. (2015). Review: Polymeric-Based 3D Printing for Tissue Engineering. *Journal of Medical and Biological Engineering*, *3*(3), 285–292. doi:10.100740846-015-0038-3 PMID:26167139

Youssef, R. F., Spradling, K., Yoon, R., Dolan, B., Chamberlin, J., Okhunov, Z., ... Landman, J. (2015). Applications of three-dimensional printing technology in urological practice. *BJU International*, *5*(5), 697–702. doi:10.1111/bju.13183 PMID:26010346

Yuan, F. S., Sun, Y. C., Wang, Y., & Lü, P. J. (2013). Accuracy evaluation of a new three-dimensional reproduction method of edentulous dental casts, and wax occlusion rims with jaw relation. *International Journal of Oral Science*, *3*(3), 155–161. doi:10.1038/ijos.2013.49 PMID:23907676

Zhu, W., OBrien, C., OBrien, J. R., & Zhang, L. G. (2014). 3D nano/microfabrication techniques and nanobiomaterials for neural tissue regeneration. *Nanomedicine (London)*, *6*(6), 859–875. doi:10.2217/nnm.14.36 PMID:24981651

ADDITIONAL READING

Castroagudin, M. R., Zhai, Y., Li, Z., Marnell, M. G., & Glavy, J. S. (2016). Cyto-3D-print to attach mitotic cells. *Cytotechnology*, *4*(4), 1641–1645. doi:10.100710616-015-9917-2 PMID:26464272

Ingber, D. E. (2016). Reverse Engineering Human Pathophysiology with Organs-on-Chips. *Cell*, *6*(6), 1105–1109. doi:10.1016/j.cell.2016.02.049 PMID:26967278

Jeandupeux, E., Lobjois, V., & Ducommun, B. (2016). 3D print customized sample holders for live light sheet microscopy. *Biochemical and Biophysical Research Communications*, *4*, 1141–1143. PMID:26093294

Numajiri, T., Nakamura, H., Sowa, Y., & Nishino, K. (2016). Low-cost Design and Manufacturing of Surgical Guides for Mandibular Reconstruction Using a Fibula. *Plastic and Reconstructive Surgery Global Open*, *7*(7), e805. doi:10.1097/GOX.0000000000000682 PMID:27536484

Pillay, V., & Choonara, Y. E. (2015). 3D Printing in Drug Delivery Formulation: You Can Dream it, Design it and Print it. How About Patent it? *Recent Patents on Drug Delivery & Formulation*, *3*(3), 192–193. doi:10.2174/1872211309666150608094744 PMID:26051151

Shinozawa, T., Yoshikawa, H. Y., & Takebe, T. (2016). Reverse engineering liver buds through self-driven condensation and organization towards medical application. *Developmental Biology*, *420*(2), 221–229. doi:10.1016/j.ydbio.2016.06.036 PMID:27364470

Zhou, Z. J., Wen, C. L., Sun, Q. M., Wang, A. P., Yan, Z. G., Liu, F., ... Li, Y. B. (2016). Morphometric measurement of the cervical spine for minimally invasive pedicle screw fixation using reverse engineering and three-dimensional reconstruction. *International Journal of Medical Robotics and Computer Assisted Surgery*. doi:10.1002/rcs.1765 PMID:27538708

KEY TERMS AND DEFINITIONS

3D Printing: Technical process of building physical objects from a three-dimensional (3D) digital model by adding many subsequent thin layers of special building material (plastic, metal, etc.).

Assistive Technology: Technology used in an assistive technology device or assistive technology service.

Assistive Technology Device: Equipment used to increase, maintain, or improve functional abilities of disabled people.

Biomedical Engineering: Application of principles and practices of engineering science to biomedical research and everyday clinical practice in health care.

Health-Related Quality of Life (HRQoL): Multi-dimensional concept covering the overall condition of a human life in areas associated with impact of the health status on quality of life: physical, mental, emotional, and social functioning.

Physical Therapy: Use of physical therapy methods, techniques and tools to promote, maintain, or restore the physical and physiological well-being of an individual with movement dysfunction.

Quality of Life (QoL): Multi-dimensional concept covering the overall condition of a human life in various areas: physical, emotional, social, political, moral, etc., their modification and enhancement.

Recovery of Function: A complete (where available) or partial return to the normal (or maximum degree) physiologic activity of an organism following disease or trauma.

Rehabilitation: Process of restoration of human functions to the full (or maximum possible) degree in patients suffering from disease or injury.

Rehabilitation Engineering: A part of biomedical engineering: application of engineering science to design, develop, adapt, test, evaluate, apply, and distribute technological solutions to problems associated with disabilities in areas of mobility, communications, hearing, vision, and cognition.

Reverse Engineering: Quick and cost-effective method of creating copies or modified versions of existing objects, unified to three subsequent stages: digital acquisition of the 3D geometric data, modification/adaptation procedures, and creation of 3D model or final product on 3D printer.

Universal Design: Concept for designing and delivering products and services usable by people with the widest possible range of functional capabilities (including people with various deficits).

This research was previously published in the Encyclopedia of Information Science and Technology, Fourth Edition edited by Mehdi Khosrow-Pour, pages 521-528, copyright year 2018 by Information Science Reference (an imprint of IGI Global).

Section 2
Civil Engineering

Chapter 4
Digital Animation for Representing Architectural Design

Roberta Spallone
Politecnico di Torino, Italy

ABSTRACT

Since the late nineties digital movies have emerged as one of the main methods for submission of architectural designs at international competitions, public presentations, and shares on websites of contemporary masters. The chapter highlights how the ability to access the fourth dimension, through the construction of a sequence of images, constitutes a specific prerogative of digital representation, which goes beyond the static constraint imposed by conventional methods of representation. The author investigates the methods, techniques, and languages of the fourth-dimensional communication of architecture. The chapter provides analysis and critiques of the case studies collected by the author from the beginning of the phenomenon and traces an ideal interpretation path, due both to the changing technologies and the emerging specific languages.

INTRODUCTION

Since the late Nineties of XX century digital movies have been emerging as one of the main methods for submission of architectural designs at international competitions, public presentations and shares on contemporary masters' websites, allowing to preview buildings, spaces and urban environments before their eventual construction.

The chapter highlights how the ability to access the fourth dimension, through the construction of a sequence of images, constitutes a specific prerogative of the digital representation, which goes beyond the static constraint imposed by conventional methods of representation codified by the Descriptive Geometry in the previous centuries.

The great complexity of architectural video production has led to the birth of new professionals who specialize in modelling, rendering, animation, graphics, montage, editing and post production and work in specialist studios. In the past, this had resulted in the generation of specific and recognizable

DOI: 10.4018/978-1-5225-7359-3.ch004

languages: the language of the design was overlaid by that of its narration, and they were not always in harmony one with the other.

The author investigates on the methods, techniques and languages of the fourth-dimensional communication of architecture, unexplored area of research thus far, already the subject of previous studies, relating them with the masters' personal poetics.

The chapter provides also analysis and critiques of several case studies collected by the author from the beginning of the phenomenon, and traces an ideal interpretation path, due both to the changing technologies and the emerging specific languages.

BACKGROUND

Digital modelling has affected the entire design process in recent years, starting with the conception through the control and communication of designs and finally to the provision of tools for building construction and then management activities. For the design stage of the process, these new modelling tools seem to have unified the constituent elements of drawing and design, reconstituting them within the etymological roots of the Latin term *designo*, the twofold meaning of which alludes to an abstract component, indicating the mental processes intrinsic to design, and another concrete component relating to the tangible operation of drawing (Garzino, Spallone, & Lo Turco, 2011).

These tools offer a possible answer to Rogers's wishes, who in 1990 said that "what is needed… is for the meaning of drawing (and therefore also the pedagogical and practical interpretation that is derived from it) not to be just the descriptive representation of an object generated by means of symbols, but for its ancient semantic value to return to it, which is summed up today by the English word design (as opposed to drawing), in which the symbols themselves contain the concept of thought, concrete intention and the design of an object" (Rogers, 1990).

Digital models have a syncretic nature which offers new opportunities to the scientific research and architectural design: tests and errors occur in a space in which our experience of problems are rendered fluid and immediate as in an architectural promenade, "3D computer models… allow more enhanced and more controlled interaction between users and models, they are able to cover the whole range of possible models in a single system of representation" (Maldonado, 2005).

Moriconi, an earlier scholar of infographics, underlined that "through the infographic support, digital drawing simulates a hypothetical reality, goes beyond the limitations of the static and allows interaction with any type of sign. With the creation of virtual images, infographic technology represents what is perhaps the most appropriate tool for interpreting the complexity of reality" (Moriconi, 2001).

According to Manovich the rise of the movie camera as a universal paradigm for the interaction with any data which is represented on three dimensions goes back to the 1980s and 90s. Indeed he affirms that "as the computer culture is gradually spatializing all representations and experiences, they become subjected to the camera particular grammar of data access. Zoom, tilt, pan and track: now we use these operations to interact with data spaces, models, objects and bodies" (Manovich, 2011).

As a new medium, the animation of digital models requires the establishment of relationships with several disciplines such as communication sciences and cinema engineering and with technologies like that of video games and it must also pass the test of comparisons with the established conventional technologies of film production. "Considering architecture as a form of art, we might learn from other artistic disciplines, such as moviemaking (cinematographic approaches, sequencing and animation), theatre (physical expression, interaction, improvisation) and music (rhythm, harmonic variation, but also digital recording and sampling). These may expand the palette of architecture (traditionally making use of drawings, models, pictures and symbols)" (Breen & Breen, 2011).

This triggers a critical discussion on the ontological nature of films, on their narrative form and on their character of exploring human emotions, and it implies the attention of videos of architecture to the perceptual effects produced by relations between persons and space.

Three dimensional digital models frequently constitute the basis for the production of particularly effective visualizations, using a variety of techniques, which can be exploited for different purposes. As Engeli noted, "the creation of specific messages relating to space requires an in-depth knowledge of the intrinsic characteristics to the different possibilities and to the aspects that are wished to emphasize.

An object that is viewed can be interpreted in different ways. The main parameters (light, materials, perspective) can bring out the nature of an object, work against it, reveal new information and add or hide aspects of the object" (Engeli, 1999).

Engeli also underlines the role of narration in the communication of a design: "when you talk about a design, the aim is to give the public an opportunity to identify with the object that is being presented. This is the main reason why it is necessary to tell stories about designs. In reality all architects perform a similar operation when they present design ideas, but usually it is not a capability that is used knowingly, despite the importance of that activity for the communication of ideas" (Engeli, 1999).

The producers themselves, who include multidisciplinary groups, composed of architects, graphic designers, modellers, producers and musicians, underline the importance of animations for the representation in architecture. In an old interview, Scott, Hampton, Alsop and Cocke, members of Squint/Opera, one of the most famous creative company, underlined the importance of videos in the presentation of a design: "I would like to encourage all architects to tell the story of their designs by using videos, because it is a tool which helps people to see a design in a different way. Our experience of architecture is dynamic: we move through spaces and it is precisely this aspect that a film is able to explore" (Oddo, 2004).

Taking the part of the stakeholders, also today the foresight of Gregotti, who observed that "who has had experience like a member of architectural competition jury panels composed by different experts, knows very well which confusion has done between the value of design and the fascination of representation considered as a painting… we will see in any year in which way video and virtual representation will act on communication in architectural competition" (Gregotti, 1995) seems to the author surely alive.

MAIN FOCUS OF THE ARTICLE

Issues, Controversies, Problems

For over fifteen years the 'spectacle of architecture' has been identified in the digital representation of space and time, especially in the animation, one of the most effective media for the prefiguration and communication of architectural design.

The author, following the development of the architectural representation from the outset through digitally constructed video believes that, with some simplification, after a first moment of the centrality of digital modelling within which paths were created or singular elements of the building were animated a following moment in which photo-realistic and hyper-realistic results were chased and another, still, characterized by the pursuit and consolidation of narrative styles, is witnessing nowadays a phase of reflection and refinement, with a return of the communication process production within the atelier of architecture, which addresses, therefore, technical choices and languages consistent with its design poetics.

The current trend, in fact, is to turn increasingly to the opportunities offered by the motion graphics, through the hybridization between static and dynamic images of different nature, photorealistic renderings tend to be singular, as they are snapshots within conceptual dynamic views, and a search for greater deepening of content takes advantage of the development of the narration on motives, concept phases and constructive solutions.

The below developed discussion, traces, through the comments of the peculiarities of several of the videos found online (not always available even today) and collected by the author over the years, an ideal path built on some case studies illustrating the developments outlined above.

As said, at the beginning of the phenomenon the centrality of digital modelling is evident in the construction of the video.

Zaha Hadid is one of the first architects who use the digital tools for the design, presentation and communication via video of the project. Central themes of her poetry, such as dynamism, fluidity, transparency, found a solution to the problems of design and representation in the digital modelling.

Also the animations of the three-dimensional models offer an opportunity to overcome the two-dimensional characteristics of her drawings and paintings and to move around buildings generated by the computer while they are still in the design phase (Schumacher, 2004). A video realized in 1999, the first that the author was able to find, in which the techniques of animation of three-dimensional model are used in the project of architecture, shows the spatial characteristics of the architectural competition proposal for the Casino and the Grand Hotel in Lugano submitted by Hadid. In the short video (0:58), produced by the computer graphics studio Neutral, which is also one of the pioneers of the specific professional field, a lava flow cools and consolidates into a new architectural form that wraps the existing buildings. The model is rendered with only two colours, red and white, in order to resume the metaphor of lava and ice, in complete harmony with Hadid's the figurative language.

In subsequent years, the improvement of the techniques and digital tools accompanies the transition from animation resolved within the conceptual modelling to an extreme tendency toward photorealism.

The availability of software used in computer-gaming, along with the launch of YouTube in 2005, promotes the proliferation and spread of design representation through video, in particular in the field of international architecture competitions, in which usually it becomes one of the mandatory requirements, in public presentations and in self-promotion of the main architectural studios through websites. The tendency to create photorealistic images aspiring to become perfect copies of reality establishes itself at this stage, in which rendering engines, textures of materials and systems for setting the lights are improving.

The video for the project by Christian de Portzamparc in the competition for the headquarters of the Region of Rhône-Alpes in Lyon (2006) describes a building in an urban scale, a complex designed as a

Figure 1. Zaha Hadid, Casino and Grand Hotel in Lugano
Source: Movie by Neutral (0:58), 1999

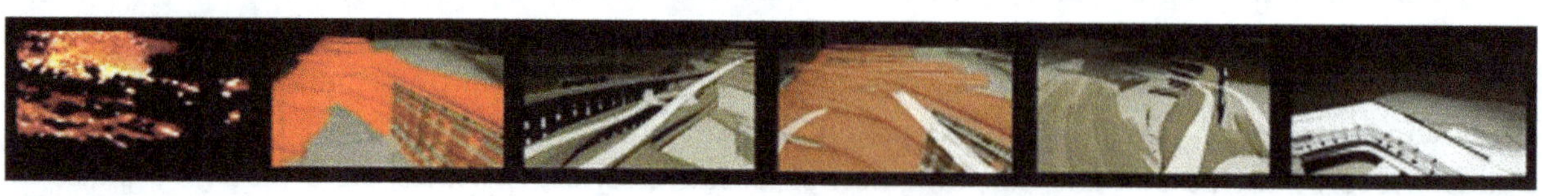

block, a place that can be enjoyed by the community in democratical and active way. The project won because of its elegance, fluidity and rigor and the particular suitability to the site (Cardani, 2006). In the video (3:12), composed by the studio Bartproject, a virtual man takes viewers into the building suggesting the functions of the different spaces through the movements of the body and emphasizing the role of the building as a business and trade centre. The constant speed of the virtual camera movement simulates the human walking. The video uses a communicative style, typical of Bartproject, which combines shooting extrapolated from the reality with a hyper-realistic rendering that overlaps transparent figures symbol of an ethereal humanity in motion. The animation shows the transparency and openness of the interior spaces - the monumental entrance, the patio, the exhibition space and the room for debate, all designed to minimize the power consumption.

Architectural competition for the *Musée des Confluences in Lyon*, won by Coop Himmelb(l)au in 2001, awards a project aimed at the needs of information-society: to make perceivable current knowledge to a broad public in a process of constant change.

The concept, named Crystal and Cloud, combines a hard space, the crystal rising towards the city, conceived as an urban space and an entrance hall for visitors, and a soft space, the cloud intended by the designers to hide the knowledge of the future (Coop Himmelb(l)au, 2003). A short movie presenting the project has been included in the Coop Himmelb(l)au website, since 2007. This was probably the basis for the longer and richer educational movie now present in the Musée des Confluences website which is also enriched by an interactive virtual visit with 360° panorama views in the principal rooms.

The longer movie (7:30), produced by Isochrom in 2006, starts exploring the area of Confluences by an aerial view, first zenith and then perspective, which shows a wide urban tissue extended to the town hall, modelled with simple volumes and rendered as a plaster model lit by natural light. As the approach is made from the river to the building, the model takes on greater realism and the movement of the camera transforms into a walk through towards and inside the building under design. A virtual person guides the observers through the main halls to the different floors and invites them to appreciate the architecture and examine the main collections by the way the shots are framed. During the tour the visitors meet "Crystal and Cloud" which characterize the shape of the building. The model is realistically, but delicately, rendered with a particular focus on bringing out the quality of the materials, transparencies, reflections and opacities and interactions between the materials and the lights. A few blobs of warm colours attributed to the entrances of some halls, arousing the curiosity to enter them, and to secondary animation elements, emerge in the interiors in cold colours, all ranging from white to grey and blue.

In the following period there is a passage from the technical virtuosism to the research and consolidation of narrative styles.

This step intended to prevent the technocratic homologation by the software, and led the video producers to create personal and recognizable narrative languages. As stated by the following three case studies, this choice anticipates the current developments.

Figure 2. Christian de Portzamparc, New Rhône-Alpes Regional headquarters in Lyon
Source: Movie by Bartproject (3:12), 2006

Figure 3. Coop Himmelb(l)au, Musée des Confluences in Lyon
Source: Movie by Isochrom (7:30), 2006

The submission by Steven Holl for the international architecture competition to design the future Musée du Louvre-Lens in the north of France, launched in early 2005, is a building in which the idea of a walk through time organizes the plan of the new museum according to a linear time increment along the rectangular frame, which is crossed by cyclical time (represented by the arcs) in short circuits and different access points.

The competition animation (4:00), made by Neutral, deepens the concept ideas, showing the aquarelle sketch drawn by Holl as the basis for the three-dimensional geometric construction which develops the idea of intersection between linear time (a parallelepiped) and cyclical time (four cylindrical rings cut to make four arcs). When the geometries become buildings, the parallelepiped appears as a hard block of bricks and the arcs as light curtains of glass. The architecture, surrounded by the green, is connected to the built city in which it is wedged. The new museum and the settings are modelled and soft rendered. The camera movement, characterized by constant speed, guides the spectators outside (by a fly trough) and inside (by a walk trough) towards the knowledge of the design. Therefore most of the animation is produced by modelling software, with consequent use of bigger time.

The proposal for the invited architectural competition for a new library in Utrecht (2008), presented by 3XN, describes a multifunctional building suitable to house a library and a cinema with residences, shops and cafes. In this way, the cultural centre could be enlivened by various activities throughout the entire day. The architects' goal is to preserve the existing idea of public life in the area, while the new building should be in function as an anthill, of which the shape resembles, characterized by multiple fronts, and should be a new meeting place for the entire city.

The 3XN claims to be deeply involved in the digital revolution (3XN, 2008) and this seems to reverberate in their choices of projectual communication, entrusted to the studio Cadpeople, demonstrating the close cooperation between the two groups. In particular, the video (3:30), resolved entirely thanks to the digital modelling, shows an original visual and communicative approach. It starts highlighting the relationship between the building and the environment, the different functions, displayed with exploded isometric, the accessibility system. Then, describing the project, a particular camera movement alternating the acceleration and stop-image is used, combined with special representation techniques that change from the conceptual, in gray scale with a few spots of primary colours, secondary animations of people, vehicles, videos, up to static photorealistic images delicately rendered.

Figure 4. Steven Holl, Musée du Louvre-Lens
Source: Movie by Neutral (4:00), 2005

Figure 5. 3XN, New library in Utrecht
Source: Movie by Cadpeople (3:30), 2008

Opposed to the essentiality carried out in the previous case study, the imaginative language, rich in contaminations between real and virtual can be found in the next case.

The animation for the presentation of the HOK Sport Olympic Stadium (2007), to be built for the Olympics and Paralympics in London, was created by the filmmakers of Squint/Opera.

The film is produced through hybridization of materials and languages, real footage and photographs with digital artefacts; the animation of three-dimensional model is reduced to a minimum in favour of applying dynamic post-production tools to the static views.

A short presentation, which simulates a communication of television news, introduces the project, firmly framed in the atmosphere of London through the city landmarks (the City Hall, the Tower Bridge and the Battersea Power Station) progressively invaded by the flags of the Olympic competition. A child spectator from the beginning to the end of the event takes the viewer into the discovery of the project elements. These are realized through geometric models of roofs, services, seating and structural system that immediately become, through animation, in other objects related to the fantasies of the child: the colourful ellipsoidal shapes of the services become flying ships, the seats in the form of gulls arrive from above, the three-dimensional trusses move to the place of foundation as they were snakes, while in the background there is the real city. Finally, the elements reassembled and made static, assume again their geometric appearance related to the environment through a model illustrating the overall functions and accessibility. A sequence in which the realistic render of the three-dimensional model is set in motion through secondary animations of public, athletes, giant screens and actions of zoom, scroll, rotate, concludes the movie.

Among the current trends of the animation for the design representation, as seen in the description of the emblematic case studies, the minimalist choices consisting in conceptual views, in animations that refer to the simple movements indicated above by Manovich, in attention to the concept communication and the artefact constructability, can be affirmed to prevail.

Moreover, the video production within the same study of architecture is witnessed in many cases, facilitated by the most common software for modelling, rendering and post-production.

Differently from the most of the architects group websites, that led by Will Alsop, which name is AllDesign, shows his projects by movies besides static renders. These movies at first realized by Squint/Opera or Virtual Artworks are nowadays produced inside the studio.

Figure 6. HOK Sport Olympic Stadium, London
Source: Movie by Squint/Opera (4:52), 2007

The choice of attributing a central role to the movies in the communication strategy of the architect's team, in tune with the graphic style of his website, stimulates more involvement by the viewers.

The video for presenting the project for the complex of three theatres in Langfang, near Beijing, in 2010, was produced by the same AllDesign studio that, in its website, claims to create movies since the best way to express their own ideas is telling a story.

The short video (2:44) introduces the narration with a recall to some images of the traditional Chinese landscape: vegetation of bamboo, flowering cherry-trees, canoes on the lakes and, finally, the wild boar, of which metamorphosis, described as the initial concept, gives shape to the project. The complex aims to fit into the natural environment with the creation of a park with ship canals and walkways leading to theatres and shady spots designed as a meeting place. The architectures, displayed in conceptual clay-render, on which the connecting blue ring raises, are part of the system of the green. Animations, fly-through, static rendering zoom and scroll and simulations of the growth of the artefacts and the green, combined with the apparition of evocative images (musical notes that come out of the buildings and moulds of traditional theatre masks) constitute the sequence.

The video for the architectural competition in 2012 for Copenhagen Arena, presented by C.F. Møllers, shows a multifunctional complex intended for sports and concerts, without losing its human scale, becoming a landmark, a building element of the landscape and an attractive element of activity. The animation (3:54), presumably produced by the same architectural studio, initially presents, in a bird eye view, the structural technologic and constructive elements and the building phases of the stadium, set within a context geometrically modelled and clay-rendered. When the video camera is lowered to the human level, its recurrent motion is the rotation around the conceptual model. The following still image is characterized by the representation as a realistic static rendering to which actions of scroll up or zoom out are slowly applied.

SOLUTIONS AND RECOMMENDATIONS

In summary, methods, techniques and figurative choices in the ambit of the architectural animation, highlighted by the selection of case studies proposed above, can be synthesized as follows.

Figure 7. AllDesign, Great Beijing Theatre and Masterplan, Langfang
Source: Movie by AllDesign (2:44), 2010

Figure 8. C.F. Møllers, Suggestion to Copenhagen Arena
Source: Movie by C.F. Møllers (3:54), 2012

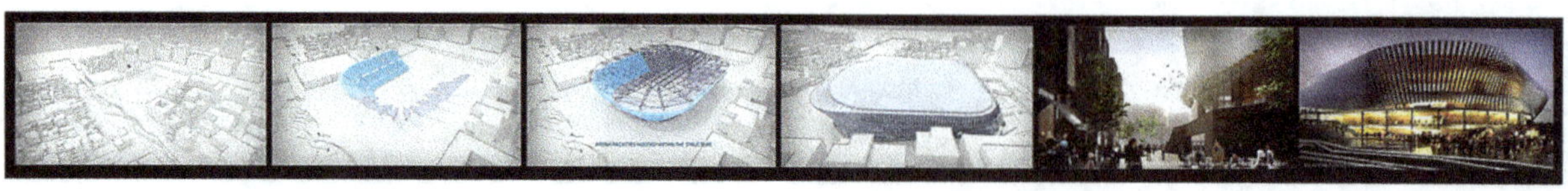

One or more geometric models of a design having an appropriate degree of detail in relation to the positions from which they are filmed, which are based on figurative convention for the architectural representation: dynamic bird eye views emphasize relationships between the building and the existing environment, while those at human eye height give a preview of the perceptual impact of the future settlements; orthogonal views of horizontal and vertical sections illustrate the space distribution and the functional aspects.

The figurative decisions reconcile the design poetics of the architects on the one hand and the narrative styles of the producers on the other, fluctuating between conceptual and hyper-realistic styles.

The camera movement is not generally at a constant speed, accelerating and slowing, sometimes moving rapidly on tracks to get close to the action or following the pace of an avatar who guides the spectator in the discovery of the place, while secondary animations, of men and vehicles in movement, monitor which projects films and fountains and fireworks overlap the primary animation or a still image of it.

The time dimension, introduced by the movement of the camera, is frequently overlapped by the representation of the natural low of time, with animations which start in the daylight and end at the night, thereby facilitating the entire daily life cycle of the object.

Animated projections are always accompanied by sounds. They can range from background music which captures the attention and helps focus the observers' attention on the noises which enliven scenes evoking feelings of empathy and the recognition of places.

Finally different video production techniques can be observed, finding their place of convergence in the digital world (Ciotti & Roncaglia, 2000): the composition of different two and three dimensional views of digital models with photographs and texts using the established technique of stop-motion combined with movement options such as fades, zooms, pans and scene changes which are practiced today using dynamic postproduction software; the modelling of building designs in their existing context, also modelled, but often in a summary manner and the formation of a long route along which a series of rendered views are created; the modelling of building designs inserted dynamically in a film of the environment using camera-tracking techniques.

FUTURE RESEARCH DIRECTIONS

Architectural videos, today with a greater presence, sometimes are accompanied, above all in the websites, by the extremely recent and occasional introduction of 360° panoramic views, which offer a modest degree of interaction, and by virtual views, which, on the contrary, offer a certain number of scenarios that can be freely selected and navigated by visitors.

CONCLUSION

The rapid development of digital technologies makes foreseeable a greater interactivity between observer and digitally simulated space, with the ability not only to freely explore the project, but also to modify it in real time. The author believes that in the same time the research on languages should advance, with the aim of increasing the wealth of knowledge provided by the animation, not only with respect to the users, but also to the designers themselves who could find another effective verification tool and project control thanks to it.

REFERENCES

Breen, J., & Breen, J. (2008). Critical Observations and Strategic Perspectives at Half-time. In K. Terzidis (Ed.), *Proceedings of First International Conference on Critical Digital: What Matter(s)?* (pp. 129-136). Cambridge, MA: Harvard University Graduate School of Design.

Cardani, E. (2006). Proposta vincente. From Lyon Confluence: Progetto, Christian de Portzamparc. *Arca, 219*, 91.

Ciotti, F., & Roncaglia, G. (2000). *Il mondo digitale. Introduzione ai nuovi media.* Roma-Bari, Italy: Laterza.

Coop Himmelb(l)au (2003). *Musée des Confluences.* Retrieved December 17, 2008, from http://www.arcspace.com/architects/coop_himelblau/musee_des_confluences/

Engeli, M. (1999). *Storie digitali. Poetiche della comunicazione.* Torino, Italy: Testo&Immagine.

Garzino, G., Spallone, R., & Lo Turco, M. (2011). Strategie digitali per modelli conoscitivi / Digital strategies for knowledge based models. In G. Garzino (Ed.), Disegno (e) in_formazione. Disegno politecnico / Drawing (and) information. Polytechnic drawing (pp. 70-111). Sant'Arcangelo di Romagna, Italy: Maggioli.

Gregotti, V. (1995). Editoriale. *Rassegna*, 61.

Himmelblau, C. (2003). Musée des Confluences. *GA Document, 73*, 12–17.

Maldonado, T. (2005). *Reale e virtuale.* Milano, Italy: Feltrinelli.

Manovich, L. (2001). *The Language of New Media.* Cambridge, MA: MIT Press.

Moriconi, C. (2001). Il modello virtuale. In A. Soletti, P. Belardi, & F. Cataliotti (Eds.), *Claudio Moriconi professione infografico* (pp. 20–30). Perugia, Italy: Università degli Studi di Perugia.

Negroponte, N. (1995). *Being digital.* New York: Alfred A. Knopf.

Oddo, F. (2004). *Squint/Opera. Nuove professionalità per raccontare l'architettura contemporanea.* Retrieved October 3, 2015, from http://architettura.it/files/20040123/index.htm

Rogers, E. N. (1990). *Gli elementi del fenomeno architettonico.* Napoli, Italy: Guida Editore.

Schumacher, P. (2004). *Hadid digitale. Paesaggi in movimento.* Torino, Italy: Testo & Immagine.

Spallone, R. (2014). Vero, falso, finto. Rappresentazioni digitali di paesaggi urbani in trasformazione fra comunicazione e seduzione [True, false, fake. Digital representations of urban landscapes subject to transformation between communication and seduction]. *DisegnareCon, 13*, 1–8.

Spallone, R., Lo Turco, M., & Sanna, M. (2009a). Representing ideas by animated digital models in architectural competitions. *Proceedings of ARCC 2009. Leadership in Architectural Research, between academia and the profession.*

Spallone, R., Lo Turco, M., & Sanna, M. (2009b). Dimension "time" to represent contemporary architectural design. *Proceedings of IDCT. Innovative Design & Construction Technologies. Building complex shapes and beyond*, 243-258.

Spallone, R., Lo Turco, M., & Sanna, M. (2010). From the place of research to the research of place: new digital technologies for simulating future architectures in building environments. *Proceedings of ARCC 2010. International Conference on Architectural Research.*

3. XN. (2008). *Digital architecture.* Retrieved May 2, 2010, from http://www.3xn.dk

ADDITIONAL READING

Levy, P. (1998). *Becoming virtual. Reality in the Digital Age.* London, England: Persius Books.

Lynn, G. (1999). *Animate form.* New York, USA: Princeton Architectural Press.

McLuhan, M. (1964). *Understanding media. The extension of man.* Berkeley, California: Gingko Press.

Sacchi, L., & Unali, M. (2003). *Architettura e cultura digitale.* Milano, Italy: Skira.

Saggio, A. (2007). *Introduzione alla rivoluzione informatica in architettura.* Roma, Italy: Carocci.

Szalapaj, P. (2005). *Contemporary Architecture and the Digital Design Process.* Massachusetts, USA: Architectural Press.

K. Terzidis (Ed.). *Proceedings of First international Conference on Critical Digital: What Matter(s)?* Cambridge, Massachusetts: Harvard University Graduate School of Design.

KEY TERMS AND DEFINITIONS

3D Modelling: In 3D computer graphics, 3D modelling is the process of developing a mathematical representation of any three-dimensional surface of object via specialized software. The product is called a 3D model.

Animation: Animation is the process of creating a continuous motion and shape change illusion by means of the rapid display of a sequence of static images that minimally differ from each other. In the architectural representations, animation can be the creation of a path inside and outside of buildings, or the putting in motion of static elements.

Compositing: Compositing is the combining of visual elements from separate sources into single images, often to create the illusion that all those elements are parts of the same scene.

Montage: Montage is a technique which uses rapid editing, special effects and music to present compressed narrative information.

Post-Production: Post-production is part of the process of filmmaking, video production and rendering, which consists in a series of editing and optimization of the product, subsequent to the camera shot or the animated sequence creation.

Rendering: Rendering is a image generated by a computer using three-dimensional modeling software, after applying materials and lights and setting a perspective view.

Secondary Animations: In architectural movies, the secondary animations consist of motions of people, vehicles, videos which overlap the animation created by the video camera movement.

This research was previously published in the Encyclopedia of Information Science and Technology, Fourth Edition edited by Mehdi Khosrow-Pour, pages 973-982, copyright year 2018 by Information Science Reference (an imprint of IGI Global).

Chapter 5
Literature Review of Augmented Reality Application in the Architecture, Engineering, and Construction Industry With Relation to Building Information

Aydin Tabrizi
University of Kansas, USA

Paola Sanguinetti
University of Kansas, USA

ABSTRACT

The rapid development and adoption of AR applications creates numerous opportunities for integrating AR with BIM and improving conventional methods used in the fields of architecture, engineering, and construction (AEC). In this chapter, the current trends in the development of AR applications and the application of AR technologies in the fields of AEC are proposed. Also, the relation between AR application and BIM in the AEC industry with the benefits of this integration and possible issues is discussed. The related examples of BIM+AR are described during the literature review. The authors believe the papers presented in this document cover the latest research trends and developments in the use of AR and its combination with BIM for AEC applications. In the future, it is expected that AR applications will be further utilized in the AEC field to enhance productivity, safety, and efficiency.

INTRODUCTION

Productivity rate in construction industry shows a general pattern of decline in comparison to other industries. This issue influences on performance and efficiency of construction projects by adding unnecessary costs, time, materials, and manpower waste Alwi, Hampson et al. (2002). In dealing with this issue, it is necessary to apply proactive approaches rather than reactive ones through using new method

DOI: 10.4018/978-1-5225-7359-3.ch005

and processes in the construction industry. During the last decade, Information and communication technologies (ICT) have been advanced significantly where the application of these technologies could improve the construction industry efficiency to some level (Park & Kim 2012). However, in order to fulfill this task in a larger scale, new methods and processes are required to develop and analyzed. This paper presents a conceptual framework to enhance construction industry efficiency via a comprehensive and proactive mechanism of Augmented Reality (AR) and Building Information Modeling (BIM) linkage. To fulfill this objective, the study begins with an extensive and critical review on AR and BIM separately in order to investigate the efficiency of each technique in the construction industry. Then the study proposes a conceptual framework of AR and BIM combination by investigating this collaboration in the enhancement of construction industry procedure.

BACKGROUND

Augmented Reality (AR) provides the means for intuitive knowledge presentation by enhancing the perceiver's situational awareness and cognitive perception of the real world. Through AR approach, virtual objects can be registered in relation to real objects where these objects can be seen in the same position and orientation of other real objects of the scene, as perceived by the user (Wang et al. 2004). In addition the real objects can be tracked and their 3D shape can be reconstructed from pictures (Azuma 1997).

AR has introduced as a technology which allows the user to see, hear, touch, smell and taste things that others cannot (Van Krevelen and Poelman 2010). It is a technology to perceive elements and objects within real world experience in a complete computational environment. It applies creatures and structures that could be used in daily activities unconsciously through interaction with others such as enabling mechanics to see instructions for repairing an unknown piece of equipment, surgeons to see ultrasound scans of organs while performing surgery on them, fire fighters to see building layouts to avoid invisible hazards and people to read reviews for each restaurant on their way (Feiner 2002). (Wang and Dunston 2007) describes AR as a tool allowing users to work with real world environment while visually receiving displays of additional computer-generated information about the item by superimposition of additional information onto the real world scene. This approach enhances the user's perception of the real environment by showing information that cannot sensed unaided.

It is expected that in the near future, an increase in the use of AR applications will occur due to the advancement of hardware and software. (McKibben and Furlonger 2009) predict that by the end of 2014 approximately 30% of workers will use some form of AR capability and after a long period of technological development and refinement, the implementation of AR applications for the general public will reach its peak. In addition, the commercial market shows a same trend by promising examples such as Project Glass as an R&D program by Google (Goldman 2012) to develop an AR head-mounted display (HMD) for enabling users to experience a truly immersive digital life. ABI Research (Hyers 2006) predicts that by the end of 2014 the revenue from the AR mobile market will reach $350 M and Juniper Research (Holden 2005) predicts that the market for AR services will reach $732 M at the same time. It is widely believed that AR technologies are maturing and that within the coming years they will be broadly adopted by industry.

Estimations show that at least 5% of total building construction costs are due to occurrence of problems in the early design process, causing insufficiency, inconsistency and omission of design-related information towards construction phase (Hwang, Thomas et al. 2009). Construction worksite layout planning relies mainly on 2D paper media where the worksite planners sketch the future layout adjacent to their real environment. This traditional approach is ineffective and prone to error because only experienced and well-trained planners are able to generate the effective layout design with paper sketch (Wang and Dunston 2007). In the architecture, engineering and construction industry (AEC), gaps between planned solutions and practical implementations, poor communications between project participants and inefficient scheduling are the main issues for the lack of sufficient information/communication technology (ICT) support and innovative business procedures (Chi, Kang et al. 2013). (Froese 2005) has categorized trends in construction ICT into three categories. The first category is stand-alone tools to assist specific work tasks such as CAD, structural analysis and estimating programs. The second category includes communication and online information sharing tools through worldwide web document management system. Finally the third category includes the potential for integrating first and second category as a cohesive model through Building Information Modeling (BIM) (Froese 2010) where project teams come together to produce comprehensive and virtual prototypes of all aspects of the construction project as the central activity.

With the progress of ICT use in the AEC industry, higher quality visualization platforms are necessary for the efficient use of shared information among involved teams. Available research studies show the attempts of construction activities simulation with feedback generation through visualizing construction information for easier understanding and data share among project participants (Kang, Anderson et al. 2007). However, this approach only enables the visualization of activities in the virtual environment without enough resemblance to the actual tasks in the real world. (Froese 2005) introduces AR approach for the generation of digital project information prior to construction with transfer onto construction site in a fully digital way where this process facilitates the comparison of the actual situation with the planned final appearance to identify the concerned items.

Building Information Modeling (BIM)

(Penttilä 2006) defines BIM as a set of interacting policies, processes and technologies that generates a methodology to manage the essential building design and project data into digital format throughout the building's life cycle. BIM creates an interdependency environment between structural, architectural and mechanical services (Dossick and Neff 2009) with digital representation of the facility's physical and functional characters to provide a shared knowledge resource for client or user to use and maintain throughout the project's life cycle (Eastman, Jeong et al. 2009).

According to the studies by (Taylor and Bernstein 2009), BIM can incorporate parametric 3D, 4D and 5D modeling where 4D includes a time dimension and 5D contains time based cost model. In addition, studies by (Aouad, Lee et al. 2006) presents the expansion of BIM into nD environment in the recent developments of AEC industry by incorporation of engineering analysis and construction business functions during the lifecycle of a building. This expansion includes scheduling, cost analysis, quality control, accessibility, safety, maintenance, sustainability and energy simulation. However, despite the recent BIM developments in the AEC industry, it still requires further progress during construction phase, specifically in relation to the daily monitoring of work and management of involved teams.

Augmented Reality (AR)

Generally, the life cycle of construction project consists of five phases as a) project feasibility analysis; b) planning/design; c) construction; d) operation; e) disposal (Halpin and Woodhead) where each phase requires the creation of physical structures and elements to create visual information for understanding and communicating the complexity of projects with consideration of relation to existing structures or elements. With recent advances in computer interface design and hardware capability, the application of AR research prototypes and test platforms have fostered in the construction industry (Yang, Tsai et al. 2011). However, AR technologies itself are not advanced enough to be applied effectively in real AEC projects and there are two main issues in this regard and they must be addressed before AR technologies can become prevalent in the AEC industry. The first issue is the limitation of available technologies to support AR approach and the second is to identify appropriate application areas for the use of AR in the construction projects. To clarify the issue, (Stanco, Battiato et al. 2011) studied the accurate alignment of virtual objects in a real world scene with regards to the user's orientation and position. Their research identified recent tracking technologies for indoor, prepared environments with accurate registration, but the lack of strong accuracies tracking techniques for outdoor environments due to the scale difference. (Yokokohji, Sugawara et al. 2000), demonstrated a system combining accelerometers and video tracking registration for analysis of indoors, but its' non-practicality for outdoor and unprepared environments.

In order to create an efficient application of AR in construction projects, it is necessary to identify the areas with the highest potential for better performance. (Balzani, Santopuoli et al. 2004), demonstrated the potential of AR as a visualization aid for underground structures and (Bruner 1966) discussed the benefits of AR for architectural assembly guidance, infrastructure field tasks, urban planning and design detailing.

Therefore, the analysis of current literature, suggests the feasibility of AR technologies to increase the likelihood of success in the AEC industry with the consideration of its suitability validation in each case.

FURTHER RESEARCH DIRECTIONS

Integration of Augmented Reality into AEC Industry

Research by (Chi, Kang et al. 2013) presents the four rapidly developing technologies in the application of AR approach to integrate the virtual world with the real world as a) Cloud computing environment; b) localization; c) portable/mobile devices and d) natural user interface (Table 1). The application of these technologies plays a significant role in the integration of AR into the AEC industry.

Cloud Computing Environment

Cloud computing has become a powerful technology with the progress in the Internet data processing speed where it has the potential to extend the use of AR and BIM applications in the AEC industry. Research by (Jardim-Goncalves and Grilo 2010) emphasizes the shifts of CloudBIM approach from stand-alone static applications to dynamic shared environment with dynamic tasks allocation and access via a network. Internet operation in real time and cloud-based computer devices helps to make the required information more accessible to the users by enhancing the AR experience with the possibility

Table 1. Architecture of AR application in 4 phases adapted from ''research trends and opportunities of augmented reality applications in aec''

Architecture of AR Application			
Data Phase	Computing Phase	Tangible Phase	Presentation Phase
Cloud Computing Environment	Localization Technologies	Portable Devices	Natural User Interface
BIM	GPS	Cheap	Gesture
Internet	RFID	Light	Motion Capture
	Barcode	Wearable	

Source: (Chi, Kang et al. 2013)

of virtual information update or change by the cloud side. Therefore, similar to the use of web browser, the involved parties can use the AR as the front-end to explore the virtual information with extra access to more variable and meaningful information in the field. According to (Kamat, Martinez et al. 2010), it can benefit construction field tasks such as schedule monitoring and building inspection where they require huge amounts of dynamic information.

Localization Technologies

(Chi, Kang et al. 2013) expect the localization technologies influence in the future development of AR applications and they introduce it as part of the core functionality of AR technology with its necessity in identifying the posture of subjects. However, the accuracy of the localization method introduces the limitations of AR applications in superimposing virtual information into the real world. To solve this issue, a research by (Zhang and Hammad 2012) discusses the development of global positioning system (GPS) and outdoor localization technologies such as radiofrequency identification (RFID), ultra-wide band (UWB) and barcoding where all combine multiple sensors to provide the accurate functionality. Despite the current developments, still all available localization approaches are limited by environmental complexity, signal quality, ranges of sensors and uncertainty where it makes it difficult for developers to maintain the accuracy level of AR and to promote the use of AR applications before improving its accuracy.

Portable and Mobile Devices

The size, weight, performance, and cost of hardware directly influence AR applications where the portability of AR depends on the level of mobility a hardware device can provide. This issue requires AR systems focus on portability and providing the ability to maneuver at outdoor environments. Research by (Yang, Tsai et al. 2011) highlights the need for all AR devices to be stable and capable of surviving in the variable environments. Until this capability is achieved, new AR applications will not become available.

Natural User Interface

(Yeh, Tsai et al. 2012) define natural user interface as intuitive control mechanism that imitate human behaviors and gestures with consideration of an increasing number of AR applications utilizing gesture controls such as Sixth-Sense as a wearable gestural interface that augments the physical world with digital information and enables the users to interact with the digital information using natural hand gestures. However, their research clarifies the dependency of AR application not only on its stability but also on the quality of the control interface. Future research requires the control interface to be more intuitive and more stable with fewer sensing limitations where it will reduce the risk of operating AR devices in dynamic environments such as construction sites.

Combination of Augmented Reality and Building Information Modeling

BIM related research has predominantly focused on enhancing communication and collaboration between stakeholders through the use of 3D modeling and 4D simulation throughout a project's life cycle (Love, Edwards et al. 2011) where BIM is utilized simply as a representation and simulation tool. The issues dealing with large quantities of data and their accessibility have hindered the use of BIM being effectively implemented on the construction site and even the issues regarding the application of BIM from design to real-time on-site construction have not explored thoroughly yet. However, the contained information within BIM should be used during construction to ensure that activities and tasks are completed on time according to schedule and to ensure the desired quality and safety standards are met (Construction 2008).

While BIM provides static and pre-defined data and information, AR can be integrated with BIM to enable the physical context of construction activities visualization with real-time monitoring of tasks. The integration of BIM with AR can provide a platform for a site management team and subcontractors to effectively interact and utilize the BIM model contained data (Hou and Wang 2011). This translates to a major opportunity presented by the use of BIM+AR technique as the ability to work at real-scale in all phases. Involved people in the building process, such as clients or prospective buyers, often have difficulties understanding and translating 2D CAD architectural representations into real 3D spatial models where they basically need to build up their individual mental models to understand the project (Donath, Beetz et al. 2001). There is also a need for a powerful and proper visualization approach to supply information to the onsite workers. BIM can make the interdependencies between work tasks more explicit with making the existing complexity more visible and manageable onsite. It can be a proactive approach where the potential negative impacts of any action can be identified earlier and mitigated avoided easily. Spatial collision analysis between trades by 3D modeling systems such as CATIA or Navisworks is an example of this approach (Wang and Love 2012). However, during the actual construction process, there is a possibility of collisions occurrence due to the change of building elements, site facilities and the movement of construction machinery. Therefore, it is challenging to introduce the onsite real-time dynamic collision detection approach to include the variations of construction sequence and schedule. Addition of AR to BIM enables real-scale representations of the existing situation to be communicated and augmented by further information where misunderstandings between planner, client and involved parties can be better avoided with savings on additional cost measures and time-delays (Donath, Beetz et al. 2001).

Therefore, BIM+AR can be proposed as the fourth era of construction IT where prior to construction, the generated digital project information can be brought onto construction site with capability of being processed in a fully digital way. Simultaneously, the site crews will have the capability of accessing the information associated with the concerned construction component or entity in a fully digital approach (Froese 2005).

Framework of Building Information Modeling and Augmented Reality Integration

(Wang, Love et al. 2013) define AR as an information aggregator to collect and consolidate information from BIM model with sensing and tracking technologies such as barcode, RFID, and GPS to identify, track and monitor individual resources. The tracking devices are mobile and ideal for onsite use to integrate BIM and AR where tags are attached to elemental components and the progress is monitored by identifying the details about the specific properties such as date, number, and text lists. Since the tags are recorded with predefined activities with the requirement of constructing a specific component and given capability to the site operator to record comments of each activity, it can be a direct link between the BIM model and AR database where both containing drawings and documents linked to a specific component and element database.

The Generation-Communication-Evaluation and Decision making (GCED) cycle was propagated by (Arayici, Coates et al. 2011) as referring to the typical onsite decision making process. In this approach a potential solution is generated before its being communicated and being aware of the potential solution, its evaluation can commence based on a set of pre-defined criteria and then decision can be made. As an example, the architects who design the building envelope interact with engineers developing the steel structures and when architects and engineers engage in discussions related to complex geometrical relationships such as facades, the GCED cycle commences (Figure 1). In the conventional approach, physical mock-ups are created considering their time-use and inaccuracy where many features and properties are lost. Another method is creation of computer-generated models prior to discussions with the consideration of their ongoing insufficiency for evaluation and collaboration purposes. Integration of BIM with AR enables the visualization of buildings 3D models with their detailed features where their properties can be visualized directly onsite to support AEC team in their communication and dynamic generation of alternative site solutions.

Similarly, in second study by (Froese 2010, Arayici, Coates et al. 2011), the framework for integration of BIM and AR is defined in four steps. First, the design and planning of construction process starts with the creation of digital prototypes in BIM model where it contains both geometric and non-geometric information. Second, the BIM model is used as the reference guide to organize the production process. Third, each involved party carries out their role according to the drawing information from the same AR based BIM model where the models are used to support effective interaction and communication. Forth, the work results are used to update the same BIM model through the function of AR annotation and commenting.

DISCUSSION

(Behzadan, Timm et al. 2008) discuss the recent developments in mobile processing units, camera quality and tracking technology where they enable AR applications to be implemented in different environments. Available mobile augmented information has various applications at the construction site including construction work planning, verification, training and safety, communication and marketing prior to construction work. Seamless interaction cycle between the real world and digital building information model developed by (Hakkarainen, Woodward et al. 2009) identifies the camera location in order to

Figure 1.

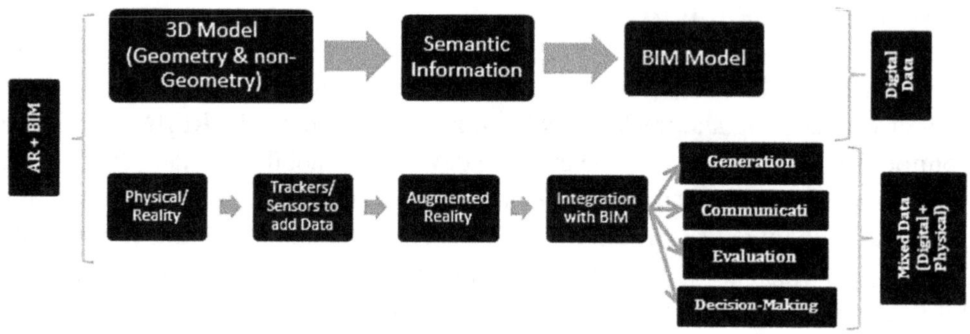

implement mobile location based visual feedback from the construction site to the BIM system with full awareness of the user's location in time and space.

(Irizarry, Gheisari et al. 2012) developed a low-cost mobile AR based tool for facility managers called InfoSPOT as a mobile AR method for accessing building information through a situation awareness approach where a real world example of a BIM was applied to test its development. The authors discovered the challenging problem of the BIM being inconsistent with the built environment and in order to solve this issue, they optimized the geometry to reduce the BIM complexity in displaying on a mobile device. Their statistical analysis indicated a generally positive response for this inexpensive solution and suggested it will assist facility managers in their routine tasks where the real-time views of space can be supplemented with vital information. They concluded that AR solutions can be easily setup and the costly hardware installations were unnecessary for a successful AR application.

(Jiao, Zhang et al. 2012) presented an AR framework integrated with a BIM and cloud computing technologies where they utilized Web3D technology to display virtual objects and resulting solution from integration with an as-built BIM model. The authors suggested the framework capability of being used in real data application where it can support collaboration in a more natural and effective way with the power of cloud services. They conducted a test case with an actual construction project and their results demonstrated the usability of the proposed framework to source virtual objects from BIM and to support collaboration. They concluded that the extension of the framework as a daily tool for construction applications require further research in the future.

(Park, Lee et al. 2012) proposed using AR for the management of defects during construction where they made an investigation in the construction industry for identifying current issues and requirements of defect management practices. Their studies presented a conceptual system framework integrating AR with BIM model for construction defect management. Based on the laboratory tests, the authors concluded that the BIM+AR system will enable managers to efficiently inspect and control their worker's job performance with allowing workers to readily confirm their works in real-time.

(Kandil, Hastak et al. 2014) proposed the integration of BIM model with AR for ductwork installation and clash detection involved in erecting the HVAC system. Even though the conflicts and clash detection can be largely identified in BIM model during design, design changes and errors or poor installation may lead to conflicts arising onsite. Using integrated AR allows a site manager to address the potential for conflicts onsite by retrieving and visualizing all the concerned properties and details in the BIM model where it can improve speed, safety and accuracy as well as reducing the cost.

CONCLUSION

Despite the limitations of building information modeling (BIM) to the design phase throughout the project lifecycle, it has started to be implemented in the construction industry in the recent years. In the meantime, augmented reality (AR) as a new and emerging technology in the construction industry deems to be a key enabler to address the current shortcomings of using BIM onsite throughout the construction process. While BIM can be used to improve the efficiency and effectiveness of design coordination, it is unable to take into account the inherent uncertainty associated with design changes and rework happens during construction and particularly in complex projects. With the rapid development and adoption of AR applications, there are numerous opportunities to integrate BIM and AR in order to visualize the information about as-built, as-planned and future progress.

In this paper, the current trends in the development of AR applications were identified and a series of examples were presented in the application of AR technologies in the AEC industry by describing how BIM+AR can be used for reasoning the interdependences of tasks, project progress monitoring, digital to physical relations, material flow tracking and visualizing design during production. However, more research and case studies are required in the future to empirically examine the proposed integrated framework in obtaining the potential productivity and performance improvements in the construction industry.

REFERENCES

Alwi, Hampson, & Mohamed. (2002). *Non value-adding activities: A comparative study of Indonesian and Australian construction projects*. Academic Press.

Aouad, G., Lee, A., & Wu, S. (2006). *Constructing the future: nD modelling*. Routledge.

Arayici, Y., Coates, P., Koskela, L., Kagioglou, M., Usher, C., & Oreilly, K. (2011). Technology adoption in the BIM implementation for lean architectural practice. *Automation in Construction*, *20*(2), 189–195. doi:10.1016/j.autcon.2010.09.016

Azuma, R. T. (1997). A survey of augmented reality. *Presence (Cambridge, Mass.)*, *6*(4), 355–385. doi:10.1162/pres.1997.6.4.355

Balzani, M., Santopuoli, N., Grieco, A., & Zaltron, N. (2004). *Laser scanner 3D survey in archaeological field: The Forum of Pompeii. International Conference on Remote Sensing Archaeology*, Beijing, China.

Behzadan, A. H., & Kamat, V. R. (2012). Enabling discovery-based learning in construction using telepresent augmented reality. *Automation in Construction*.

Behzadan, A. H., Timm, B. W., & Kamat, V. R. (2008). General-purpose modular hardware and software framework for mobile outdoor augmented reality applications in engineering. *Advanced Engineering Informatics*, *22*(1), 90–105. doi:10.1016/j.aei.2007.08.005

Bruner, J. S. (1966). *Towards a theory of instruction*. Harvard University Press.

Caudell, T. P., & Mizell, D. W. (1992). *Augmented reality: An application of heads-up display technology to manual manufacturing processes. System Sciences, 1992. Proceedings of the Twenty-Fifth Hawaii International Conference on*. IEEE. 10.1109/HICSS.1992.183317

Chi, H.-L., Kang, S.-C., & Wang, X. (2013). Research trends and opportunities of augmented reality applications in architecture, engineering, and construction. *Automation in Construction*, *33*, 116–122. doi:10.1016/j.autcon.2012.12.017

Construction, M.-H. (2008). Building Information Modeling Trends SmartMarket Report. New York: Academic Press.

Donath, D., Beetz, J., Grether, K., Petzold, F., & Seichter, H. (2001). *Augmented reality techniques for design and revitalisation in existing built environments*. Buffalo, NY: Angenommener Beitrag zur ACADIA.

Dong, S., Feng, C., & Kamat, V. R. (2012). Sensitivity analysis of augmented reality-assisted building damage reconnaissance using virtual prototyping. *Automation in Construction*.

Dossick, C. S., & Neff, G. (2009). Organizational divisions in BIM-enabled commercial construction. *Journal of Construction Engineering and Management*, *136*(4), 459–467. doi:10.1061/(ASCE)CO.1943-7862.0000109

Eastman, C., Jeong, Y.-S., Sacks, R., & Kaner, I. (2009). Exchange model and exchange object concepts for implementation of national BIM standards. *Journal of Computing in Civil Engineering*, *24*(1), 25–34. doi:10.1061/(ASCE)0887-3801(2010)24:1(25)

Feiner, S. K. (2002). Augmented reality: A new way of seeing. *Scientific American*, *286*(4), 48–55. do i:10.1038cientificamerican0402-48 PMID:11905108

Froese, T. (2005). Impact of emerging information technology on information management. *International Conference on Computing in Civil Engineering*, Cancun, Mexico. 10.1061/40794(179)162

Froese, T. M. (2010). The impact of emerging information technology on project management for construction. *Automation in Construction*, *19*(5), 531–538. doi:10.1016/j.autcon.2009.11.004

Goldman, D. (2012). *Google unveils 'Project Glass' virtual-reality glasses*. Money.

Hakkarainen, M., Woodward, C., & Rainio, K. (2009). Software architecture for mobile mixed reality and 4D BIM interaction. *Proc. 25th CIB W78 Conference*.

Halpin & Woodhead. (1998). *Construction management*. Wiley.

Holden, W. (2005). *Juniper research*. White Paper.

Hou, L., & Wang, X. (2011). Experimental Framework for Evaluating Cognitive Workload of Using AR System for General Assembly Task. *Proceedings of the 28th International Symposium on Automation and Robotics in Construction*.

Hwang, B.-G., Thomas, S. R., Haas, C. T., & Caldas, C. H. (2009). Measuring the impact of rework on construction cost performance. *Journal of Construction Engineering and Management*, *135*(3), 187–198. doi:10.1061/(ASCE)0733-9364(2009)135:3(187)

Hyers, K. (2006). *US mobile broadcast video market: Five predictions*. ABI Research.

Irizarry, J., Gheisari, M., Williams, G., & Walker, B. N. (2012). InfoSPOT: A mobile Augmented Reality method for accessing building information through a situation awareness approach. *Automation in Construction*.

Jardim-Goncalves, R., & Grilo, A. (2010). SOA4BIM: Putting the building and construction industry in the Single European Information Space. *Automation in Construction*, *19*(4), 388–397. doi:10.1016/j.autcon.2009.11.009

Jiao, Y., Zhang, S., Li, Y., Wang, Y., & Yang, B. (2012). Towards cloud Augmented Reality for construction application by BIM and SNS integration. *Automation in Construction*.

Kamat, V. R., Martinez, J. C., Fischer, M., Golparvar-Fard, M., Peña-Mora, F., & Savarese, S. (2010). Research in visualization techniques for field construction. *Journal of Construction Engineering and Management*, *137*(10), 853–862. doi:10.1061/(ASCE)CO.1943-7862.0000262

Kandil, Hastak, & Dunston. (2014). Creating 3D web-based game environment using BIM models for virtual on-site visiting of building HVAC systems. *Bridges, 10*.

Kang, J. H., Anderson, S. D., & Clayton, M. J. (2007). Empirical study on the merit of web-based 4D visualization in collaborative construction planning and scheduling. *Journal of Construction Engineering and Management*, *133*(6), 447–461. doi:10.1061/(ASCE)0733-9364(2007)133:6(447)

Love, P. E., Edwards, D. J., Han, S., & Goh, Y. M. (2011). Design error reduction: Toward the effective utilization of building information modeling. *Research in Engineering Design*, *22*(3), 173–187. doi:10.100700163-011-0105-x

McKibben, D., & Furlonger, D. (2009). *Magic quadrant for operational risk management software for financial services*. Gartner Research. Available: http://www. gartner. com/technology/home. jsp

Papagiannakis, G., Singh, G., & Magnenat-Thalmann, N. (2008). A survey of mobile and wireless technologies for augmented reality systems. *Computer Animation and Virtual Worlds*, *19*(1), 3–22. doi:10.1002/cav.221

Park, C.-S., & Kim, H.-J. (2012). A framework for construction safety management and visualization system. *Automation in Construction*.

Park, C.-S., Lee, D.-Y., Kwon, O.-S., & Wang, X. (2012). A Framework for Proactive Construction Defect Management using BIM, Augmented Reality and Ontology-based Data Collection Template. *Automation in Construction*.

Penttilä, H. (2006). *Describing the changes in architectural information technology to understand design complexity and free-form architectural expression*. ITcon.

Shin, D. H., Park, J., Woo, S., & Jang, W.-S. (2012). Representations for imagining the scene of non-existing buildings in an existing environment. *Automation in Construction*.

Stanco, F., Battiato, S., & Gallo, G. (2011). *Digital Imaging for Cultural Heritage Preservation: Analysis, Restoration, and Reconstruction of Ancient Artworks*. CRC Press.

Taylor, J. E., & Bernstein, P. G. (2009). Paradigm trajectories of building information modeling practice in project networks. *Journal of Management Engineering*, *25*(2), 69–76. doi:10.1061/(ASCE)0742-597X(2009)25:2(69)

Van Krevelen, D., & Poelman, R. (2010). A survey of augmented reality technologies, applications and limitations. *International Journal of Virtual Reality*, *9*(2), 1.

Wang & Dunston (2007). *Design, strategies, and issues towards an augmented reality-based construction training platform*. ITcon.

Wang, X., & Love, P. (2012). BIM+ AR: Onsite information sharing and communication via advanced visualization. *Computer Supported Cooperative Work in Design (CSCWD), 2012 IEEE 16th International Conference on*. IEEE.

Wang, X., Love, P. E., Kim, M. J., Park, C.-S., Sing, C.-P., & Hou, L. (2013). A conceptual framework for integrating building information modeling with augmented reality. *Automation in Construction*, *34*, 37–44. doi:10.1016/j.autcon.2012.10.012

Wong, J. K., Li, H., Wang, H., Huang, T., Luo, E., & Li, V. (2012). Toward low-carbon construction processes: The visualisation of predicted emission via virtual prototyping technology. *Automation in Construction*.

Yang, M.-D., Chao, C.-F., Huang, K.-S., Lu, L.-Y., & Chen, Y.-P. (2012). Image-based 3D scene reconstruction and exploration in augmented reality. *Automation in Construction*.

Yang, T.-J., Tsai, Y.-M., & Chen, L.-G. (2011). Smart display: A mobile self-adaptive projector-camera system. *Multimedia and Expo (ICME), 2011 IEEE International Conference on*. IEEE.

Yeh, K.-C., Tsai, M.-H., & Kang, S.-C. (2012). On-site building information retrieval by using projection-based augmented reality. *Journal of Computing in Civil Engineering*, *26*(3), 342–355. doi:10.1061/(ASCE)CP.1943-5487.0000156

Yokokohji, Y., Sugawara, Y., & Yoshikawa, T. (2000). Accurate image overlay on video see-through HMDs using vision and accelerometers. *Virtual Reality (Waltham Cross)*.

Zhang, C., & Hammad, A. (2012). Improving lifting motion planning and re-planning of cranes with consideration for safety and efficiency. *Advanced Engineering Informatics*, *26*(2), 396–410. doi:10.1016/j.aei.2012.01.003

KEY TERMS AND DEFINITIONS

Augmented Reality: A live direct or indirect view of a physical and real-world environment where the elements are augmented by computer-generated input such as sound, video, graphics, etc.

AEC: Stands for architectural, engineering and construction field.

Building Information Modeling (BIM): A process where it involves the generation and management of digital representations of physical and functional characteristics of items or places.

Literature Review: A text of a scholarly paper including the current knowledge and findings alongside the theoretical and methodological contributions to a particular topic.

This research was previously published in the Encyclopedia of Information Science and Technology, Fourth Edition edited by Mehdi Khosrow-Pour, pages 983-993, copyright year 2018 by Information Science Reference (an imprint of IGI Global).

Section 3
Computer Engineering

Chapter 6
Architecture of an Open-Source Real-Time Distributed Cyber Physical System

Stefano Scanzio
CNR-IEIIT, Italy

ABSTRACT

Cyber physical systems are based on a number of nodes connected through a communication network, which can interact with the environment. In this chapter, a completely open-source architecture of a cyber physical system based on off-the-shelf components will be presented. Its main characteristics are high real-time capabilities and the use of both wired and IEEE 802.11 wireless technologies for communication. The Linux operating system installed on common personal computers and communication technologies derived from the IT world make the proposed architecture highly customizable, inexpensive, and performing. Moreover, the presence of a time synchronization service allows the sharing of time between nodes. Specific software and techniques, some based on synchronized nodes, are used to increase determinism and reliability in both wired and wireless extensions.

INTRODUCTION

Cyber physical systems (CPSs) (Guan et al., 2016) are based on a number of nodes connected through a communication network, which can interact with the environment. Application contexts that foresee the use of CPSs are Industry 4.0, building automation and smart grids (Yu & Xue, 2016). In such systems, nodes are connected through a wired or a wireless network, but commonly a typical configuration relies on the combination of both. In many applications, such as those regarding manufacturing systems, determinism on data exchanges between nodes of the system is the most important requirement.

On the other hand, in recent years, open-source software and applications are approaching CPS market. Open-source provides a low cost and effective method to implement inexpensive, performing and bug free applications. The need of open-source components for many application contexts derives from their capability to be highly customizable. This characteristic impacts directly on performance

DOI: 10.4018/978-1-5225-7359-3.ch006

to preform actuations or sensing physical quantities, in terms of determinism and latency. The use of open-source software for real embedded systems has been described in the book (Cibrario Bertolotti, & Hu, 2015). Other examples of use of open-source software will be provided in the next sections. In the PC world, the Linux operating system is an example of widely adopted open-source platform, which offered long-term support and development. These last features are really important for systems such as CPS, which are updated and replaced infrequently. The Linux operating system is often chosen by researchers, users and developers. Moreover, the code of the operating system, and of its software suite, is well documented, allowing to expert users a really high degree of control of any behavioral aspect. From the network point of view, Ethernet is the "de facto" standard for wired networks. Regarding wireless, a number of technologies exist, but the most common, fast and easily interoperable with Ethernet is IEEE 802.11, also known as Wi-Fi. For cost reasons and ease of configuration and installation, these technologies are the best candidates for a CPS based on open-source components. It is worth pointing out that, while the majority of protocols for industrial automation rely on Ethernet, possible candidates in industrial CPS for wireless extensions are IEEE 802.11, Bluetooth or wireless sensor networks based on the IEEE 802.15.4 standard (Lu et al., 2016).

In this chapter, a possible implementation of a CPS based on the Linux operating system is presented. The proposed architecture exploits the RTAI or XENOMAI hard real-time schedulers to guarantee the required degree of determinism of nodes. It makes use of synchronization protocols, in both the wired and the wireless extensions, to provide all the nodes a common view of time. Transmission latency can be reduced in the wired network by using hard real-time protocol stacks (such as, e.g., RTnet) and channel access methods as the time division multiple access (TDMA). Regarding Wi-Fi, redundancy techniques based on the transmission of the very same data packet on two non-overlapping networks allow to reduce both the number of packets lost and latency. Finally, we will show how the proposed architecture simplifies the integration of popular industrial protocols (e.g., EtherCAT or Modbus TCP) within the communication system.

BACKGROUND

CPSs usually interact with the surrounding environment through a wired or a wireless network. Sensing applications used to measure physical values are characterized by few timing requirements. Usually, a network wide notion of the time is maintained through specific synchronization protocols. Common time allows network nodes to associate a time to measured values. It is the case of wireless sensor networks (WSN) (Zaman, Ragab, & Abdullah, 2012) or monitoring systems. A large class of CPS applications is control systems (CSs). In CSs, remote actuators have to be continuously managed by a control unit. In its control-loop, the CS cyclically makes use of data acquired by remote sensors to properly command actuators. The main requirements of CSs are the determinism of the communication network and its reliability. In practice, a network packet has to reach the receiving node within a predefined deadline. In hard real-time (HRT) systems deadlines cannot be exceed, while in soft real-time (SRT) systems they can be exceed with low probability, i.e., the deadline miss ration has to be bounded. For wired networks, a number of protocols, technologies and node architectures, typically originated from the industrial world, are able to ensure HRT constraints. Nowadays, only SRT is possible for wireless networks. This is mainly due to the non-exclusive use of the communication medium (i.e., the ether) by nodes involved in the communication. Nevertheless, a number of countermeasures appeared in scientific literature to increase determinism and reliability of wireless protocols, especially for Wi-Fi. Finally, in such type of networks, nodes have to send packets with the correct timing constraints, to ensure the determinism required by the application. To this extent, RT properties of nodes have to be improved by using, e.g., a RT operating system. In the following, design guidelines and references will be provided for the implementation of a distributed wired/wireless CPS, with completely open-source components, RT capabilities and conventional hardware.

OPEN-SOURCE ARCHITECTURE FOR A REAL-TIME CPS

Architecture

A reference network architecture of a CPS is schematically presented in Figure 1. A *wired backbone* interconnects *wired nodes*. In the proposed open-source architecture, it consists in standard Ethernet cables and switches. Some *wireless extensions* are connected to the wired backbone. They allow node mobility and cable replacement for such applications in which cabling is hard or cables are prone to wear and tear (e.g., in robotic arms). Wi-Fi has been chosen as the reference wireless technology for wireless extensions because it is faster if compared with other WSNs technologies, it is completely interoperable at the data link layer with Ethernet, and its basic components (i.e., access points (APs) and Wi-Fi adapters) are available off-the-shelf at relatively low cost. APs are used as interconnection elements between the wired backbone and wireless extensions. Two types of Wi-Fi nodes can coexist in the same CPS: conventional *Wi-Fi nodes* and redundant *Red. Wi-Fi nodes* equipped with two or more wireless adapters. The latter type of node is used for SRT applications with demanding latency and reliability

Figure 1. Reference architecture of a CPS

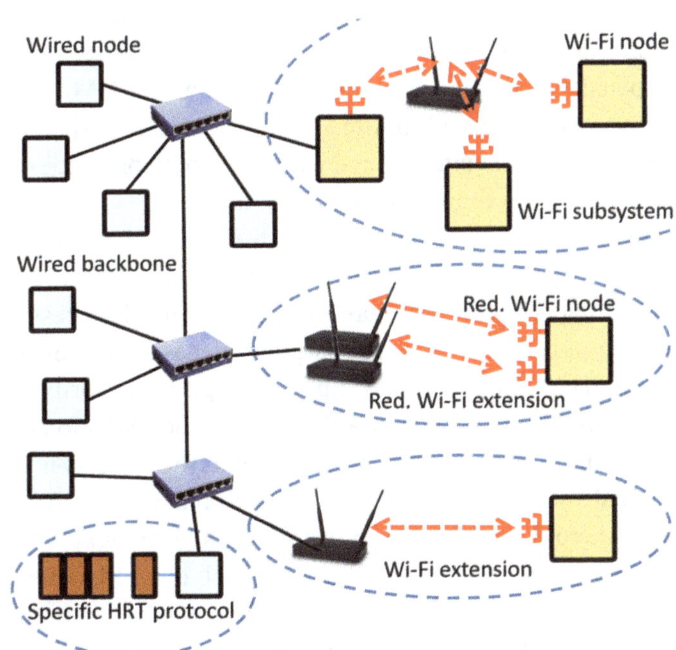

requirements. Some nodes can be equipped with both wired and wireless adapters. An example is the border node of a *Wi-Fi subsystem* (e.g., the interface node of an industrial machinery produced by a specific vendor) which is connected to the wired backbone to communicate with other components of the CPS and internally exploits wireless communication for the connection with moving elements. Finally, some specific network segments may require a high degree of determinism. In this case, *specific HRT network protocols* can ensure the expected performance. Both wired and wireless nodes of the CPS are personal computers (PCs) or embedded systems running the Linux operating system.

Wired Backbone

In the proposed architecture, wired backbone is based on the IEEE 802.3 Ethernet standard. Switches are used to interconnect wired nodes. As a matter of fact, the level of determinism in terms of latency of switched Ethernet is enough for the majority of the CPS application contexts. Nevertheless, switches introduce indeterminism on delivery time, because packets queued inside them experience a delay that depends on network traffic. To avoid this problem, a synchronization protocol can be used to provide a common notion of time to nodes. As a consequence, if network packets containing a command arrive to the receivers sufficiently in advance, synchronized actuations can be performed by different nodes, even if the underlying communication network is not deterministic enough. Synchronization protocols will be analyzed in detail later. If the network packet containing the actuation command cannot be sent sufficiently in advance, because the actuation command is based on other sensors data that must be adequately new, network determinism has to be improved. A common and effective remedy is traffic prioritization. Most important traffic has to be tagged with the higher priority, as define in IEEE P802.1p standard, to increase Quality of Service (QoS). Such solution is adopted by the EtherNet/IP industrial protocol. Further advantages can be achieved by using specifically modified switches. For example, in TTEthernet high priority traffic is sent on specific time intervals. TTEthernet ensures that in such intervals other traffic cannot interfere with the high priority one. TTEthernet preserves the compatibility with standard Ethernet (i.e., traffic is routed with conventional Ethernet rules outside the reserved intervals). Unfortunately, as other HRT protocols, TTEthernet requires specific and expensive hardware. A software-based, less expensive and less deterministic solution, is TDMA. In TDMA, synchronized nodes subdivide the communication network in time slots. Each node is then allowed to transmit only in its assigned slots, preventing possible interference with other network nodes. The main disadvantage of this technique is that all nodes have to implement TDMA.

Quite counter-intuitively, in relatively small networks, the major source of indeterminism is not the network, but is the node itself. A simplified schema of a Linux-based PC is reported in Figure 2.

In a PC, concurrently, a number of processes make use of the central processing unit (CPU). Between these processes, the control application (CA) is in charge to manage the transmission and reception of the relevant data of the CPS. The transmission of a packet (e.g., to perform an actuation on a remote node) by means of the CA is usually triggered by a timer or by the reception of another packet, which for instance contains a new sensed datum. In both cases, when the CPU is in use by others process, a context switch operation to substitute the running process with that of the CA is needed. The additional delay to save the context of the old process and to restore the CA context worsen the HRT capability of the node, if compared with the more favorable case in which the CPU is idle when the CA process is triggered. To send the packet, the CA that runs in user-space makes use of the *system calls* provided by the operating system. System calls software is executed in kernel-space. The transition from user to

Figure 2. Simplified schema of a PC wired node

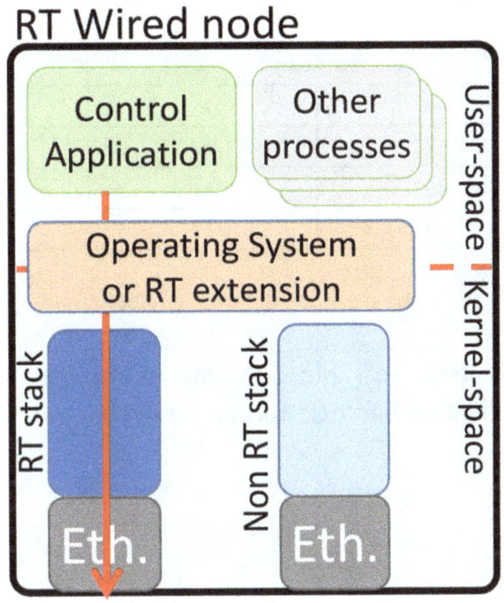

kernel spaces requires another context switch operation, which leads to further indeterminism. Moreover, the use of shared resources by means of concurrent processes decreases determinism (e.g., the access to the Ethernet adapter or to the main memory is performed through the shared system bus). Additionally, the reaction time of the system to an interrupt depends on pending interrupts not yet managed by the CPU. A first remedy is the installation of specific Linux kernel patches with the aim to reduce the time intervals in which the operating system cannot be preempted by other ready processes, such as the CA. The most known is RT patch. Among its most important features, it minimizes the time in which the CPU is not sensible to interrupts and it allows the Linux kernel to assign a priority to each interrupt (i.e., a major priority can be assigned to Ethernet adapter and timing interrupts). Although RT patch allows to increase determinism, the resulting system can only be considered SRT. A more effective counter-measure relies on the use of HRT extensions, such as RTAI or XENOMAI. Normal processes and the operating system itself are scheduled by these extensions as non-real-time processes. HRT processes are not scheduled by the Linux kernel but by RTAI or XENOMAI. In this way, they do not suffer of the interference caused by other non-real-time processes. Unfortunately, to guarantee HRT performance, CA software cannot make use of system calls, and device drivers have to be programmed exploiting specific RTAI or XENOMAI libraries.

The most recent article of performance comparison between standard Linux kernel, RT patch and RTAI is (Cereia, Cibrario Bertolotti, & Scanzio, 2011). Some results of that work are reported in Table 1. It analyzes the jitter (i.e., the deviation from the desired actuation time) in a system without additional load and with an intensive I/O load of 500 IRQs per second. In the latter configuration, results confirm the impossibility to use a standard Linux kernel in RT applications (maximum jitter is 1151 μs), and the effectiveness of RT patch and RTAI (maximum jitter is 10.4 and 8.0 μs, respectively).

When packets have to be sent timely, also driver and protocol stack implementations must have RT properties. A commonly used HRT protocol stack for wired adapters is RTnet, which ensures a bounded delay on packets delivery. Jitters of standard Linux and RTnet were analyzed in (Cena, Scanzio, Va-

Table 1. Comparison between standard Linux kernel, RT patch and RTAI

	Jitter (μs)			
	No Load		I/O Load	
	σ	Max	σ	Max
Std. Linux kernel	5.72	1434.0	5.75	1151.0
RT patch	0.12	1.8	0.63	10.4
RTAI	0.12	2.5	0.32	8.0

Data source: (Cereia, Cibrario Bertolotti, & Scanzio, 2011).

lenzano, & Zunino, 2011). As reported in Table 2, the maximum jitter of the standard driver in the I/O load condition is for many application contexts too high (1910.6 μs). On the other side, with RTnet, jitter remains bounded to 38 μs.

Wi-Fi Extensions

Wi-Fi extensions are really useful in many application contexts. Unfortunately, they cannot guarantee the determinism of cables. In particular, the number and type of wireless nodes in proximity of the CPS is usually out of the control of the application designer. Choosing the band and the transmission channel of the Wi-Fi system in a frequency not exploited by other wireless networks is the first best practice. Sometimes this solution is not possible, because in a specific environment users have not control over the surrounding networks. With the widespread diffusion of mobile devices, the elimination of these interfering nodes is becoming even more complicated. Moreover, electromagnetic disturbances are another source of packets loss. TDMA techniques have been extensively exploited in Wi-Fi to prevent the indeterminism of the CSMA/CA method used by nodes to manage channel access (Sevani, Raman, & Joshi, 2014). As for the wired case, synchronized nodes are allowed to transmit only in their assigned slots. Unfortunately, even with TDMA, sensitivity to channel disturbances is equal to normal Wi-Fi. In addition, to obtain a sufficient degree of determinism, these methods require that all the wireless nodes follow TDMA rules. Even if TDMA and the other listed techniques can noticeably improve channel quality, they cannot deal with protracted disturbances or heavy interferences.

An effective solution is seamless redundancy, which is based on frequency diversity. The architecture of a *Red. Wi-Fi node* implementing seamless redundancy and based on a PC is reported in Figure 3. The most significant difference with a typical Wi-Fi node is the presence of two (or more) Wi-Fi adapters.

Table 2. Comparison between standard and RTnet protocol stack

	Jitter (μs)			
	No Load		I/O Load	
	σ	Max	σ	Max
Standard	11.2	72.3	35.2	1910.6
RTnet	6.0	38.0	4.0	33.6

Data source: (Cena, Scanzio, Valenzano, & Zunino, 2011).

Figure 3. Simplified schema of a PC-based Red Wi-Fi node

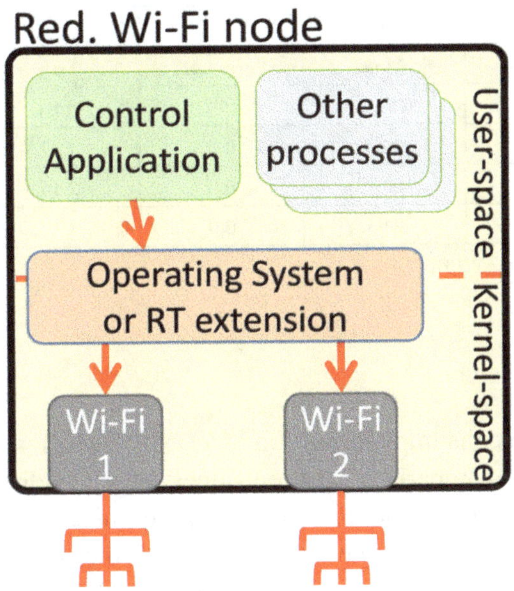

With seamless redundancy, two copies of the same message are transmitted on distinct wireless channels. The receiver node takes care of discarding duplicates, and it forwards to the application layer only the first message copy arrived to destination. The probability that a disturbance or an interference affects all the channels involved in the parallel transmission is relatively small. Seamless redundancy for Wi-Fi was firstly proposed by (Rentschler & Laukemann, 2012) and it was based on the Parallel Redundancy Protocol (PRP), which is a protocol aimed at ensuring end-to-end redundancy for wired Ethernet networks. Its first adaptation to Wi-Fi relied on specific devices named RedBoxes that implement the standard PRP protocol over Ethernet. Wi-Fi equipment was connected to RedBoxes to analyze PRP performance in a wireless network. Seamless redundancy based on PRP is an end-to-end solution, i.e., between sender and receiver nodes. In Wi-Red technique (Cena, Scanzio, Valenzano, & Zunino, 2014), seamless redundancy was applied at the link level. For instance, in the communication path between two Wi-Fi nodes, the first link is between the node and the AP, while the second is between the AP and the second node. With Wi-Red, further improvements are possible: by implementing specific techniques to avoid the sending of packets already arrived through other channels, or by exploiting heuristics for scheduling the traffic on the two interfaces.

Results published in scientific literature (Cena, Scanzio, & Valenzano, 2016) confirm that significant improvements can be achieved in real installations in terms of packet losses and transmission latencies. Table 3 compares performances of two single channels with those obtained with seamless redundancy. Results, obtained sending periodically (every 100 ms) unicast and multicast traffic, are in terms of average latency, maximum latency and percent of packet lost much better than those of the single channel. Even if, at present, a HRT driver implementation similar to RTnet for Wi-Fi adapters is not available, the use of HRT extensions such as RTAI or XENOMAI is recommend to improve determinism as much as possible.

Table 3. Comparison between Wi-Fi transmissions on individual channels and on redundant channels

	Unicast			Multicast		
	μ	Max	Perc. Lost	μ	Max	Perc. Lost
	(ms)	(ms)	(%)	(ms)	(ms)	(%)
Channel A	0.92	100.77	0.0	1.57	106.30	0.51
Channel B	0.95	100.92	0.0	2.16	92.71	1.11
Channel A+B	0.71	8.93	0.0	1.12	44.35	0.02

Data source: (Cena, Scanzio, & Valenzano, 2016).

Clock Synchronization

Synchronization makes possible the implementation of TDMA and offers some essential services at the application level. To permit the comparison between values sensed by different nodes, sensed data must be associated with a temporal information. In addition, synchronization enables different nodes to perform simultaneous actuations, addressing the indeterminism of the communication network by sending packets in advance. Synchronization protocols are based on the exchange of specific sequences of packets, and on timestamps obtained on sending and/or reception of a subset of these packets. Timestamps are used by synchronization protocols for the regulation of nodes time by means of specific clock regulation algorithms (Mongelli & Scanzio, 2016). Their precision is directly related to timestamps accuracy. If hardware timestamping is not supported by the adapter, timestamps are typically acquired at the first instruction of the Interrupt Service Routine (ISR), executed to serve the interrupt raised by the network adapter, which follows the reception of a packet. Synchronization quality in the case of software timestamps is in the order of tens microseconds, while for hardware timestamps a typical accuracy is in the order of some nanoseconds. For wired networks, the “de facto” standard is the master-slave IEEE 1588 protocol (Cena, Cibrario Bertolotti, Scanzio, Valenzano, & Zunino, 2013). The IEEE 1588 protocol is based on a tree synchronization hierarchy, where a number of slave nodes synchronize with a master node. A slave node can act as a master for its sub-tree. For Wi-Fi, specific synchronization protocols, namely the *timing measurement* and *timing advertisement*, have been included in the 2012 version of IEEE 802.11 standard, but at the time of writing they are not implemented in most of the available off-the-shelf Wi-Fi adapters, and they have some disadvantages (Cena, Scanzio, Valenzano, & Zunino, 2015). The porting of IEEE 1588 to Wi-Fi has been analyzed in (Mahmood, Exel, & Sauter, 2014). The AP acts as a slave for the wired backbone and as a master for the Wi-Fi extension. This solution is really effective, and, above all, it makes possible the sharing of the same synchronization protocol to all the nodes of the CPS. Unfortunately, implementation requires modifications to the AP. The Reference Broadcast Infrastructure Synchronization (RBIS) protocol (Cena et al., 2015) does not require any modification to Wi-Fi equipment. As a consequence, it can be directly implemented in software on conventional hardware. RBIS requires a Wi-Fi node as a master, i.e., the master is not the AP. This last requirement limits the possible network topologies. In fact, the master node, to synchronize with the wired backbone, has two network adapters: a wired interface connected to the backbone and a Wi-Fi interface connected to the AP. On the other side, RBIS enables the connection of Wi-Fi subsystems to the wired backbone, as represented in Figure 1. Synchronization error achieved by RBIS on a real implementation based on a conventional PC is less than 3.3 μs (Cena et al., 2015).

Other RT Protocols

Usually, actuations and sensing have to be performed within a given deadline. To this extent, some network segments can rely on specific and more deterministic protocols, especially if the CPS is used for CSs with safety requirements. A number of HRT protocols exist, each with its own peculiarities. EtherCAT (Scanzio, 2012) is an example of really high deterministic protocol. In EtherCAT, master node can be implemented on a conventional PC. This feature and the availability of open-source software make it particularly suitable to be used in Linux-based CPS. Slave nodes, those performing sensing and actuations, are implemented in hardware to increase determinism. EtherCAT has its own synchronization protocol that guarantees a synchronization quality smaller than 50 ns (Cena, Cibrario Bertolotti, Scanzio, Valenzano, & Zunino, 2012), and which can be easily integrated with IEEE 1588 through specific EtherCAT slave devices. The most powerful open-source EtherCAT master software for Linux is EtherLab. A description of EtherLab and some guidelines for its use and configuration can be found in (Scanzio, 2012), while its performance was evaluated in (Cereia et al., 2011). An EtherCAT master implemented on a Linux-based conventional PC requires a very high determinism on sending Ethernet frames. For this reason, the use of HRT extensions as RTAI or XENOMAI is highly recommended.

SOLUTIONS AND RECOMMENDATIONS

A completely open-source architecture for CPS based on Linux, off-the-shelf network devices and standard PCs has been proposed. Determinism of nodes is achieved with HRT extensions (i.e., RTAI or XENOMAI) or with a SRT patch named RT patch. Moreover, specific HRT protocol stacks, such those provided by RTnet, can be used for wired connections. Depending on the application requirements, the proposed architecture has different levels of capability and scalability. For wired networks, it may rely on conventional Ethernet, but TDMA techniques can be used to increase determinism. Specific HRT protocols such EtherCAT can be exploited to further increase determinism. Regarding wireless extensions, if conventional Wi-Fi is not enough, TDMA is a first improvement, which can be further enhanced with techniques based on seamless redundancy. However, to the current state-of-the-art, Wi-Fi can be considered only for SRT systems, in which deadlines can be exceeded with low probability (Cena, Scanzio, & Valenzano, 2016). Finally, a synchronization service is transparently available in all nodes of the CPS. It is based on the IEEE 1588 standard for wired segments, and on recent researches for Wi-Fi extensions.

FUTURE RESEARCH DIRECTIONS

For many years, a constant research activity was targeted to improve determinism of wired networks. Its outcomes were a suite of HRT protocols that rely on Ethernet, and deterministic software and operating systems for nodes. Many of these software (and hardware) are commercial products, some are open-source, which are the most used in research. Nowadays, research activity on CPS wired networks is aimed to increase performance and to seamless integrate all network components (Bangemann, Riedl, Thron, & Diedrich, 2016). Regarding determinism, HRT extensions installed on PCs have counter-intuitively worst performance than in the past, because today's CPUs are multi-core and their target is to maximize throughput instead of determinism. The evolution of the hardware requires the implementation of new algorithms and software. The use of big-data analysis and machine learning techniques is a way to improve the quality of some services, such as for synchronization protocols. Nodes virtualization allows to decrease configuration and management effort in large CPS, and it is a very promising research area, which opens many interesting investigation aspects regarding especially determinism. An even more promising research area concerns wireless. CPS has a really high demand of deterministic communication over the ether. Seamless redundancy is a new remedy, but not a definitive solution. Very large improvements can be achieved on existing techniques, and the opportunity to propose new effective methods is very high. A really heterogeneous set of wireless technologies is currently available in the market, with very orthogonal characteristics (e.g., IEEE 802.11, Bluetooth, IEEE 802.15.4, WirelessHART, LoRa, etc.). For each of these technologies, techniques to increase determinism and reliability have to be specifically implemented to obtain the maximum improvement (Lu et al., 2016). Finally, other services such localization need to be improved and included in wireless networks of the future, in order to better support node mobility.

CONCLUSION

The proposed architecture for the implementation of a CPS is based on open-source software, on standard hardware and on the Linux operating system. Other architectural solutions are possible, especially if commercial products are considered. The latter are typically easier to install and configure, but their integration in the overall architecture is more complex. The proposed open-source solution is very scalable and customizable, especially because software components are easily modifiable and highly configurable (i.e., source code is open). The reference wireless technology for wireless extensions discussed in this chapter is Wi-Fi. It offers a number of advantages in terms of both integration with Ethernet networks and throughput. Nowadays, general purpose systems (e.g., PC equipped with Linux) and off-the-shelf devices (e.g., Ethernet and Wi-Fi adapters, or APs and switches) can be the basic components of high performing CPS. This is a great achievement. Moreover, future research activities are very promising and aimed to replace cable with wireless technology for an increasing number of applications.

REFERENCES

Bangemann, T., Riedl, M., Thron, M., & Diedrich, C. (2016). Integration of Classical Components Into Industrial Cyber–Physical Systems. *Proceedings of the IEEE, 104*(5), 947–959. doi:10.1109/JPROC.2015.2510981

Cena, G., Cibrario Bertolotti, I., Scanzio, S., Valenzano, A., & Zunino, C. (2012). Evaluation of EtherCAT Distributed Clock Performance. *IEEE Transactions on Industrial Informatics*, *8*(1), 20–29. doi:10.1109/TII.2011.2172434

Cena, G., Cibrario Bertolotti, I., Scanzio, S., Valenzano, A., & Zunino, C. (2013). Synchronize your watches: Part I: General-purpose solutions for distributed real-time control. *IEEE Industrial Electronics Magazine*, *7*(1), 18–29. doi:10.1109/MIE.2012.2232354

Cena, G., Scanzio, S., & Valenzano, A. (2016). Seamless Link-Level Redundancy to Improve Reliability of Industrial Wi-Fi Networks. *IEEE Transactions on Industrial Informatics*, *12*(2), 608–620. doi:10.1109/TII.2016.2522768

Cena, G., Scanzio, S., Valenzano, A., & Zunino, C. (2011). Evaluation of the real-time properties of open-source protocol stacks. In *IEEE 16th Conference on Emerging Technologies & Factory Automation (ETFA)* (pp. 1-8). 10.1109/ETFA.2011.6059003

Cena, G., Scanzio, S., Valenzano, A., & Zunino, C. (2015). Implementation and Evaluation of the Reference Broadcast Infrastructure Synchronization Protocol. *IEEE Transactions on Industrial Informatics*, *11*(3), 801–811. doi:10.1109/TII.2015.2396003

Cereia, M., Cibrario Bertolotti, I., & Scanzio, S. (2011). Performance of a Real-Time EtherCAT Master Under Linux. *IEEE Transactions on Industrial Informatics*, *7*(4), 679–687. doi:10.1109/TII.2011.2166777

Cibrario Bertolotti, I., & Hu, T. (Eds.). (2015). *Embedded Software Development: The Open-Source Approach*. Boca Raton, FL: CRC Press.

Guan, X., Yang, B., Chen, C., Dai, W., & Wang, Y. (2016). A comprehensive overview of cyber-physical systems: from perspective of feedback system. *IEEE/CAA Journal of Automatica Sinica, 3*(1), 1-14.

Lu, C., Saifullah, A., Li, B., Sha, M., Gonzalez, H., Gunatilaka, D., ... Chen, Y. (2016). Real-Time Wireless Sensor-Actuator Networks for Industrial Cyber-Physical Systems. *Proceedings of the IEEE, 104*(5), 1013–1024. doi:10.1109/JPROC.2015.2497161

Mahmood, A., Exel, R., & Sauter, T. (2014). Delay and Jitter Characterization for Software-Based Clock Synchronization Over WLAN Using PTP. *IEEE Transactions on Industrial Informatics*, *10*(2), 1198–1206. doi:10.1109/TII.2014.2304413

Mongelli, M., & Scanzio, S. (2016). A neural approach to synchronization in wireless networks with heterogeneous sources of noise. *Ad Hoc Networks*, *49*, 1–16. doi:10.1016/j.adhoc.2016.06.002

Rentschler, M., & Laukemann, P. (2012). Towards a reliable parallel redundant WLAN black channel. In *9th IEEE International Workshop on Factory Communication Systems (WFCS)* (pp. 255-264). Lemgo, Germany. 10.1109/WFCS.2012.6242573

Scanzio, S. (2012). SoftPLC-Based Control: A Comparison between Commercial and Open-Source EtherCAT Technologies. In M. Khan & A. Ansari (Eds.), *Handbook of Research on Industrial Informatics and Manufacturing Intelligence: Innovations and Solutions* (pp. 440–463). Hershey, PA: Information Science Reference. doi:10.4018/978-1-4666-0294-6.ch019

Sevani, V., Raman, B., & Joshi, P. (2014). Implementation-Based Evaluation of a Full-Fledged Multihop TDMA-MAC for WiFi Mesh Networks. *IEEE Transactions on Mobile Computing*, *13*(2), 392–406. doi:10.1109/TMC.2012.251

Yu, X., & Xue, Y. (2016). Smart Grids: A Cyber–Physical Systems Perspective. *Proceedings of the IEEE, 104*(5), 1058–1070. doi:10.1109/JPROC.2015.2503119

Zaman, N., Ragab, K., & Abdullah, A. (2012). *Wireless Sensor Networks and Energy Efficiency: Protocols, Routing and Management*. Hershey, PA: IGI Global. doi:10.4018/978-1-4666-0101-7

ADDITIONAL READING

Cena, G., Cibrario Bertolotti, I., Scanzio, S., Valenzano, A., & Zunino, C. (2013). Synchronize Your Watches: Part II: Special-Purpose Solutions for Distributed Real-Time Control. *IEEE Industrial Electronics Magazine*, *7*(2), 27–39. doi:10.1109/MIE.2013.2248431

Cena, G., Scanzio, S., Valenzano, A., & Zunino, C. (2014). Ethernet for Control Automation Technology. In R. Zurawski (Ed.), *Industrial Communication Technology Handbook* (2nd ed., pp. 18-1–18-27). Boca Raton, FL: CRC Press.

Cena, G., Scanzio, S., Vitturi, S., & Zunino, C. (2011). Industrial Ethernet and TCP/IP-Based Systems. In B. G. Liptak & H. Eren (Eds.), Instrument Engineers' Handbook, Volume 3: Process Software and Digital Networks, Fourth Edition (pp. 636-653). Boca Raton, FL: CRC Press.

Cucinotta, T., Mancina, A., Anastasi, G. F., Lipari, G., Mangeruca, L., Checcozzo, R., & Rusina, F. (2009). A Real-Time Service-Oriented Architecture for Industrial Automation. *IEEE Transactions on Industrial Informatics*, *5*(3), 267–277. doi:10.1109/TII.2009.2027013

Djukic, P., & Mohapatra, P. (2012). Soft-TDMAC: A Software-Based 802.11 Overlay TDMA MAC with Microsecond Synchronization. *IEEE Transactions on Mobile Computing*, *11*(3), 478–491. doi:10.1109/TMC.2011.56

Mongelli, M., & Scanzio, S. (2014). Approximating Optimal Estimation of Time Offset Synchronization With Temperature Variations. *IEEE Transactions on Instrumentation and Measurement*, *63*(12), 2872–2881. doi:10.1109/TIM.2014.2320400

Perahia, E., & Stacey, R. (Eds.). (2013). *Next Generation Wireless Lans: 802.11n and 802.11ac* (2nd ed.). New York, NY: Cambridge University Press. doi:10.1017/CBO9781139061407

Wu, Y. C., Chaudhari, Q., & Serpedin, E. (2011). Clock Synchronization of Wireless Sensor Networks. *IEEE Signal Processing Magazine*, *28*(1), 124–138. doi:10.1109/MSP.2010.938757

KEY TERMS AND DEFINITIONS

Cyber Physical System: A system that integrates computer and networking to control a physical process.

IEEE 802.11 (Wi-Fi): A standard for the implementation of high-throughput wireless local area networks.

IEEE 802.3 (Ethernet): A collection of standards for the definition of Ethernet. Ethernet is the most known wired network technology.

Real-Time Networks: Communication networks with demanding requirements in terms of latency and determinism.

Seamless Redundancy: Inclusion of duplicate components, not strictly necessary, to improve reliability.

Synchronization Protocols: Network protocols for the distribution of a common time reference to the nodes of a communication network.

Time Division Multiple Access (TDMA): A channel access method to manage a mutual exclusive access of transmitting nodes to a network.

Wi-Red: A seamless redundancy technique applied to Wi-Fi to improve determinism and reliability of IEEE 802.11 wireless networks.

This research was previously published in the Encyclopedia of Information Science and Technology, Fourth Edition edited by Mehdi Khosrow-Pour, pages 1227-1237, copyright year 2018 by Information Science Reference (an imprint of IGI Global).

Chapter 7
Consistency Is Not Enough in Byzantine Fault Tolerance

Wenbing Zhao
Cleveland State University, USA

ABSTRACT

The use of good random numbers is crucial to the security of many mission-critical systems. However, when such systems are replicated for Byzantine fault tolerance, a serious issue arises (i.e., how do we preserve the integrity of the systems while ensuring strong replica consistency?). Despite the fact that there exists a large body of work on how to render replicas deterministic under the benign fault model, the solutions regarding the random number control are often overly simplistic without regard to the security requirement, and hence, they are not suitable for practical Byzantine fault tolerance. In this chapter, the authors present a novel integrity-preserving replica coordination algorithm for Byzantine fault tolerant systems. The central idea behind our CD-BFT algorithm is that all random numbers to be used by the replicas are collectively determined, based on the contributions made by a quorum of replicas, at least f+1 of which are not faulty.

INTRODUCTION

In Byzantine fault tolerance (BFT), a core concern is to ensure the consistency of replicas despite malicious attacks from the clients and compromised replicas (Zhao, 2014). This is accomplished by totally ordering incoming requests and by rendering the replica's operations deterministic (Zhang et al., 2011). In the presence of application non-determinism, such as the access of local clocks, replicas are rendered deterministic by forcing all non-faulty replicas to use the same values either supplied by the primary or computed deterministically. While this approach works well for some applications, such as a replicated file system, doing so could lead to the compromise of the service integrity for applications that rely on the use of random numbers.

For example, consider an Internet application that relies on the use of session-ids for stateful interactions between the server and its clients. As pointed out in (Dorrendorf, Gutterman, & Pinkas, 2007), if the session-id of an active session can be predicted, the client's session with the server could be hijacked,

DOI: 10.4018/978-1-5225-7359-3.ch007

which could lead to the leak of confidential information regarding the client, such as name, address, and the order history, or unauthorized orders (if the one-click option for placing orders is enabled). The session-id may be predicted by searching the limited entropy space if weak random bits are used in an application. For example, the authors of (Dorrendorf, Gutterman, & Pinkas, 2007) reverse-engineered a version of Tomcat (a popular Java Servlet Engine) and the related operations in a Window's based Java Virtual Machine. They could attack the system by performing about 251 searches in finding an active session-id.

Therefore, it is critical not to weaken the strength of the random bits essential for the integrity of their operations when replicating these systems for Byzantine fault tolerance. For a sound coordination algorithm, it is essential to enable each replica to access its own entropy source and maintain its independence in such operations. However, this desire is in conflict with the basic requirement for state machine replication (Schneider, 1990), which mandates that the replicas must be deterministic or rendered deterministic to maintain strong replica consistency. The conflicting requirements for security and replication must be reconciled to avoid the dilemma of either favoring security over high availability by not performing state machine replication of the systems, or trading security for high availability by removing the randomness of the systems in order to perform state machine replication.

In this chapter, we present a novel replica coordination algorithm, referred to as the Collective-Determination BFT algorithm, or CD-BFT algorithm in short, towards the reconciliation of the conflicting requirements for security and for strongly consistent replication. The central idea behind this algorithm is that all random numbers to be used by the replicas are collectively determined, and furthermore, the determination is based on the contributions made by a quorum of replicas, at least one of which is correct.

In the CD-BFT algorithm, the replicas first reach a Byzantine agreement on the set of contributions from replicas, and then apply a deterministic algorithm, such as the bitwise exclusive-or operation (Young & Yung, 2004), to compute the final random value. The freshness of the random numbers generated is application dependent. Our approach does not alter the freshness of the random numbers. If a pseudo-random number generator is used, it should be periodically reseeded from a good entropy source.

BACKGROUND

An arbitrary fault is often referred to as a Byzantine fault. The term was introduced in (Lamport, Shostak, & Pease, 1982) to highlight a specific faulty behavior that a Byzantine faulty process may disseminate conflicting information to other processes. For example, a compromised process might exhibit such Byzantine faulty behavior. Byzantine fault tolerance refers to the capability of tolerating Byzantine faults in a system. It can be achieved by introducing sufficient redundancy into the system and by using a sophisticated replica coordination algorithm that can cope with Byzantine faulty replicas and clients. A basic requirement for such an algorithm is to ensure that all server replicas agree on the total ordering of the requests received despite the existence of Byzantine faulty replicas and clients. Such an agreement is often referred to as Byzantine agreement (Lamport, Shostak, & Pease, 1982).

Recently, a number of efficient BFT algorithms (Castro & Liskov, 2002; Kotla et al., 2007; Yin et al., 2003) have been proposed. Our CD-BFT algorithm is derived from the PBFT algorithm and we use the same system model as that in (Castro & Liskov, 2002). The PBFT algorithm operates in an asynchronous distributed environment. The safety property of the algorithm, *i.e.,* all correct replicas agree on the total ordering of requests, is ensured without any assumption of synchrony. However, for the algorithm to make progress towards a Byzantine agreement (*i.e.*, liveness), certain synchrony is needed, for example, a reasonable assumption is that the message transmission and processing delay has an asymptotic upper bound. This bound is dynamically explored in the algorithm in that each time a view change occurs, the timeout for the new view is doubled.

How to ensure strong replica consistency in the presence of replica non-determinism has been of research interest for a long time (Castro & Liskov, 2002; Castro, Rodrigues, & Liskov, 2003; Powell, 1991; Slember & Narasimhan; 2006), most of which are for fault tolerant systems using the benign fault model. However, while the importance of the use of good random numbers has long been recognized in building secure systems (Viega & McGraw, 2002), we have yet to see substantial research work on how to preserve the randomized operations necessary to ensure the system integrity in a fault tolerant system. For the type of systems where the use of random numbers is crucial to their service integrity, the benign fault model is obviously inadequate and the Byzantine fault model must be employed if fault tolerance is required.

Some form of replica non-determinism, in particular those related to timestamp operations, has been studied in the context Byzantine fault tolerant systems (Castro & Liskov, 2002; Castro, Rodrigues, & Liskov, 2003). However, as we will argue in the next section, the existing approach is vulnerable to the presence of colluding Byzantine faulty replicas and clients. We also studied the replica non-determinism issue with the emphasis of classification of non-determinism types and the systematic handling of various types of replica non-determinism (Zhang et al., 2011).

RECONCILIATION OF SECURITY AND REPLICATION REQUIREMENTS

Pitfalls in Controlling Replica Randomness

In this section, we analyze a few well-known approaches that one may attempt to use to ensure replica consistency in the presence of replica randomness. We show that they are not robust against Byzantine faulty replicas and clients. We reiterate here the importance of using random bits for security- related operations. Any attempt to weaken or remove such randomness from the system for the purpose of replication may compromise the system integrity. Furthermore, for Byzantine fault tolerance, it is essential *not* to place undue trust to any single replica, such as the primary replica, because it is virtually impossible to verify in real-time if the random number proposed by a replica is truly random, and consequently, if that particular replica is compromised, the integrity of the entire replicated system will be lost.

For systems that use a pseudo-random number generator, the replica consistency can be easily achieved by using the same seed value to initialize the generator. One might attempt to use the sequence number assigned to the request as the seed. Even though this approach is perhaps the simplest way to render replicas deterministic (since no extra communication step is needed and no extra information is to be included in the control messages for total ordering of requests), it takes the randomness away from the system. Consequently, a Byzantine faulty client can easily guess the seed and predict the random numbers. A seemingly more robust approach is to use the timestamp as the seed to the pseudo-random number generator. As shown in (Viega & McGraw, 2002; Young & Yung, 2004), the use of timestamp does not offer more robustness to the system because it a Byzantine faulty client can easily guess the seed.

The only option remaining seems to be the use of a truly random number to seed a strong pseudo-random number generator (or to obtain random numbers entirely from a high entropy source). We note that the elegant mechanism described in (Castro & Liskov, 2002) cannot be used in this case because backups have no means to verify whether the number proposed by the primary is taken from a strong random number generator seeded periodically using a high-entropy source, or is generated according to a deterministic algorithm. If the latter is the case, the Byzantine faulty primary could continue colluding with Byzantine faulty clients without being detected. A slight improvement over this scheme is to proactively rotate the primary. However, this improved scheme does not solve the problem: while the faulty replica serves as the primary, it can use a predictable seed and/or a weak random number generator without being detected, and hence, the system integrity cannot be guaranteed during this period.

Therefore, the most effective way in countering such threats seems to collectively determine the random numbers, based on the contributions from a quorum of replicas so that Byzantine faulty replicas cannot influence the final outcome.

The Collective-Determination BFT Algorithm

The normal operation of the CD-BFT algorithm is illustrated in Figure 1. The algorithm is executed by a set of 3f + 1 replicas to tolerate up to f Byzantine faulty replicas. One of the replicas is designated as the primary while the remaining replicas are backups. Each replica is assigned a unique id i, where i varies from 0 to 3f. For view v, the replica whose id i satisfies i = v mod (3f + 1) would serve as the primary. The view starts from 0. For each view change, the view number is increased by one and a new primary is selected.

All messages exchanged among the replicas, and those between the replicas and the clients are protected by an authenticator (Castro & Liskov, 2002) (for multicast messages), or by a message authentication code (MAC) (for point-to-point communications). An authenticator is formed by a number of MACs, one for each target of the multicast. We assume that the replicas and the clients each has a public/private key pair, and the public keys are known to everyone. These keys are used to generate symmetric keys needed to produce/verify authenticators and MACs. To ensure freshness, the symmetric keys are periodically refreshed by the mechanism described in (Castro & Liskov, 2002). We assume that the adversaries have limited computing power so that they cannot break the security mechanisms described above.

A compromised replica may replace a high entropy source to which it uses to seed its random number generator with a deterministic algorithm, and convey such an algorithm and the random numbers collectively determined via a covert channel to its colluding clients. We assume that a faulty replica cannot explicitly piggyback such information with a reply through the normal communication channel, which could be ensured by using an application-level gate- way, or a privacy firewall as described by Yin et al. (2003). However, we assume that the bandwidth of the covert channel is very limited such that a faulty replica cannot transmit confidential state information to its colluding clients *in real time*, more specifically, given the size of the secret data L, the refresh period T, and the covert channel bandwidth R, we assume T < LR. For example, if the random number is of size 32 bits, and the covert channel bandwidth is 0.1 bit/second, the refresh period should be made less than 320 seconds. How to limit the bandwidth of covert channels is still under intense research (Gianvecchio & Wang, 2007; Shah, Molina, & Blaze, 2006), and it is out of the scope of this chapter.

When ordering a request, the primary determines the order of the request (*i.e.,* assigns the next sequence number to the request), and queries the application for the type of operation associated with the request. If the operation involves with a random number as input, the primary activates the mechanism for the CD-BFT algorithm. The primary then obtains its share of random number by extracting from its

Figure 1. Normal operation of the CD-BFT algorithm

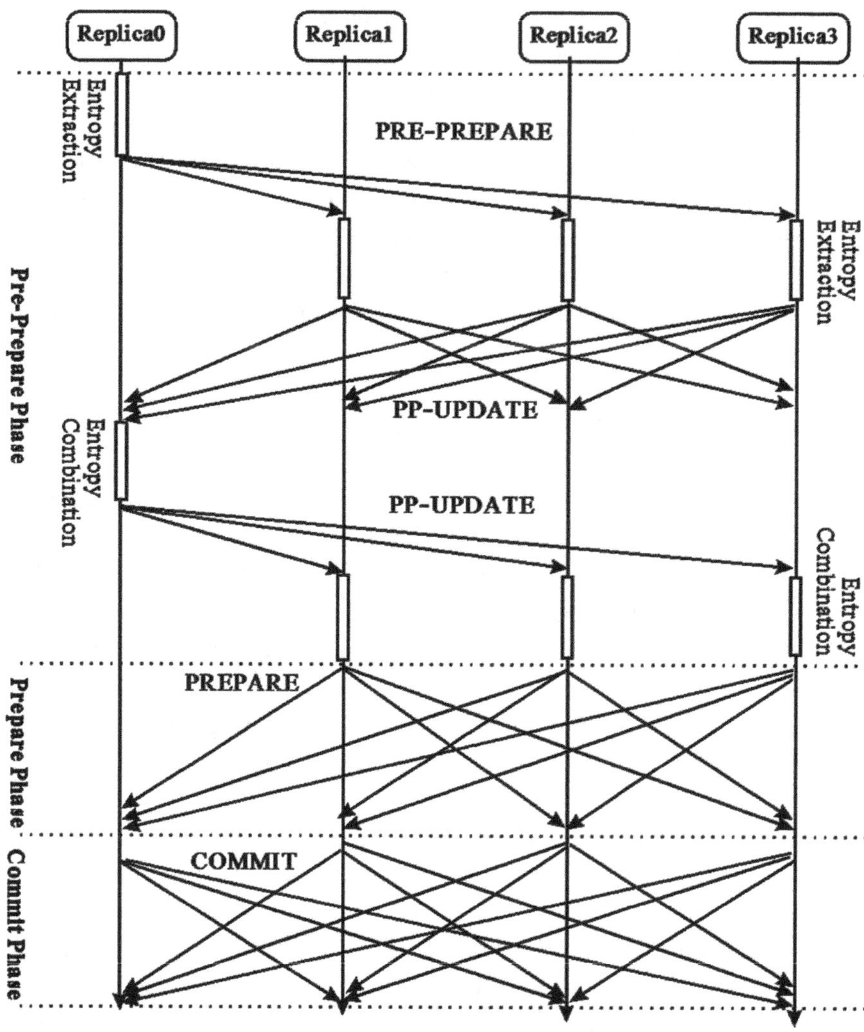

own entropy source, and piggybacks the share with the PRE-PREPARE message multicast to all backups. The pre-prepare message has the form <PRE-PREPARE, v, n, d, R_p>α_p, where v is the view number, n is the sequence number assigned to the request, d is the digest of the request, R_p is the random number generated by the primary, and α_p is the authenticator for the message.

When a backup receives a PRE-PREPARE message, it performs the usual procedure such as the verification of the authenticator before it accepts the message. It also checks if the request will indeed trigger a randomized operation, to prevent a faulty primary from putting unnecessary loads on correct replicas (which might lead to a denial of service attack). If the PRE-PREPARE message is acceptable, the replica creates a PRE-PREPARE certificate for storing the relevant information, generates a share of random number from its entropy source, and multicasts to all replicas a PP-UPDATE message, in the form <PP-UPDATE, v, n, i, R_i, d>α_i, where i is the sending replica identifier, R_i is the random number share from replica i.

When the primary has collected 2f PP-UPDATE messages, it combines the random numbers received according to a deterministic algorithm in the entropy combination step, as shown in Figure 1, and builds a PP-UPDATE message with slightly different content than those sent by backups. In the PP-UPDATE message sent by the primary, the R_i component is replaced by a set of 2f + 1 tuples containing the random numbers contributed by replicas (including its own share), S. Each tuple in the set S has the form <R, i>. The replica R_i identifier is included in the tuple to ease the verification of the set at backups.

On receiving a PP-UPDATE message, a backup accepts the message and stores the message in its data structure provided that the message has a correct authenticator, it is in view v and it has accepted a PRE-PREPARE message to order the request with the digest d and sequence number n. A backup proceeds to the entropy combination step provided that it has accepted a PP-UPDATE message from the primary and 2f PP-UPDATE messages sent by the replicas referenced in the set S. The backup requests a retransmission from the primary for any missing PP-UPDATE message.

When it finishes the entropy combination, a backup multicasts a prepare message. The message has the form $< PREPARE, v, n, i, d'>\alpha_i$, where d′ is the digest of the request concatenated by the combined random number. The replica then waits for 2f valid PREPARE messages from different replicas (possibly including the message sent or would have been sent by itself), after which, it multicasts a commit message to all replicas in the form $<COMMIT, v, n, i, d'>\alpha_i$. When a replica receives 2f+1 valid COMMIT messages, it decides on the sequence number and the collectively determined random number. At the time of delivery to the application, both the request and the random number are passed to the application.

In Figure 1, the duration of the entropy extraction and combination steps have been intentionally exaggerated for clarify. In practice, the entropy combination can be achieved by applying a bitwise exclusive-or operation on the set of random numbers collected, which is very fast. The cost of entropy extraction depends on the scheme used. Some schemes, such as the TrueRand method (Lacy et al., 1993), allow very fast entropy extraction. TrueRand works by gathering the underlying randomness from a computer by measuring the drift between the system clock and the interrupts-generation rate on the processor.

DISCUSSION

For clarity, when describing the CD-BFT algorithm, we have assumed that each remote method invocation involves only a single random number. In practice, an invocation might need to access a number of random numbers. Our algorithm can be trivially modified to accommodate this need. Let n be the number of random numbers needed for an operation. In the CD-BFT algorithm, during the pre-prepare phase, each replica obtains n random numbers from its entropy source and includes them in the PRE-PREPARE and PP-UPDATE messages. The entropy combination step would now combine n random numbers instead of 1. The remaining steps are identical to those in the original algorithm.

If the entropy consumption rate exceeds the entropy generation rate of the entropy source, which is the case for many practical systems, a pseudo-random number generator is normally used. If such systems are replicated for Byzantine fault tolerance, unless the pseudo-random number generator is frequently reseeded, it is not wise to collectively determine the seed to the random number generator so that the sequence of pseudo-random numbers can be generated deterministically without replica coordination. This is because if a compromised replica leaks the collectively-determined seed to some faulty clients (an adversary can take the time to leak the seed number through a very low-bandwidth covert channel, if the reseeding is not carried out frequently), they can predict all the future random numbers produced by the generator, which would compromise the integrity of the system (*i.e.*, the backward security will be lost). For better security, the replicas should instead seed their pseudo-random number generators independently and collectively determine *every* random number generated. It is also advised that each replica uses a different (strong) pseudo-random number generator for enhanced security. In fact, even if all correct replicas use the same relatively weak random number generator that can be broken by searching 2^E space, our algorithm can expand the search space significantly to $2^{(f+1)E}$ because the collectively determined random number is based on contributions from at least f + 1 correct replicas.

FUTURE RESEARCH DIRECTIONS

It would be interesting to do a comparison study of our approach and the threshold cryptography (Desmedt, 1994; Rabin, 1998; Shoup, 2000). Threshold cryptography can be readily used to perform many secure operations across several replicas without ever exposing the security key, and hence, it is more robust against malicious attacks because it works even in the presence of high-bandwidth covert channels between a compromised replica and its colluding clients. However, threshold cryptograph is not without its limitations shown below. How to over come these limitations would be very interesting research topics.

- Threshold cryptograph is extremely computationally expensive. We have experimented with a Java-based threshold cryptography framework on our test-bed. For a 1024 bit-long key, the key shares generation operation takes over 1 second, and the key shares combination operation takes over 2 seconds. This might be too expensive for most Web-based applications, which require soft real-time responses. It is worthwhile to investigate if a shorter key is sufficient depending on the security requirement of an application.
- To achieve proactive threshold cryptography, the key shares must be frequently refreshed, which involves not only expensive computations, but substantial message exchanges among the replicas as well. How this step impacts the overall system performance should be evaluated.
- Threshold cryptography assumes the availability of a trusted dealer to distribute the initial shares. How to implement such a trusted dealer would be an interesting research challenge.

Nevertheless, threshold cryptography has been integrated into some Byzantine fault tolerance systems, such as in Byzantine fault tolerant distributed hash tables (Young et al., 2009), and Byzantine fault tolerant distributed storage (Cachin and Tessaro, 2005).

CONCLUSION

The use of good random numbers is crucial to the security of many mission-critical systems. However, when such systems are replicated for Byzantine fault tolerance, a serious issue arises, i.e., how do we preserve the integrity of the systems while ensuring strong replica consistency? Despite the fact that there exists a large body of work on how to render replicas deterministic under the benign fault model, the solutions regarding the random number control are often overly simplistic without regard to the security requirement, and hence, they are not suitable for practical Byzantine fault tolerance. In this chapter, we have presented a novel integrity-preserving replica coordination algorithm for Byzantine fault tolerant systems. The central idea behind our CD-BFT algorithm is that all random numbers to be used by the replicas are collectively determined, based on the contributions made by a quorum of replicas, at least f+1 of which are not faulty.

REFERENCES

Cachin, C., & Tessaro, S. (2006, June). Optimal resilience for erasure-coded Byzantine distributed storage. In *Proceedings of the International Conference on Dependable Systems and Networks*, (pp. 115-124). IEEE. 10.1109/DSN.2006.56

Castro, M., & Liskov, B. (2002). Practical Byzantine fault tolerance and proactive recovery. *ACM Transactions on Computer Systems*, *20*(4), 398–461. doi:10.1145/571637.571640

Castro, M., Rodrigues, R., & Liskov, B. (2003). BASE: Using abstraction to improve fault tolerance. *ACM Transactions on Computer Systems*, *21*(3), 236–269. doi:10.1145/859716.859718

Desmedt, Y. (1994). Threshold cryptography. *European Transactions on Telecommunications*, *5*(4), 449–457. doi:10.1002/ett.4460050407

Dorrendorf, L., Gutterman, Z., & Pinkas, B. (2007). Cryptanalysis of the random number generator of the windows operating system. *Proceedings of the 14th ACM Conference on Computer and Communications Security*, 476–483.

Gianvecchio, S., & Wang, H. (2007). Detecting covert timing channels: an entropy-based approach. *Proceedings of the 14th ACM Conference on Computer and Communications Security*, 307–316. 10.1145/1315245.1315284

Gutterman, Z., & Malkhi, D. (2005). Hold your sessions: An attack on Java session-id generation. *Lecture Notes in Computer Science*, *3376*, 44–57. doi:10.1007/978-3-540-30574-3_5

Kotla, R., Alvisi, L., Dahlin, M., Clement, A., & Wong, E. (2007). Zyzzyva: Speculative byzantine fault tolerance. *Proceedings of 21st ACM Symposium on Operating Systems Principles*, 45-58.

Lacy, J., Mitchell, D., & Schell, W. (1993). Cryptolib: Cryptography in software. *Proceedings of the 4th USENIX Security Symposium*, 1–17.

Lamport, L., Shostak, R., & Pease, M. (1982). The Byzantine generals problem. *ACM Transactions on Programming Languages and Systems*, *4*(3), 382–401. doi:10.1145/357172.357176

Powell, D. (Ed.). (1991). *Delta-4: A Generic Architecture for Dependable Distributed Computing.* Springer-Verlag. doi:10.1007/978-3-642-84696-0

Rabin, T. (1998). A simplified approach to threshold and proactive RSA. *Proceedings of the 18th Annual International Cryptology Conference*, 89-104. 10.1007/BFb0055722

Schneider, F. (1990). Implementing fault-tolerant services using the state machine approach: A tutorial. *ACM Computing Surveys*, *22*(4), 299–319. doi:10.1145/98163.98167

Shah, G., Molina, A., & Blaze, M. (2006). Keyboards and covert channels. *Proceedings of the 15th USENIX Security Symposium*, 59–75.

Shoup, V. (2000). Practical threshold signatures. *Lecture Notes in Computer Science*, *1097*, 207–220. doi:10.1007/3-540-45539-6_15

Slember, J., & Narasimhan, P. (2006). Living with nondeterminism in replicated middleware applications. *Proceedings of the ACM/IFIP/USENIX 7th International Middleware Conference*, 81–100. 10.1007/11925071_5

Viega, J., & McGraw, G. (2002). *Building Secure Software*. Addison-Wesley.

Yin, J., Martin, J.-P., Venkataramani, A., Alvisi, L., & Dahlin, M. (2003). Separating agreement from execution for byzantine fault tolerant services. *Proceedings of the ACM Symposium on Operating Systems Principles*, 253–267. 10.1145/945445.945470

Young, A., & Yung, M. (2004). *Malicious Cryptography: Exposing Cryptovirology*. Wiley.

Young, M., Kate, A., Goldberg, I., & Karsten, M. (2010, June). Practical robust communication in dhts tolerating a byzantine adversary. In *Proceedings of the IEEE 30th International Conference on Distributed Computing Systems* (pp. 263-272). IEEE. 10.1109/ICDCS.2010.31

Zhang, H., Zhao, W., Moser, L. E., & Melliar-Smith, P. M. (2011). Design and Implementation of a Byzantine Fault Tolerance Framework for Non-Deterministic Applications. *IET Software*, *5*(3), 342–356. doi:10.1049/iet-sen.2010.0013

Zhao, W. (2014). *Building dependable distributed systems*. John Wiley & Sons. doi:10.1002/9781118912744

ADDITIONAL READING

Adya, A., Bolosky, W. J., Castro, M., Cermak, G., Chaiken, R., Douceur, J. R., ... & Wattenhofer, R. P. (2002). FARSITE: Federated, available, and reliable storage for an incompletely trusted environment. ACM SIGOPS Operating Systems Review, 36(SI), 1-14.

Aiyer, A. S., Alvisi, L., Clement, A., Dahlin, M., Martin, J. P., & Porth, C. (2005). BAR fault tolerance for cooperative services. *Operating Systems Review*, *39*(5), 45–58. doi:10.1145/1095809.1095816

Amir, Y., Coan, B., Kirsch, J., & Lane, J. (2008, June). Byzantine replication under attack. In Dependable Systems and Networks With FTCS and DCC, 2008. DSN 2008. IEEE International Conference on (pp. 197-206). IEEE. 10.1109/DSN.2008.4630088

Amir, Y., Danilov, C., Dolev, D., Kirsch, J., Lane, J., Nita-Rotaru, C., . . . Zage, D. (2006, June). Scaling Byzantine Fault-Tolerant Replication toWide Area Networks. In Dependable Systems and Networks, 2006. DSN 2006. International Conference on (pp. 105-114). IEEE. 10.1109/DSN.2006.63

Amir, Y., Danilov, C., Dolev, D., Kirsch, J., Lane, J., Nita-Rotaru, C., ... Zage, D. (2008). Steward: Scaling Byzantine Fault-Tolerant Replication to Wide Area Networks. *IEEE Transactions on Dependable and Secure Computing*, 80–93.

Bessani, A. N., da Silva Fraga, J., & Lung, L. C. (2006, April). BTS: A Byzantine fault-tolerant tuple space. In *Proceedings of the 2006 ACM symposium on Applied computing* (pp. 429-433). ACM. 10.1145/1141277.1141377

Chai, H., Melliar-Smith, P. M., Moser, L. E., Zhang, H., & Zhao, W. (2013). Toward Trustworthy Coordination of Web Services Business Activities. *IEEE Transactions on Services Computing*, *6*(2), 276–288. doi:10.1109/TSC.2011.57

Chun, B. G., Maniatis, P., & Shenker, S. (2008, June). Diverse Replication for Single-Machine Byzantine-Fault Tolerance. In *USENIX Annual Technical Conference* (pp. 287-292).

Clement, A., Wong, E. L., Alvisi, L., Dahlin, M., & Marchetti, M. (2009, April). Vol. 9, pp. 153–168). Making Byzantine Fault Tolerant Systems Tolerate Byzantine Faults. In NSDI.

Cowling, J., Myers, D., Liskov, B., Rodrigues, R., & Shrira, L. (2006, November). HQ replication: A hybrid quorum protocol for Byzantine fault tolerance. In *Proceedings of the 7th symposium on Operating systems design and implementation* (pp. 177-190). USENIX Association.

Distler, T., & Kapitza, R. (2011, April). Increasing performance in byzantine fault-tolerant systems with on-demand replica consistency. In *Proceedings of the sixth conference on Computer systems* (pp. 91-106). ACM. 10.1145/1966445.1966455

Driscoll, K., Hall, B., Sivencrona, H., & Zumsteg, P. (2003). Byzantine fault tolerance, from theory to reality. In Computer Safety, Reliability, and Security (pp. 235-248). Springer Berlin Heidelberg. doi:10.1007/978-3-540-39878-3_19

Garcia, R., Rodrigues, R., & Preguiça, N. (2011, April). Efficient middleware for byzantine fault tolerant database replication. In *Proceedings of the sixth conference on Computer systems* (pp. 107-122). ACM. 10.1145/1966445.1966456

Hendricks, J., Ganger, G. R., & Reiter, M. K. (2007). Low-overhead byzantine fault-tolerant storage. *Operating Systems Review*, *41*(6), 73–86. doi:10.1145/1323293.1294269

Kapitza, R., Behl, J., Cachin, C., Distler, T., Kuhnle, S., Mohammadi, S. V., ... Stengel, K. (2012, April). CheapBFT: resource-efficient byzantine fault tolerance. In *Proceedings of the 7th ACM european conference on Computer Systems* (pp. 295-308). ACM.

Kotla, R., & Dahlin, M. (2004, June). High throughput Byzantine fault tolerance. In Dependable Systems and Networks, 2004 International Conference on (pp. 575-584). IEEE. 10.1109/DSN.2004.1311928

Veronese, G. S., Correia, M., Bessani, A. N., & Lung, L. C. (2009, September). Spin One's Wheels? Byzantine Fault Tolerance with a Spinning Primary. In Reliable Distributed Systems, 2009. SRDS'09. 28th IEEE International Symposium on (pp. 135-144). IEEE. 10.1109/SRDS.2009.36

Yin, J., Martin, J. P., Venkataramani, A., Alvisi, L., & Dahlin, M. (2002, June). Byzantine fault-tolerant confidentiality. In *Proceedings of the International Workshop on Future Directions in Distributed Computing* (pp. 12-15).

Zhang, H., Chai, H., Zhao, W., Melliar-Smith, P. M., & Moser, L. E. (2012). Trustworthy coordination of Web services atomic transactions. *IEEE Transactions on Parallel and Distributed Systems*, *23*(8), 1551–1565. doi:10.1109/TPDS.2011.292

Zhang, Y., Zheng, Z., & Lyu, M. R. (2011, July). BFTCloud: A byzantine fault tolerance framework for voluntary-resource cloud computing. In *Proceedings of the IEEE International Conference on Cloud Computing* (pp. 444-451). IEEE. 10.1109/CLOUD.2011.16

Zhao, W. (2008, December). Integrity-preserving replica coordination for byzantine fault tolerant systems. In *Proceedings of the 14th IEEE International Conference on Parallel and Distributed Systems* (pp. 447-454). IEEE. 10.1109/ICPADS.2008.45

Zhao, W. (2009). Design and implementation of a Byzantine fault tolerance framework for Web services. *Journal of Systems and Software*, *82*(6), 1004–1015. doi:10.1016/j.jss.2008.12.037

Zhao, W. (2015). Optimistic Byzantine fault tolerance. International Journal of Parallel, Emergent and Distributed Systems, 1-14 (preprint).

Zhao, W., & Zhang, H. (2009). Proactive service migration for long-running Byzantine fault-tolerant systems. IET software, 3(2), 154-164.

KEY TERMS AND DEFINITIONS

Entropy: It is a measure of uncertainty, or randomness.

Message Authentication Code (MAC): A MAC is produced by a keyed secure hash function on a message. It is used to ensure the integrity of the message such that if a message protected by a MAC is tampered, it can be detected by comparing the MAC included with in the message and the recomputed MAC.

Pseudo-Random Number Generator: It is an algorithm used to generate a sequence of numbers that approximate the properties of random numbers. The algorithm depends on one or a small set of initial numbers, usually referred to as the seed to the generator.

Quorum: A quorum of a set consists of the minimum number of components to perform a predefined function. In Byzantine fault tolerance replication, 2f+1 is needed to form a quorum in a set of 3f+1 replicas.

Random Number: It is a number generated by some process that cannot be reproduced or predicted.

Security: The security of a system refers to its capability of protecting itself from harm, such as external attacks. More specifically, a secure system is one that guarantees confidentiality, integrity, and the availability of the system.

Strong Replica Consistency: The states of the replicas of a process should remain identical at the end of the processing of each request.

System Integrity: The integrity of a system refers to the capability of performing correctly according to the original specification of the system under various adversarial conditions.

Threshold Cryptography: Basic cryptographic operations such as encryption, decryption, signature generation, and verification are performed by a group of processes without reconstructing the shared secret.

This research was previously published in the Encyclopedia of Information Science and Technology, Fourth Edition edited by Mehdi Khosrow-Pour, pages 1238-1247, copyright year 2018 by Information Science Reference (an imprint of IGI Global).

Section 4
Electrical Engineering

Chapter 8
Mechanisms of Electrical Conductivity in Carbon Nanotubes and Graphene

Rafael Vargas-Bernal
Instituto Tecnológico Superior de Irapuato, Mexico

ABSTRACT

There is enormous interest in carbon nanomaterials due to their exceptional physical properties, from the perspective of science and engineering of materials applied to the electronics industry. Significant progress has been made towards understanding the mechanisms of electrical conductivity of carbon nanotubes and graphene. However, scientists around the world continue studying these mechanisms to exploit them fully in different electronic applications with a high technological impact. This chapter discusses the mechanisms of electrical conductivity of both nanomaterials, analyzes the present implications, and projects its importance for future generations of electronic devices. In particular, it is important to note that different mechanisms may be identified when these nanomaterials are used individually, when they are incorporated as fillers in composite materials or hybrid materials, or even when they are doped or functionalized. Finally, other electrical variables with important role in electrical conductivity of these materials are also explored.

INTRODUCTION

In the search for alternative materials to semiconductor materials used commonly in electronics such as silicon, germanium, gallium arsenide, gallium phosphide, etc., researchers around the world have been developing carbon-based materials with ideal electrical properties to operate with high efficiency in nanoelectronics. Carbon nanotubes (CNTs) and graphene represent two technological options for these innovative materials, which can be used either individually, or in composite or hybrid materials as electrical filler. They offer electrical properties such as high electrical conductivity and high dielectric permittivity, which can be tuned by synthesis, doping, functionalization, etc. These qualities can be exploited in applications such as interconnects, electronic devices such as field-effect transistors, batteries, fuel

DOI: 10.4018/978-1-5225-7359-3.ch008

cells, supercapacitors (Yusoff, 2015), electrodes for touch screens (Zheng, 2015), flexible transparent memory circuits, materials for electrostatic discharge (ESD) and electromagnetic interference (EMI) shielding (Vargas-Bernal, 2015c), etc.

This chapter will review the most important electrical transport mechanisms associated with the electrical conductivity of carbon nanotubes and graphene, since these can be used in individual way or within composite or hybrid materials, with the aim of discovering the origin of their extraordinary electrical properties than have been used, are being used, and will be used in diverse technological applications. The effect of a set of technical variables related with electrical behavior of carbon nanotubes and graphene, and associated with the electrical conductivity such as band gap, intrinsic mobility, percolation threshold, electrical conductivity, and dielectric permittivity, are also discussed.

BACKGROUND

Electrical conduction can be defined as the movement of electrical carriers through a transmission medium. A transmission medium is a material substance that transmits or guides through of itself electromagnetic waves. This movement of carriers generates an electrical current in response to an electrical field. Moreover, in each type of material, different mechanisms of electrical conduction are presented. For example, electrons are electrical carriers in metals, and the Ohm's law is the mathematical relationship used to determine the mathematical expression between the electrical current (*I*) and the applied potential difference (*V*) between a pair of ends of the material (Bird, 2014):

$$\mathrm{I} = \frac{V}{R} = VG, \tag{1}$$

where *R* and *G* are electrical resistance and electrical conductance, respectively. Thus, one or more electrons from each atom can move freely within the metal, since they are loosely bound to the atom in the higher level of the valence band. These electrons are incorporated to the conduction band as electrical carriers due to the potential difference, and therefore, an electrical current is generated. An electrical current is a flow of electrical charge carried out regularly by moving electrons through a medium.

Electrical conductivity (σ) also called specific conductance can be defined as the ability of a material for conducting an electrical current. In three-dimensional conductor materials, the electrical conductance can be mathematically expressed as:

$$\mathrm{G} = \frac{A}{\rho L} = \frac{Wt}{\rho L} = \frac{Wt\sigma}{L}, \tag{2}$$

where *A* is the cross-sectional area, *L* is the length, *W* is the width, *t* is the thickness, and, ρ and σ are electrical resistivity and electrical conductivity of the material, respectively. Two different types of electrical conductivities can be found in materials: surface conductivity and bulk conductivity. Surface conductivity or sheet conductance quantifies the electrical conductance of thin films with uniform thickness nominally. This represents the rate between the electrical conductivity of the material, and the thickness of the thin film. Therefore, it is mathematically expressed as:

$$G_s = \frac{t}{\rho} = \sigma t, \qquad (3)$$

whose units are square per Ohm or Siemen square or denoted by sq/Ω or □/Ω or S·sq or S·□, which is dimensionally equal to an Siemen. Bulk conductance, specific electrical conductance, or volume conductivity (σ) is expressed in units of Siemens per meter (S/m).

Materials can be electrically classified in accordance with their conductivities as conductive, static conductive, or static dissipative (Grady, 2011). The surface conductivity regimes for each are approximately greater than 10^{-4} sq/Ω, 10^{-6}–10^{-4} sq/Ω, and 10^{-12}-10^{-6} sq/Ω, respectively. Moreover, the corresponding volume conductivity regimes are > 0.1, 10^{-3}–10^{-1}, and 10^{-9}–10^{-3} S/m. Two applications can be identified in accordance with the value of conductivity: electromagnetic interference (EMI) shielding and electrostatic dissipation (ESD). EMI shielding uses materials with high conductivity, while ESD requires materials with low conductivity.

In materials such as insulators and semiconductors, there is an energy range called forbidden band or band gap, where electron energy states cannot exist between the top of the valence band and the bottom of the conducting band.

MAIN FOCUS OF THE ARTICLE

Carbon nanotubes and graphene has a structure of conjugated system, where a system of connected *p*-orbitals with delocalized electrons in atoms, presents alternating single and multiple bonds, which in general may lower the overall energy of the system and increase stability. Two *p*-orbitals form a π-bond. A π-bond is a covalent chemical bond, where two lobes of one atomic orbital are overlapped to other two lobes of the other atomic orbital involved. Therefore, the π-electrons do not belong to a single bond or atom, but rather to a group of atoms (Jug, 2001). The conjugation can be viewed as the overlap of one *p*-orbital with another across a sigma bond. A simple model of the energy levels can be considered as a quantum mechanical problem of a one-dimensional particle, representing the movement of a π-electron along a long conjugated chain of carbon atoms as found in carbon nanotubes and/or graphene. In this model, the lowest possible absorption energy corresponds to the energy difference between the highest occupied molecular orbital (HOMO), and the lowest unoccupied molecular orbital (LUMO). Almost all electronic transitions in conjugated π-systems are carried from a bonding molecular orbital (MO) to an antibonding MO (π to π^*), but electrons from non-bonding Lone pairs (pair of valence electrons that are not shared with another atom) can also be promoted to a π-system MO (n to π^*) in charge-transfer complexes. A HOMO to LUMO transition is carried out by an electron if it is allowed by the selection rules for electromagnetic transitions. Thus, the electrical conductivity is guaranteed at using carbon nanotubes and/or graphene in electronic devices. Carbon nanotubes can achieve electrical conductivities between 100 and 200E03 S/m, while graphene can achieve an electrical conductivity between 1,738 S/m and 100E06 S/m.

The hexagonal lattice, found in the carbon nanomaterials, possesses the longest mean free path of any known material, in the order of microns. Therefore, a ballistic transport is presented in these materials, since the distance that an electron can travel freely without bumping into anything, or having its path disrupted by scattering is large, which reduces their electrical resistance even at room temperature.

Graphene is an allotrope of carbon, formed by a simple sheet of graphite, which is one-atom thick; with carbon atoms arranged in a regular hexagonal pattern (see Figure 1). It has an extremely low weight, since, a graphene sheet with an area of one square meter; this material only weights 0.77 grams. As a semi-metallic material, this material is a semiconductor, and also it has high electron mobility at room temperature. Graphene has no bandgap, since its conduction and valence bands fulfill the Dirac points. It is possible to induce a small bandgap in graphene by doping. A material without band gap can convert all wavelengths of light to electrons, with energy levels that even did not found in semiconductor materials, and thus, graphene is a wonderful candidate for use in photovoltaic cells. Graphene has high energy density and/or highest current density (a million times that of copper), that favors the storage capacity and rate of charging/discharging, which make it suitable for batteries and electrical supercapacitors. It has the highest intrinsic mobility (100 times more than silicon) between 15,000 $cm^2 \cdot V^{-1} \cdot s^{-1}$ and 200,000 $cm^2 \cdot V^{-1} \cdot s^{-1}$ nearly unaffected by the temperature (10K and 100K), a fact that hints at defect scattering, being the dominant scattering mechanism (Tanaka, 2014). Graphene has a unique combination of properties that is ideal for next-generation electronics, including mechanical flexibility, high electrical conductivity, and chemical stability.

The graphene presents a very high electrical conductivity, thanks to its zero-overlap semi-metallic behavior, with holes and electrons as charge carriers. The electrons in carbon atoms are distributed in the following way: 2 in the inner shell and 4 in the outer shell. Three outer electrons are connected to other 3 carbon atoms, and the fourth electron is left free to be exploited in the electrical conductivity. These free electrons are called pi (π) electrons and they are located in both sides of the graphene sheet.

Figure 1. Different types of graphene sheets in accordance with their chiral indices (n,m) and chiral angle (ψ)

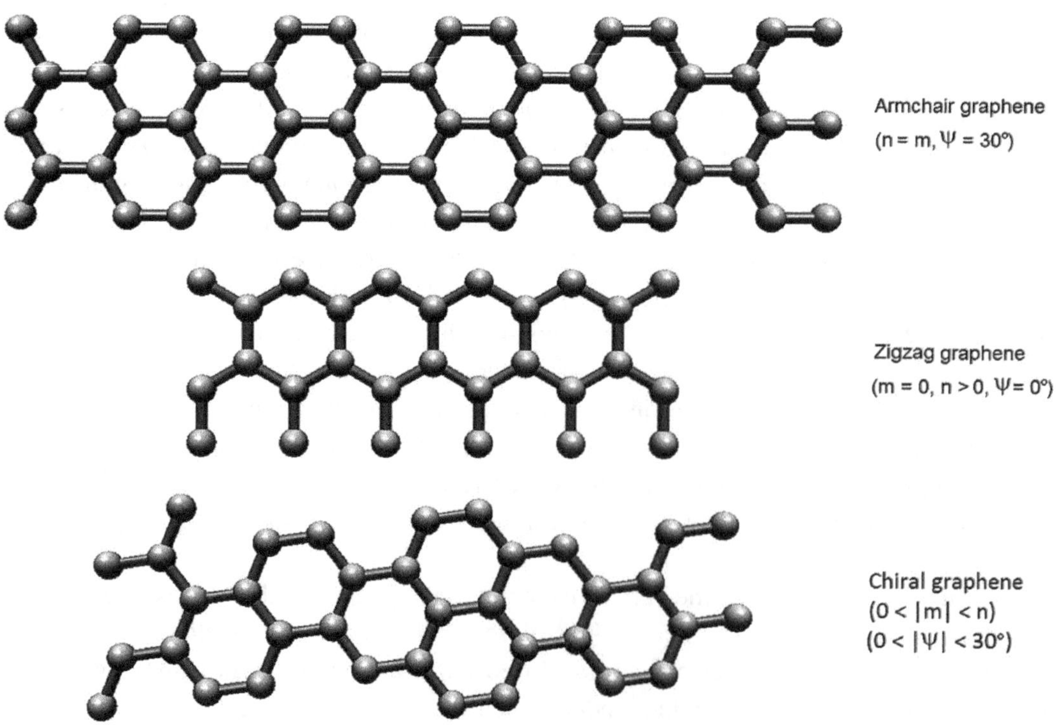

Moreover, these electrons overlap, and improve the carbon-to-carbon bonds in graphene, leading to electronic properties established by the bonding and anti-bonding, of the orbitals associated with them.

Three different types of sheets of graphene can be identified: armchair, zigzag and chiral, in accordance with the chiral indices, as is illustrated in Figure 1 (Vargas-Bernal, 2015b). Zigzag graphene nanoribbons have a metallic behavior, while armchair graphene nanoribbons can have behavior either metallic or semiconducting. This last depends of the number (N) of atoms across the width where k is the number of nanoribbons presented by the material, and therefore, when N=3k-1 a metallic behavior is obtained, if N=3k a semiconducting behavior is presented, and finally, if N=3k+1, a moderate semiconducting behavior is achieved (see Figure 2).

Another important parameter which defines the electrical properties of graphene is its bandgap. Graphene has no bandgap, since its conduction and valence bands fulfill the Dirac points. The Dirac points are six locations in momentum space, on the edge of the Brillouin zone, divided into two non-equivalent sets of three points. The two sets are labeled as K and K', as is illustrated in Figure 3. Electrons through the graphene propagate according to the tight-binding model (electronic band structure), where they are energetically dispersed as a traveling wave within the medium. The mathematical relationship for determining 2D energy dispersion of the electrons in π bands of graphene is given as (Das, 2015):

$$\varepsilon_{2D} = \pm\gamma_0\sqrt{1+4cos\left(\frac{\sqrt{3}k_x a}{2}\right)cos\left(\frac{k_y a}{2}\right)+4cos^2\left(\frac{k_y a}{2}\right)} \quad (4)$$

where γ_0 = 3.033 eV is the nearest-neighbor overlap energy. The ε-k (energy versus wave vector) relationship of the electrical carriers is linear for lowest energies near the six corners of the 2D hexagonal

Figure 2. Electrical behavior of the graphene related with the number of atoms across the width of the nanoribbon of graphene and the wave vector k

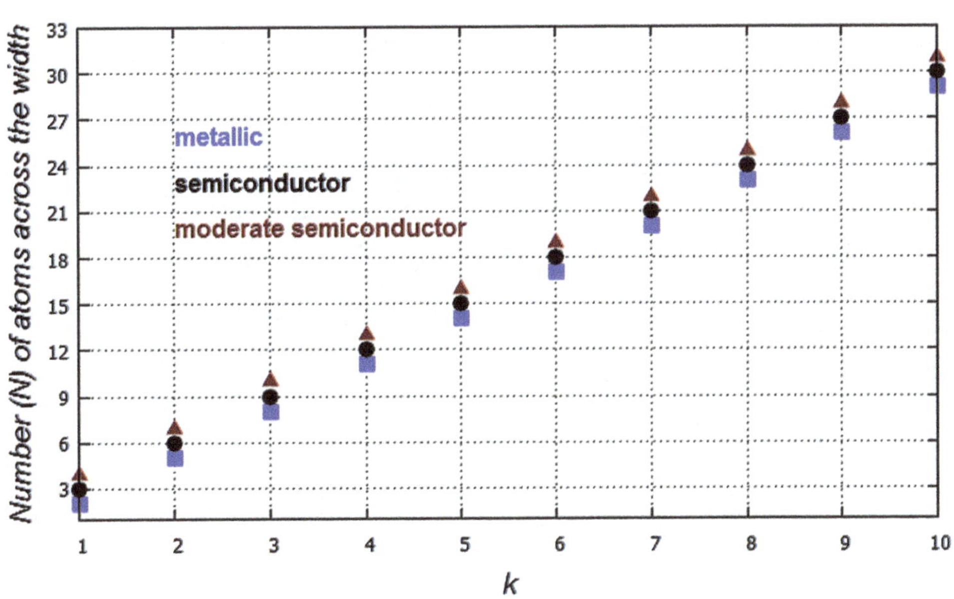

Brillouin zone, which leads to zero effective mass for electrons and holes, as shown in Figure 3 (Shuai, 2012).

A carbon nanotube is an allotrope of carbon that can be conceptualized as a layer of graphite with one-atom-thick called graphene, which is wrapped as a cylinder. One, two or more layers of graphene can be concentrically wrapped giving place to a single-wall nanotubes (SWNTs), double-wall nanotubes (DWNTs) or multi-wall nanotubes (MWNTs). The wrapping of a graphene sheet can be identified by means of the indices (n, m), where n and m denote integer numbers representing the number of unit vectors along directions in the honeycomb crystal lattice. When $m = 0$, the carbon nanotubes are called zigzag nanotubes, if $n = m$ they are called armchair nanotubes, and in otherwise, they are called chiral nanotubes, as depicted in Figure 4.

The diameter of an ideal carbon nanotube is determined from its (n, m) chiral indices as:

$$\mathrm{d} = \frac{\sqrt{3}a_{CC}}{\pi}\sqrt{n^2 + nm + m^2} \tag{5}$$

where $a_{CC} = 0.142$ nm is the carbon-carbon bond length, as is illustrated in Figure 5. The electrical properties of the carbon nanotubes are directly proportional to the value of the diameter, that is, great values of diameter imply high conductivity and vice versa. The chiral angle is defined as:

$$\psi = cos^{-1}\left(\frac{2n + m}{2\sqrt{n^2 + nm + m^2}}\right) \tag{6}$$

It allows distinguishing three classes of carbon nanotubes with different electrical properties: armchair ($n{=}m$, $\psi = 30°$) with a metallic behavior, zigzag ($m{=}0$, $n > 0$, $\psi{=}0°$) with a semiconducting behavior, and chiral ($0<|m|<n$, $0<|\psi|<30°$) with a moderate semiconducting behavior, as shown in Figure 6.

Figure 3. Band structure of graphene (ε-k relationship of graphene nanoribbons)

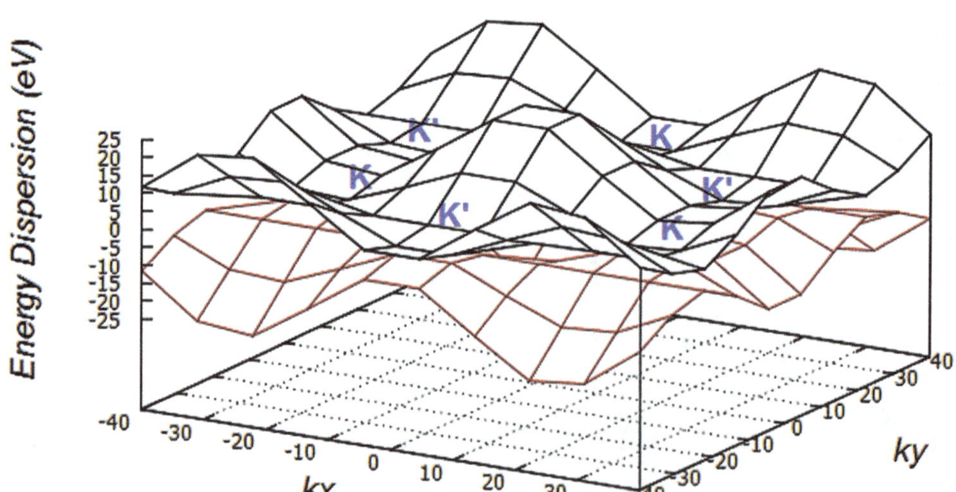

Figure 4. Different types of carbon nanotubes according with chiral indices

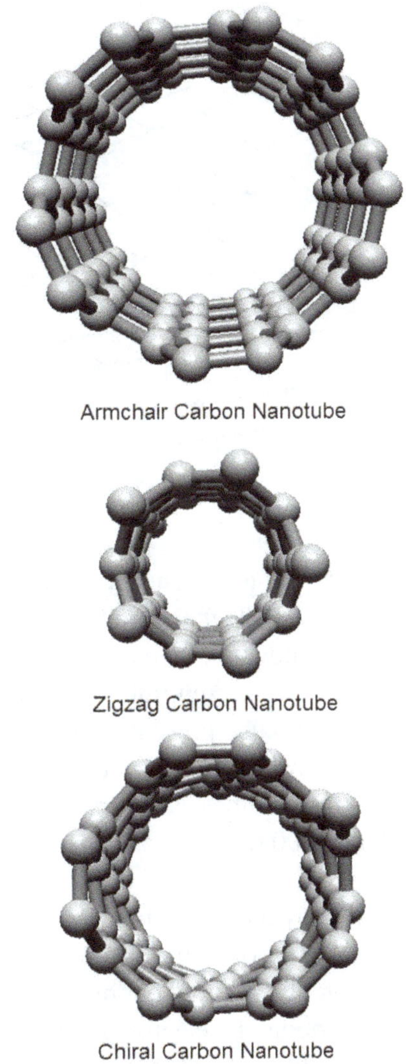

Figure 5. Estimation of the diameter of carbon nanotubes according to the chiral indices

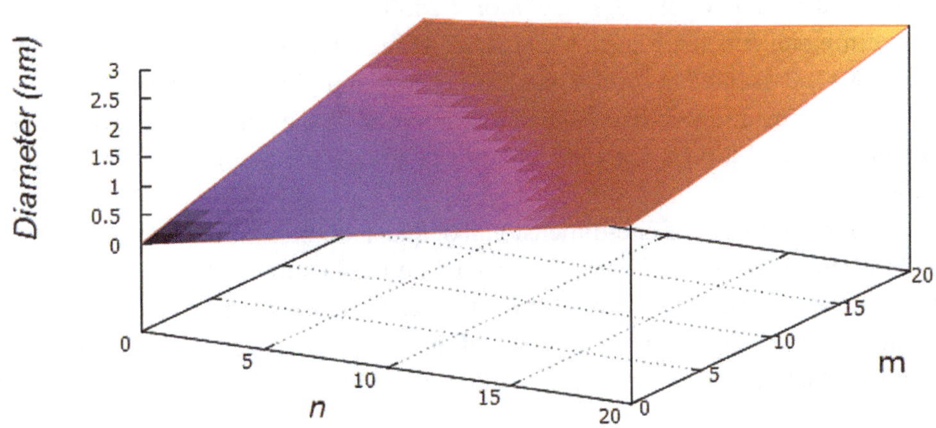

Figure 6. Electrical classification of the carbon nanotubes according to the chiral angle and indices

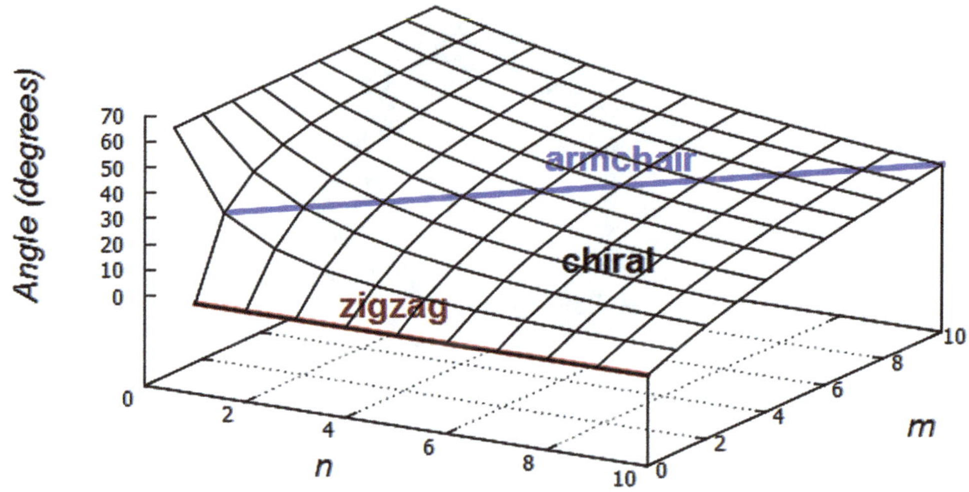

In accordance with Kataura's graph, the band gap energy in a carbon nanotube is inversely proportional to the diameter of the nanotube. In a SWNT, their band gap goes from 0 to 2 eV and their electrical conductivity presents metallic or semiconducting behavior. For a given (n, m) nanotube, when $n = m$, the nanotube has a metallic behavior; if n-m is a multiple of 3, then the nanotube is semiconducting, with a small bandgap, and otherwise the nanotube presents a moderate semiconducting behavior with band gap (Vargas-Bernal, 2012). The chirality has strong influence on electrical properties in nanotubes with small diameter. Theoretically, carbon nanotubes can carry an electrical current density of the order of 4E13 A/m^2, whose value is more than 1,000 times greater than that of metals, such as copper, which reduce the effect of electromigration.

In carbon nanotubes, electrons travel only along the tube's axis, and therefore, these materials are referred to as one-dimensional conductors. At room temperature, the dominant scattering mechanism of transport of electrons implies the emission of optical phonons and it is called ballistic transport. When in a carbon nanotube and graphene nanoribbon, the electrons in transit are not scattered with too many phonons and the devices have about 100 nm long, then the transport of electrons depends on the nanoribbon edge structure and the electron energy. The maximum electrical conductance of a single-walled carbon nanotube is estimated as two times (sheet with two sides) the value than that of a single ballistic quantum channel, that is, $2G_0$ where $G_0 = 2e^2/h$ or $7.748091734625\times10^{-5}$ S, where e is the electrical charge of the electron and h is the Planck's constant. At high temperatures, a different mechanism of transport of electrons is presented in carbon nanotubes called hopping. Hopping is a thermally activated process in which an electron moves from one site to another. The electron loses all information about its phase in the process, and it does not exist, coherence among the amplitudes for finding an electron at different lattice sites.

Electrical transport properties of one-dimensional or quasi-1D nanomaterials such as quantum wires, nanofibers, and carbon nanotubes are been subjected of intense scrutiny with the aim of introducing these materials into electronic applications. Semiconducting carbon nanotubes are used in nano-electronic devices, while metallic carbon nanotubes are applied in nano-interconnects (Das, 2015). The electrical transport in semiconducting carbon nanotubes can be ballistic (tunneling) (see Figure 7), or diffusive

Figure 7. Electrical ballistic transport in carbon nanomaterials

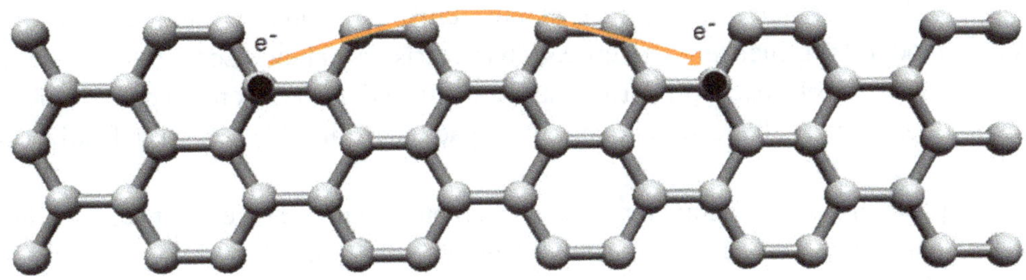

(scattering) (see Figure 8). In carbon nanomaterials, tunneling can be defined as a quantum mechanical phenomenon, where an electron tunnels through or cross a potential energy barrier (space without matter), which it classically could not be surmount. Mechanism of diffusion based on scattering appears when defects, impurities or lattice vibrations are found in the carbon nanomaterials. Diffusion is the movement of electrons from a region of high concentration to a region of low concentration, where variables such as pressure or temperature are involved.

During ballistic transport, carbon nanomaterials do not dissipate heat (Frank, 1998). Electrical conduction in carbon nanomaterials is quantized due to their one-dimensionality, and the number of allowed electronic states is limited. In other words, charge carriers are transmitted through discrete conduction channels without experiencing scattering due to impurities, local defects or lattice vibrations (Vargas-Bernal, 2015a). In addition, the electrons do not find electrical resistance, and no energy dissipation occurs in the conduction channel. For armchair carbon nanotubes, there are two sub-bands that cross the Fermi level, and to semiconducting carbon nanotubes, these sub-bands do not cross the Fermi level. Thus, there are two conducting channels and each band accommodates two electrons of opposite spin (White, 1998). Conduction of electrons in armchair carbon nanotubes experience an effective disorder that is averaged over the tube's circumference, which increases the electron's mean free paths with nanotube's diameter, and therefore, the ballistic transport in the carbon nanotube is guaranteed. The electrical conductivity of individual carbon nanotubes under ballistic conduction achieves the value of 100E06 S/m for single-walled and 3.33E06 S/m for multi-walled carbon nanotubes, respectively (Li, 2007).

Functionalization of carbon nanomaterials is necessary to interface it with other moieties in order to expand the scope of its electrical/electronic applications (Sreeprasad, 2013). Unfortunately, this also modulates its electrical properties, and therefore, a wide knowledge related with the effect of carrier scattering, carrier concentration, charge polarity, quantum-capacitance enhanced doping, energy levels,

Figure 8. Electrical diffusive transport in carbon nanomaterials

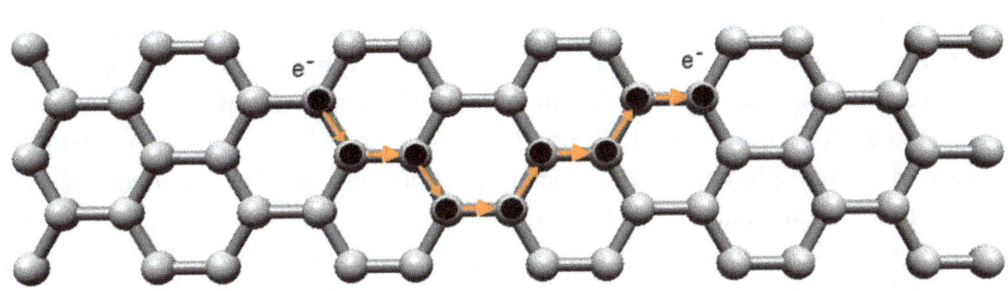

transport mechanisms, and orbital hybridization of energy-bands, must be developed. In (Sreeprasad, 2013), the authors have included factors such as covalent bonds, adsorption, π-π bonds, and lattice incorporation to model its influence on electrical characteristics. They discovered that three different mechanisms governing the electrical properties can be identified: (1) conversion of carbon's hybridized state, (2) dipole interactions enhanced via quantum capacitance, and (3) orbital hybridization with an interfacing molecule.

The use of dopants in carbon nanomaterials such as metals increases the electrical conductivity and its electrical current capacity. Ag nanowires have been used to generate hybrid fibers with graphene oxide (Xu, 2013) for applications in high-performance materials, chemical supercapacitors, ultralight aerogels, sensors and catalysts.

Nanocomposites for electrical applications and based on carbon nanotubes or graphene and polymers, must achieve a certain percolation threshold or a minimum volume fraction, to form a cluster of carbon nanotubes or graphene sheets connected, which will operate as electrical pathways, between a pair of electrodes, to carry out the electrical transport of electrical carriers (Hu, 2014). In this case, the electrical conductivity implies a tunneling of electrons (non-ohmic behavior), one by one, through the polymer found between two neighboring carbon nanotubes o two neighboring graphene nanoribbons.

Carbon nanomaterials can be used as electrical fillers in composite materials or hybrid materials for electronic applications. These materials must be dispersed in polymeric matrix to avoid the formation of agglomerates of fillers. The electrical transport mechanism in these materials is due to tunneling between electrical fillers, since polymers are regularly electrical insulators (Wen, 2012). When aligned carbon nanomaterials are used in a composite material based on polymer (Gong, 2015), anisotropic electrical conductivity in the longitudinal and transverse directions is obtained due to: 1) nanomaterials are distortable, and the Van der Waals forces are presented, and 2) there is structural distortion of the fillers. In addition, aligned graphene in composite materials based on polymers can provide high dielectric constants of over 14000 with 3 wt% of graphene at 1 KHz thanks to the charge accumulation favored by the aligning (Yousefi, 2014).

Surfactants can be used to exfoliate and disperse carbon nanomaterials to be used in composites based in polymers (Mittal, 2015). These additives allow reduce the percolation threshold (minimum percentage to form an interconnected network between electrical fillers), and therefore, a high electrical conductivity is easily achieved, either using high boiling point solvents or low boiling point solvents.

SOLUTIONS AND RECOMMENDATIONS

Different mechanisms are contributing to the electrical conductivity; a complete analysis must be realized to determine the percentage of its influence on electrical properties of the carbon nanomaterials. In real applications, ballistic and diffusive transports are completely influenced by different perturbations from fabrication and operation of the carbon nanomaterials and their composites or hybrid materials. Models of conductivity electrical must be developed to understand the types of transport presented in aligned carbon nanotubes (Yousefi, 2014; Gong, 2015), segregated structures (Du, 2011), compacted materials composed of multiple carbon nanomaterials (Marinho, 2012), and mixes of graphene and carbon nanotubes in composite and hybrid materials.

Until now, in the case of composite materials, diverse strategies have been proposed to reduce the percentage of the CNTs and/or graphene (electrical fillers) to be incorporated in composites based on polymers or metals with the aim of obtaining extraordinary electrical properties (Min, 2010, Jomaa, 2015). The most critical factor implied in the reduction of the filler is the reduction of the percolation threshold or the minimum value of carbon nanotubes and/or graphene used to achieve the electrical conductivity in composite o hybrid materials. Among these techniques that reduce the quantity of filler used in a composite or a hybrid material, can be mentioned: surfactant treating, acid treating, and chemical functionalization. The surfactants lower the surface tension of carbon nanomaterials, which preventing the formation of aggregates of fillers, and thus, an increasing of electrical conductivity is obtained. The cleaning of carbon nanomaterials by means of acids increases electrical conductivity of them, decreasing the concentration of impure materials found on surface of these materials. However, an excessive time in acid can damage the structure of the fillers. A better dispersion can be achieved with functionalization since it changes the hybridization of the carbon nanomaterials, and thus, electrical properties are also changed.

FUTURE RESEARCH DIRECTIONS

This chapter has presented an overview of the most significant advances in the research on electrical transport mechanism that can be exploited in carbon nanomaterials such as graphene and carbon nanotubes. More research must be realized to discover the effect of the ambient on electrical conductivity of both carbon nanomaterials as was developed in (Roch, 2015). A lot of work must be realized to model electrical transport in carbon nanomaterials that have been functionalized to be used in electronic applications.

Carbon nanomaterials will continue offering a highly disruptive technology to be exploited in the interconnection of electronic devices, and the fabrication of electronic devices based either in individual elements or as composite and hybrid materials. These materials will be used in the next generation of electrical wires, with the aim of replacing traditional metal ones based on copper or aluminum (Lekawa-Raus, 2014).

CONCLUSION

In this article, different mechanisms involved in the electrical conductivity of carbon nanotubes and graphene, either working individually or within a composite or hybrid materials, have been discussed. In the case of composite or hybrid materials can be distinguished two mechanisms: the formation of a conducting network interconnecting, carbon nanotubes and/or graphene sheets, and the hopping of the electrons within carbon nanomaterials. Carbon nanomaterials can present ballistic transport (perfect material) or diffusive transport (when defects, impurities and lattice vibrations are found). New mechanisms are presented when carbon nanomaterials are functionalized: hybridized states of the carbon atoms, dipole interactions based on quantum capacitance, and orbital hybridization in the interface.

ACKNOWLEDGMENT

This work is part of a project that is being financed by CONACYT-Mexico (project 152524) and supported by PRODEP and ITESI (project 284.15-PD).

REFERENCES

Bird, J. (2014). *Electrical Circuit: Theory and Technology* (5th ed.). New York: Routledge.

Das, D., & Rahaman, H. (2015). *Carbon Nanotube and Graphene Nanoribbon Interconnects*. Boca Raton, FL: CRC Press.

Du, J., Zhao, L., Zeng, Y., Zhang, L., Li, F., Liu, P., & Liu, C. (2011). Comparison of Electrical Properties between Multi-Walled Carbon Nanotube and Graphene Nanosheet/High Density Polyethylene Composites with a Segregated Network Structure. *Carbon*, *49*(4), 1094–1110. doi:10.1016/j.carbon.2010.11.013

Frank, S., Poncharal, P., Wang, Z. L., & de Heer, W. A. (1998). Carbon Nanotube Quantum Resistors. *Science*, *280*(5370), 1744–1746. doi:10.1126cience.280.5370.1744 PMID:9624050

Gong, S., Zhu, Z. H., & Meguid, S. A. (2015). Anisotropic Electrical Conductivity of Polymer Composites with Aligned Carbon Nanotubes. *Polymer*, *56*, 498–506. doi:10.1016/j.polymer.2014.11.038

Grady, B. P. (2011). *Carbon Nanotube-Polymer Composites: Manufacture, Properties, and Applications*. Hoboken, NJ: John Wiley & Sons. doi:10.1002/9781118084380

Hu, K., Kulkarni, D. D., Choi, I., & Tsukruk, V. V. (2014). Graphene-Polymer Nanocomposites for Structural and Functional Applications. *Progress in Polymer Science*, *39*(11), 1934–1972. doi:10.1016/j.progpolymsci.2014.03.001

Jomaa, M. H., Masenelli-Varlot, K., Seveyrat, L., Lebrun, L., Dib Jawhar Dib, M. C., Beyou, E., & Cavaillé, J. Y. (2015). Investigation of Elastic, Electrical and Electromechanical Properties of Polyurethane/Grafted Carbon Nanotubes Nanocomposites. *Composites Science and Technology*, *121*, 1–8. doi:10.1016/j.compscitech.2015.10.019

Jug, K., Hilbert, P. C., & Shail, S. (2001). σ-π Energy Separation in Modern Electronic Theory for Ground States of Conjugated Systems. *Chemical Reviews*, *101*(5), 1477–1500. doi:10.1021/cr990328e PMID:11710230

Lekawa-Raus, A., Patmore, J., Kurzepa, L., Bulmer, J., & Koziol, K. (2014). Electrical Properties of Carbon Nanotube based Fibers and Their Use in Electrical Wiring. *Advanced Functional Materials*, *24*(24), 3661–3682. doi:10.1002/adfm.201303716

Li, Q. W., Li, Y., Zhang, X. F., Chikkannanavar, S. B., Zhao, Y. H., Dangelewicz, A. M., ... Zhu, Y. T. (2007). Structure-Dependent Electrical Properties of Carbon Nanotube Fibers. *Advanced Materials*, *19*(20), 3358–3363. doi:10.1002/adma.200602966

Marinho, B., Ghislandi, M., Tkalya, E., Koning, C. E., & de With, G. (2012). Electrical Conductivity of Compacts of Graphene, Multi-Wall Carbon Nanotubes, Carbon Black, and Graphite Powder. *Powder Technology*, *221*, 351–358. doi:10.1016/j.powtec.2012.01.024

Min, C., Shen, X., Shi, Z., Chen, L., & Xu, Z. (2010). The Electrical Properties and Conducting Mechanisms of Carbon Nanotube/Polymer Nanocomposites: A Review. *Polymer-Plastics Technology and Engineering*, *49*(12), 1172–1181. doi:10.1080/03602559.2010.496405

Mittal, G., Dhand, V., Rhee, K. Y., Park, S.-J., & Lee, W. R. (2015). A Review on Carbon Nanotubes and Graphene as Fillers in Reinforced Polymer Nanocomposites. *Journal of Industrial and Engineering Chemistry*, *21*, 11–25. doi:10.1016/j.jiec.2014.03.022

Roch, A., Greifzu, M., Talens, E. R., Stepien, L., Roch, T., Hege, J., ... Leson, A. (2015). Ambient Effects on the Electrical Conductivity of Carbon Nanotubes. *Carbon*, *95*, 347–353. doi:10.1016/j.carbon.2015.08.045

Shuai, Z., Wang, L., & Song, C. (2012). Theory of Charge Transport in Carbon Electronic Materials. Springer. doi:10.1007/978-3-642-25076-7

Sreeprasad, T. S., & Berry, V. (2013). How Do the Electrical Properties of Graphene change with Its Functionalization? *Small*, *9*(3), 341–350. doi:10.1002mll.201202196 PMID:23169614

Tanaka, K., & Iijima, S. (Eds.). (2014). *Carbon Nanotubes and Graphene* (2nd ed.). Kidlington: Oxford, Elsevier.

Vargas-Bernal, R. (2015a). Performance Analysis of Interconnects based on Carbon Nanotubes for AMS/RF IC Design. In M. Fakhfakh, E. Tlelo-Cuautle, & M. H. Fino (Eds.), *Performance Optimization Techniques in Analog, Mixed-Signal, and Radio-Frequency Circuit Design* (pp. 336–363). Hershey, PA: IGI Global. doi:10.4018/978-1-4666-6627-6.ch014

Vargas-Bernal, R. (2015b). Advances in Computational Modeling of Electronic Devices based on Graphene. *IEEE Journal on Emerging and Selected Topics in Circuits and Systems*, *5*(1), 109–116. doi:10.1109/JETCAS.2015.2398219

Vargas-Bernal, R. (2015c). Performance Analysis Electromagnetic Interference Shielding based on Carbon Nanomaterials used in AMS/RF IC Design. In M. Fakhfakh, E. Tlelo-Cuautle, & M. H. Fino (Eds.), *Performance Optimization Techniques in Analog, Mixed-Signal, and Radio-Frequency Circuit Design* (pp. 268–294). Hershey, PA: IGI Global. doi:10.4018/978-1-4666-6627-6.ch011

Vargas-Bernal, R., & Herrera-Pérez, G. (2012). Carbon Nanotube- and Graphene based Devices, Circuits and Sensors for VLSI Design. In VLSI Design, (pp. 41-66). Rijeka, Croatia: Intech.

Wen, M., Sun, X., Su, L., Shen, J., Li, J., & Guo, S. (2012). The Electrical Conductivity of Carbon Nanotube/Carbon Black/Polypropylene Composites prepared through Multistage Stretching Extrusion. *Polymer*, *53*(7), 1602–1610. doi:10.1016/j.polymer.2012.02.003

White, C. T., & Todorov, T. N. (1998). Carbon Nanotubes as Long Ballistic Conductors. *Nature*, *393*(6682), 240–242. doi:10.1038/30420

Xu, Z., Liu, Z., Sun, H., & Gao, C. (2013). Highly Electrically Conductive Ag-doped Graphene Fibers as Stretchable Conductors. *Advanced Materials*, *25*(23), 3249–3253. doi:10.1002/adma.201300774 PMID:23640822

Yousefi, N., Sun, X., Lin, X., Shen, X., Jia, J., Zhang, B., ... Kim, J. K. (2014). Highly Aligned Graphene/Polymer Nanocomposites with Excellent Dielectric Properties for High-Performance Electromagnetic Interference Shielding. *Advanced Materials*, *26*(31), 5480–5487. doi:10.1002/adma.201305293 PMID:24715671

Yusoff, A. R. M. (Ed.). (2015). Graphene-based Energy Devices. Wiley-VCH.

Zheng, Q., & Kim, J.-K. (2015). *Graphene for Transparent Conductors: Synthesis, Properties and Applications*. New York, NY: Springer. doi:10.1007/978-1-4939-2769-2

ADDITIONAL READING

Mahmoodi, M., Arjmand, M., Sundararaj, U., & Park, S. (2012). The Electrical Conductivity and Electromagnetic Interference Shielding of Injection Molded Multi-Walled Carbon Nanotube/Polystyrene Composites. *Carbon*, *50*(4), 1455–1464. doi:10.1016/j.carbon.2011.11.004

Matsumoto, K. (Ed.). (2015). Frontiers of Graphene and Carbon Nanotubes: Devices and Applications, Tokyo, Kanto: Springer. doi:10.1007/978-4-431-55372-4

Sadasivuni, K. K., Ponnamma, D., Kim, J., & Thomas, S. (Eds.). (2015). Graphene-based Polymer Nanocomposites in Electronics, Heidelberg, Baden-Württemberg, Springer. doi:10.1007/978-3-319-13875-6

Shafraniuk, S. (2015). *Graphene: Fundamentals, Devices, and Applications*. Boca Raton, Florida: Pan Stanford Publishing. doi:10.1201/b18258

Tiwari, A., & Syväjärvi, M. (Eds.). (2015). *Graphene Materials: Fundamentals and Emerging Applications*. Hoboken, New Jersey: John Wiley & Sons. doi:10.1002/9781119131816

KEY TERMS AND DEFINITIONS

Ballistic Conduction: or **Ballistic Transport**: The transport of electrons in a medium having negligible electrical resistivity or the highest electrical conductivity.

Band Gap: An energy difference (in electron volts) between the top of the valence band and the bottom of the conduction band in insulators and semiconductors.

Carbon Nanomaterials: Nanostructures of carbon such as fullerenes, carbon nanotubes, nanofibers and graphene with unique physicochemical properties with multiple technological applications.

Carbon Nanotubes: Allotropes of carbon with a cylindrical nanostructure of length-to-diameter of up to 132,000,000:1, which have unusual properties and valuable for nanotechnology, electronics, optics and other fields of materials science and technology.

Diffusive Transport: The movement of electrons from a region of high concentration to a region of low concentration where variables such as pressure or temperature are involved.

Electrical Conductivity: The physical property that quantifies how strongly a given material opposes the flow of electrical current. A low resistivity indicates a material that readily allows the movement of electrical charge.

Electronic Device: Device that accomplishes its purpose controlling the flow of electrons applied to digital electronics, analog electronics, microelectronics, optoelectronics, or integrated circuits.

Graphene: A two-dimensional, crystalline allotrope of carbon whose atoms are densely packed in a regular sp^2-bonded atomic-scale chicken wire (hexagonal) pattern composed by a one-atom thick layer of graphite.

Interconnect: A path of material that connects two elements or components in an integrated circuit, through which electrical current is transported.

Quantum Tunneling or Tunneling: A quantum mechanical phenomenon, where an electron tunnels through a band gap, which it classically could not surmount.

This research was previously published in the Encyclopedia of Information Science and Technology, Fourth Edition edited by Mehdi Khosrow-Pour, pages 2673-2684, copyright year 2018 by Information Science Reference (an imprint of IGI Global).

Section 5

Environmental Science and Agriculture

Chapter 9
Carbon Capture From Natural Gas via Polymeric Membranes

Nayef Mohamed Ghasem
UAE University, UAE

Nihmiya Abdul Rahim
UAE University, UAE

Mohamed Al-Marzouqi
UAE University, UAE

ABSTRACT

Polymeric membrane is a promising energy and an active alternative for conventional CO2 absorption column. The type of absorption liquid and operating parameters plays an efficient role in the ultimate absorption/stripping performance using gas-liquid membrane contactor. The gas flow rate has a significant effect on CO2 absorption performance; by contrast, it has no effect on stripping performance. Further, the CO2 absorption performance in membrane contactor could be enhanced by high liquid flow rates. The gas-liquid contact time was a key factor in enhancing the stripping flux at low temperature while liquid phase boundary layer thickness and associated mass transfer resistance is important at elevated temperature. By controlling the liquid phase velocity and the length of module at low temperature, better stripping performance can be achieved. The effect of liquid temperature on absorption performance in gas-liquid is not straightforward, since the liquid temperature cooperatively influences several factors.

INTRODUCTION

With the rapid development of technology, there is an increasing demand for fuels. Natural gas is an environmentally friendly, renewable, and clean energy source. It is also the third largest proportion in energy structure throughout the world after coal and oil. The composition of the raw natural gas extracted from producing wells depends on the type, depth, and location of the underground deposit and the geology of the area. Natural gas consists primarily of methane as the prevailing element but it also contains considerable amounts of light and heavier hydrocarbons as well as contaminating compounds

DOI: 10.4018/978-1-5225-7359-3.ch009

of CO_2, N_2, Hg, He, H_2S and etc. The presence of acid gases such as CO_2 and H_2S can cause corrosion of pipeline and equipment and they present a major safety risk. Also they reduce the energy content of the gas and affect the selling price of the natural gas. Further in Liquefied Natural Gas (LNG) processing plant, while cooling the natural gas to a very low temperature, the CO_2 can be frozen and block pipeline systems and cause transportation drawback. Consequently, natural gas produced at the wellhead must be processed, i.e., cleaned, before it can be safely delivered to the high-pressure, long-distance pipelines that transport the product to the consuming public.

The traditional method for CO_2 separation is amine scrubbing. Although high product yields and purities can be obtained, the disadvantage of this method is its high energy consumption, especially during stripper, in combination with high liquid losses due to evaporation of the solvent in the stripper (Naim et al., 2012). In addition, as liquid and gas streams cannot be controlled independently the occurrence of flooding, foaming, channeling and entrainment of the absorption liquid also limits the process. Membrane technology is a promising method to replace the conventional absorption technology. It has a high energy efficiency, is easy to scale-up because of its modular design and it has a high area-to-volume ratio. A limitation can be found in the permeability-selectivity tradeoff relation. Gas–liquid membrane contactor (GLMC) combines the advantages of membrane technology with those of absorption liquid (Ze et al., 2014). In a GLMC the microporous membrane acts as a fixed interface between the feed gas and the absorption liquid without dispersing one phase into another and this decoupling of the gas and liquid phase prevents any momentum transfer occurring across the phase boundary. As a consequence, the operation problems and constraints take place in conventional absorption technology can be resolved. Further the employment of microporous membrane elucidates the permeability-selectivity tradeoff relation drawback challenged in membrane technology. The performance of GLMC as CO_2 absorber and stripper depend upon several factors such as type of membrane, type of absorption liquid, module configurations and process parameters. Understanding the optimistic attributes of these factors on CO_2 separation performance of GLMC is vital important to develop the GLMC that gives the outstanding CO_2 absorption/stripping performance. The focus of this work is to illustrate the potential for the energy efficient and effective separation of CO_2/CH_4 gas mixture via lean solvent and regenerating of the rich solvent through absorption/stripping mechanism taking place in a hollow fiber GLMC process.

BACKGROUND

GLMCs have attracted great interest over the past decade as CO_2 absorber and stripper. In the absorber, CO_2 diffuses from the feed gas through the porous membrane and is then absorbed in the flowing liquid. Then this CO_2 rich liquid circulated form the absorber to the stripper membrane contactor module in which stripped CO_2 will be carried by sweep gas (Figure 1).

In the GLMC, gas and liquid flow on the different side of the microporous membrane and membrane acts only as a barrier between two phases without dispersing one phase to another. In general, when hydrophobic microporous membranes are used in membrane contactors, the gas-liquid interface is immobilized at the opening of the pores of microporous membrane by careful control of the pressure difference between the two phases. For applications in gas–liquid absorption/desorption, the driving force is based on the concentration gradient. The gas molecules to be separated diffuse from the concentrated phase to the gas liquid interface via the membranes pores the then contacts the diluted phase on the other

Figure 1. Membrane gas absorption/stripper process

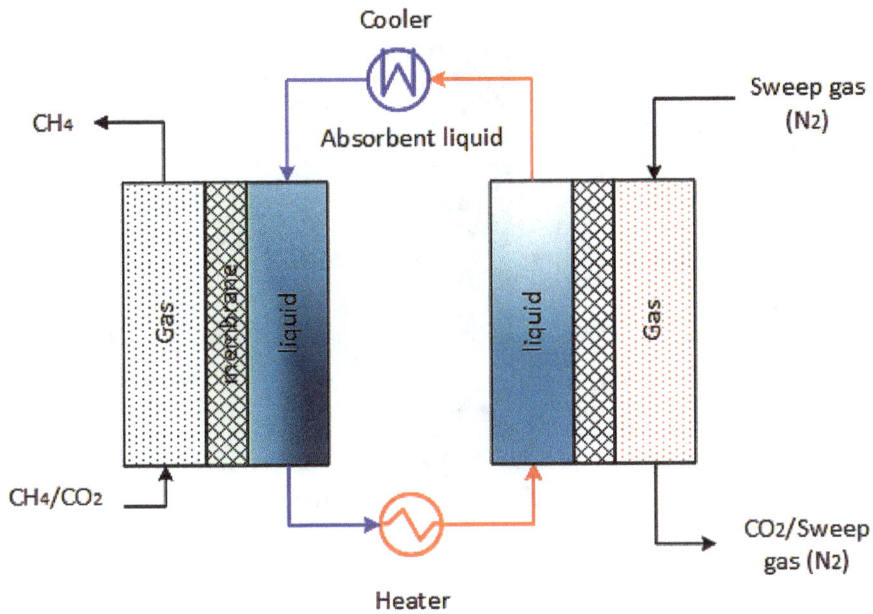

side. For instance, in the case of CO_2/CH_4 separation, as shown in Figure 2, CO_2 molecules diffuses from the feed gas side through the membrane and is then absorbed in the selective absorption liquid.

In GLMC the mass transfer process consists three steps in series: the transfer from one phase to the membrane surface, transfer within membrane pores and transfer from other phase interface to the bulk. Figure 3 shows the concentration profiles that are formed when species "i" is transferred from the gas phase to the liquid and if hydrophobic membranes are used. In order to describe this mass transfer process between two phases through porous membrane a resistance in series model based on film theory can be used. The resistances to the mass transfer encountered in both cases are those offered by the boundary layers and the membrane and can be drawn, as in Figure 4, by considering an electrical analogy.

Figure 2. CO_2 molecule transport through microporous membrane

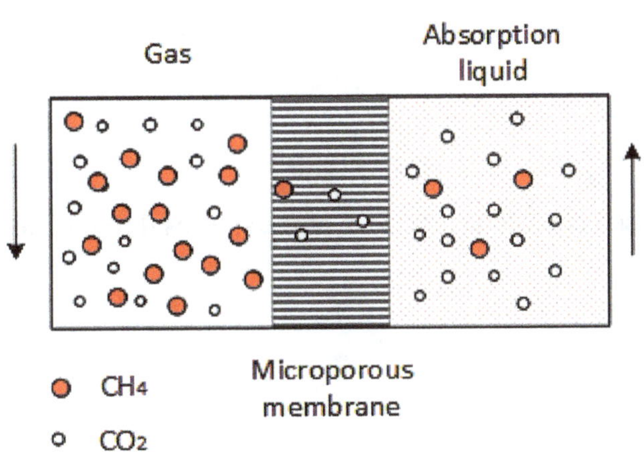

Figure 3. Concentration profiles formed when species "i" is transferred from one phase to another phase through hydrophobic membrane

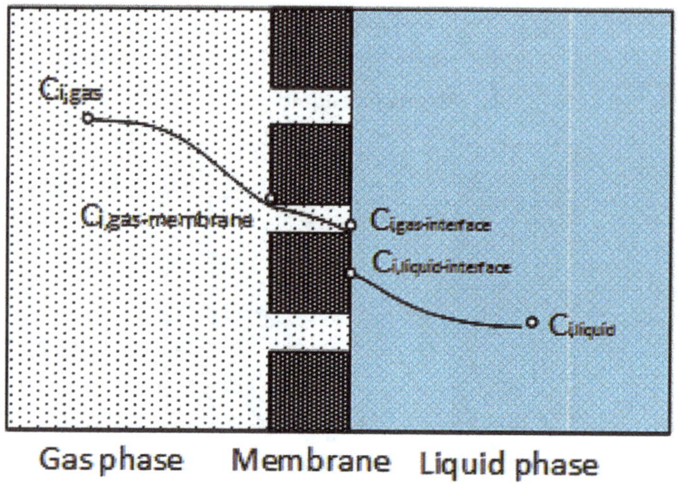

Figure 4. Resistance in series model

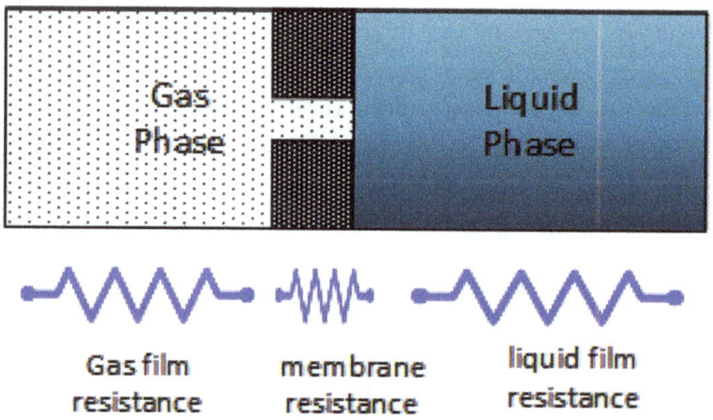

As shown in the Figure 4, three resistances exist in the resistance in series model:

- Gas film resistance.
- Liquid film resistance.
- Membrane resistance.

A general expression used to calculate the flux of the species through the membrane is:

$$J = K_{ov}(C_1 - C_2) \quad [1]$$

with

$$\frac{1}{K_{ov}} = \frac{1}{k_G}\frac{d_o}{d_i} + \frac{1}{k_m} + \frac{1}{mEk_L} \qquad [2]$$

hence

$$K_{ov} = f(k_G, k_m, k_L, m, E) \qquad [3]$$

where J is flux. C_1 and C_2 are concentration of CO_2 in the two phases. K_{ov} is overall mass transfer coefficient. k_G, k_m and k_L are individual mass transfer coefficient of gas phase, membrane and liquid phase respectively. M is physical solubility and E is enhancement factor if any chemical reaction present.

Hence overall mass transfer coefficient in GLMC, be determined by three individual mass transfer coefficients, solubility and enhancement factor which in turn be influenced by so many factors. Firstly, membrane mass transfer coefficient influenced by so many membrane inherent properties such as membrane pore size, porosity and tortuosity which on the other hand controlled by adopted fabrication method . Further membrane mass transfer coefficient also subject to whether the membranes pores are gas filled (ideal non wetted mode) or liquid filled (non-preferred wetted mode). Next, the gas and liquid mass transfer coefficients as well depend in GLMC module configurations and adopted process parameters during separation process such as gas and liquid flow rates, temperature etc. Solubility and enhancement factor depends on the type of solvent used. So detail understanding of these properties is vital important for complete understanding of GLMC. Qi and Cussler were the first to propose the idea of CO_2 absorption by sodium hydroxide in a HFMC. Afterwards GLMC as CO_2 absorber has been comprehensively studied by several researchers and exciting experimental and theoretical results have been reported (Ghasem et al., 2011, 2012a, 2012b, 2013, 2014). Currently, efficient and effective CO_2 stripping from liquid absorbents by using membrane contactor has become the target of many researchers in the field and few experimental results have been reported [Rahim et al., 2014, 2015].

MAIN FOCUS OF THE ARTICLE

The schematic diagram in Figure 5 shows the experimental setup used to study the potential of custom made hollow fiber GLMC for separation of CO_2/CH_4 gas mixture via lean solvent and regeneration of the rich solvent. Homemade and characterized 28% PVDF hollow fiber membranes (Ghasem e al., 2012, 2013) were used to prepare the GLMC modules. Shell and tube type modules were constructed using Perspex tube as shell (transparent) and hollow fiber membranes as tubes. The details of the membrane contactor modules are given in Table 1. Firstly the CO_2 absorption (by using fresh absorption liquids) and stripping (by using manually CO_2 loaded absorption liquids) performance of GLMC was tested separately. Then the absorption experiments were conducted by recycling the absorption liquid several times with and without the stripping between each recycle to investigate the potential of using GLMC for continuous mode operation.

To study the absorption performance 10/90 vol% CO_2/CH_4 gas mixture was applied to the shell side of the membrane contactor module at different gas inlet flow rate (10 to 100 ml/min) by using mass flow controllers (Alicat Scientific). Different 0.5 M aqueous solutions of primary (MEA), secondary (DEA), hindered (AMP) amines and amine salts (PG) were used as absorption liquid. The liquid feed stream supplied to lumen side of the module at variable flow rate of 10-40 ml/min by using peristaltic pump (Masterflex® L/S®). Also the liquid feed stream was heated for different temperatures; range from room temperature 20 °C to 60 °C by using feedback control heater (WiseStir®).Exit gas concentration was measured using gas chromatography (Shimadzu, Japan). Also the long term absorption performance of absorption liquids was examined by running absorption experiments for 200 minutes. Moreover for PG the ratio between potassium hydroxide and glycine were altered in order to study the effect of solution pH on absorption/stripping performance. Further the recyclability of absorption liquids were studied by recycling the absorption liquid from absorber module with and without stripping between each cycle.

To study the stripping performance the liquid feed stream either coming from the absorber module or manual CO_2 preloaded aqueous solutions (different initial CO_2 loading) were pumped to the stripping module. Pure nitrogen, as a sweep gas, was made to flow through the module shell side at different gas inlet flow rate (100 to 600 ml/min) by using mass flow controllers (Alicat Scientific). The liquid feed stream supplied to lumen side of the module at variable flow rate of 10-50 ml/min by using peristaltic pump (Masterflex® L/S®). Also the liquid feed stream was heated for different temperatures, range from

Table 1. Specifics of the gas liquid membrane contactor module

Fiber	28% PVDF
Module length (mm)	260
Module outer dia. (mm)	15
Module inner dia. (mm)	11
Fiber outer dia. (mm)	1.5
Fiber inner dia. (mm)	0.42
No of fibers	10

Figure 5. CO_2 absorption/stripping experiment setup

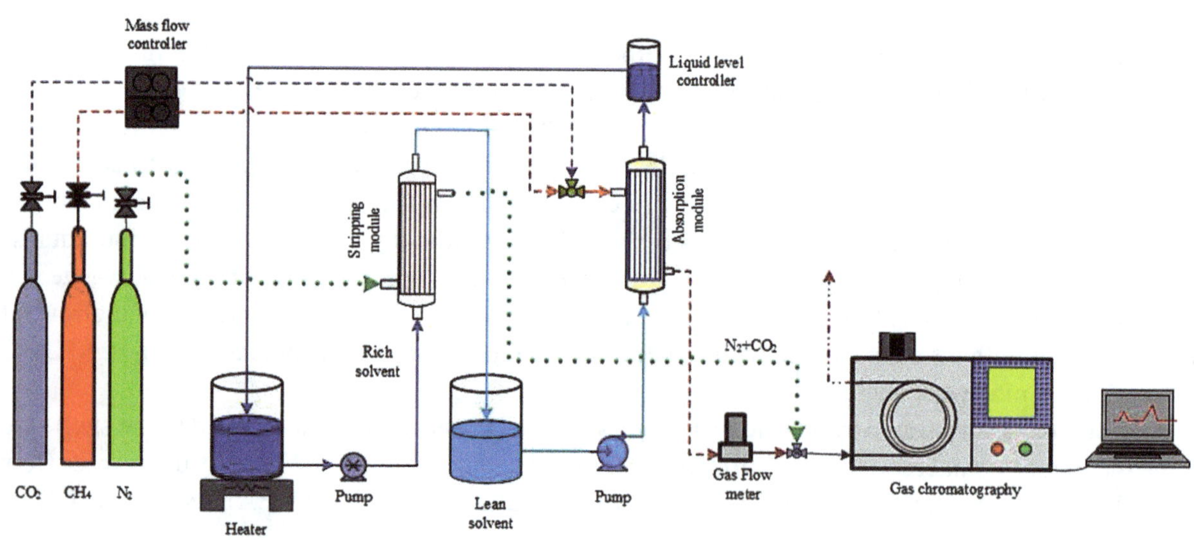

room temperature 25 °C to 90 °C by using feedback control heater (WiseStir®). A counter-current flow mode was applied for the gas and liquid phases. Variable pressure difference range from 0.01 to 0.05 MPa was applied between the liquid stream and gas stream by using control valve in order to avoid the formation of bubbles on the liquid side and to study the pressure effect. The system was in operation for 30 min to achieve a steady state condition before taking samples. The liquid phase CO_2 concentration at the inlet and outlet of the stripper module was measured by using double chemical titration method (described in the succeeding section), to determine stripping flux and efficiency. Also the sweep gas exit concentration was measured by using gas chromatography (Shimadzu, Japan) in order to make sure the titration results. The experimental CO_2absorption/stripping efficiency was calculated as:

$$\eta(\%) = \left(\frac{v_i C_i - v_o C_o}{v_i C_i} \right) \times 100 \tag{4}$$

Where C_i and C_o (mol/m^3) are the feed gas CO_2 concentration for absorption and liquid phase CO_2 concentration for stripping at the inlet and outlet of the membrane module respectively.v_i and v_o (m^3/min) are the feed gas flow rate for absorption and liquid flow rate for stripping at the inlet and outlet of the membrane module respectively

The CO_2 absorption/stripping flux of the module can be calculated as:

$$J_{co_2} = \frac{\left(v_i C_i - v_o C_o \right)}{A_i} \tag{5}$$

where J_{CO2} (mol/m^2 s) is the CO_2 flux and A_i (m^2) is inner surface of the hollow fiber membranes.

Measurement of CO_2 Loading in Liquid Sample

CO_2 concentration in the aqueous solution was determined by double chemical titration method (Seo et al., 1996) An excess amount of 1.0 M NaOH solution was added to the known amount of CO_2 rich liquid sample, converting dissolved CO_2 into the nonvolatile ionic species.Then an excess amount of 1.0 M $BaCl_2$ solution was added to the solution. The solution then stirred and heated for 3 hour at temperature of 700C. Absorbed CO_2 precipitates as $BaCO_3$. The excess NaOH was titrated with 1.0 M HCl solution using phenolphthalein as the indicator. After this titration, using Methyl Orange as the indicator, HCl was added to measure the amount of $BaCO_3$. The volume of HCl added to neutralize the basic species in the solution was used to calculate the CO_2 loading.

The CO_2 loading was calculated according to the following equation:

$$\alpha = \frac{v_{HCl}}{2 \times v_{sample}} \tag{6}$$

where α = CO_2 loading in mol of CO_2/volume of solution, v_{HCl} = volume of HCl request to neutralize the $BaCO_3$ in ml, v_{sample} = volume of sample taken for analysis in ml.

SOLUTIONS AND RECOMMENDATIONS

The potential of custom made hollow fiber GLMC for separation of CO_2/CH_4 gas mixture via lean solvent and regeneration of the rich solvent was tested by using different aqueous solutions of primary (MEA), secondary (DEA), hindered (AMP) amines and amine salts (PG) as absorption liquids. The experimental results proved GLMC is promising energy efficient and effective alternative for conventional CO_2 absorption/stripping technology and the type of absorption liquid and operating parameters plays a vital role in eventual absorption/stripping performance given away by GLMC.

CO_2 Absorption Performance of GLMC

The CO_2 absorption experiments in GLMC revealed the absorption performance order of studied absorbent liquids is PG>MEA> AMP>DEA> regardless of operating conditions adopted (Figure 6). The aqueous amine-based solutions are usually adopted CO_2 absorption liquids and MEA is the most commonly used amongst them. The structures of alkanolamines include primary, secondary, ternary amines containing at least one OH and amine group such as MEA, DEA and MDEA. The reactivity of amines to CO_2 follows the order primary, secondary and ternary amines. So as obtained MEA has greater abortion performance than the DEA. In addition to these amines, the steric hindrance amines such as AMP also proposed by several researchers. This is because that the steric character reduces the stability of the formed carbamate and easy regenerate the solution. The absorption performance of AMP is better than the DEA as shown in figures. Amino acid salts (AAS) aqueous solutions attract great research interest in recent years as CO_2 absorbent liquid in GLMC because of its prominent characters such as their better affinities towards CO_2 than alkanolamines and their high surface tension because of its ionic nature. Hence mostly used PG absorption performance was compared with other absorbent liquids and its absorption performance

Figure 6. CO_2 absorption performance of various absortion liquids in GLMC (0.5M aqeuous solutions at room temperature flow in lumen side at10 ml/min; gas mixture 9%CO_2/91%CH_4 flows in shell side)

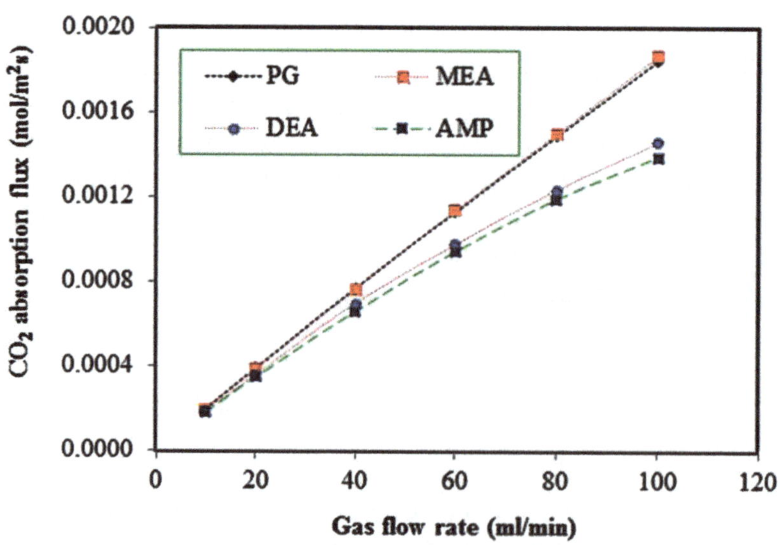

was greater than the MEA aqueous solutions. As well even after 8 hours long run the PG did not wet the membrane. Whereas the amine solutions cause membrane wetting averagely after five six hours of continuous running, this is due to its low surface tension.

Moreover from the absorption experiments it was observed the gas flow rate have a significant effect on CO_2 absorption performance in GLMC. Although as shown in Figure 6 the CO_2 absorption flux increased with gas flow rate the CO_2 removal efficiency decreased with increased gas flow rate. For instance for PG the removal efficiency reduced from 99% to 94% with the increased of gas flow rate from 10 ml/min to 100 ml/min. This can be attributed to decrease in contact time and increase in driving force for mass transfer. Increasing the gas flow rate decreases the residence time of the gas phase in the membrane contactor and hence contact time of the gas phase with liquid. On the other hand the increase in the gas velocity results in the reduction of the boundary layer and the improvement of the total mass-transfer rate. Further it was observed for any absorption liquid the CO_2 absorption performance in GLMC could be enhanced by high liquid flow rates. This enhancement in absorption performance was significant for the absorption liquids which have poor absorption performance than the absorption liquids which have good absorption performance. For PG the removal efficiency increased only by 3.3%, whereas for DEA the removal efficiency increased by 17% with increase of liquid flow rate from 10 ml/min to 50 ml/min. This is attributed to higher liquid velocity leads to a lower CO_2 concentration in the liquid phase, which in turn results in a higher CO_2 concentration gradient between gas and liquid phase phases. Also increasing the liquid velocity turns the fluid flow form laminar to turbulent this reduces the liquid phase boundary layer.

The effect of temperature on CO_2 absorption performance in GLMC is attributed to collective effect of solubility (physical absorption), chemical reaction (chemical absorption), diffusion and evaporation of absorbent. Further increase in temperature would decrease the viscosity of the solution which is favorable character of GLMC absorbent liquid. It is well known the favored chemical reaction rate and diffusion rate with temperature would enhance the absorption performance. While decrease in CO_2 solubility and an increase in evaporation of absorbent (wetting) with temperature would reduce the absorption performance. Experimentally it was observed MEA and shows no any significant effect of temperature on absorption performance contrary to the results obtained with AMP and DEA. This may be due to for MEA and PG the favored chemical reaction and diffusion with temperature were compensated by the reduced solubility. Also these absorbent have good reactivity even at room temperature (around 95% removal), so the enrichment in absorption performance with temperature was not highlighted. Whereas for AMP and DEA the favored effect of temperature on chemical reaction and diffusion were higher than the reduced effect temperature on CO_2 solubility and so the enrichment in absorption performance with temperature was significant. However all solvents shows some flattening off at elevated temperatures. Because the wetting caused by elevated temperature cannot be neglected in long term operation. Further the thermal degradation of membrane material also should be considered. So, solvent temperature is a crucial factor to be controlled very carefully for long term operation performance.

CO_2 Stripping Performance of GLMC

CO_2 stripping performance of PVDF hollow fiber GLMC was compared between primary (MEA), secondary (DEA), hindered (AMP) amines and amine salts (PG). The liquid solutions were manually preloaded with CO_2 for saturation level in order to obtain higher CO_2 flux for comparing purposes In terms of regeneration AMP and PG perform well as absorbent liquid comparing to MEA and DEA. The results in

Figure 7 show the effect of different aqueous solution on stripping flux as function of temperature. The regeneration energy of the absorbents was directly related with the heat of reaction and showed higher values when the binding force between amines and CO_2 molecules was larger. Therefore, absorbents that formed carbamate required larger quantities of heat, and the absorption heats of the MEA and DEA were higher than those of AMP and PG. So AMP and PG shows better stripping performance compare to MEA and DEA. Regardless of type of solvent the CO_2 stripping flux and efficiency increases with temperature rapidly. Because as the reaction rate increases with temperature, the formation of carbamate, which is the final product of CO_2 reaction, become unstable and the energy consumed in regeneration become smaller.

The studies on sweep gas flow rates exposed the gas flow rate have no any significant effect on stripping flux and efficiency regardless of type of solvent. However, when operating at high rich solution temperature, the low sweep gas flow rate, allow the vapor molecules to easily enter through the pores and wet the membrane. This reduces the effective long term operation of the membrane contactor module. So, moderate sweep gas flow rate gives better performance comparing to low sweep gas flow rate. Liquid flow rate shows some notable criteria on CO_2 stripping flux and efficiency. At low temperatures the increase in liquid flow rate reduces the stripping efficiency. In contrast at high temperature the increase in liquid flow rate increases the stripping efficiency. Two phenomenon govern the stripping efficiency. The residence time (contact time between gas and liquid phase) and liquid phase boundary layer thickness. At low temperature contact time overtaken by the boundary layer thickness. Lower liquid velocity tend to higher residence time, which lets dissolved CO_2 to shift to gas–liquid interface and results in increasing the driving force of mass transfer. In contrast at high temperature due to low solubility more CO_2 end to be released. So rather than residence time boundary layer thickness is important. Because the liquid flow rate reduces the boundary layer thickness and increase the mass transfer coefficient, at high temperature high flow rates gives the high removal efficiency.

Figure 7. Effect of liquid phase temperature on CO_2 stripping flux for four different solvents (saturation initial CO_2 loading, v_l= 50ml/min, v_g = 600 ml/min and $P_l - P_g = 0.5 \times 10^5$ Pa)

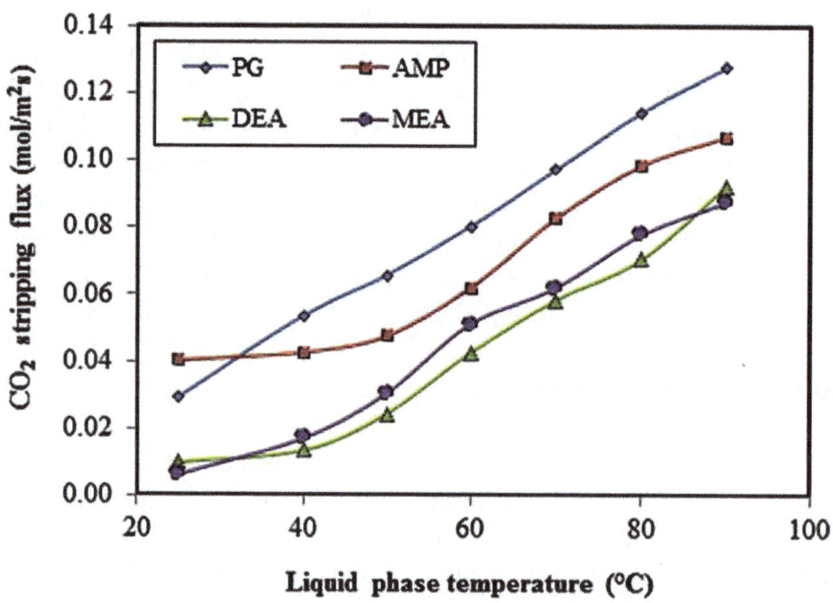

The liquid pressure also has great influence in the stripping performance. The considerable increase in CO_2 flux with liquid pressure can be attributed to the increase in driving force for desorption as a result of an increase in CO_2 concentration. Many researchers proved the CO_2 absorption flux increase drastically with gas phase pressure. So vice versa by increasing the liquid side pressure the CO_2 desorption flux can be increased. But, even though, hydrophobic PVDF hollow fiber membrane can resist wetting, applying higher pressure in the liquid side can gradually cause wetting. It is well known that partial wetting of membrane can increase mass transfer resistance significantly. So the liquid side pressure should be maintained below the break through pressure.

The effect of absorption liquid pH on CO_2 absorption/stripping performance in gas liquid membrane contactor module was investigated by using asymmetric solutions of PG (i.e. solutions containing different molar amounts of amino acid (glycine) and base (potassium hydroxide)). At lower pH the CO_2 reaction equilibrium shifts towards the release of CO_2 enhancing the partial pressure of CO_2. So higher molar ratios of amino acid shows better stripping and reduced absorption performance (Figure 8). Subsequently in continuous operation of absorption and followed up by stripping, the stripping performance was controlled by reduced absorption performance because of resultant low initial CO_2 concentrations. Hence for the net effect the molar ratio has to be optimized to achieve both good absorption and subsequent stripping performance.

FUTURE RESEARCH DIRECTIONS

GLMC due to attested several advantages has become the promising alternative for conventional CO_2 absorption/stripping process. Although the GLMC for CO_2 separation has been extensively studied

Figure 8. Effect of different asymmetric solutions on overall efficiency

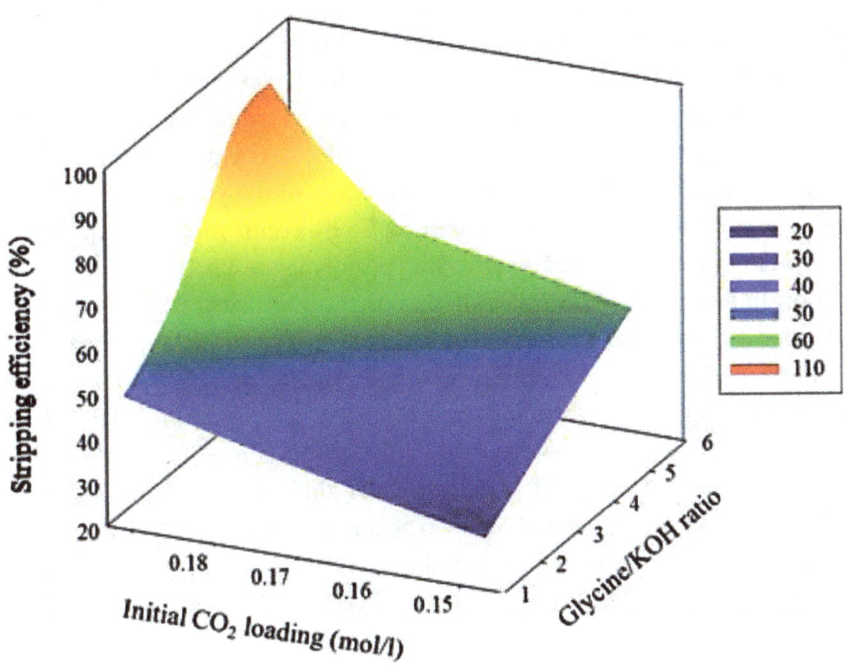

there is still a long way ahead before this technique to completely replace the existing CO_2 separation technology. The main hitch is long term stability of GLMC absorption process due to membrane wetting. The key is employing surface modified super hydrophobic membranes and high surface tension absorption liquids in GLMC absorption/stripping processes. Currently, various techniques such as surface grafting, bore filling grafting, coating/interfacial polymerization and in-situ polymerization are being investigated to improve the surface hydrophobicity. Some works are going on using ionic liquid membranes in GLMC applications. Further as a consequences of seek out for new absorption liquids that not only have high surface tension but also can be regenerable in efficient way, AAS and ionic liquids have attracted great research interest. Most importantly AASs have favorable biodegradation properties, which make the disposal of these solvents easier and with lower environmental impacts. In spite of all these work still the membrane wetting at elevated temperatures is arguing. Moreover, despite the fact that solvent regeneration is responsible for the major cost component in gas separation processes due to energy consumption, the studies on CO_2 stripping using GLMC have started recently and there are only few reports documented in the open literature. Hence future research directions should focus on super hydrophobic membranes, absorption liquids that can be regenerated in an energy efficient way and eventually suitable membrane-absorbent combination, therefore, GLMC will thrive as a perfectly energy efficient and effective CO_2 absorption/stripping technology, which can replace the current CO_2 separation technology completely.

CONCLUSION

The experimental results proved GLMC is promising energy efficient and effective alternative for conventional CO_2 absorption/stripping technology and the type of absorption liquid and operating parameters plays a vital role in eventual absorption/stripping performance given away by GLMC. Studied AAS (PG) shows both good absorption and stripping performance. The gas flow rate has a significant effect on CO_2 absorption performance where as it has no effect in stripping performance. Further the CO_2 absorption performance in GLMC could be enhanced by high liquid flow rates. This enhancement was significant for the absorption liquids which have poor absorption performance than the absorption liquids which have good absorption performance. In contrast, the stripping performance enhancement with liquid flow rate depends on liquid temperature. Because the gas–liquid contact time was a key factor to enhance the stripping flux at low temperature while liquid phase boundary layer thickness and associated mass transfer resistance is important at elevated temperature. So by controlling the liquid phase velocity and the length of module at low temperature better stripping performance can be achieved. The effect of liquid temperature on absorption performance in GLMC is not straightforward. Since the liquid temperature cooperatively influence the several factors which determines the absorption performance positively and negatively, it should be handled in care depending on the system used. However by increasing the rich solution temperature the stripping performance can be improved preferably. To improve the stripping performance enhanced CO_2 desorption (based on pH-shift) was studied. In this study the pH values were changed by using asymmetric solutions of PG. Lower pH shifts the reaction equilibrium towards the release of CO_2 and shows better stripping and reduced absorption performance. Thus for the net effect the molar ratio has to be optimized to achieve both good absorption and subsequent stripping performance.

ACKNOWLEDGMENT

The authors would like to acknowledge the financial support provided by the research and graduate study of the United Arab Emirates University (UAEU) and National Research Fund (NRF). Grant number 31N168-UPAR (9) 2013.

REFERENCES

Ghasem, N., Al-Marzouqi, M. H., & Rahim, N. A. (2012a). Effect of polymer extrusion temperature on poly (vinylidene fluoride) hollow fiber membranes: Properties and performance used as gas–liquid membrane contactor for CO_2 absorption. *Separation and Purification Technology*, *99*, 91–103. doi:10.1016/j.seppur.2012.07.021

Ghasem, N., Al-Marzouqi, M. H., & Rahim, N. A. (2014). Absorption of CO_2 Form Natural Gas via Gas-liquid PVDF Hollow Fiber Membrane Contactor and Potassium Glycinate as Solvent. *Jurnal Teknologi*, *69*(9), 121–126. doi:10.11113/jt.v69.3409

Ghasem, N. M., Al-Marzouqi, M., & Zhu, L. P. (2012b). Preparation and properties of polyethersulfone hollow fiber membranes with o-xylene as a additive used in membrane contactors for CO_2 absorption. *Separation and Purification Technology*, *92*, 1–10. doi:10.1016/j.seppur.2012.03.005

Ghasem, N. M., Al-Marzouqi, M. H., & Duaidar, A. (2011). Effect of quenching temperature on the performance of poly (vinylidene fluoride) microporous hollow fiber membranes fabricated via thermally induced phase separation technique on the removal of CO_2 from CO_2 gas mixture. *International Journal of Greenhouse Gas Control*, *5*(6), 1550–1558. doi:10.1016/j.ijggc.2011.08.012

Ghasem, N. M., Al-Marzouqi, M. H., & Rahim, N. A. (2013). Modeling of CO_2 absorption in a membrane contactor considering solvent evaporation. *Separation and Purification Technology*, *110*, 1–10. doi:10.1016/j.seppur.2013.03.008

Naim, R., & Ismail, A. F. (2013). Effect of polymer concentration on the structure and performance of PEI hollow fiber membrane contactor for CO2 stripping. *Journal of Hazardous Materials*, *250*, 354–361. doi:10.1016/j.jhazmat.2013.01.083 PMID:23474409

Rahim, N. A., Ghasem, N., & Al-Marzouqi, M. H. (2014). Stripping of CO_2 from different aqueous solvents using PVDF hollow fiber membrane contacting process. *Journal of Natural Gas Science and Engineering*, *21*, 886–893. doi:10.1016/j.jngse.2014.10.016

Rahim, N. A., Ghasem, N., & Al-Marzouqi, M. H. (2015). Absorption of CO_2 from natural gas using different amino acid salt solutions and regeneration using hollow fiber membrane contactors. *Journal of Natural Gas Science and Engineering*, *26*, 108–117. doi:10.1016/j.jngse.2015.06.010

Seo, D.-J., & Hong, W.-H. (1996). Solubilities of carbon dioxide in aqueous mixtures of diethanolamine and 2-amino-2-methyl-1-propanol. *Journal of Chemical & Engineering Data*, *41*(2), 258–260. doi:10.1021/je950197o

Ze, Z., & Sx, J. (2014). Hollow fiber membrane contactor absorption of CO_2 from the flue gas: Review and perspective. *Global NEST Journal*, *16*, 355–374.

ADDITIONAL READING

Drioli, E., Criscuoli, A., & Curcio, E. (2011). *Membrane Contactors: Fundamentals, Applications and Potentialities: Fundamentals, Applications and Potentialities* (Vol. 11). Elsevier.

Ghasem, N. M., & Al-Marzouqi, M. (2014). Modeling and Experimental Study of Gas-Liquid Membrane Contactor. Encyclopedia of Information Science and Technology, Third Edition (10 Volumes) July, 2014. Chapter 537, pages 5442-5453. DOI: doi:10.4018/978-1-4666-5888-2

Li, J.-L., & Chen, B.-H. (2005). Review of CO_2 absorption using chemical solvents in hollow fiber membrane contactors. *Separation and Purification Technology*, *41*(2), 109–122. doi:10.1016/j.seppur.2004.09.008

Mansourizadeh, A., & Ismail, A. F. (2009). Hollow fiber gas–liquid membrane contactors for acid gas capture: A review. *Journal of Hazardous Materials*, *171*(1-3), 38–53. doi:10.1016/j.jhazmat.2009.06.026 PMID:19616376

Mansourizadeh, A., & Ismail, A. F. (2011). CO_2 stripping from water through porous PVDF hollow fiber membrane contactor. *Desalination*, *273*(2-3), 386–390. doi:10.1016/j.desal.2011.01.055

Mulder, M. (1997). *Basic Principles of Membrane Technology*. Kluwer Academic Publishers.

Naim, R., Ismail, A. F., Cheer, N. B., & Abdullah, M. S. (2014). Polyvinylidene fluoride and polyetherimide hollow fiber membranes for CO_2 stripping in membrane contactor. *Chemical Engineering Research & Design*, *92*(7), 1391–1398. doi:10.1016/j.cherd.2013.12.001

Rahbari-Sisakht, M., Ismail, A. F., Rana, D., & Matsuura, T. (2013). Carbon dioxide stripping from diethanolamine solution through porous surface modified PVDF hollow fiber membrane contactor. *Journal of Membrane Science*, *427*, 270–275. doi:10.1016/j.memsci.2012.09.060

Simioni, M., Kentish, S. E., & Stevens, G. W. (2011). Membrane stripping: Desorption of carbon dioxide from alkali solvents. *Journal of Membrane Science*, *378*(1-2), 18–27. doi:10.1016/j.memsci.2010.12.046

Yan, S., He, Q., Zhao, S., Zhai, H., Cao, M., & Ai, P. (2015). CO_2 removal from biogas by using green amino acid salts: Performance evaluation. *Fuel Processing Technology*, *129*, 203–212. doi:10.1016/j.fuproc.2014.09.019

KEY TERMS AND DEFINITIONS

AASS: Amino acid salt solution. Chemical solvent derived by mixing amino acid and alkaline hydroxides.

CO_2 Absorption: Operation used in removing acid gas from other gases using liquid solvent.

CO_2 Stripping: Operation used in removing absorbed acid gas from absorbent liquid.

GLMC: Gas Liquid membrane contactor. A device that have bundle of fibers enclosed on a shell in which gas and liquid flow without dispersing with each other.

Hydrophobic: The not-affinity to water.

Natural Gas: Gas consists with more than seventy percent methane.

PVDF: Polyvinylidene fluoride, polymer used in fabrication.

Wetting: Membrane pores filled with liquid instead of gas.

This research was previously published in the Encyclopedia of Information Science and Technology, Fourth Edition edited by Mehdi Khosrow-Pour, pages 3043-3055, copyright year 2018 by Information Science Reference (an imprint of IGI Global).

Chapter 10
Enhancing the Resiliency of Smart Grid Monitoring and Control

Wenbing Zhao
Cleveland State University, USA

ABSTRACT

In this chapter, the authors present the justification and a feasibility study of applying the Byzantine fault tolerance (BFT) technology to electric power grid health monitoring. They propose a set of BFT mechanisms needed to handle the PMU data reporting and control commands issuing to the IEDs. They report an empirical study to assess the feasibility of using the BFT technology for reliable and secure electric power grid health monitoring and control. The authors show that under the LAN environment, the overhead and jitter introduced by the BFT mechanisms are negligible, and consequently, Byzantine fault tolerance could readily be used to improve the security and reliability of electric power grid monitoring and control while meeting the stringent real-time communication requirement for SCADA operations.

INTRODUCTION

Smart grid is one of the hottest research areas in recent years. The development of smart grid is partially driven by the fact that the traditional data communication infrastructure for electric power grid can no longer meet the needs of new developments (Wang, Xu, & Khanna, 2011):

- The recent deregulation would allow many independent parties to enter the utility industry by offering alternative channels for electric power generation, distribution, and trade. This inevitably demands timely, reliable and secure information exchanges among these parties (Bose, 2005).
- The current data communication infrastructure lacks the support for large-scale real-time coordination among different electric power grid health monitoring and control systems, which could have prevented the 2003 massive blackout incident in North America (Birman et al., 2005).

DOI: 10.4018/978-1-5225-7359-3.ch010

- The use of modern computer networking technology could also revolutionize the everyday electric power grid operations, as shown by the huge benefits of substation automation and the use of Phasor Measurement Units (PMUs) for electric power grid health monitoring (Melliopoulos, 2007).

However, the openness and the ease of information sharing and cooperation brought by smart grid also increased the likelihood of cyber attacks on the electric power grid, as demonstrated recently by an experiment conducted by the US Department of Energy's Idaho Lab (CNN, 2007). To address such vulnerability, intrusion detection and intrusion tolerance techniques must be used to enhance the current and future data communication infrastructure for the electric power grid. Byzantine fault tolerance is a fundamental technique to achieve the objective (Castro & Liskov, 2002; Zhao, 2014a).

In this chapter, we focus our discussions on the security and reliability of smart grid health monitoring and control. We elaborate in detail the need for Byzantine fault tolerance and the challenges of applying Byzantine fault tolerance into this problem domain. In particular, we investigate experimentally the feasibility of using such sophisticated technology to meet potentially very stringent real-time requirement for the health monitoring and control of smart grid, while ensuring high degree of reliability and security of the system.

BACKGROUND

A Byzantine faulty process may behave arbitrarily. In particular, it may disseminate conflicting information to different components of a system, which constitutes a serious threat to the integrity of a system (Lamport, Shostak, & Pease, 1982). Because a Byzantine faulty process may also choose not to send a message, or refuse to respond to requests, it can exhibit crash fault behavior as well. Consider the scenario that multiple PMUs periodically report their measurement results to a controller for electric power grid health monitoring. When it detects an abnormally, the controller may wish to issue specific control instructions to the actuating devices, such as Intelligent Electronic Devices (IEDs) (Hossenlopp, 2007) located at the same substation as those PMUs to alleviate the problem. Due to the critical role played by the controller, it must be replicated to ensure high availability. Otherwise, the controller would become a single-point of failure. The main components and their interactions are illustrated in Figure 1.

However, the controller replicas, the PMUs, and the IEDs, might be compromised under cyber attacks. Consider the following two scenarios:

- A Byzantine faulty PMU could potentially send inconsistent data to different controller replicas. Without proper coordination among the controller replicas, the state of the replicas might diverge in the former case, which would lead to inconsistent decisions among the replicas.
- A compromised controller replica could send conflicting commands to different IEDs. Without a sound mechanism at each IED, a malicious command might be executed in the latter case, which could lead to the destruction of a generator or a transmission line, as reported by CNN (2007).

Byzantine fault tolerance (BFT) refers to the capability of a system to tolerate Byzantine faults (Lamport, Shostak, & Pease, 1982). If BFT is used, the cyber attacks illustrated above could be defeated provided that the number of compromised controller replicas, f, is below a threshold, and the number of

Figure 1. The interaction of substation devices (PUMs and IEDs) and the controller replicas

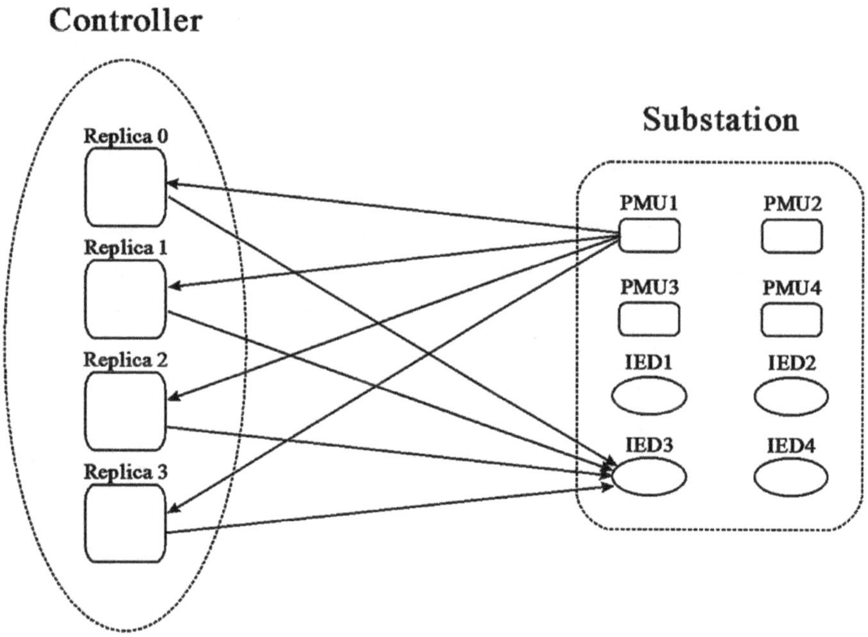

non-faulty PMUs and IEDs are sufficient for the normal operation of the substation. For the client-server system shown in Figure 1, BFT can be achieved by using 3f + 1 replicas to tolerate up to f faulty replicas and by ensuring all non-faulty replicas to execute the same set of requests in the same order. The latter means that the server replicas must reach an agreement on the set of requests and their relative ordering despite the presence of Byzantine faulty replicas and clients. Such an agreement is often referred to as a Byzantine agreement (Lamport, Shostak, & Pease, 1982). The Byzantine agreement among the replicas ensures that a faulty client (i.e., a PMU) cannot cause the divergence of the state of non-faulty controller replicas. Furthermore, a Byzantine agreement must be reached among all non-faulty replicas on each command issued by the controller for reasons to be explained in the next section. Before an IED can accept the command, it must wait until it has collected at least f + 1 identical command from different replicas.

We should note that Byzantine fault tolerance has been a hot research area in many other areas, such as Web services (Merideth et al., 2005; Zhao, 2009) and data storage systems (Rhea et al., 2003). Even though the works are in a different context, many insights are useful for BFT controls in electric power grid applications. In particular, the mechanisms designed to cope with the interaction of a replicated object and the un-replicated external entities reported in (Merideth et al., 2005) have been partially incorporated in this work.

BYZANTINE FAULT TOLERANT MONITORING AND CONTROL MECHANISMS

In this work, we choose to use PBFT, a well-known Byzantine agreement algorithm developed by Castro and Liskov (2002). The PBFT algorithm is designed to support client-server applications running in an asynchronous distributed environment with the Byzantine fault model. The implementation of the algorithm contains two parts. At the client side, a lightweight library is in charge of sending the client's request to the primary replica, retransmitting the request to all server replicas on the expiration of a retransmission timer (to cope with the primary faults and network faults), and collecting and voting on the corresponding replies. The main PBFT algorithm is executed at the server side by a set of 3f + 1 replicas to tolerate up to f faulty replicas. One of the replicas is designated as the primary while the remaining are backups. Furthermore, all messages are protected by a digital signature, or an authenticator (Castro & Liskov, 2002) so that a faulty replica or client cannot tamper with the messages and cannot impersonate as another non-faulty replica or client.

The normal operation of the (server-side) PBFT algorithm involves three phases: pre-prepare, prepare, and commit. In the beginning of the pre-prepare phase, the primary multicasts a pre-prepare message containing the client's request, the current view number and a sequence number assigned to the request to all backups. A backup verifies the request message and the ordering information. If the backup accepts the message, it multicasts to all other replicas a prepare message containing the ordering information and the digest of the request being ordered. This starts the prepare phase. A replica waits until it has collected 2f prepare messages from different replicas (including the message it has sent if it is a backup) that match the pre-prepare message before it multicasts a commit message to other replicas, which starts the commit phase. The commit phase ends when a replica has received 2f matching commit messages from other replicas. At this point, the request message has been totally ordered and it is ready to be delivered to the server application if all previous requests have already been delivered. If the primary or the client is faulty, the Byzantine agreement on the ordering of a request might not be reached, in which case, a new view is initiated, triggered by a timeout on the current message being ordered. A different primary is designated in a round-robin fashion for each new view installed.

For electric power grid health monitoring and control, however, the above BFT algorithm cannot be used directly, because normally the controller replicas collect input from the PMUs and the control commands are issued to IEDs. Furthermore, the updates from PMUs are one-way messages in that the PMUs normally do not wait for an explicit response from the controller. On the other hand, IEDs are acting as the server role when it receives the control commands from the controller replicas. Table 1 provides a summary of the actions taken by the controller replicas in response to receiving a report from a PMU and to sending of a command to an IED.

On collecting PMU data, the controller replicas engage in a Byzantine agreement for each input message as usual, as shown in Figure 2. However, the message delivery procedure must be modified. When a replica reaches a Byzantine agreement on the message, and it has delivered all previously ordered messages, it invokes the callback function provided by the controller to deliver this message. On the return of the up-call, no message is sent back to the PMU.

Upon issuing a control command, a controller replica does not directly send the command to the target IED. Instead, a round of Byzantine agreement on the command message is conducted, as shown in Figure 3. The procedure is very similar to that of PMU input message ordering, except that the pre-prepare message is triggered by the issuing of a control command rather than the receiving of a client's request, and the command is sent to the target IED when the Byzantine agreement is reached, instead of delivering a request. As mentioned in the previous Section, the target IED must not accept a control command immediately because the command might have been sent by a faulty controller replica. By waiting until it has received f + 1 identical command from different controller replicas, it can guarantee that at least one of them is sent by a non-faulty replica, because at most f replicas can be faulty according to our assumption.

Figure 2. Normal operation of the BFT algorithm for PMU report handling

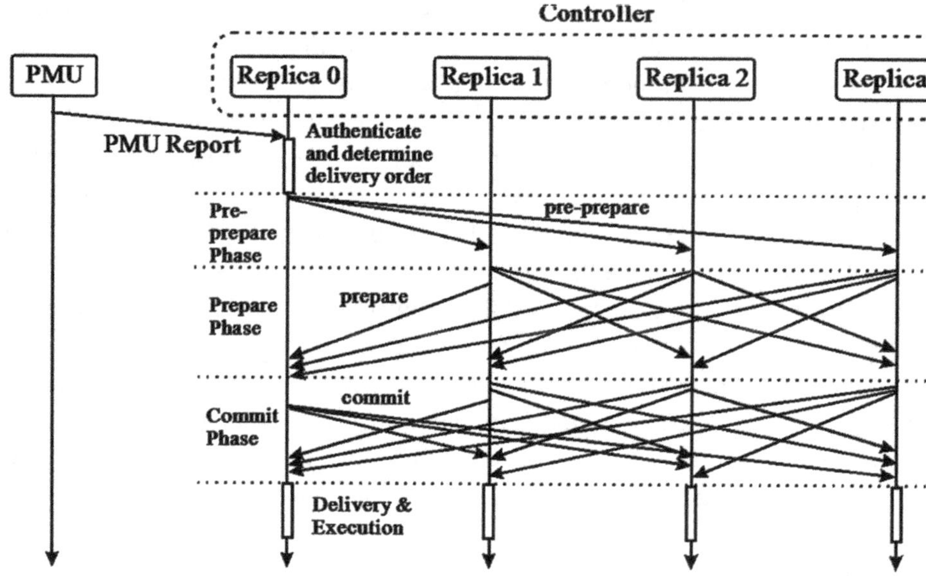

Figure 3. Byzantine agreement (normal operation) on the control command at the controller replicas and the acceptance of the command at an IED

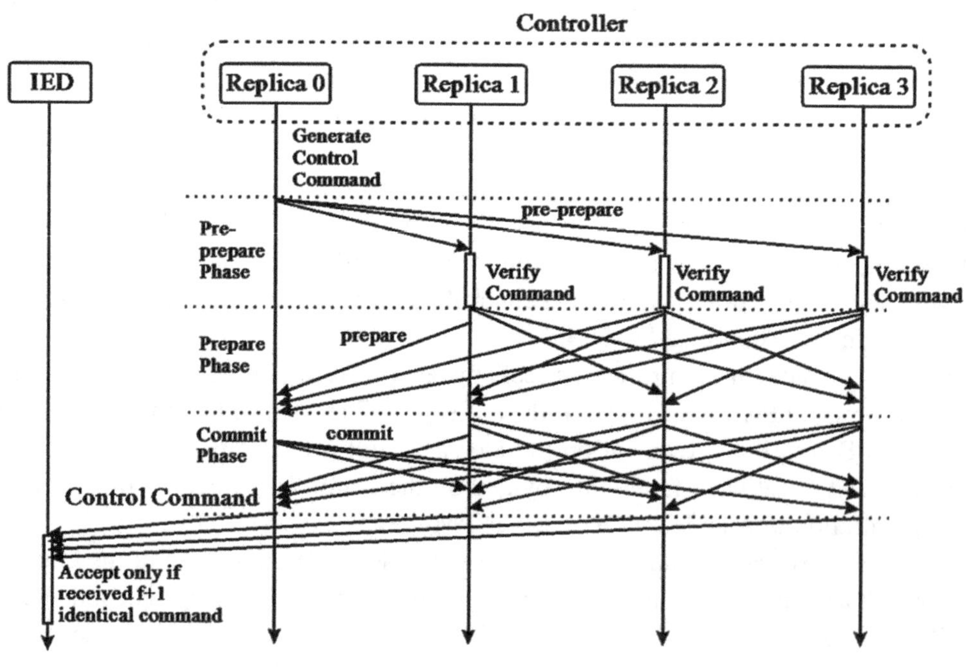

If the replicas operate completely deterministically and in lock-step, the round of Byzantine agreement for the commands to the IEDs is not necessary. However, it is virtually impossible to guarantee lock-stepped execution of the replicas across a network. If the control command contains information such as the time to execute the command, the commands issued by different replicas would contain different

Table 1. A summary of the actions taken by the controller replicas in response to receiving a report from a PMU and to sending of a command to an IED

Event	Actions Taken
On receiving a report from PMU	Byzantine agreement on the report No message is sent back to PMU
On sending of a command to IED	Byzantine agreement on the command prior to sending IED must collect f+1 consistent commands from different replicas

timestamps, which would make it impossible for the IEDs to authenticate and compare the commands for acceptance. Therefore, in general, it is necessary for the replicas to reach an agreement on the command to be issued to the IEDs. Here, we assume that a backup replica is able to verify if the command proposed by the primary is valid. A backup would initiate a view change if it deems that the command from the primary is invalid. If a backup replica cannot verify the command issued by the primary, more sophisticated mechanisms must be used, as reported in (Zhang et al., 2011).

EXPERIMENTAL ASSESSMENT

The implementation of our Byzantine fault tolerance framework is based on the PBFT library developed by Castro and Liskov (2002). We incorporated the changes necessary for electric power grid monitoring and control as mentioned in the previous Section.

The test-bed consists of 12 PCs in a local-area network (LAN) connected by a 100Mbps Ethernet. Four of the PCs in the LAN are equipped with Pentium IV 2.8GHz processors and 256MB memory, and the remaining PCs in the LAN each has a single Pentium III 1GHz processor. All PCs on the LAN run the Red Hat 8.0 Linux. The remote PC has one Pentium IV 3.2GHz processor running CentOS 4.5 Linux.

The main objective of the performance evaluation is to assess whether or not the Byzantine fault tolerance mechanisms are efficient enough to meet the real-time communication requirement for power grid health monitoring and control. Consequently, we characterize the response time and jitter of the Byzantine fault tolerant system.

The test application simulates the electric power grid health monitoring and control scenario as shown in Figure 1. The controller is replicated in the 4 Pentium IV PCs (one replica per PC) and the PMUs and IEDs are run on the remaining 8 Pentium III PCs (a pair of PMU and IED on each PC). During the experiments, up to 8 concurrent PMU-IED pairs are used.

A PMU (as the client) periodically reports its measured data to the replicated controller according to the IEEE 1344 standard. Upon each PMU message received, the controller replicas generate a command and send it to the corresponding IED (collocated on the same node as the reporting PMU). Note that this is done purely for the purpose of performance characterization and might not match the practical usage scenarios. The payload of each PMU report is 14 bytes long. The payload of each control command is set to 128 bytes long.

Furthermore, the PCs in our test-bed are not equipped with high-resolution GPS devices, preventing us from directly measuring one-way latencies for PMU reports and control command notifications. Instead, we measure the round-trip time from the sending of a PMU report to the receiving of a command in response to the report at a collocated IED.

To gain insight on the jitters of networking and Byzantine agreement processing delays, we measure the intervals between two consecutive sending of PMU reports at each PMU and the intervals of consecutive deliveries of the PMU reports at each controller replica, and compare the probability density functions (PDFs) of the sending intervals and the delivery intervals. The PDFs provide a much more detailed and accurate picture on the predictability of the arrival rate of the PMU reports at the controller replicas than using the mean values and standard deviations. For similar reasons, the PDF is used to capture and present the round-trip times. In each run, 10,000 samples are taken.

Figure 4 shows the experimental results under the normal operation condition. The number of concurrent PMU-IED pairs varies from 1 to 8. The PDFs for the sending intervals measured at the PMUs are shown in Figure 4(a). The interval between two consecutive sending is controlled by the nanosleep() API provided by Linux. Even though the target interval is 10 milliseconds, the actual intervals vary slightly (with peak value of about 11.6 milliseconds). If there is no jitter in networking and Byzantine agreement processing, the PDFs of the delivery intervals measured at the controller replicas should be identical to those of the sending intervals. The PDFs of the delivery intervals shown in Figure 4(b) indicate that there is noticeable jitter. However, the jitter is small enough to sustain a 60Hz PMU sampling rate for all scenarios tested (up to 8 concurrent PMU-IED pairs), which is often regarded as the most demanding SCADA requirement (Johnston, 2005). Furthermore, Figure 4(c) shows that the round-trip time is in the sub- millisecond range, again for all scenarios measured, which is more than adequate to ensure urgent sensing data delivered and control command acted upon.

When the primary controller replica is faulty, it may take asignificant amount of time (e.g., 2 seconds) for a view change to complete. During this period of time, the controller is basically out-of-service. To address this issue, the controller should periodically send contingency control commands to the IEDs. If an IED does not receive a control command in time, it should resort to the contingency command. We emphasize that this situation, even though not desirable, is far better than the IED executing a command sent by a malicious controller, which can lead to the destruction of critical components of the power grid.

Figure 4. The measured PDFs with various number of PMU-IED pairs. (a) The report-sending interval at the PMU. (b) The delivery interval at the controller replicas. (c) The round-trip time from sending a PMU report and the receiving of a control command

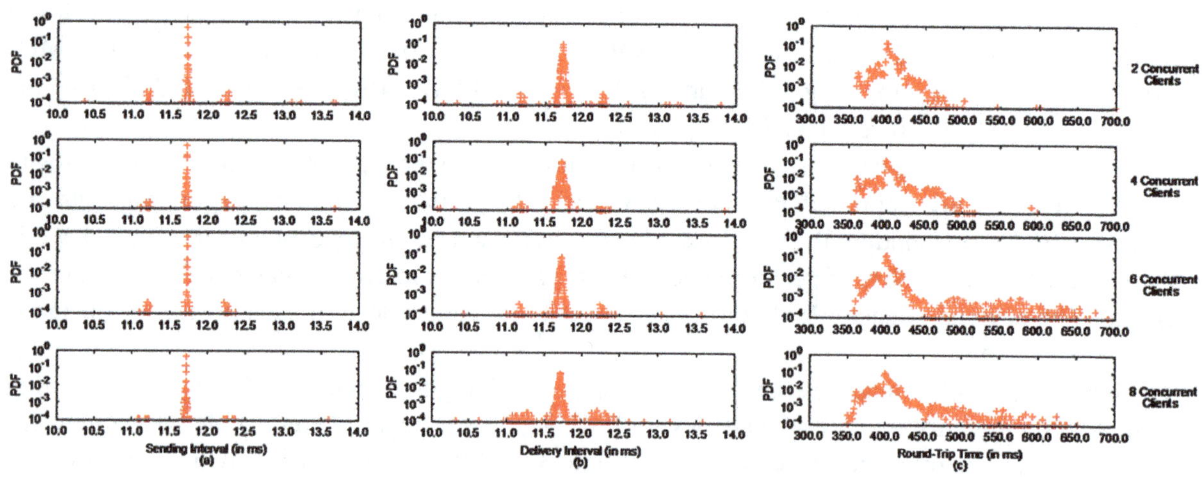

FUTURE RESEARCH DIRECTIONS

There is a large body of research work on how to restructure the current data communication infrastructure for electric power grid, such as (Zhang, Wang, & Xiang, 2015; Birman et al., 2005; Bose, 2005; Hossenlopp, 2007; Melliopoulos, 2007). The SCADA security issues have also attracted worldwide attention (http://sandia.gov/scada/). However, the work that targets both the security and reliability aspects of the infrastructure is rarely seen. Our work appears to be the first to assess if it is possible to apply the Byzantine fault tolerance technology for electric power grid health monitoring and control. We plan to carry out more in-depth investigations. In particular, the performance of the BFT technology in the wide-area network environment should be carefully evaluated. Furthermore, we also plan to explore context-ware and adaptive fault tolerance (Buys et al., 2011) to improve the performance and robustness of our framework. Another research direction is to improve the performance of the proposed framework by adapting our replication protocol specific for SACADA interactions. We call this line of research application-aware Byzantine fault tolerance (Zhao, 2014a; Zhao, 2014b).

The importance of stable sampling rate for networked sensing and control is addressed in (Liberatore, 2006). In (Liberatore, 2006), Liberatore proposed a playback-based method to increase the predictability of the sampling rate. This method seems to be a good candidate to be integrated with the proposed framework to address the potential jittering problem in the wide-area networks.

CONCLUSION

In this chapter, we presented the justification and a feasibility study of applying the Byzantine fault tolerance technology to electric power grid health monitoring. We proposed and implemented the BFT mechanisms needed to handle the PMU data reporting and control commands issuing to the IEDs. We carried out an empirical study to assess the feasibility of using the BFT technology for reliable and secure electric power grid health monitoring and control. We show that under the LAN environment, the overhead and jitter introduced by the BFT mechanisms are negligible, and consequently, Byzantine fault tolerance could readily be used to improve the security and reliability of electric power grid monitoring and control while meeting the stringent real-time communication requirement for SCADA operations.

While the brief out-of-service time (typically in 1-2 seconds) during a view change can be a concern for electric power grid health monitoring and control, additional mechanism, such as the playback scheme proposed in (Liberatore, 2006) could be used to alleviate the problem. In any case, the BFT sensing and control ensures that a PMU report from a compromised PMU cannot cause the state divergence of the correct controller replica, and a control command from a compromised control replica is never accepted by a correct actuating device such as an IED.

REFERENCES

Birman, K. P., Chen, J., Hopkinson, E. M., Thomas, R. J., Thorp, J. S., Renesse, R. V., & Vogels, W. (2005). Overcoming communications challenges in software for monitoring and controlling power systems. *Proceedings of the IEEE*, *93*(5), 1028–1041. doi:10.1109/JPROC.2005.846339

Bose, A. (2005). Improved communication/computation infrastructure for better monitoring and control. In *Proceedings of the IEEE Power Engineering Society General Meeting* (pp. 2715–2716). 10.1109/PES.2005.1489755

Buys, J., De Florio, V., & Blondia, C. (2011) Towards context-aware adaptive fault tolerance in SOA applications. In *Proceedings of the 5th ACM international conference on Distributed event-based system*, (pp. 63-74). 10.1145/2002259.2002271

Castro, M., & Liskov, B. (2002). Practical Byzantine fault tolerance and proactive recovery. *ACM Transactions on Computer Systems*, *20*(4), 398–461. doi:10.1145/571637.571640

CNN. (2007). *Staged cyber attack reveals vulnerability in power grid*. Retrieved from http://www.cnn.com/2007/US/09/26/power.at.risk/index.html

Hossenlopp, L. (2007). Engineering perspectives on iec 61850. *IEEE Power and Energy Magazine*, *5*(3), 45–50. doi:10.1109/MPAE.2007.365819

Johnston, R. (2005). *Obtaining high performance phasor measurements in a geographically distributed status dissemination network* (Master's thesis). Washington State University.

Lamport, L., Shostak, R., & Pease, M. (1982). The Byzantine gen- erals problem. *ACM Transactions on Programming Languages and Systems*, *4*(3), 382–401. doi:10.1145/357172.357176

Liberatore, V. (2006). Integrated playback, sensing and network control. *Proceedings of the 25th IEEE International Conference on Computer Communications*, 1–12.

Melliopoulos, A. P. S. (2007). Substation automation: Are we there yet? *IEEE Power and Energy Magazine*, *5*(3), 28–30. doi:10.1109/MPAE.2007.365815

Merideth, M., Iyengar, A., Mikalsen, T., Tai, S., Rouvellou, I., & Narasimhan, P. (2005). Thema: Byzantine-fault-tolerant middleware for web services applications. *Proceedings of the IEEE Symposium on Reliable Distributed Systems*, 131–142. 10.1109/RELDIS.2005.28

Rhea, S., Eaton, P., Geels, D., Weatherspoon, H., Zhao, B., & Kubiatowicz, J. (2003). Pond: the oceanstore prototype. *Proceedings of the 2nd USENIX Conference on File and Storage Technologies* (Vol. 3, pp. 1-14).

Wang, W., Xu, Y., & Khanna, M. (2011). A survey on the communication architectures in smart grid. *Computer Networks*, *55*(15), 3604–3629. doi:10.1016/j.comnet.2011.07.010

Zhang, H., Zhao, W., Moser, L. E., & Melliar-Smith, P. M. (2011). Design and implementation of a Byzantine fault tolerance framework for non-deterministic applications. *IET Software*, *5*(3), 342–356. doi:10.1049/iet-sen.2010.0013

Zhang, Y., Wang, L., & Xiang, Y. (2015). Power System Reliability Analysis With Intrusion Tolerance in SCADA Systems. *IEEE Transactions on Smart Grid.*

Zhao, W. (2009). Design and implementation of a Byzantine fault tolerance framework for Web services. *Journal of Systems and Software*, *82*(6), 1004–1015. doi:10.1016/j.jss.2008.12.037

ADDITIONAL READING

Aiyer, A. S., Alvisi, L., Clement, A., Dahlin, M., Martin, J. P., & Porth, C. (2005). BAR fault tolerance for cooperative services. *Operating Systems Review*, *39*(5), 45–58. doi:10.1145/1095809.1095816

Amir, Y., Coan, B., Kirsch, J., & Lane, J. (2008, June). Byzantine replication under attack. In Dependable Systems and Networks With FTCS and DCC, 2008. DSN 2008. IEEE International Conference on (pp. 197-206). IEEE. 10.1109/DSN.2008.4630088

Amir, Y., Danilov, C., Dolev, D., Kirsch, J., Lane, J., Nita-Rotaru, C., ... Zage, D. (2008). Steward: Scaling Byzantine Fault-Tolerant Replication to Wide Area Networks. *IEEE Transactions on Dependable and Secure Computing*, 80–93.

Bessani, A. N., Alchieri, E. P., Correia, M., & Fraga, J. S. (2008, April). DepSpace: A Byzantine fault-tolerant coordination service. *Operating Systems Review*, *42*(4), 163–176. doi:10.1145/1357010.1352610

Casey, P. R., Jaber, N., & Tepe, K. E. (2011, October). Design and implementation of a cross-platform sensor network for smart grid transmission line monitoring. In *Proceedings* of *the IEEE International Conference on Smart Grid Communications* (pp. 285-290). IEEE. 10.1109/SmartGridComm.2011.6102334

Castro, M., & Liskov, B. (2000, October). Proactive recovery in a Byzantine-fault-tolerant system. In Proceedings of the 4th conference on Symposium on Operating System Design & Implementation-Volume 4 (pp. 19-19). USENIX Association.

Castro, M., & Liskov, B. (2001, July). Byzantine fault tolerance can be fast. In Dependable Systems and Networks, 2001. DSN 2001. International Conference on (pp. 513-518). IEEE. 10.1109/DSN.2001.941437

Chai, H., Melliar-Smith, P. M., Moser, L. E., Zhang, H., & Zhao, W. (2013). Toward Trustworthy Coordination of Web Services Business Activities. *IEEE Transactions on Services Computing*, *6*(2), 276–288. doi:10.1109/TSC.2011.57

Chun, B. G., Maniatis, P., & Shenker, S. (2008, June). Diverse Replication for Single-Machine Byzantine-Fault Tolerance. In *USENIX Annual Technical Conference* (pp. 287-292).

Clement, A., Wong, E. L., Alvisi, L., Dahlin, M., & Marchetti, M. (2009, April). Vol. 9, pp. 153–168). Making Byzantine Fault Tolerant Systems Tolerate Byzantine Faults. In NSDI.

Cowling, J., Myers, D., Liskov, B., Rodrigues, R., & Shrira, L. (2006, November). HQ replication: A hybrid quorum protocol for Byzantine fault tolerance. In *Proceedings of the 7th symposium on Operating systems design and implementation* (pp. 177-190). USENIX Association.

Distler, T., & Kapitza, R. (2011, April). Increasing performance in byzantine fault-tolerant systems with on-demand replica consistency. In *Proceedings of the sixth conference on Computer systems* (pp. 91-106). ACM. 10.1145/1966445.1966455

Driscoll, K., Hall, B., Sivencrona, H., & Zumsteg, P. (2003). Byzantine fault tolerance, from theory to reality. In Computer Safety, Reliability, and Security (pp. 235-248). Springer Berlin Heidelberg. doi:10.1007/978-3-540-39878-3_19

Garcia, R., Rodrigues, R., & Preguiça, N. (2011, April). Efficient middleware for byzantine fault tolerant database replication. In *Proceedings of the sixth conference on Computer systems* (pp. 107-122). ACM. 10.1145/1966445.1966456

Hendricks, J., Ganger, G. R., & Reiter, M. K. (2007). Low-overhead byzantine fault-tolerant storage. *Operating Systems Review*, *41*(6), 73–86. doi:10.1145/1323293.1294269

Kapitza, R., Behl, J., Cachin, C., Distler, T., Kuhnle, S., Mohammadi, S. V., ... Stengel, K. (2012, April). CheapBFT: resource-efficient byzantine fault tolerance. In *Proceedings of the 7th ACM European conference on Computer Systems* (pp. 295-308). ACM.

Kotla, R., & Dahlin, M. (2004, June). High throughput Byzantine fault tolerance. In *Proceedings of the International Conference on Dependable Systems and Networks* (pp. 575-584). IEEE.

Veronese, G. S., Correia, M., Bessani, A. N., & Lung, L. C. (2009, September). Spin One's Wheels? Byzantine Fault Tolerance with a Spinning Primary. In *Proceedings of the 28th IEEE International Symposium on Reliable Distributed Systems* (pp. 135-144). IEEE. 10.1109/SRDS.2009.36

Yin, J., Martin, J. P., Venkataramani, A., Alvisi, L., & Dahlin, M. (2002, June). Byzantine fault-tolerant confidentiality. In *Proceedings of the International Workshop on Future Directions in Distributed Computing* (pp. 12-15).

Zhang, H., Chai, H., Zhao, W., Melliar-Smith, P. M., & Moser, L. E. (2012). Trustworthy coordination of Web services atomic transactions. *IEEE Transactions on Parallel and Distributed Systems*, *23*(8), 1551–1565. doi:10.1109/TPDS.2011.292

Zhang, Y., Zheng, Z., & Lyu, M. R. (2011, July). BFTCloud: A byzantine fault tolerance framework for voluntary-resource cloud computing. In *Proceedings of the IEEE International Conference on Cloud Computing*, (pp. 444-451). IEEE. 10.1109/CLOUD.2011.16

Zhao, W. (2009). Design and implementation of a Byzantine fault tolerance framework for Web services. *Journal of Systems and Software*, *82*(6), 1004–1015. doi:10.1016/j.jss.2008.12.037

Zhao, W. (2014a). *Building dependable distributed systems*. John Wiley & Sons. doi:10.1002/9781118912744

Zhao, W. (2014b). Application-Aware Byzantine Fault Tolerance. In *Proceedings of the 12th IEEE International Conference on Dependable, Autonomic and Secure Computing* (pp. 45-50). IEEE. 10.1109/DASC.2014.17

Zhao, W. (2015). Optimistic Byzantine fault tolerance. *International Journal of Parallel, Emergent and Distributed Systems*, 1-14 (preprint).

Zhao, W., & Zhang, H. (2009). Proactive service migration for long-running Byzantine fault-tolerant systems. IET software, 3(2), 154-164.

KEY TERMS AND DEFINITIONS

Byzantine Fault Tolerance: It refers to the capability of a system to tolerate Byzantine faults.

Intelligent Electronic Device (IED): It is an actuating device that is capable of receiving commands from a controller. Example IEDs include protective relaying devices, and voltage regulators.

Jitter: It refers to the deviation from the periodicity of a sequence of events or signals.

Normal Operation: It refers to the operation of an algorithm during a period that either there is no fault, or the faults do not disrupt its operation. For example, when a backup replica crashes, the PBFT algorithm would still operate as normal.

Phasor Measurement Unit (PMU): It is a device that measures the electrical waves in an electric power grid. The measurements must be synchronized with a global clock, such as a GPS.

Sampling Rate: It is defined as the number of samples taken per unit of time.

SCADA: It is short for Supervisory Control and Data Acquisition. It is a type of industrial control system that monitor and control industrial processes that exist physically.

View Change: It refers to the configuration change of the group of replicas that engage in Byzantine fault tolerance. When the primary is suspected of being faulty, a new view is initiated so that a different replica is elected as the primary for the new view.

This research was previously published in the Encyclopedia of Information Science and Technology, Fourth Edition edited by Mehdi Khosrow-Pour, pages 3056-3065, copyright year 2018 by Information Science Reference (an imprint of IGI Global).

Chapter 11
E-Waste, Chemical Toxicity, and Legislation in India

Prashant Mehta
National Law University Jodhpur, India

ABSTRACT

In present digital age, we constantly upgrade or replace our numerous electronic devices due to continuous technological advances and short product life cycles. With increasing "market penetration" in developing countries, "replacement market" in developed countries, and "high obsolescence rate," a large pile of e-waste is generated either internally or it is generated in developed countries and often ends up for recycling in developing countries. The current practices of e-waste management and poor awareness in India is posing a huge challenge to the environment regulators, governments, and policy makers as much work needs to be done at ground level to achieve sustainable results This chapter provides a comprehensive overview of India's current e-waste scenario, analyzes hazardous metals and considers environmental and health risks posed by them, understands existing legal framework and strategic interventions, and explores immediate technical solutions to manage and minimize its impact on all.

INTRODUCTION

The industrial revolution was a period of dynamic change and dramatic innovation in the history of human society (Ayers, 1999). Across the world, societies are constantly reinventing to manage revolutionary changes that have radically transformed the lifestyle of people. Some of these changes are subtle and barely noticeable, while other changes are blatant and abrupt, like advances in Information and Communication Technology (ICT) and widespread use of Electrical and Electronics Equipment (EEE), which has made human civilization to grow in a more efficient manner.

Following economic liberalization in 1991, the Indian ICT industry has been one of the major drivers of economic progress both in terms of volume and applications. It has assumed the role of providing a forceful leverage to the socio, economic, and technological growth of a developing society (Joseph, 2007). However consumption and production processes of these complex electronic devices are unsustainable, pose a serious challenge to environment and human health, making e-waste one of the largest

DOI: 10.4018/978-1-5225-7359-3.ch011

growing waste streams (Lundgren, 2012a). With waste market getting increasingly global, such waste is illegally exported to crude e-waste recycling hotspots in Asian countries, such as China, India, and Pakistan, and in some African countries, like Ghana and Nigeria (Castillo, 2011). Such illegal trade in e-waste is camouflaged and conducted under the pretext of obtaining 'reusable' equipment or 'donations' from developed nations.

E-waste comprises of ICT and EEE products that are not fit for their originally intended use. It includes computers, its accessories (monitors, printers, keyboards, and central processing units), typewriters, mobile phones and chargers, remotes, compact discs, headphones, batteries, LCD/Plasma TVs, air conditioners, refrigerators, and other household appliances (Lalchandani, 2010). The increasing 'market penetration' in the developing countries, 'replacement market' in the developed countries (Borthakur & Sinha, 2013), coupled with rapid developments, innovation, miniaturization, and replacement resulted into higher rate of obsolescence of electronics products. It is estimated that in 2014 world-wide 41.8 million metric tonnes (Mt) e-waste was generated and most of it was not collected and treated in environmentally sound manner (Baldé, Wang, Kuehr, & Huisman, 2015). Most of this either end up with municipal waste in landfills or unauthorized recycling yard (Greenpeace Press Report, 2008).

As noted by UNEP in 2005, "Every year, 20 to 50 million tonnes of e-waste is generated world-wide, which could bring serious risks to human health and the environment" (Schwarzer, Giuliani, Kluser, & Peduzzi, 2005). Even though there is no clear data on the quantity of e-waste generated and disposed of each year in India, it is estimated 70 percent of e-waste handled in India is imported. It also estimates that between the years of 2007-2020, domestic television e-waste will double, computer e-waste is expected to increase five-fold, while cell phones will increase eighteen times (Disabled World, 2015). Thus knowledge society of 21st century is creating its own toxic footprint which is most debated issue amongst the environmentalists, environment regulators, world-wide environment forums, governmental, and non-governmental agencies, and policy makers.

BACKGROUND OF STUDY

Solid waste management, which is already a mammoth task in India, is becoming more complicated by the invasion of e-waste, which has complex characteristics as it differs chemically and physically from urban or industrial waste. Each wave of technology creates a set of waste previously unknown by humans (Sikdar & Vaniya, 2014) making e-waste management a big issue in both developed and developing countries.

The current practices of e-waste management in developing countries suffer from a many drawbacks such as informal recycling, inadequate legislation, low public awareness of the hazardous nature of e-waste, use of obsolete methods, and inadequate emphasis on the employee's protection (Cobbing, 2008), is jeopardizing people's health and environment (Smith, Sonnenfeld, & Naguib Pellow, 2006a).

Having reviewed literature from various other studies conducted in India and abroad, and understanding the magnitude of this problem, it is time for India to critically review its management of e-waste, to work towards a strategy to create the necessary infrastructure, and mechanisms to support sustainable and environmentally friendly e-waste management besides sensitizing consumers, waste recyclers, and future decision makers on issues like e-waste characteristics, its trans-boundary movement recycling technology, social, and environmental considerations, and toxic effect on health.

DEFINITION OF E-WASTE

Even though there is no standard definition for e-waste, some of the reported definitions of e-waste in literature are mentioned below:

According to the Basel Convention, "Wastes are substances or objects, which are disposed of or are intended to be disposed of, or are required to be disposed of by the provisions of national laws" (Text of Basel Convention, 2014).

According to Basel Action Network (BAN), "E-waste includes a wide and developing range of electronic appliances ranging from large household appliances, such as refrigerators, air-conditioners, cell phones, stereo systems, and consumable electronic items to computers discarded by their users" (Puckett, Byster, Westervelt, Gutierrez, Davis, Hussain, & Dutta, 2002), (Gaidajis, Angelakoglou, & Aktsoglou, 2010).

As per European Directive 2002/96/EC, "Waste electrical and electronic equipment (WEEE), including all components, subassemblies, and consumables, which are part of the product at the time of discarding" (European Parliament, 2003, January 27), (Jain, 2008), (European Parliament, 2012, July 4).

As per European Directive 75/442/EEC, Article I(a), "Any substance or object which the holder discards or is required to discard in compliance with the national legislative provisions". Further it includes all components, subassemblies, and consumables which are part of the product at the time of discarding (Borthakur & Singh, 2012).

According to Organisation for Economic Co-operation and Development (OECD), "Any household appliance consuming electricity and reaching its life cycle end", also referred as composite waste (OECD, 2007).

These differences in definitions, of what constitutes e-waste, have the potential to create disparities in both the quantification of e-waste generation and the identification of e-waste flows across nations. The lack of a precise definition of e-waste is one of the key issues that need to be addressed on an international level (Lundgren, 2012b).

COMPONENTS OF E-WASTE

E-waste is classified as hazardous waste (Tsydenova & Bengtsson, 2011a), and it imposes many challenges on the recycling industry (Smith, Sonnenfeld, & Naguib Pellow, 2006b). Modern electronics can contain up to 60 different elements; many are valuable, some are hazardous (Third World Network, 1991) and some are both. Several rare elements are also used (Frazzoli, Orisakwe, Dragone, & Mantovani, 2010). The types and amounts of metals used in electronics products vary with evolution of technology. The most complex mix of hazardous substances is usually present in the printed wiring boards (PWBs) that contain valuable metals like copper, silver, gold, palladium, and platinum, brominated flame retardants used in connectors, cathode ray tubes and LCD contain heavy metals like lead and barium, switches and flat screens contain mercury, older capacitors and transformers contain poly chlorinated biphenyl's (PCB's), poly vinyl chloride (PVC) coated copper cables and casing, plastics from computer hardware that release highly toxic dioxins and furans (Sum, 1991). The fraction including iron, copper, aluminum, gold, and other metals in e-waste is over 60 percent, plastics account for about 30 percent, and the hazardous pollutants comprise only about 2.70 percent of waste (Widmer, Oswald-Krapf, Sinha-Khetriwal, Schnellmann, & Bo¨ni, 2005) besides alloys that mostly decreases metal's recyclability.

TOXIC ELEMENTS IN E-WASTE

E-waste contains thousands of components made of deadly chemicals, heavy metals, flame retardants, and potentially hazardous substances whose main routes of human exposure are through inhalation, dust ingestion, dermal exposure, and oral intake. Metal toxicity causes breathing difficulties, respiratory irritation, coughing, choking, pneumonitis, tremors, neuropsychiatric problems, convulsions, coma and even death (Yu, Welford, & Hills, 2006 a). Some toxic chemicals found in e-waste are analyzed below.

1. Beryllium (Be) is used as copper-beryllium alloys in computer motherboards, relays, and connectors (Taylor, Ding, Ehler, Foreman, Kaszuba, & Sauer, 2003). Beryllium refining produces fumes, dust, and oxides which are both acutely and chronically toxic to humans. If inhaled in large concentration, it causes acute lung disease, breathing discomfort, coughing, rapid heartbeat, and death in extreme cases. Its compounds are carcinogenic in nature (IARC, 1993b) and studies have shown that people can still develop beryllium disease (beryllicosis) after many years of last exposure.
2. Cadmium (Cd) is a toxic heavy metal found naturally in very low concentration (Salomons & forstner, 1984). It is used in switches, rechargeable (Ni-Cd) batteries, stabilizers, resistors, and corrosion-resistant alloys. It is released into the environment as powdered dust during crushing and milling of plastics, CRTs, and circuit boards. It is potentially a long-term cumulative poison associated with deficits in cognition, learning, behavior, and neuromotor skills in children, causes severe pain in the joints and spine (Itai-itai disease), affects kidneys and softens bones (osteomalacia and osteoporosis) in humans. There is evidence of the role of cadmium and beryllium in carcinogenicity (Strickland, & Kensler, 1995), (Pruss-Ustun & Corvalan, 2006).
3. Hexavalent Chromium (Cr-VI) is used to protect metal housings in a computer from corrosion. It is very reactive and soluble in water, making it more mobile in environment (Mukherjee, 1998). Its corrosive nature cause skin allergies (dermatitis), damage DNA, liver, kidneys, pulmonary congestion, edema, bronchial maladies including asthmatic bronchitis, and lung cancer (IARC, 1990a).
4. Lead (Pb) as lead oxide comes from breaking of CRT, lead powder is released while removing solder form microchips, and lead fumes comes from high temperature smelting processes exposing the workers (Schutz, Olsson, Jensen, Gerhardsson, Borjesson, Mattsson, & Skerfving, 2005). In unlined landfills, lead would dissolve in leachate or mix with ground water leading to contamination. It is neurotoxin that exerts toxic effects on the central nervous system (organic affective syndrome), peripheral nervous systems (motor neuropathy), the hemopoietic system (anemia), the genitourinary system (capable of causing damage to all parts of nephron), and the male and female reproductive systems (Harrington, Aw, & Baker, 2003). It affects mental development in children with impaired cognitive function, behavioral disturbances, attention deficits, hyperactivity, conduct problems, and lower IQ.
5. Mercury (Hg) is used as lighting device which illuminates most flat panel monitors. Workers can inhale of mercury vapour or dust which released while breaking and burning of circuit boards and switches. It affects kidneys, immune system, damage to the genitourinary system (tubular dysfunction), central and peripheral nervous systems, reduced fertility, and impairs fetus growth. When inorganic mercury spreads out in the water, it is transformed into toxic methylated mercury by microbial activity, which bio-accumulates, biomagnifies in living organisms, and concentrates through the food chain, particularly by fish (Hu, & Speizer, 2001), (WHO, 1989).

6. Brominated Flame Retardants (BFRs) like Polybrominated biphenyl (PBB), Polybrominated diphenyl ether (PBDE), and Tetrabromobisphenol-A (TBBPA) are chemically persistent organic pollutants (POP) along with toxic antimony trioxide which is used as flame retardants in electronic devices. They release carcinogenic brominated dioxins and furans as gases during fire. PBDE used in transformers and capacitors is bioaccumulative, impair brain function, and can cause liver and malfunctioning of endocrine system (thyroid damage). TBBPA used in printed circuit boards contains bromine that can leach into landfills. Dust on computer cabinets contains BFRs.
7. Polyvinyl Chloride (PVC) is found in circuit boards, cabinets, and insulation on cables. It is hazardous because contains upto 56 percent chlorine which are precursors to polychlorinated di-benzo-p-dioxins and di-benzo-furans (classified as POP under Stockholm Convention) during incineration along with large quantities of hydrogen chloride gas, which when inhaled may leads to cancer, respiratory problems, affect reproductive, and immune system.
8. Polycyclic Aromatic Hydrocarbons (PAH) is generated from e-waste recycling activities and have potential impacts on soil, vegetation, and human health include breathing difficulties, respiratory irritation, coughing, choking, pneumonitis, tremors, neuropsychiatric problems, convulsions, coma and even death (Yu, Welford, & Hills, 2006 b). Epidemiological studies in the past on occupational exposure to PAH, provides sufficient evidence of the role of PAH in the induction of skin and lung cancers (Stewart & Kliehues, 2003).
9. Cobalt (Co) is extensively used in integrated circuits, semi-conductors, magnetic recording media, thin metallic films, and rechargeable batteries. It is mainly absorbed from the pulmonary and the gastrointestinal tracts and cause allergic dermatitis, rhinitis, vomiting, thyroid damage, and impaired vision. Cobalt dust may cause an asthma-like disease with symptoms ranging from cough, shortness of breath, and dyspnea to decreased pulmonary function, nodular fibrosis, permanent disability, and death.

In addition to the hazardous materials e-waste also contain a large number of valuable precious metals like Gold (Au), Silver (Ag), Platinum (Pt), and Palladium (Pd) in concentrations 40 to 50 times richer than there naturally occurring deposits. These precious metals are extracted using hydrometallurgical processes, where valuable metals first leached into acid or alkali solutions and then they are concentrated by using various methods like precipitation, cementation, solvent extraction etc.

E- WASTE DISPOSAL AND RECYCLING PRACTICES IN INDIA

E-wastes disposal is a big problem faced by many countries including India. It is estimated that, by 2020, India could see nearly 500 percent rise in the number of old computers being dumped (Schluepa, et.al., 2009). E-waste disposal process work in two ways:

1. By removing the hazardous items;
2. By separating recyclable materials.

The three main groups of substances that may be released during recycling and material recovery, and may pose significant human and environmental health risks are:

1. Original constituents of EEE like lead and mercury;
2. Added substance during recovery processes like cyanide; and
3. Hazardous byproducts formed by incineration of e-waste like PAH.

In India, e-waste collection, transportation, segregation, dismantling, recycling, and disposal is done by unorganized small enterprise that is difficult to regulate. They employ untrained labours (more so children and women) who work in poorly-ventilated or enclosed areas without appropriate equipment and technical expertise. The process includes manual disassembly, melting, acid extraction of metals from complex mixtures, and extruding plastics.

Further incineration of printed circuit boards for desoldering and removal of chips exposes workers to fumes of metals, particularly those in solder (often lead and tin), and other hazardous substances that can be potentially released (Tsydenova, & Bengtsson, 2011b) which pollutes the surrounding air. Inhalation and dust ingestion impose a range of potential occupational hazards including silicosis (Lepawsky, & McNabb, 2010). It exposes them to dangerous slow-poisoning chemicals on a regular basis impacting their health. Computer wastes that are land-filled produces contaminated leachate which eventually pollutes the groundwater, whereas acids and sludge obtained from melting computer chips, if disposed on the ground causes acidification of soil and irreversible damage to environment.

LEGAL FRAMEWORK IN INDIA

The environmentally sound management of used EEE imports, recycling, and their disposal has become the subject of serious discussion, debate, and significant challenge among the government, organizations, environmental groups, and component manufacturers in India. The department related Parliamentary Standing Committee on Science & Technology, under Ministry of Environment & Forests (MoEF) which is also responsible for environmental legislation, in its 192nd report on the 'Functioning of the Central Pollution Control Board (CPCB), which plays important role in drafting guidelines, has concluded that e-waste is going to be a big problem in the future due to modern life style, increase in the living standards of people, and augmentation of economic growth.

The rules and regulations for waste control in India are primarily listed under the aegis of Environmental Protection Act 1986. Despite a wide range of environmental legislation in India there are no specific laws or guidelines for electronic or computer waste (Devi, Shobha, & Kamble 2004). Electronic waste is included under List-A and List-B of Schedule-3 of the Hazardous Wastes (Management & Handling Rules), 1989, as amended in 2000 and 2003. In 2007, separate guidelines on e-waste management were implemented, but they were voluntary and had limited impact. These guidelines include details such as e-waste composition; recycle, re-use, and recovery potential of items of economic value, identification of possible hazardous contents in e-waste, treatment, disposal options, and the environmentally sound e-waste treatment technologies (Rajya Sabha Unstarred Question No. 1887, 2009). However these rules primarily dealt with industrial waste and lack elements to deal with complex nature of e-waste.

Subsequently in 2008, these rules were amended to include toxic content and made registration mandatory for recyclers. The provision of environmental protection is delegated among various states in India. Following Supreme Court directions (Writ Petition (Civil) No. 657, 1995), the states have notified a set of hazardous waste laws and built a number of hazardous waste disposal facilities in the last ten years. However, the Controller and Auditor General (CAG) report found that over 75 percent of state bodies were not implementing these laws (Writ Petition (Civil) No. 10, 1995), giving rise to sloppy enforcement of e-waste related legislation.

India is also a signatory to the Basal Convention (under UNEP) on the control of Transboundary Movement of hazardous wastes and their disposal but officially opposes enforcement of BAN Amendment (Basel Action Network, 2011). The regulations banning the importation of hazardous waste for disposal are weak and an imported consignment of electronic scrap still comes into the country, as they are not properly classified as plastic scrap or mixed waste. In 2014, India generated 17 lakh tonnes of e-waste increasing growing at rate of 4-5 percent per year. With such exponential growth, the Indian government finally woke up and responded by framing the E-waste (Management and Handling Rules), 2016 with the aim to make re-cycling of e-wastes environmentally friendly. To begin with, the rules put India along with a select club of nations like the United States and many European nations to have legislation to regulate and manage e-waste. These rules recognized the producer's liability for recycling and reducing e-waste in the country. It will apply to every producer, consumer, and bulk consumer involved in manufacture, sale, purchase, and processing of electronic equipment or components. It also brought disposal of CFL lights under its preview.

While the rules seem impressive on paper, environmentalist argues that there is total oversight of the ground situation. These rules ignore the unorganized sectors where 90 percent of the e-waste is generated. Also there is lack of a safe e-waste recycling infrastructure in the organized sector where only a fraction of the e-waste (estimated 10 percent) finds its way to recyclers due to absence of an efficient take back scheme for consumers. The ministry is giving the producers of EEE a breathing period of one year to set up their collection centers (Kumar & Shah, 2014) and develop technical guidelines for the environmentally sound management of e-wastes. Thus computer, mobile handset, and consumer goods manufacturers, will be required to come up with e-waste collection centers or introduce 'take back' systems. Therefore over reliance on the capacities of the informal sector poses severe risks to the environment and human health.

STRATEGIC INTERVENTION

A smart e-waste management system have to access current e-waste situation, recognize that e-wastes is complex mixture of hazardous substances, reduce the generation of e-waste through smart manufacturing and maintenance, reuse till functioning of EEE, and finally recycle those components that cannot be repaired or refurbished. Recycling and reuse of materials are the next level of potential options to reduce e-waste (Ramachandra & Saira, 2004). Based on current situational analysis in India, following strategic intervention is proposed:

1. Extended Producers Responsibility be introduced that involves collection and disposal of e-waste in environmentally sound manner or else stringent financial penalties must be imposed by state governments.
2. Update the legal and institutional framework for e-waste management including effective enforcement of laws, regulations, and standards. Introduce strict liability clause in proposed rules for effectiveness.

3. Defining responsibilities of prime stake holders at the level of government, supply chain, consumers, and develop a comprehensive policy that address all issues ranging from production, trade to final disposal, including technology transfers for the recycling of electronic waste.
4. Raise public awareness, advocate for e-waste management across all stakeholders through public-private-partnerships linkages.
5. Introduce a concept of 'e-waste exchange' as an independent market instrument offering services for sale and purchase of e-waste.
6. Tighten import norms and custom procedure at port of entry. Maintain statistical records of imported of EEE goods for further analysis including their final disposal.
7. Create a facilitative environment for investment in e-waste handling and disposal infrastructure by creating modern e-waste recycling facility and provide tax incentives to make it more effective.
8. Introduce advance recycling fees. Also setup and operationalise an e-waste fund to benefit those working within this industry.

FUTURE RESEARCH DIRECTIONS

Developing national approach to handle e-waste, strengthening regulatory environment, designing new methods to increase waste collection, integrated modelling concepts to build waste recycling capacities, and building awareness among people can be carried out on larger scale across India. Also study of impact on toxic heavy metals individually as well as their cumulative effect on ecosystem needs to be further explored. Further quantities of waste material that moves between countries, waste flows within a country and between countries and hazardous substance emissions associated with manual recycling process, social, and its health impact on children and women can also be further investigated. A comparative study of various EEE product categories can made.

CONCLUSION

From the discussion above, it is aptly clear that India faces an enormous task of handling and disposing growing piles of e-waste and its impact on human health (more so of women and children) and natural environment has increased manifold. Therefore policy level interventions should include strong e-waste regulation, tight control on import and export of e-wastes, and facilitation in development of recycling infrastructure. Lack of strict enforcement of legislation is also worrisome situation.

It requires building of public awareness, establishment of institutional infrastructures (including e-waste collection, transportation, treatment, storage, recovery, and disposal) at national and regional levels. Furthermore product end-of-life management should be made a priority during design of new electronic products using green materials, innovation in product technology, life cycle analysis, public outreach, and so on. Beyond conservation of raw materials and energy, there are additional environmental benefits of recycling, such as reduced land disturbance, water use, air emissions, and waste generation which in turn improves efficiency and environment. In conclusion it is time for us to look deep and ahead today as tomorrow it will be too late to act.

REFERENCES

Ayers, S. (1999). *The cultural impact of computer technology*. Yale-New Haven Teachers Institute. Retrieved February 29, 2016 from http://www.yale.edu/ynhti/curriculum/units/1999/7/99.07.07.x.html

Baldé, C. P., Wang, F., Kuehr, R., & Huisman, J. (2015). The global e-waste monitor - 2014. United Nations University, IAS-SCYCLE.

Basel Action Network (BAN). (2011). *Toxic trade news: "Cochin Port a safe conduit for imported e-waste". "Most aspects of e-waste not regulated in U.S., Va."; "Research identifies U.S electronic wastes as likely source of toxic jewellery imports from China"; "178 countries agree to allow the ban on exports of toxic wastes to developing countries to become law"*. Retrieved March 1, 2016 from http://www.ban.org/

Borthakur, A., & Singh, P. (2012). Electronic waste in India: Problem and Policies. *International Journal of Environment Science*, *3*, 354–362.

Borthakur, A., & Sinha, K. (2013). Generation of electronic waste in India: Current scenario, dilemmas and stakeholders. *African Journal of Environmental Science and Technology*, *7*(9), 899–910. doi:10.5897/AJEST2013.1505

Castillo, M. (2011, January 14). *Electronic Waste: Where does it go and what happens to it?* Retrieved March 1, 2016 from http://techland.time.com/2011/01/14/electronic-waste-where-does-it-go-and-what-happens-to-it/

Cobbing, M. (2008). *Toxic Tech: Not in our backyard. Uncovering the hidden flows of e-waste*. Report from Greenpeace International, Amsterdam. Retrieved March 1, 2016 from http://www.greenpeace.org/raw/content/belgium/fr/press/reports/toxic-tech.pdf

Devi, B. S., Shobha, S. V., & Kamble, R. K. (2004). E-Waste: The Hidden harm of Technological Revolution. *Journal IAEM*, *31*, 196–205.

Disabled World. (2015). *Electronic waste: Medical & health issues*. Retrieved March 3, 2016 from http://www.disabled-world.com/health/ewaste.php

Disposal of e-waste. (2009). Rajya Sabha Unstarred Question No. 1887.

European Parliament. (2003, January 27). Directive 2002/96/EC of the European Parliament and of the Council on waste electrical and electronic equipment (WEEE). *Official Journal of the European Union, L, 37*, 24–38.

European Parliament. (2012, July 4). Directive 2012/19/EU of the European Parliament and of the Council on waste electrical and electronic equipment (WEEE). *Official Journal of the European Union, L, 197*, 38–71.

Frazzoli, C., Orisakwe, O. E., Dragone, R., & Mantovani, A. (2010). Diagnostic health risk assessment of electronic waste on the general population in developing countries scenarios. *Environmental Impact Assessment Review*, *30*(6), 388–399. doi:10.1016/j.eiar.2009.12.004

Gaidajis, G., Angelakoglou, K., & Aktsoglou, D. (2010). E-waste Environmental Problem and Current Management. *Journal of Engineering Science and Technology, Review*, *3*(1), 193–199.

Greenpeace. (2008). *Take Back Blues, An assessment of e-waste takes back in India.* Retrieved February 29, 2016 from www.greenpeace.org/india/press/reports/recycling-ofelectronic-waste

Guidance Manual for the Implementation of the Organisation for Economic Co-operation and Development (OECD). (2007). *Recommendation C(2004)100 on Environmentally Sound Management (ESM) of Waste, Section 5.2, 12.* Retrieved March 2, 2016 from http://www.oecd.org/env/waste/39559085.pdf

Harrington, J. M., Aw, T. C., & Baker, E. L. (2003). Occupational and environmental health and safety. In A. W. David, M. C. Timothy, D. F. John, & J. B. Edward (Eds.), *Oxford Textbook of Medicine* (4th ed.; Vol. 1, pp. 956–960). New York: Oxford University Press.

Hu, H., & Speizer, F. E. (2001). Specific environmental and occupational hazards. In Harrison's Principles of Internal Medicine (15th ed.; vol. 2, pp. 2591-2592). McGraw-Hill Inc.

International Agency for Research on Cancer (IARC). (1990). Chromium. IACR Monograph, 49, 677.

International Agency for Research on Cancer (IARC). (1993). Beryllium and Beryllium Compounds. International Agency for Research on Cancer (IARC) Monograph, 58, 444.

Jain, A. (Ed.). (2008). *E-waste: Implications, regulations, and management in India and current global best practices*. New Delhi: Teri Press.

Joseph, K. (2007). Electronic waste management in India - Issues and Strategies. *Proceedings of the Eleventh International Waste Management and Landfill Symposium.*

Kumar, R., & Shah, J. D. (2014). Current Status of Recycling Waste Printed Circuit Boards in India. *Journal of Environmental Protection*, 9-16.

Lalchandani, N. (2010, April 24). E-scare. *The Times of India*.

Lepawsky, J., & McNabb, C. (2010). Mapping international flows of electronic waste. *Canadian Geographer*, *54*(2), 177–195. doi:10.1111/j.1541-0064.2009.00279.x

Lundgren, K. (2012a). The Global Impact of e-waste: Addressing the Challenge. International Labour Office, Programme on Safety and Health at Work and the Environment (Safe Work), Sectoral Activities Department (Sector). Geneva: ILO.

Lundgren, K. (2012b). The Global Impact of e-waste: Addressing the Challenge. International Labour Office, Programme on Safety and Health at Work and the Environment (Safe Work), and Sectoral Activities Department (SECTOR). Geneva: ILO, ISBN 978-92-2-126897-0 (print)

Mukherjee, A. B. (1998). Chromium in the environment of Finland. *The Science of the Total Environment*, *217*(1-2), 9–19. doi:10.1016/S0048-9697(98)00163-6 PMID:9695169

Pruss-Ustun, A., & Corvalan, C. (2006). *Preventing disease through healthy environments: Towards an estimate of environmental burden of disease*. WHO Publication.

Puckett, J., Byster, L., Westervelt, S., Gutierrez, R., Davis, S., Hussain, A., & Dutta, M. (2002). *Exporting Harm - The High-Tech Trashing of Asia; The Basel Action Network (BAN) Silicon Valley Toxics Coalition*. Seattle, WA: SVTC.

Ramachandra, T. V., & Saira, V. K. (2004, March). Environmentally sound options for waste management, *Envis Journal of Human Settlements*. Retrieved March 2, 2016 from http://www.ces.iisc.ernet.in/energy/paper/ewaste/ewaste.html

Salomons, W., & Frostner, U. (1984). Metals in hydrocycle. Springer-Verlag. doi:10.1007/978-3-642-69325-0

Schluepa, M., Hageluekenb, C., Kuehrc, R., Magalinic, F., Maurerc, C., Meskersb, C., . . . Wang, F. (2009). *Recycling: From e-waste to resources, Sustainable Innovation and Technology Transfer Industrial Sector Studies (Nairobi and Bonn, UNEP and STeP)*. Retrieved March 2, 2016 from http://www.unep.org/pdf/Recycling_From_e-waste_to_resources.pdf

Schutz, A., Olsson, M., Jensen, A., Gerhardsson, L., Borjesson, J., Mattsson, S., & Skerfving, S. (2005). Lead in finger bone, whole blood, plasma and urine in lead-smelter workers: Extended exposure range. *International Archives of Occupational and Environmental Health, 78*(1), 35–43. doi:10.100700420-004-0559-5 PMID:15750821

Schwarzer, S., Giuliani, A., De Bono, G., Kluser, S., & Peduzzi, P. (2005). *"E-waste: the hidden side of IT equipment's manufacturing and use". Early Warnings on Emerging Environmental Threats, No. 5, United Nations Environment Programme (UNEP)*. GRID Europe.

Sikdar, M., & Vaniya, S. (2014). The New Millennium and Emerging Concerns. *International Journal of Scientific and Research Publications, 4*(2).

Smith, T., Sonnenfeld, D. A., & Naguib Pellow, D. (Eds.). (2006). Challenging the chip: Labor rights and environmental justice in the global electronics industry. Philadelphia, PA: Temple University Press.

Stewart, B. W., & Kliehues, P. (2003). *World Cancer Report*. Lyon: IARC Press.

Strickland, P. T., & Kensler, T. W. (1995). Chemical and physical agents in our environment. In Clinical Oncology. Churchill Livingston Inc.

Sum, E. Y. L. (1991). The recovery of metals from electronic scrap. *Journal of Metallurgy, 43*, 53–61.

Taylor, T. P., Ding, M., Ehler, D. D., Foreman, T. M., Kaszuba, J. P., & Sauer, N. N. (2003). Beryllium in the environment: A Review. *Journal of Environmental Science and Health. Part A, Toxic/Hazardous Substances & Environmental Engineering, 38*(2), 439–469. doi:10.1081/ESE-120016906 PMID:12638707

Text of the Basel Convention. (2014). *The Control of Transboundary Movements of Hazardous Wastes and Their Disposal*. United Nations Environment Programme (UNEP). Retrieved March 1, 2016 from http://www.basel.int/portals/4/basel%20convention/docs/text/baselconventiontext-e.pdf

Third World Network. (1991). Toxic Terror: Dumping of Hazardous Wastes in the Third World, Penang, Malaysia. Author.

Tsydenova, O., & Bengtsson, M. (2011a). Chemical hazards associated with treatment of waste electrical and electronic equipment. *Waste Management (New York, N.Y.)*, *31*(1), 45–58. doi:10.1016/j.wasman.2010.08.014 PMID:20869229

Tsydenova, O., & Bengtsson, M. (2011b). Chemical hazards associated with treatment of waste electrical and electronic equipment. *Waste Management (New York, N.Y.)*, *31*(1), 45–58. doi:10.1016/j.wasman.2010.08.014 PMID:20869229

Widmer, R., Oswald-Krapf, H., Sinha-Khetriwal, D., Schnellmann, M., & Bo¨ni, H. (2005). Global perspectives on e-waste. *Environmental Impact Assessment Review*, *25*(5), 436–458. doi:10.1016/j.eiar.2005.04.001

World Health Organization (WHO). (1989). Mercury, Environmental Health Criteria, 86. WHO.

Yu, J., Welford, R., & Hills, P. (2006 a). Industry responses to EU WEEE and ROHS Directives: Perspectives from China. *Corporate Social Responsibility and Environmental Management*, *13*(5), 286–299. doi:10.1002/csr.131

Yu, J., Welford, R., & Hills, P. (2006 b). Industry responses to EU WEEE and ROHS Directives: Perspectives from China. *Corporate Social Responsibility and Environmental Management*, *13*(5), 286–299. doi:10.1002/csr.131

ADDITIONAL READING

Brigden, K., Labunska, I., Santillo, D., & Allsopp, M. (2005). *Recycling of electronic wastes in China & India: Workplace & environmental contamination*. Greenpeace International.

Bruke, M. (2007). The Gadget Scrap, Chemistry World, 44-48. Retrieved April 1, 2016 from www.chemistryworld.org

Fornalczyk, A., Willner, J., Francuz, K., & Cebulski, J. (2013). E-waste as a source of valuable metals. *Archives of Materials Science and Engineering*, *63*(2), 87–92.

Gupta, V., & Kumar, A. (2014). E-waste status and management in India. *Journal of Information Engineering and Applications*, *4*(9), 41–48.

Khetriwal, D. S., Luepschen, C., & Kuehr, R. (Eds.). (2013). *Solving the e-waste problem: An interdisciplinary compilation of international e-waste research*. United Nations University.

Krishna, R., & Saha, S. (2015). Study paper on e-waste management. Retrieved March 30, 2016 from http://tec.gov.in/pdf/Studypaper/e%20waste%20management_11.08.pdf

Namias, J. (2013). The future of electronic waste recycling in the United States: Obstacles and domestic solutions. Retrieved March 31, 2016 from http://www.seas.columbia.edu/earth/wtert/sofos/Namias_Thesis_07-08-13.pdf

KEY TERMS AND DEFINITIONS

Dioxins and Furan: These are persistent environmental pollutants (POPs), formed as unintentional by-product during e-waste incineration to recover valuable metals.

Electronic Waste (E-Waste): It refers to EEE waste, including all components, subassemblies, and consumables which are part of the product destined for reuse, resale, salvage, recycling or disposal at the time of discarding.

Environmental Hazard: A substance, state or event which has the potential to threaten the surrounding natural environment that adversely affects people's health (Pollution and Natural disasters).

Environmental Sustainability: It is defined as could be defined as a condition of balance, resilience, and interconnectedness that allows human society to satisfy its needs while neither exceeding the capacity of its supporting ecosystems to continue to regenerate the services necessary to meet those needs nor by our actions diminishing biological diversity.

Metal Poisoning: Toxic metals in certain form and dose sometimes imitate the action of an essential element in the body; interfere with the metabolic processes that cause illness (metal poisoning).

Occupational Hazard: It is risk accepted as a consequence of a particular occupation and they encompass chemical, biological, psychosocial, and physical hazards.

Recycling: It is the process of converting waste materials into reusable objects; dismantling, separating fractions, and recovering material in order to reduce the consumption of fresh raw materials, energy usage, air pollution (from incineration) and water pollution (from land filling) from e-waste after the lifespan of the equipment.

This research was previously published in the Encyclopedia of Information Science and Technology, Fourth Edition edited by Mehdi Khosrow-Pour, pages 3066-3076, copyright year 2018 by Information Science Reference (an imprint of IGI Global).

Chapter 12
Green IT and the Struggle for a Widespread Adoption

Edward T. Chen
University of Massachusetts – Lowell, USA

ABSTRACT

Increasing pollution levels, consumption of electricity as well as other natural resources, and the continuous buildup of outdated computer systems in landfills are plaguing the computer systems industry. Green information technology (IT) is a sensible solution providing multiple resources and alternatives for day-to-day computer use that could reduce the negative impact on our environment without reducing the effectiveness and capabilities of the technology. This chapter describes the struggle of adoption and provides basic concepts and sustainable solutions of green IT for businesses and individuals.

INTRODUCTION

Since the inception of computers, both for business and personal purposes, there have been multiple environmental issues that resulted from this technology. The complex electronics require significant electricity to operate them, large amounts of energy to keep them cool for prolonged usage, and various chemicals and resources to construct them. Notably, within the last decade, there has been a movement building for the ecologically responsible construction, use and disposal of computer systems and their components, including monitors, batteries and printer cartridges. This initiative is commonly known as Green Information Technology (IT), or Green IT.

As both consumable and enterprise level computing products grows, a need for sustainability arises. A balance between the energy consumption and the provided services is required to ensure the environment can survive the influx of billions and billions of devices. Concepts like the Internet of Things, Big Data, smart devices and phones, and complex business analytics for corporations all drive the need for more connected devices. These devices consume more electricity than ever before and data runs the planet (Murugesan & Gangadharan, 2012; Subburaj, Kulkarni, & Jia, 2014).

The Green IT (green information technology) is the practice of environmentally sustainable computing (McLaughlin, 2013). The lack of regulations, standardizations, and standard operating procedures

DOI: 10.4018/978-1-5225-7359-3.ch012

has left this notion out of the mainstream and under the radar of many organizations' information technology (IT) implementations. Several ideas at different levels have been proposed over the years. Its current adoption rate is not enough for sustainability. G-Readiness framework combines properties, processes, and components that are well defined and measurable to ensure success in the greening of IT (Molla, Cooper, Corbitt, Deng, Peszynski, Pittayachawan, & Teoh, 2008). Large technology companies have designed, patented, and implemented as a way to offer a differentiated service and a competitive advantage through green IT. Some of their innovations have the potential to be replicated for further successes (Murugesan & Gangadharan, 2012).

BACKGROUND

Though there is not a general consensus on the exact definition of Green IT (also referred to as green computing, green information and communication technologies (ICT), or ICT sustainability), the most commonly accepted definition was coined by San Murugesan, an outspoken university professor, in his 2008 article entitled "Harnessing Green IT: Principles and Practices". Murugesan defined green IT as "the study and practice of designing, manufacturing, using and disposing of computers, servers, and associated subsystems… efficiently and effectively with minimal or no impact on the environment" (Murugesan, 2008). Multiple efforts can be made, both from individual home users as well as those of entire businesses, to reduce the negative impact on the environment from the technology they are using.

The hardware, software, and components that make up technology are always changing and evolving. Some components like computer processors, are gaining the ability to process information faster while the integrated circuits is getting smaller. Other devices gain new features with each new release and make the older model seem outdated or no longer usable. This perception is particularly accurate with personal technology such as laptops, phones, and tablets. Unused excess hardware accumulates in staggering quantities. In the corporate world, technology is advancing faster than the needs of many businesses. Data centers are filled with high-powered servers and storage devices, which run 24/7/365 in a production environment. Attractive and enticing price points combined with clever marketing presentations convince companies that the deployment of these systems is necessary to solve their IT and IS (information systems) problems (Nguyen, Cheriet, Lemay, Reijs, Mackarel, & Pastrama, 2012).

According to Gartner Research, there are 2 billion computers in use today. They predict that the number of devices and things, items such as thermostats, refrigerators, cars, and other non-traditional computing hardware and sensors, on the Internet could surpass 40 billion by the year 2020 (Akhgar, Pattinson, & Dastbaz, 2015). Greenpeace estimates that if the Internet were a country, it would fall between Japan and Russia, or 5th place, in overall electricity consumption in the world (Cook & Pomerantz, 2015). 50% of the world's population owns a cellular telephone. This number is only going to go up as emerging countries begin to rely on the same technology as First World countries. Tablets are expected to outpace computers in sales and use before the end of this year (Akhgar, Pattinson, & Dastbaz, 2015). The amount of technology in use and the amount of technology that has been cast aside present two challenges for the concept of green IT: reducing energy consumptions of current hardware and finding ways to safely recycle previous hardware that is no longer in use. Stated in a different way, it is solving the two problems of how to reduce CO_2 emissions and how to lower e-waste (Ahmad & Ranka, 2016; Elliot, 2007).

A study was conducted in 2009 to investigate why the lack of growth with implementing and supporting green IT initiatives and standards. It surveyed Chief Information Officers (CIOs) and other IS professionals to find the "barriers" that keep green IT from being implemented. The results of the survey show no surprises, citing a lack of business leadership; the unknown costs versus cost savings for green IT solutions, and the absence of value by turning to green alternatives (Dedrick, 2010). Also uncovered through the same survey was the importance of government incentives or regulations mandating converting to green IT. Without formal direction or instructions to do so, companies are not eager to start the perceived arduous process to switch.

INFORMATION TECHNOLOGY'S GREEN PROBLEM

Energy consumption is a major aspect of IT, with the methods to produce electricity still largely powered by the depletion of fossil fuels such as coal and oil. It is estimated that the carbon dioxide (CO_2) produced by a single desktop computer over its lifetime is 1,096 kilograms (Thomson & van Belle, 2015). This consumption and pollution is amplified by the increasing occurrence of non-efficient software and coding, requiring computers to take longer periods to process finite tasks. A 2009 disputed study found that the average search using the popular search engine Google produced approximately 7 grams of CO_2 and required roughly half of the amount of energy needed to boil a kettle of water (Swaine, 2009; Warman, 2009). A portion of the total electricity consumed by IT is to power the computers and their components, while an even larger portion (30%) is used to cool the computers and their related hardware, particularly in the data centers associated with these information systems (IS) that house the computers and their related components, such as servers and power supplies (Murugesan, 2008; Nguyen, 2012).

The majority of businesses routinely require that their data centers remain cooled at all times at temperatures less than 70 degrees Fahrenheit (21 degrees Celsius) and one study found that only 7% of data centers in the world run at or above a temperature of 75 degrees Fahrenheit (23.8 degrees Celsius) (Mitchell, 2013). The majority of energy spent to cool these facilities utilizes air cooling as opposed to other more efficient methods of cooling and require continuous operation to maintain the set temperature, regardless of whether the IT is even in use. It is estimated that many processors sit idle between 85 to 95% of the day, requiring nearly the same amount of electricity as when active (Shah, 2012). Energy consumption utilized by IT is continually increasing as IT becomes more prevalent in society, with electricity usage by data centers in the United States rising 74% between 2000 and 2010 (van Bussel, Smitter, & Vandepas, 2015).

Computer production requires extensive resources and chemicals, with their associated production facilities consuming large amounts of energy for daily operations as well as climate control. Materials such as lead, mercury, cadmium and hexavalent chromium, combined with large amounts of water, aluminum and plastic, are required to make these machines and their intricate components. With today's pressing need for IT, this technology is more necessary than ever before to ensure each company and household remains capable of performing all manners of tasks and communications both within their community and throughout the world. Despite these high production levels of state-of-the-art equipment, many networks built to last 33 years are seeing themselves disposed of after only 3 to 5 years of use, mainly due to a knee-jerk reaction by their owners to consistently replace and update their 'obsolete' systems for the supposedly required next generations of technology (Ahmad & Ranka, 2016; Ogden, 2013).

These discarded systems are then largely disposed of inappropriately due to minimal disposal regulations and recycling options for IT, resulting in approximately 80% of the systems being dumped in nearby landfills or exported to developing countries such as China and Pakistan where the disposal regulations are less stringent. In 2008, it was estimated that two-thirds of the estimated 870 million personal computers that will be made within the next five years will end up discarded in landfills (Murugesan, 2008). In 2006, the global production of IT-related waste was estimated at 20 to 50 million tons per year (van Bussel et al., 2015). These inefficient practices regarding IT are contributing towards global warming, the expedited depletion of natural resources, and increasing the waste in landfills, leading to the pollution of both the land as well as water sources in the vicinity.

SOLUTIONS AND RECOMMENDATIONS

Despite this dire situation, many options exist to implement green IT and make computing less of a hazard on the environment. The first practice the majority of companies and households can perform is power management, ranging from simply turning off their computers when not in use to providing a power management features to the IT to automatically reduce its electrical load following a period of inactivity. This feature, known as sleep mode, can reduce costs by 60 to 70% and now comes standard on most new computer systems after lengthy pressure was exerted on the manufacturing industry to standardize power-saving features in their products (Murugesan, 2008). Another prevalent shift in the IT industry is the development of ecological hardware that requires much lower levels of electricity to operate as well as attempts to increase the lifespan of the equipment. An example of this innovative new technology is the Atom series of energy-efficient processors designed by Intel. They are ultra-thin and lightweight and their use in IT reduces the space and energy required on the IS. In addition, the fit-PC by Compulab is a series of fanless personal computers that are smaller and require less power than traditional IT, often able to function for long periods of time on batteries and requiring lower-power processors such as the Intel Atom.

Monitors that display the IT images have improved significantly in technology and availability, greatly reducing energy costs. Light-emitting diode (LED) displays are now very comparable in cost and beneficial due to their use of significantly less electricity than liquid crystal display (LCD) monitors as well as their predecessor, the cathode ray tube (CRT) display. In 2014, the German ion research company GSI Helmholtz had created the world's most energy-efficient supercomputer, the L-CSC, in Frankfurt as evidenced by the 'Green 500' report released in November of 2014. Despite being incredibly energy-efficient, the computer was also rated the fourth fastest computer in Germany upon its release (Phys.org, 2014), demonstrating that green computers can be just as powerful and effective as standard models. In addition, efficient coding of software and other applications can significantly reduce the time required for computing processes by increasing the software processing speed on each computer, reducing the electricity consumption as well as energy required to cool the IS. The increasing development and availability of green IT hardware and software are helping provide efficient alternatives to traditional IT equipment used throughout the past 30 years.

There is a wide variety of cooling system methods that can be used to maintain a requisite temperature in the spaces that house computer systems. In addition to traditional air conditioning, alternative methods such as air-side economizers for facilities and liquid cooling systems for PC's are showing themselves to be viable options for keeping equipment cool and, in many cases, requiring a fraction of the cost of air conditioning while being several hundred times more efficient. Despite these benefits, it is understood that a massive overhaul of a company's cooling systems, as well as the required investment that would accompany it, are not simple fixes that can be implemented overnight. Despite these hurdles, the primary solution to high cooling costs that is immediate and is being recommended by leading green companies (such as FedEx, Raytheon and Northrop Grumman) as well as the American Society of Heating, Refrigerating and Air-Conditioning Engineers (ASHRAE) is to simply raise the temperature in data centers. Rather than require a temperature of no warmer than 70 degrees Fahrenheit (F), multiple experts state that all companies should be safely able to raise the temperature of its data centers to 75 degrees F and possibly even as high as 80.6 degrees F to maintain the climate required to protect the IS while simultaneously reducing the energy demands of the facility cooling systems (Mitchell, 2013).

Finally, the most effective solution to reducing energy, hardware, and software costs for all entities utilizing IT seems to be virtualizing equipment. Virtualization is the practice of consolidating servers, desktops, and any other types of equipment to fewer pieces of equipment that can still handle the load requirements. Server virtualization is common in which many virtual servers are hosted on a smaller number of more powerful servers. Instead of having 20 servers throughout the country, entities can invest in three larger server facilities that can still provide full network coverage throughout the company while significantly reducing facility operating, cooling and maintenance costs (Pandi & Somasundaram, 2016). Northrop Grumman is an example of this type of virtualization, eliminating 4,000 physical servers and combining 19 data centers and 81 smaller server rooms into only three facilities (Mitchell, 2013).

Desktop virtualization works in many ways by the same principles. Thin clients are stateless and fan-less computer desktop terminals with no hard drives. They simply link in to the data center to access all desktop capabilities while only drawing one-fifth of the electricity of a traditional desktop computer (Murugesan, 2008; Nguyen, 2012). Even virtual chillers are becoming a frequently-utilized concept, providing climate controls to multiple facilities within the area at a fraction of the cost each facility would require to cool through individual systems. These concentrated efforts may take considerable planning and expense to arrange in the majority of circumstances, but the initiatives pay off for years in the future with exponentially lower energy and facilities costs to the companies (Pandi & Somasundaram, 2016).

LIMITATIONS

Despite the obvious benefits and wide array of alternatives for green IT that are available to companies now more than ever before, the main deterrent is that there are very minimal laws of compliance and regulation regarding green IT. Instead of strict legislation requiring specific changes to be implemented, the majority of initiatives and alternative options available within the United States and beyond are solely advisory. The United States Environmental Protection Agency (EPA) founded the Energy Star program in 1992, giving it a significant upgrade in 2006 to include computers and other IT, but it is still mainly an optional program and not a requirement for all appliances and equipment to have the green star logo depicting environment-conscious operation. As of 2008, only 26 of the 50 United States had established a statewide recycling program for the proper disposal of older computers. The Green Electronics Council created the Electronic Product Environmental Assessment Tool (EPEAT) as a method to assist prospective IT buyers through the ranking of available for purchase equipment by more than 50 criteria topics, assigning scores based on how green the technology is (Ahmad & Ranka, 2016).

In 2007, President George W. Bush signed Executive Order number 13423, which required all federal agencies to utilize EPEAT when purchasing new computer systems and requiring all vendors contracted by the Federal government to utilize EPEAT in their system purchases. President Obama recently modified this requirement in March of 2015 with Executive Order 13693, stating that all government procurements of electronics should "meet or exceed specifications, standards, or labels recommended by (the) EPA" (Moodie, 2015). This legislation is expected to reduce the prominence of the EPEAT ratings system by instead requiring companies wishing to do business with the federal government to follow one of the many recommended ratings systems, rather than specifically the most well-known, EPEAT, but at least the directive to procure green equipment remains a priority.

The Restriction of Hazardous Substances (ROHS) in Electrical and Electronic Equipment Directive makes significant restrictions on the European Union market for equipment containing specific amounts of certain hazardous substances (Murugesan, 2008). Since August of 2006, Greenpeace International has produced their quarterly Guide to Greener Electronics, ranking several companies on their overall use (or lack thereof) of green IT. Then in 2007, several manufacturers of computer equipment (including Dell, HP, Microsoft, and Intel) formed the Green Grid, a consortium dedicated to making data centers and information systems more ecologically efficient.

Four general drivers for a company to choose to utilize green IT are economic, regulatory, market opportunity, and finally influence from social, cultural and political pressures (Thomson & van Belle, 2015). Despite the wide variety of laws and regulations aimed to transform companies into becoming more environmentally-conscious, the majority remain overwhelmingly optional for companies to follow. Thus, the primary motivator for companies to go green is economic, strongly appealing primarily to their own self-interests (Bohas & Poussing, 2016). Implementing green methods reduces energy costs. Green practices can extend the life of the IT and decreases the costs of disposal of obsolete equipment. In addition, following green practices will set a positive example for other companies to follow and can greatly increase the public image and reputation of the firm, known as "Green corporate image" (Bohas & Poussing, 2016).

However, many businesses look at green IT through a purely economic mindset. Despite the benefits to the business and the environment, fiscal incentives or government subsidies will have to be involved to aid the businesses in the costly transition towards greener IT (Mitchell, 2013). Through the use of several of the advisory organizations listed above, there is a growing voice for green practices with increasing pressures being placed on both the companies as well as the governments of the world to enact stricter regulations to enable change. There is also a small but growing market opportunity for companies that specialize in green IT consulting to help companies shift to cost effective and ecological green practices. This will likely help increase the prevalence of green computing in the years to come.

Another major benefit of green IT is that for the majority of the process, managerial practices should not have to change significantly in the business settings. The process of shifting systems to greener alternatives will require close coordination and decision-making between the management of each business and the IS personnel involved. Following these shifts, the systems will largely function in a way that does not change the day-to-day aspects of each normal business process. The employees will likely just see a greater emphasis on turning off their computers at the end of the day and other similar green initiatives that are both understandable and significant practices that businesses can implement to help their environment.

Other than the higher costs that are still regularly associated with newer technology, it does not appear to be any subsequent technological problems or complications that will arise as a result of green IT. Alternatively, green methods are designed and implemented specifically to reduce the existing energy problems that have been prevalent for decades. They are consistently being seen as reliable solutions to the disadvantages of outdated technology.

FUTURE RESEARCH DIRECTIONS

The Internet of Things (IoT) promises a lot of innovation through sensors operating and interoperating together. It has already solved problems that previously having no operational solutions. IoT brings with it immense computing capabilities and endless possibilities that combine science fiction with an improved quality of life. The need for a seeming endless amount of storage to keep the big data collected from over 40 billion IoT devices. This alone facilitates the immediate need for green IT. However, this technology and this level of capability can also drive green IT initiatives.

As Big Data becomes a term that continues to grow in usage, the amount of data is also growing. The more data that needs to be stored, the more natural resources it will take to power the servers, keep them cool, and house them in large facilities. One innovative team decided to see if solving green IT can be accomplished by reducing the abundant amount of data that is currently burning up more energy than necessary. While this study still needs further development, it helps to further the cause here of helping to define how to make green IT more accessible to the general IT public. It categorizes green IT into six core components. The first component is longevity of the hardware. How well was the product produced? What is the expected life span? The second component is software optimization, allowing for the hardware to consume less resource to compile and run the software. Third one is power management in finding the right formula and standard to shut off computer resources when not in use. The fourth component is the recycling of materials and not just recycling computers but finding alternative uses for the hardware. Fifth component on the list is telecommuting. An indirect benefit from fewer cars on the road will reduce carbon dioxide emissions. And finally, the sixth component is energy-efficient computing or low-power IT (van Bussel, Smitter, & van de Pas, 2015).

Current and previous research has been focused mainly on trying to better understand how green IT can be codified for implementation. Longitudinal studies have looked at ways that Green IT could be standardized, means to quantify form over function, and even look at the individuals who will need to support it and their impressions from a psychological approach. If green IT is to surpass its current adoption rate, stronger research needs to be conducted by pairing technology with green philosophies. For example, implementing power-saving functionality at the firmware level of a network switch can save energy for the entire network. Sensors abound from the Internet of Things revolution should make collected data meaningful for experiments and publish new findings and standards easier to reach target audiences (Kaushik & Vidyarthi, 2016).

CONCLUSION

Through a series of missteps or misinterpretations, green IT is still struggling to find its place within the technology field. Studies have been conducted; outcomes have been published; and successes have been documented; but something remains missing in this vital area that blends innovations with ecological awareness and respect. The G-readiness framework is probably the most concrete way for a common business to assess the potential of green IT for them based on the factors most important to keeping the business operational. And while the framework might seem daunting at first, there is a structure to it that forces a common sense and logical approach to reduce energy and look at IT in a smarter way.

Companies have adopted the virtualization of servers. Virtualization is not considered a win for green IT, nor did it help to mainstream the concept. Instead, IT professionals everywhere have no idea that when they learn how to run multiple services virtually, they reduce their carbon footprints. Those same IT professionals will be needed to collaborate with their business counterparts to ensure they can provide innovative solutions to problems as they arise. This will be the cornerstone to the implementation of green IT within a business.

The green IT community is still small and many of the same people have been working together on the different studies and experiments trying to garner a stronger and larger following. Social media should be exploited to ensure the next generation of IT professionals and operators know how to think green right from the beginning. It's the Millennials that will be faced with the green IT challenge, a challenge to sustain a healthy planet through the Internet of Things and the 50 billion devices and zettabytes worth of data that comes with it. Green IT must become the new normal before that time. It is no longer an option, and it is no longer something for someone else to worry about.

Currently, there are not enough green IT regulations implemented through the governments of the world, including that of the United States. They would require at least a small percentage of its implementation by the business sector to kick start the unanimous shift towards fully green computing. Further, until commercial entities are required to comply with green IT practices, the personal-use market will also not attain compliance and yet another decade could likely pass resulting in very minimal improvements. Until the requirements are enacted, the majority of businesses will simply not switch over. Before they are able to devote the requisite time and energy to make this essential shift, they want to see that green IT practices can benefit them significantly.

The concept of Green IT is steadily increasing in prominence and the benefits such as reduced energy costs are making green IT practices very attractive to all businesses. They desire a sense of longevity and a cost-effective IS infrastructure. The omnipresence and convenience of IT in all venues of life factored in with the environmental benefits from computing. The reduced greenhouse gas emissions evidenced from telecommuting, online education enterprises and video teleconferencing, will keep the IT industry strong and a permanent fixture in today's society. Green IT will eventually become the standard to follow, but the implementation will be much slower if major elements do not become mandatory through the use of effective legislation and increasing numbers of entities willing to set the correct industry examples.

REFERENCES

Ahmad, I., & Ranka, S. (2016). *Handbook of energy-aware and green computing*. CRC Press.

Akhgar, B., Pattinson, C., & Dastbaz, M. (2015). *Green Information Technology: A Sustainable Approach*. Waltham, MA: Elsevier, Inc.

Bohas, A., & Poussing, N. (2016). An empirical exploration of the role of strategic and responsive corporate social responsibility in the adoption of different green IT strategies. *Journal of Cleaner Production*, *122*, 240–251. doi:10.1016/j.jclepro.2016.02.029

Cook, G., & Pomerantz, D. (2015). *Clicking Clean: A Guide to Building the Green Internet*. Washington, DC: Greenpeace, Inc.

Dedrick, J. (2010). Green IS: Concepts and issues for information systems research. *Communications of the Association for Information Systems*, *27*(11), 173–183.

Elliot, S. (2007) Environmentally sustainable ICT: a critical topic for IS research? *Pacific Asia Conference on Information Systems, 114*, 100-112.

Kaushik, A., & Vidyarthi, D. P. (2016). A green energy model for resource allocation in computational grid using dynamic threshold and GA. *Sustainable Computing: Informatics and Systems, 9*, 42–56.

McLaughlin, E. (2013). *Definition of green IT (green information technology)*. Retrieved March 30, 2016 from http://searchcio.techtarget.com/definition/green-IT-green-information-technology

Mitchell, R. (2013). IT gets its green on. *Computerworld*, 26–29.

Molla, A. (2009). Organizational motivations for Green IT: exploring Green IT matrix and motivation models. *Pacific Asia Conference on Information Systems*.

Molla, A., Cooper, V., Corbitt, B., Deng, H., Peszynski, K., Pittayachawan, S., & Teoh, S. Y. (2008). E-readiness to G-readiness: developing a green information technology readiness framework. *Australasian Conference on Information Systems, 35*, 669-678.

Moodie, A. (2015). *Is the Obama Administration lowering the bar on green electronics?* Retrieved March 20, 2016 from http://www.theguardian.com/sustainable-business/2015/apr/23/obama-epeat-epa-technology-green-electronics-council

Murugesan, S. (2008). Harnessing Green IT: Principles and practices. *IT Professional, 10*(1), 24–33. doi:10.1109/MITP.2008.10

Murugesan, S., & Gangadharan, G. (2012). *Harnessing green IT: principles and practices*. Chichester, UK: John Wiley and Sons. doi:10.1002/9781118305393

Nguyen, K. K., Cheriet, M., Lemay, M., Reijs, V., Mackarel, A., & Pastrama, A. (2012). Environmental-aware virtual data center network. *Computer Networks, 56*(10), 2538–2550. doi:10.1016/j.comnet.2012.03.008

Ogden, J. (2013). Blog: Green IT – a key to long-term savings. *CIO Asia*. Retrieved June 20, 2016 from http://www.cio-asia.com/tech/data-center/blog-green-it--a-key-for-long-term-savings/?page=1

Pandi, K. M., & Somasundaram, K. (2016). Energy efficient in virtual infrastructure and green cloud computing: A review. *Indian Journal of Science & Technology, 9*(11), 1–8. doi:10.17485/ijst/2016/v9i11/89399

Phys.org. (2014). *German Supercomputer is a world champion in saving energy*. Retrieved March 10, 2016 from http://phys.org/news/2014-11-german-supercomputer-world-champion-engergy.html

Shah, R. (2012). Eco-friendly IT: Greener approach to IT. *Journal of Management and Business Research, 2*(2), 4–25.

Subburaj, S., Kulkarni, S., & Jia, L. (2014). Green IT: Sustainability by aligning business requirements with IT resource utilization. *International Journal of Communication Networks and Distributed Systems, 12*(1), 30–46. doi:10.1504/IJCNDS.2014.057986

Swaine, J. (2009). *Two Google Searches 'Produce Same CO_2 as Boiling a Kettle*. Retrieved March 30, 2016 from http://www.telegraph.co.uk/technology/google/4217055/Two-Google-searches-produce-same-CO2-as-boiling-a-kettle.html

Thomson, S., & van Belle, J. (2015). Antecedents of Green IT adoption in South African higher education institutions. *The Electronic Journal Information Systems Evaluation*, *18*(2), 172–186.

Van Bussel, G., Smitter, N., & Vandepas, J. (2015). Digital archiving, Green IT and environment. Deleting data to manage critical effects of the data deluge. *The Electronic Journal Information Systems Evaluation*, *18*(2), 187–198.

Warman, M. (2009). *Green I.T.: How many Google searches does it take to boil a kettle?* Retrieved March 30, 2016 from http://www.telegraph.co.uk/technology/google/4241791/Green-I.T.-how-many-Google-searches-does-it-take-to-boil-a-kettle.html

ADDITIONAL READING

Ahmad, I., & Ranka, S. (2016). *Handbook of energy-aware and green computing. Chapman & Hall/ CRC Computer and Information Science Series*. CRC Press.

Bohas, A., & Poussing, N. (2016). An Empirical Exploration of the Role of Strategic and Responsive Corporate Social Responsibility in the Adoption of Different Green IT Strategies. *Journal of Cleaner Production*, *122*, 240–251. doi:10.1016/j.jclepro.2016.02.029

Kaushik, A., & Vidyarthi, D. P. (2016). A green energy model for resource allocation in computational grid using dynamic threshold and GA. *Sustainable Computing: Informatics and Systems*, *9*, 42–56.

Pandi, K. M., & Somasundaram, K. (2016). Energy efficient in virtual infrastructure and green cloud computing: A review. *Indian Journal of Science & Technology*, *9*(11), 1–8. doi:10.17485/ijst/2016/v9i11/89399

KEY TERMS AND DEFINITIONS

Carbon Footprint: The total amount of greenhouse gas emissions usually expressed in tons of carbon dioxide (CO2).

Data Center: A physical or virtual infrastructure used to house a large group of networked computer servers for the remote storage, processing, or distribution of large amounts of data.

Energy Consumption: Amount of energy consumed in a process or system, or by an individual, organization or country.

E-Waste: Any old, end-of-life or discarded electronic or electrical devices or their components.

Green IT: Green information technology is the practice of environmentally sustainable and responsible use of computers and related resources.

This research was previously published in the Encyclopedia of Information Science and Technology, Fourth Edition edited by Mehdi Khosrow-Pour, pages 3077-3085, copyright year 2018 by Information Science Reference (an imprint of IGI Global).

Chapter 13
Identification of Green Procurement Drivers and Their Interrelationship Using Fuzzy TISM and MICMAC Analysis

Surajit Bag
Tega Industries South Africa Pty Ltd., South Africa

ABSTRACT

The objective of the chapter is to identify the leading green procurement drivers and identify the interrelationships using fuzzy total interpretive structural modeling and MICMAC approach. For the purpose of this study, 25 drivers were identified from existing literature that influence green procurement practices. These are finally refined through experts' opinions. The final fuzzy model consists of fuzzy relationships between one-to-one criteria. The key drivers which emerged from fuzzy TISM and MICMAC analyses are government policy and regulations, total quality environmental management, management support, management review, continuous education of employees, cross-functional team building, organization culture, and green process and technology. The green procurement model may assist supply chain practitioners in better decision making and successful implementation of green procurement programs.

INTRODUCTION

Green procurement is sourcing products and services that cause minimal adverse environmental impacts. It incorporates human health, social and environmental concerns into the search for high quality products and services at competitive prices. Green procurement is generally considered a mammoth task by the procurement managers. Recently focus has been given by regulatory bodies to apply pressure on firms for implementing green programs. Environment protection bodies are regularly organizing seminars and conferences to educate and train managers in such greening initiatives. In some countries the government has developed green specifications for items and mandatory part of tender requirement for public procurement. However, the green procurement programs are still under nascent phase in most developing countries.

DOI: 10.4018/978-1-5225-7359-3.ch013

The present research is motivated based on the study of Azevedo et al., (2011) where they have pointed potential future research area in exploring the enablers and barriers influencing companies in taking green procurement decisions. Secondly, Appolloni et al., (2014) conducted a review on green procurement considering the time frame between 1996 and 2013 but does not highlight the inter-relationships between the drivers of green procurement practices and they have also kept it under one of future research directions. They have also mentioned the need for strong qualitative and quantitative research to support the progress of green procurement.

The objective of the current study is to identify the leading drivers that influence green procurement programs and determine the interactions among the identified drivers. This chapter is structured into four additional sections. The next section presents the background of the study which helps to identify the green procurement drivers. The third section introduces Fuzzy TISM. Finally, conclusions, limitations and directions of future research are presented.

BACKGROUND

In this section an attempt has been taken to briefly explain the key drivers of green procurement.

Government Policy and Regulations

Governments are among the largest consumers in an economy. The public sector on average spends 45%-65% of their budgets on procurement. Given this substantial purchasing power, governments have enormous leverage to stimulate and drive markets for sustainable production and consumption when they make a determined effort to purchase 'green' products and services. Adopting such an approach is a smart form of procuring goods and service – it not only improves the efficiency of public procurement but also uses the public market power to bring about significant environmental and socioeconomic benefits. Supply chain management operates within a regulatory framework set by National Government and extended by provinces and local governments to specific policies, legislation and regulations. In South Africa for instance important legislation influencing this function includes the Public Finance Management Act (1999), Preferential Procurement Policy Framework Act (2000), Preferential Procurement Framework Regulations (2001) and National Treasury Regulations (2005). The Municipal Finance Management Act (MFMA) of 2003 governs the financial and supply chain management functions of Local Government. In developing green procurement policies, local government would need to ensure that these policies: are aligned with their existing Supply chain management regulatory frameworks; avoid a clash between the Preferential Procurement regulations and environmental principles or criteria in the policy; incorporate green procurement in all dimensions of the supply chain management cycle; and institutionalize green procurement within the existing structures set out by the regulatory framework. Government policy and regulations positively influences green procurement (Min & Galle 1997; Diabat & Govindan 2011; Hassini et al., 2012; Bag., 2014; Appolloni et al., 2014)

Total Quality Environmental Management

Firms with successful TQEM programs will have more formal mechanisms for interacting with suppliers. Business units with successful TQEM programs exhibit a greater degree of competitive focus and strategic sourcing. In every step of the manufacturing process there will be quality check to avoid rejections and wastage and this will assist in saving natural resources. TQEM positively influences green procurement (Khidir et al., 2010; Diabat & Govindan 2011; Ageron et al., 2012; Dubey et al., 2013; Bag., 2014; Bag & Anand., 2014; Dubey et al., 2014).

Management Support

Management support is important in success of any projects and specially for green procurement programs where strategic decisions are mainly involved. Management support have positively influenced green procurement (Min & Galle 1997; Zhu et al., 2008; Arslan 2010; Bag., 2014).

Management Review

Management review periodically is necessary to check the progress of green procurement programs and see that timelines are met. The review will capture the bottlenecks, critical paths and develop strategies to find out ways to complete the activities at economical cost within the timeline. As per expert opinion management review positively influence green procurement.

Continuous Education of Employees

Organizations practicing green procurement must have the transformation and diversity manager to carry out the necessary trainings of employees. Training will assist employees in gaining knowledge and deeper understanding of green procurement and its importance in supply chain management. Moreover, the training budget must be utilized carefully in proper training and must be aligned with the company mission and vision so that organization ultimately benefits in the long run. As per expert opinion continuous education of employees positively influences green procurement.

Cross Functional Team Building

In a manufacturing firm there are people from planning, procurement, production, quality assurance, logistics and other functions. Since green procurement involves close coordination with all related supply chain functions therefore it is imperative that organization form a green procurement committee comprising people from all functions, i.e. a cross-functional team to drive the green procurement project. This committee will be responsible for generating weekly progress reports, communicating to internal and external environment, maintain records of consumption of resources, reduction in energy usage and associated cost savings. As per expert opinion cross functional team building positively influences green procurement.

Organization Culture

Empowering junior managers and creating innovation culture among employees will facilitate green procurement practices. Ultimately the organisational culture, structure and process must support the green procurement practices. As per expert opinion organizational culture positively influences green procurement.

Green Process and Technology

Introducing 'green' tendering criteria can influence the marketplace and result in new entrants in the field of environmental technologies and products - potentially resulting in increased competition and reduced prices. Firms must remove the obsolete machinery and convert to green process and technology (renewable source of energy, reducing utility costs, environmental innovation and source reduction) to drive the green procurement programs. Green process and technology positively influences green procurement (Green et al., 2000; Hassini et al., 2012; Bag., 2014).

Information System Infrastructure

IT Eco-Efficiency and IT Eco-Innovation is important dimension of information system infrastructure required to drive organization wide green programs. Additionally, SAP, ERP, RFID, and data analytics are important tools to derive the MIS reports for assessing the progress of green procurement projects. IT infrastructure positively influence green procurement (Green et al., 1998; Hervani et al., 2005; Bag 2014).

Green Design

Driving the green procurement project starts in the design stage to ensure that products comply with restrictions on specified chemical substances in parts and materials, while complying with obligations for labeling, information provision and energy-saving standards for finished products. The green design should comply with environmental norms and regulations such as WEEE, RoHS, REACH, EU Directive on Packaging and Packaging Waste. Green Design positively influence green procurement (Mavi et al., 2013).

Re-Use, Re-Engineering, and Recycling of Products and Materials

The 3 Rs' forms the basis of any green procurement programs and enhance cost savings and savings of natural resources. 3Rs' positively influence green procurement (Mavi et al., 2013).

Supplier Flexibility

It is critical for suppliers to adopt flexibility and supply alternate material as per revised bill of material in green procurement projects. The supply risk must be minimized by supplying the alternate material in time at economical costs. Supplier flexibility positively influence green procurement (Mavi et al., 2013; Bag, 2016b)

Suppliers' Capability to Innovate

It is imperative that supplier has the capability of innovation by demonstrating knowledge and willingness by coming with new eco-products at cheaper costs. It positively influences green procurement (Min & Galle., 2001; Chiou et al., 2011; Ageron et al., 2012).

Supply Risk Management

Green procurement adopt some of the best practices which automatically minimizes the supply risks associated with traditional supply chain management. It positively influences green procurement (Ageron et al., 2012; Bag 2014; Bag & Anand., 2014; Dubey et al., 2014).

Trust Building in Suppliers

Trust is a soft factor associated with green procurement. It positively influences green procurement (Ageron et al., 2012; Ji et al., 2014).

Low Supplier Lead Time

Green procurement prerequisites best operational practices such as vendor managed inventory, just in time approach and lean manufacturing to optimize costs. Automatically the supplier lead time becomes low in such cases and suppliers are able to deliver material as per delivery schedule. This involves proper strategy building and careful monitoring. Therefore, supplier lead time influences green procurement (Ageron et al., 2012; Bag. 2014; Bag & Anand., 2014; Dubey et al., 2014).

Customer Satisfaction

Customer satisfaction is the outcome of good green procurement practices and measured as a success parameter in any green programs. It positively influences green procurement (Diabat & Govindan, 2011; Ageron et al., 2012; Bag., 2014; Bag & Anand., 2014; Dubey et al., 2014).

Annual Savings From Green Procurement Practices

Organizations derive both tangible and intangible benefits from green procurement practices. The direct savings from green procurement practices are significant and highly motivates the procurement managers. As per expert opinion annual savings from green purchasing positively influences green procurement.

Annual Saving of Natural Resources

The annual savings of natural resources motivates managers in driving green procurement projects. The key areas where reduction can be achieved are: consumption of coal used to fire boiler, diesel oil in forklifts, steam for running hydraulic systems and compressor, energy for utility and paper for printing will lead to significant savings of natural resources. As per expert opinion savings of natural resources positively influences green procurement.

Procurement Excellence

Adopting best world class sustainability practices such as green procurement leads to procurement excellence. Therefore, procurement excellence is the outcome and considered as a variable for measuring success of green procurement (Bag, 2016a).

DATA ANALYSIS

The current study intends to develop green procurement theory based on fuzzy total interpretive structural modeling (Fuzzy TISM) approach. Fuzzy TISM is an advanced method and designed in a manner to capture both the statements of respondents as well as logic and interpretation. TISM has greater explanatory power than other established inductive approaches.

Fuzzy Total Interpretive Structural Modeling (FUZZY TISM)

The steps of Fuzzy TISM have been followed as per Khatwani et al., (2015). The data analysis is segregated into following sub-sections:

Start the Decision Making Process

In the current study, the responses of five experts from manufacturing sector have been gathered. The input has been utilized to refine the drivers and further used in modeling.

Selection of Criteria and Sub Criteria of Green Procurement

This is the second step in Fuzzy TISM modeling approach. For the purpose of this study, twenty-five drivers were identified from literature which influences green procurement practices and finally refined through experts' opinion. The final twenty sub criteria are presented in the previous section.

In the next step the process of gathering responses for fuzzy TISM has been explained in details.

Gathering Responses and Calculation of Aggregated SSIM

For the purpose of study, the responses of five procurement managers from manufacturing sector has been collected for evaluation of interrelationship among selected drivers. These experts are well experienced with the green procurement practices. A joint meeting was organized with these five experts and the research problem and methodology was presented and explained to them in 30 minutes' duration. Further they were requested to fill up the VAXO matrix. The entire process of filling up the VAXO matrix separately by these five experts took almost one hour. From the submitted five separate VAXO matrices, finally the aggregated SSIM matrix was calculated by applying mode. The aggregated SSIM matrix is shown in Table 1 (Appendix).

Calculation of Fuzzy Reachability Matrix

The fuzzy reachability matrix is generated from aggregated fuzzy SSIM matrix which was presented in the above stage. The fuzzy reachability matrix is shown in Table 2 (Appendix).

Calculation of Final Fuzzy Reachability Matrix

The final fuzzy reachability matrix is derived from step above and presented in Table 3 (Appendix). From the final fuzzy reachability matrix, the dependence power (X) and driving power (Y) is calculated based on summing of columns and rows and calculating the crisp value. The crisp value of each element is used as an input to develop fuzzy MICMAC analysis.

Driving Power and Dependence Power Matrix (MICMAC) Based on Fuzzy Reachability Matrix From Table 3

See Figure 1.

Discussions Based on MICMAC Analysis Based on Fuzzy Reachability Matrix

Cluster 1 - Autonomous Variables: These variables have a weak drive power and weak dependence power. In this cluster we have six variables. Total Quality Environmental management (E2), Management Support (E3), Management Review (E4), Cross functional team building (E6), Organization Culture (E7) and Information System Infrastructure (E9).

Cluster 2 - Dependence Variables: These variables have a weak drive power but strong dependence power. In this cluster we have twelve variables. Green process and Technology (E8), Green Design (E10), Re-use, Re-engineering and Recycling of products and materials (E11), Supplier Flexibility (E12), Suppliers' capability to Innovate (E13), Supply risk management (E14), Trust building in suppliers (E15), Low supplier lead time (E16), Customer Satisfaction (E17), Annual Savings from green procurement practices (E18), Annual Saving of natural resources (E19) and Procurement Excellence (E20).

Cluster 3 - Linkage Variables: These variables have a strong drive power as well as strong dependence power. Linkage variables are very sensitive and unstable. Any action on these variables will trigger an effect on other variables and also a feedback on themselves. In this cluster we have no variables.

Figure 1. MICMAC analysis based on crisp values

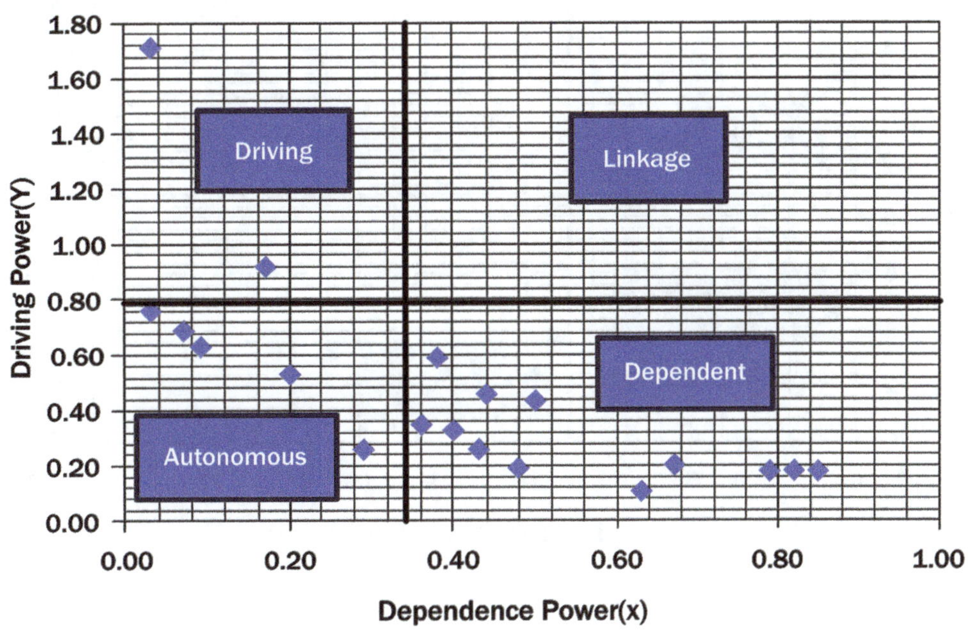

Cluster 4 - Driving Variables: These variables have a strong drive power but weak dependence power. In this cluster we have two variables. Government policy and Regulations (E1) and Continuous education of employees (E5).

Defuzzified Reachability Matrix

The Defuzzified reachability matrix with fuzzy linguistic terms VH, H, as 1 and rest as 0 is shown in Table 4 (Appendix).

Defuzzified Reachability Matrix and Transitivity Check

The paired comparisons are translated in the form of reachability Matrix. The matrix has been also checked for transitivity rule. A final reachability matrix, post transitivity check is presented in Table 5 (Appendix).

Driving Power and Dependence Power Matrix (MICMAC) Based on Defuzzified Reachability Matrix From Table 5

See Figure 2.

Figure 2. MICMAC analysis based on Table 6

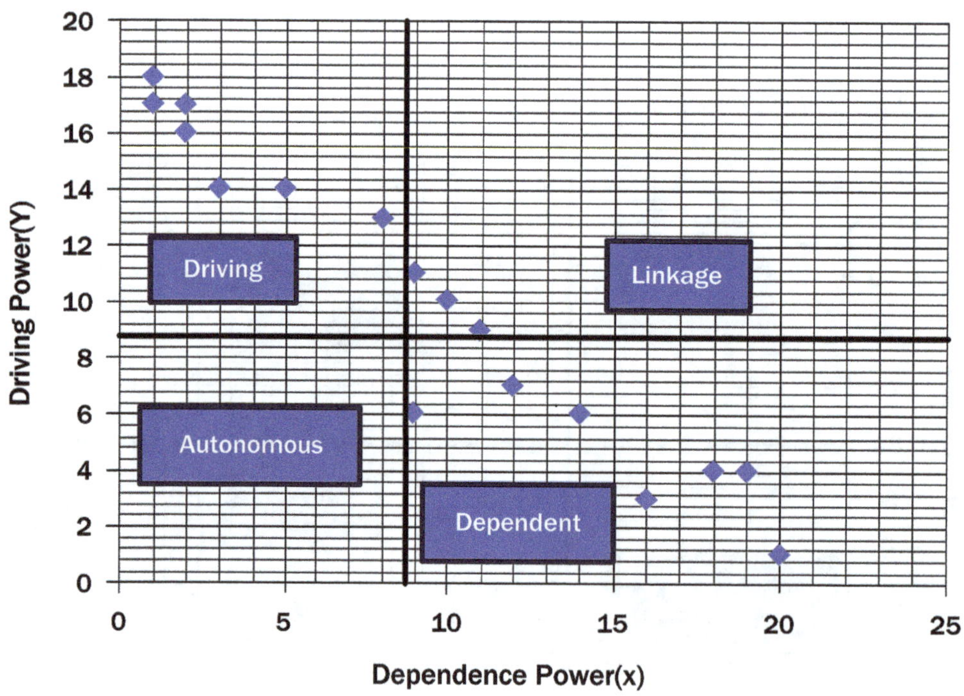

Discussion on MICMAC Analysis Based on Defuzzified Reachability Matrix

Cluster 1 - Autonomous Variables: These variables have a weak drive power and weak dependence power. In this cluster we have no variables.

Cluster 2 - Dependent Variables: These variables have a weak drive power but strong dependence power. In this cluster we have nine variables. Information System Infrastructure (E9), Suppliers' capability to Innovate (E13), Supply risk management (E14), Trust building in suppliers (E15), Low supplier lead time (E16), Customer Satisfaction (E17), Annual Savings from green procurement practices (E18), Annual Saving of natural resources (E19), Procurement Excellence (E20).

Cluster 3 - Linkage Variables: These variables have a strong drive power as well as strong dependence power. Linkage variables are very sensitive and unstable. Any action on these variables will trigger an effect on other variables and also a feedback on themselves. In this cluster we have three variables. Green Design (E10), Re-use, Re-engineering and Recycling of products and materials (E11) and Supplier Flexibility (E12).

Cluster 4 - Driving Variables: These variables have a strong drive power but weak dependence power. In this cluster we have eight variables. Government policy and Regulations (E1), Total Quality Environmental management (E2), Management Support (E3), Management Review (E4), Continuous education of employees (E5), Cross functional team building (E6), Organization Culture (E7), and Green process and Technology (E8).

Level Partition on Reachability Matrix

The final reachability matrix obtained in Table 5 is now partitioned into different levels. After the first iteration, the driver classified to level 1 is discarded and the partitioning procedure is repeated on the remaining drivers to determine the level 2. These iterations are continued until the level of each driver has been determined and presented in Table 6 (Appendix).

TISM Diagraph

The connective and interpretive information contained in the interpretive direct interaction matrix and diagraph is used to derive the TISM. The nodes in the diagraph are replaced by the interpretation of elements placed in boxes. The interpretation of the cells of interpretive direct interaction matrix is depicted by the side of the respective links in the structural model. The final TISM model after removing the transitive links is presented in Figure 3.

CONCLUSION

In the current study author proposes Fuzzy TISM method for identifying the interrelationships among elements influencing green procurement. Due to incorporation of fuzziness in TISM the decision makers have the flexibility in assigning the level of influence of pair wise elements. Apart from wider flexibility Fuzzy TISM also enhance the quality of decision making. The findings from the TISM model show that Customer Satisfaction is in level 1, Annual Savings from green procurement practices, Annual Saving of natural resources and Procurement Excellence are in level 2, Supply risk management is in level 3, Infor-

Figure 3. TISM diagraph

mation System Infrastructure and Low supplier lead time is in level 4, Suppliers' capability to Innovate and Trust building in suppliers is in level 5, Supplier Flexibility is in level 6, Re-use, Re-engineering and Recycling of products and materials is in level 7, Green Design is in level 8, Green process and Technology is in level 9, Continuous education of employees, Cross functional team building and Organization Culture is in level 10, Total Quality Environmental management and Management Review is in level 11 and Government policy and Regulations and Management Support are in the bottom level.

The driving factors which emerged from Fuzzy TISM and MICMAC analyses are Government policy and Regulations (E1), Total Quality Environmental management (E2), Management Support (E3), Management Review (E4), Continuous education of employees (E5), Cross functional team building (E6), Organization Culture (E7), and Green process and Technology (E8). These elements are the key players which must be considered while planning green procurement programs.

LIMITATIONS AND DIRECTIONS OF FUTURE RESEARCH

The present research has certain limitations. The model is developed purely based on interview with five procurement management experts from the manufacturing sector. In future studies the author proposes to statistically validate the model using big sample size and further extend theory of green procurement.

REFERENCES

Ageron, B., Gunasekaran, A., & Spalanzani, A. (2012). Sustainable supply management: An empirical study. *International Journal of Production Economics*, *140*(1), 168–182. doi:10.1016/j.ijpe.2011.04.007

Appolloni, A., Sun, H., Jia, F., & Li, X. (2014). Green Procurement in the private sector: A state of the art review between 1996 and 2013. *Journal of Cleaner Production*, *85*, 122–133. doi:10.1016/j.jclepro.2014.08.106

Arslan, T., Yilmaz, V., & Aksoy, H. K. (2010). Structural Equation Model for Environmentally Conscious Purchasing Behavior. *International Journal of Environmental of Research*, *6*(1), 323–334.

Azevedo, S. G., Carvalho, H., & Machado, V. C. (2011). The influence of green practices on supply chain performance: A case study approach. *Transportation Research Part E, Logistics and Transportation Review*, *47*(6), 850–871. doi:10.1016/j.tre.2011.05.017

Bag, S. (2014). *Developing a GSCM model for the Indian rubber goods manufacturing sector (Unpublished Doctoral Dissertation). University of Petroleum & Energy Studies*. Available at http://ils.ddn.upes.ac.in:8001/cgi-bin/koha/opac-authoritiesdetail.pl?authid=13645

Bag, S. (2016). Green strategy, supplier relationship building and supply chain performance: Total interpretive structural modelling approach. *International Journal of Procurement Management*, *9*(4), 398–426. doi:10.1504/IJPM.2016.077702

Bag, S. (2016b). Flexible procurement systems is key to supply chain sustainability. *Journal of Transport and Supply Chain Management*, *10*(1), 1–9. doi:10.4102/jtscm.v10i1.213

Bag, S., & Anand, N. (2014). Modeling Green Supply Chain Management framework using ISM and MICMAC analysis. *African Journal of Business Management*, *8*(22), 1053.

Chiou, T. Y., Chan, H. K., Lettice, F., & Chung, S. H. (2011). The influence of greening the suppliers and green innovation on environmental performance and competitive advantage in Taiwan. *Transportation Research Part E, Logistics and Transportation Review*, *47*(6), 822–836. doi:10.1016/j.tre.2011.05.016

Diabat, A., & Govindan, K. (2011). An analysis of the drivers affecting the implementation of green supply chain management. *Resources, Conservation and Recycling*, *55*(6), 659–667. doi:10.1016/j.resconrec.2010.12.002

Dubey, R., & Ali, S. (2014). Identification of flexible manufacturing system dimensions and their interrelationship using total interpretive structural modelling and fuzzy MICMAC analysis. *Global Journal of Flexible Systems Management*, *15*(2), 131–143. doi:10.100740171-014-0058-9

Dubey, R., Bag, S., Ali, S., & Venkatesh, V. (2013). Green purchasing is key to superior performance: An empirical study. *International Journal of Procurement Management*, *6*(2), 187–210. doi:10.1504/IJPM.2013.052469

Green, K., Morton, B., & New, S. (1998). Green purchasing and supply policies: Do they improve companies environmental performance? *Supply Chain Management: An International Journal*, *3*(2), 89–95. doi:10.1108/13598549810215405

Green, K., Morton, B., & New, S. (2000). Greening organizations purchasing, consumption, and innovation. *Organization & Environment*, *13*(2), 206–225. doi:10.1177/1086026600132003

Hassini, E., Surti, C., & Searcy, C. (2012). A literature review and a case study of sustainable supply chains with a focus on metrics. *International Journal of Production Economics*, *140*(1), 69–82. doi:10.1016/j.ijpe.2012.01.042

Hervani, A. A., Helms, M. M., & Sarkis, J. (2005). Performance measurement for green supply chain management. *Benchmarking. International Journal (Toronto, Ont.)*, *12*(4), 330–353.

Ji, P., Ma, X., & Li, G. (2014). Developing green purchasing relationships for the manufacturing industry: An evolutionary game theory perspective. *International Journal of Production Economics*.

Khatwani, G., Singh, S. P., Trivedi, A., & Chauhan, A. (2015). Fuzzy-TISM: A fuzzy extension of TISM for group decision making. *Global Journal of Flexible Systems Management*, *16*(1), 97–112. doi:10.100740171-014-0087-4

Khidir ElTayeb, T., Zailani, S., & Jayaraman, K. (2010). The examination on the drivers for green purchasing adoption among EMS 14001 certified companies in Malaysia. *Journal of Manufacturing Technology Management*, *21*(2), 206–225. doi:10.1108/17410381011014378

Mavi, R. K., Kazemi, S., Najafabadi, A. F., & Mousaabadi, H. B. (2013). Identification and assessment of logistical factors to evaluate a green supplier using the fuzzy logic DEMATEL method. *Polish Journal of Environmental Studies*, *22*(2), 445–455.

Min, H., & Galle, W. P. (1997). Green purchasing strategies: Trends and implications. *International Journal of Purchasing and Materials Management*, *33*(2), 10–17. doi:10.1111/j.1745-493X.1997.tb00026.x

Min, H., & Galle, W. P. (2001). Green purchasing practices of US firms. *International Journal of Operations & Production Management, 21*(9), 1222–1238. doi:10.1108/EUM0000000005923

Zhu, Q., Sarkis, J., Cordeiro, J. J., & Lai, K. H. (2008). Firm-level correlates of emergent green supply chain management practices in the Chinese context. *Omega, 36*(4), 577–591. doi:10.1016/j.omega.2006.11.009

ADDITIONAL READING

Berry, C. (2011). *The sustainable procurement guide: procuring sustainably using BS 8903*. BSi.

Theron, C., & Dowden, M. (2014). *Strategic Sustainable Procurement: Law and Best Practice for the Public and Private Sectors*. Do Sustainability.

KEY TERMS AND DEFINITIONS

Green Procurement: Green Procurement is procuring of products and services that cause minimal adverse environmental impacts. It incorporates human health, social and environmental concerns into the search for high quality products and services at competitive prices.

Total Interpretive Structural Modeling (TISM): Fuzzy TISM is an extension of Interpretive structural modeling (ISM) technique. ISM methodology, transforms unclear, poorly articulated models of systems into clear, well-defined models. ISM uses experts to judge the variables, and the relations among the variables are interpreted. ISM depends on the experts' knowledge and familiarity with the firm, its operations, and its industry. ISM generates deep knowledge of the subject and is greatly helpful for practitioners.

This research was previously published in the Encyclopedia of Information Science and Technology, Fourth Edition edited by Mehdi Khosrow-Pour, pages 3086-3102, copyright year 2018 by Information Science Reference (an imprint of IGI Global).

APPENDIX

Table 1. Aggregated SSIM matrix

	E 20	E 19	E 18	E 17	E 16	E 15	E 14	E 13	E 12	E 11	E 10	E 9	E 8	E 7	E 6	E 5	E 4	E 3	E 2	E 1
E 1	O(No)	V(VH)	V(VH)	O(No)	O(No)	O(No)	O(No)	O(No)	O(No)	V(VH)	V(VH)	V(L)	V(VH)	O(No)	O(No)	O(No)	O(No)	O(No)	V(VH)	
E 2	V(VH)	V(VH)	V(VH)	V(VH)	V(L)	V(H)	V(VH)	V(H)	V(H)	V(VH)	V(VH)	V(H)	V(VH)	O(No)	V(VH)	V(H)	O(No)	O(No)		
E 3	V(VH)	V(VH)	V(VH)	V(H)	V(L)	V(H)	V(H)	V(H)	V(H)	V(VH)	V(VH)	V(VH)	V(VH)	V(H)	V(VH)	V(VH)	V(H)			
E 4	V(VH)	V(VH)	V(VH)	V(H)	V(H)	V(L)	V(H)	V(L)	V(L)	V(H)	V(H)	V(L)	V(H)	V(L)	V(H)	V(H)				
E 5	V(VH)	V(H)	V(H)	V(L)	O(No)	O(No)	O(No)	O(No)	O(No)	V(H)	V(H)	O(No)	V(H)	O(No)	V(L)					
E 6	V(VH)	V(VH)	V(VH)	V(L)	V(L)	V(L)	V(H)	V(L)	V(H)	V(VH)	V(VH)	V(L)	V(VH)	O(No)						
E 7	V(VH)	V(VH)	V(VH)	V(VH)	V(L)	V(VH)	V(VH)	V(VH)	V(VH)	V(VH)	V(VH)	V(VH)	V(VH)							
E 8	V(VH)	V(VH)	V(VH)	V(VH)	V(VH)	V(VH)	V(VH)	V(H)	V(H)	V(VH)	V(VH)	V(H)								
E 9	V(H)	V(L)	V(L)	V(L)	V(L)	O(No)	V(H)	O(No)	O(No)	V(L)	V(L)									
E 10	V(H)	V(VH)	V(VH)	V(VH)	V(H)	V(H)	V(H)	V(H)	V(H)	V(VH)										
E 11	V(VH)	V(VH)	V(VH)	V(VH)	V(H)	V(H)	V(H)	V(H)	V(H)											
E 12	V(H)	V(H)	V(L)	O(No)	V(VH)	V(H)	V(VH)	V(VH)												
E 13	V(VH)	V(VH)	V(VH)	V(H)	V(H)	V(L)	V(VH)													
E 14	V(VH)	O(No)	O(No)	V(H)	A(VH)	A(VH)														
E 15	V(VH)	V(H)	V(H)	O(No)	V(VH)															
E 16	V(H)	V(L)	V(L)	V(L)																
E 17	A(VH)	A(H)	A(H)																	
E 18	X(VH)	X(VH)																		
E 19	X(VH)																			
E 20																				

Source: Author own compilation

Table 2. Fuzzy reachability matrix

	E1	E2	E3	E4	E5	E6	E7	E8	E9	E10	E11	E12	E13	E14	E15	E16	E17	E18	E19	E20
E 1	1	VH	No	No	No	No	No	VH	L	VH	VH	No	No	No	No	No	No	VH	VH	No
E 2	No	1	No	No	H	VH	No	VH	H	VH	VH	H	H	VH	H	L	VH	VH	VH	VH
E 3	No	No	1	H	VH	VH	H	VH	VH	VH	VH	H	H	H	H	L	H	VH	VH	VH
E 4	No	No	No	1	H	H	L	H	L	H	H	L	L	H	L	H	H	VH	VH	VH
E 5	No	No	No	No	1	L	No	H	No	H	H	No	No	No	No	No	L	H	H	VH
E 6	No	No	No	No	No	1	No	VH	L	VH	VH	H	L	H	L	L	L	VH	VH	VH
E 7	No	No	No	No	No	No	1	VH	VH	VH	VH	VH	VH	VH	VH	L	VH	VH	VH	VH
E 8	No	No	No	No	No	No	No	1	H	VH	VH	H	H	VH	VH	VH	VH	VH	VH	VH
E 9	No	No	No	No	No	No	No	No	1	L	L	No	No	H	No	L	L	L	L	H
E 10	No	No	No	No	No	No	No	No	No	1	VH	H	H	H	H	H	VH	VH	VH	H
E 11	No	No	No	No	No	No	No	No	No	No	1	H	H	H	H	H	VH	VH	VH	VH
E 12	No	No	No	No	No	No	No	No	No	No	No	1	VH	VH	H	VH	No	L	H	H
E 13	No	No	No	No	No	No	No	No	No	No	No	No	1	VH	L	H	H	VH	VH	VH
E 14	No	No	No	No	No	No	No	No	No	No	No	No	No	1	No	No	H	No	No	VH
E 15	No	No	No	No	No	No	No	No	No	No	No	No	No	VH	1	VH	No	H	H	VH
E 16	No	No	No	No	No	No	No	No	No	No	No	No	No	VH	No	1	L	L	L	H
E 17	No	No	No	No	No	No	No	No	No	No	No	No	No	No	No	No	1	No	No	No
E 18	No	No	No	No	No	No	No	No	No	No	No	No	No	No	No	No	H	1	VH	VH
E 19	No	No	No	No	No	No	No	No	No	No	No	No	No	No	No	No	H	VH	1	VH
E 20	No	No	No	No	No	No	No	No	No	No	No	No	No	No	No	No	VH	VH	VH	1

Source: Author own compilation

Table 3a. Final fuzzy reachability matrix

	E1	E2	E3	E4	E5	E6	E7	E8	E9	E10	E11
E 1	(1,1,1)	(0.75,1.0,1.0)	(0,0,0.25)	(0,0,0.25)	(0,0,0.25)	(0,0,0.25)	(0,0,0.25)	(0.75,1.0,1.0)	(0.25,0.5,0.75)	(0.75,1.0,1.0)	(0.75,1.0,1.0)
E 2	(0,0,0.25)	(1,1,1)	(0,0,0.25)	(0,0,0.25)	(0.5,0.75,1.0)	(0.75,1.0,1.0)	(0,0,0.25)	(0.75,1.0,1.0)	(0.5,0.75,1.0)	(0.75,1.0,1.0)	(0.75,1.0,1.0)
E 3	(0,0,0.25)	(0,0,0.25)	(1,1,1)	(0.5,0.75,1.0)	(0.75,1.0,1.0)	(0.75,1.0,1.0)	(0.5,0.75,1.0)	(0.75,1.0,1.0)	(0.75,1.0,1.0)	(0.75,1.0,1.0)	(0.75,1.0,1.0)
E 4	(0,0,0.25)	(0,0,0.25)	(0,0,0.25)	(1,1,1)	(0.5,0.75,1.0)	(0.5,0.75,1.0)	(0.25,0.5,0.75)	(0.5,0.75,1.0)	(0.25,0.5,0.75)	(0.5,0.75,1.0)	(0.5,0.75,1.0)
E 5	(0,0,0.25)	(0,0,0.25)	(0,0,0.25)	(0,0,0.25)	(1,1,1)	(0.25,0.5,0.75)	(0,0,0.25)	(0.5,0.75,1.0)	(0,0,0.25)	(0.5,0.75,1.0)	(0.5,0.75,1.0)
E 6	(0,0,0.25)	(0,0,0.25)	(0,0,0.25)	(0,0,0.25)	(0,0,0.25)	(1,1,1)	(0,0,0.25)	(0.75,1.0,1.0)	(0.25,0.5,0.75)	(0.75,1.0,1.0)	(0.75,1.0,1.0)
E 7	(0,0,0.25)	(0,0,0.25)	(0,0,0.25)	(0,0,0.25)	(0,0,0.25)	(0,0,0.25)	(1,1,1)	(0.75,1.0,1.0)	(0.75,1.0,1.0)	(0.75,1.0,1.0)	(0.75,1.0,1.0)
E 8	(0,0,0.25)	(0,0,0.25)	(0,0,0.25)	(0,0,0.25)	(0,0,0.25)	(0,0,0.25)	(0,0,0.25)	(1,1,1)	(0.5,0.75,1.0)	(0.75,1.0,1.0)	(0.75,1.0,1.0)
E 9	(0,0,0.25)	(0,0,0.25)	(0,0,0.25)	(0,0,0.25)	(0,0,0.25)	(0,0,0.25)	(0,0,0.25)	(0,0,0.25)	(1,1,1)	(0.25,0.5,0.75)	(0.25,0.5,0.75)
E 10	(0,0,0.25)	(0,0,0.25)	(0,0,0.25)	(0,0,0.25)	(0,0,0.25)	(0,0,0.25)	(0,0,0.25)	(0,0,0.25)	(0,0,0.25)	(1,1,1)	(0.75,1.0,1.0)
E 11	(0,0,0.25)	(0,0,0.25)	(0,0,0.25)	(0,0,0.25)	(0,0,0.25)	(0,0,0.25)	(0,0,0.25)	(0,0,0.25)	(0,0,0.25)	(0,0,0.25)	(1,1,1)
E 12	(0,0,0.25)	(0,0,0.25)	(0,0,0.25)	(0,0,0.25)	(0,0,0.25)	(0,0,0.25)	(0,0,0.25)	(0,0,0.25)	(0,0,0.25)	(0,0,0.25)	(0,0,0.25)
E 13	(0,0,0.25)	(0,0,0.25)	(0,0,0.25)	(0,0,0.25)	(0,0,0.25)	(0,0,0.25)	(0,0,0.25)	(0,0,0.25)	(0,0,0.25)	(0,0,0.25)	(0,0,0.25)
E 14	(0,0,0.25)	(0,0,0.25)	(0,0,0.25)	(0,0,0.25)	(0,0,0.25)	(0,0,0.25)	(0,0,0.25)	(0,0,0.25)	(0,0,0.25)	(0,0,0.25)	(0,0,0.25)
E 15	(0,0,0.25)	(0,0,0.25)	(0,0,0.25)	(0,0,0.25)	(0,0,0.25)	(0,0,0.25)	(0,0,0.25)	(0,0,0.25)	(0,0,0.25)	(0,0,0.25)	(0,0,0.25)
E 16	(0,0,0.25)	(0,0,0.25)	(0,0,0.25)	(0,0,0.25)	(0,0,0.25)	(0,0,0.25)	(0,0,0.25)	(0,0,0.25)	(0,0,0.25)	(0,0,0.25)	(0,0,0.25)
E 17	(0,0,0.25)	(0,0,0.25)	(0,0,0.25)	(0,0,0.25)	(0,0,0.25)	(0,0,0.25)	(0,0,0.25)	(0,0,0.25)	(0,0,0.25)	(0,0,0.25)	(0,0,0.25)
E 18	(0,0,0.25)	(0,0,0.25)	(0,0,0.25)	(0,0,0.25)	(0,0,0.25)	(0,0,0.25)	(0,0,0.25)	(0,0,0.25)	(0,0,0.25)	(0,0,0.25)	(0,0,0.25)
E 19	(0,0,0.25)	(0,0,0.25)	(0,0,0.25)	(0,0,0.25)	(0,0,0.25)	(0,0,0.25)	(0,0,0.25)	(0,0,0.25)	(0,0,0.25)	(0,0,0.25)	(0,0,0.25)
E 20	(0,0,0.25)	(0,0,0.25)	(0,0,0.25)	(0,0,0.25)	(0,0,0.25)	(0,0,0.25)	(0,0,0.25)	(0,0,0.25)	(0,0,0.25)	(0,0,0.25)	(0,0,0.25)
Dependence Power	(1,1,5.75)	(1.75,2,6.5)	(1,1,5.75)	(1.5,1.75,6.50)	(2 75,3.50,8)	(3.25,4.25,8.5)	(1.75,2.25,7)	(5.75,7.5,11)	(4.25,6,10.25)	(6.75,9,12.25)	(7.5,10,13)
Crisp Value	0.03	0.07	0.03	0.07	0.17	0.20	0.09	0.38	0.29	0.44	0.50

Source: Author own compilation

Table 3b. Final fuzzy reachability matrix

	E12	E13	E14	E15	E16	E17	E18	E19	E20	Driving Power	Crisp Value
E 1	(0,0,0.25)	(0,0,0.25)	(0,0,0.25)	(0,0,0.25)	(0,0,0.25)	(0,0,0.25)	(0.75,1.0,1.0)	(0.75,1.0,1.0)	(0,0,0.25)	(5.75,7.50,10.75)	1.71
E 2	(0.5,0.75,1.0)	(0.5,0.75,1.0)	(0.75,1.0,1.0)	(0.5,0.75,1.0)	(0.25,0.5,0.75)	(0.75,1.0,1.0)	(0.75,1.0,1.0)	(0.75,1.0,1.0)	(0.75,1.0,1.0)	(10.50,14.25,16.75)	0.70
E 3	(0.5,0.75,1.0)	(0.5,0.75,1.0)	(0.5,0.75,1.0)	(0.5,0.75,1.0)	(0.25,0.5,0.75)	(0.5,0.75,1.0)	(0.75,1.0,1.0)	(0.75,1.0,1.0)	(0.75,1.0,1.0)	(11.5,15.75,19.25)	0.76
E 4	(0.25,0.5,0.75)	(0.25,0.5,0.75)	(0.5,0.75,1.0)	(0.25,0.5,0.75)	(0.5,0.75,1.0)	(0.5,0.75,1.0)	(0.75,1.0,1.0)	(0.75,1.0,1.0)	(0.75,1.0,1.0)	(8.50,12.50,16.50)	0.69
E 5	(0,0,0.25)	(0,0,0.25)	(0,0,0.25)	(0,0,0.25)	(0,0,0.25)	(0.25,0.5,0.75)	(0.5,0.75,1.0)	(0.5,0.75,1.0)	(0.75,1.0,1.0)	(4.75,6.75,11.25)	0.92
E 6	(0.5,0.75,1.0)	(0.25,0.5,0.75)	(0.5,0.75,1.0)	(0.25,0.5,0.75)	(0.25,0.5,0.75)	(0.25,0.5,0.75)	(0.75,1.0,1.0)	(0.75,1.0,1.0)	(0.75,1.0,1.0)	(7.75,11,14.25)	0.53
E 7	(0.75,1.0,1.0)	(0.75,1.0,1.0)	(0.75,1.0,1.0)	(0.75,1.0,1.0)	(0.25,0.5,0.75)	(0.75,1.0,1.0)	(0.75,1.0,1.0)	(0.75,1.0,1.0)	(0.75,1.0,1.0)	(10.25,13.5,15.25)	0.63
E 8	(0.5,0.75,1.0)	(0.5,0.75,1.0)	(0.75,1.0,1.0)	(0.75,1.0,1.0)	(0.75,1.0,1.0)	(0.75,1.0,1.0)	(0.75,1.0,1.0)	(0.75,1.0,1.0)	(0.75,1.0,1.0)	(9.25,12.25,14.75)	0.59
E 9	(0,0,0.25)	(0,0,0.25)	(0.5,0.75,1.0)	(0,0,0.25)	(0.25,0.5,0.75)	(0.25,0.5,0.75)	(0.25,0.5,0.75)	(0.25,0.5,0.75)	(0.5,0.75,1.0)	(3.50,5.50,9.50)	0.26
E 10	(0.5,0.75,1.0)	(0.5,0.75,1.0)	(0.5,0.75,1.0)	(0.5,0.75,1.0)	(0.5,0.75,1.0)	(0.75,1.0,1.0)	(0.75,1.0,1.0)	(0.75,1.0,1.0)	(0.5,0.75,1.0)	(7,9.50,13.25)	0.46
E 11	(0.5,0.75,1.0)	(0.5,0.75,1.0)	(0.5,0.75,1.0)	(0.5,0.75,1.0)	(0.5,0.75,1.0)	(0.75,1.0,1.0)	(0.75,1.0,1.0)	(0.75,1.0,1.0)	(0.75,1.0,1.0)	(6.50,8.75,12.50)	0.43
E 12	(1,1,1)	(0.75,1.0,1.0)	(0.75,1.0,1.0)	(0.5,0.75,1.0)	(0.75,1.0,1.0)	(0,0,0.25)	(0.25,0.5,0.75)	(0.5,0.75,1.0)	(0.5,0.75,1.0)	(5,6.75,12.50)	0.35
E 13	(0,0,0.25)	(1,1,1)	(0.75,1.0,1.0)	(0.25,0.5,0.75)	(0.5,0.75,1.0)	(0.5,0.75,1.0)	(0.75,1.0,1.0)	(0.75,1.0,1.0)	(0.75,1.0,1.0)	(5.25,7,10.75)	0.33
E 14	(0,0,0.25)	(0,0,0.25)	(1,1,1)	(0,0,0.25)	(0,0,0.25)	(0.5,0.75,1.0)	(0,0,0.25)	(0,0,0.25)	(0.75,1.0,1.0)	(2.25,2.75,7.25)	0.11
E 15	(0,0,0.25)	(0,0,0.25)	(0.75,1.0,1.0)	(1,1,1)	(0.75,1.0,1.0)	(0,0,0.25)	(0.5,0.75,1.0)	(0.5,0.75,1.0)	(0.75,1.0,1.0)	(4.25,5.5,9.5)	0.26
E 16	(0,0,0.25)	(0,0,0.25)	(0.75,1.0,1.0)	(0,0,0.25)	(1,1,1)	(0.25,0.5,0.75)	(0.25,0.5,0.75)	(0.25,0.5,0.75)	(0.5,0.75,1.0)	(3,4.25,8.75)	0.19
E 17	(0,0,0.25)	(0,0,0.25)	(0,0,0.25)	(0,0,0.25)	(0,0,0.25)	(1,1,1)	(0,0,0.25)	(0,0,0.25)	(0,0,0.25)	(1,1,5.75)	0.20
E 18	(0,0,0.25)	(0,0,0.25)	(0,0,0.25)	(0,0,0.25)	(0,0,0.25)	(0.5,0.75,1.0)	(1,1,1)	(0.75,1.0,1.0)	(0.75,1.0,1.0)	(3,3.75,8)	0.18
E 19	(0,0,0.25)	(0,0,0.25)	(0,0,0.25)	(0,0,0.25)	(0,0,0.25)	(0.5,0.75,1.0)	(0.75,1.0,1.0)	(1,1,1)	(0.75,1.0,1.0)	(3,3.75,8)	0.18
E 20	(0,0,0.25)	(0,0,0.25)	(0,0,0.25)	(0,0,0.25)	(0,0,0.25)	(0.75,1.0,1.0)	(0.75,1.0,1.0)	(0.75,1.0,1.0)	(1,1,1)	(3.25,4,8)	0.18
Dependence Power	(5,7,11.50)	(5.5,7.75,12)	(9.25,12.5,15.50)	(6.25,8.25,12.50)	(6.5,9.5,13.50)	(9.5,13.5,16.75)	(11.75,16,17.75)	(12,16.25,18)	(12.75,17,18.5)		
Crisp Value	0.36	0.40	0.63	0.43	0.48	0.67	0.79	0.82	0.85		

Table 4. Defuzzified reachability matrix

	E1	E2	E3	E4	E5	E6	E7	E8	E9	E10	E11	E12	E13	E14	E15	E16	E17	E18	E19	E20
E 1	1	1	0	0	0	0	0	1	0	1	1	0	0	0	0	0	0	1	1	0
E 2	0	1	0	0	1	1	0	1	1	1	1	1	1	1	1	0	1	1	1	1
E 3	0	0	1	1	1	1	1	1	1	1	1	1	1	1	1	0	1	1	1	1
E 4	0	0	0	1	1	1	0	1	0	1	1	0	0	1	0	1	1	1	1	1
E 5	0	0	0	0	1	0	0	1	0	1	1	0	0	0	0	0	0	1	1	1
E 6	0	0	0	0	0	1	0	1	0	1	1	1	0	1	0	0	0	1	1	1
E 7	0	0	0	0	0	0	1	1	1	1	1	1	1	1	1	0	1	1	1	1
E 8	0	0	0	0	0	0	0	1	1	1	1	1	1	1	1	1	1	1	1	1
E 9	0	0	0	0	0	0	0	0	1	0	0	0	0	1	0	0	0	0	0	1
E 10	0	0	0	0	0	0	0	0	0	1	1	1	1	1	1	1	1	1	1	1
E 11	0	0	0	0	0	0	0	0	0	0	1	1	1	1	1	1	1	1	1	1
E 12	0	0	0	0	0	0	0	0	0	0	0	1	1	1	1	1	0	0	1	1
E 13	0	0	0	0	0	0	0	0	0	0	0	0	1	1	0	1	1	1	1	1
E 14	0	0	0	0	0	0	0	0	0	0	0	0	0	1	0	0	1	0	0	1
E 15	0	0	0	0	0	0	0	0	0	0	0	0	0	1	1	1	0	1	1	1
E 16	0	0	0	0	0	0	0	0	0	0	0	0	0	1	0	1	1	0	0	1
E 17	0	0	0	0	0	0	0	0	0	0	0	0	0	0	0	0	1	0	0	0
E 18	0	0	0	0	0	0	0	0	0	0	0	0	0	0	0	0	1	1	1	1
E 19	0	0	0	0	0	0	0	0	0	0	0	0	0	0	0	0	1	1	1	1
E 20	0	0	0	0	0	0	0	0	0	0	0	0	0	0	0	0	1	1	1	1

Source: Author own compilation

Table 5. Defuzzified reachability matrix and transitivity check

	E1	E2	E3	E4	E5	E6	E7	E8	E9	E10	E11	E12	E13	E14	E15	E16	E17	E18	E19	E20
E1	1	1	0	0	1*	1*	0	1	1*	1	1	1*	1*	1*	1*	1*	1*	1	1	1*
E2	0	1	0	0	1	1	1*	1	1	1	1	1	1	1	1	1*	1	1	1	1
E3	0	0	1	1	1	1	1	1	1	1	1	1	1	1	1	1*	1	1	1	1
E4	0	0	0	1	1	1	0	1	1*	1	1	1*	1*	1	1*	1	1	1	1	1
E5	0	0	0	0	1	0	0	1	1*	1	1	1*	1*	1*	1*	1*	1*	1	1	1
E6	0	0	0	0	0	1	0	1	1*	1	1	1	1*	1	1*	1*	1*	1	1	1
E7	0	0	0	0	0	0	1	1	1	1	1	1	1	1	1	1*	1	1	1	1
E8	0	0	0	0	0	0	0	1	1	1	1	1	1	1	1	1	1	1	1	1
E9	0	0	0	0	0	0	0	0	1	0	0	0	0	1	0	0	1*	1*	1*	1
E10	0	0	0	0	0	0	0	0	0	1	1	1	1	1	1	1	1	1	1	1
E11	0	0	0	0	0	0	0	0	0	0	1	1	1	1	1	1	1	1	1	1
E12	0	0	0	0	0	0	0	0	0	0	0	1	1	1	1	1	1*	1*	1	1
E13	0	0	0	0	0	0	0	0	0	0	0	0	1	1	0	1	1	1	1	1
E14	0	0	0	0	0	0	0	0	0	0	0	0	0	1	0	0	1	0	0	1
E15	0	0	0	0	0	0	0	0	0	0	0	0	0	1	1	1	1*	1	1	1
E16	0	0	0	0	0	0	0	0	0	0	0	0	0	1	0	1	1	1*	1*	1
E17	0	0	0	0	0	0	0	0	0	0	0	0	0	0	0	0	1	0	0	0
E18	0	0	0	0	0	0	0	0	0	0	0	0	0	0	0	0	1	1	1	1
E19	0	0	0	0	0	0	0	0	0	0	0	0	0	0	0	0	1	1	1	1
E20	0	0	0	0	0	0	0	0	0	0	0	0	0	0	0	0	1	1	1	1

Source: Author own compilation

Table 6. Level partitioning

Elements	Reachability Set	Antecedent Set	Intersection Set	Level
E1	1,2,5,6,8,9,10,11,12,13,14,15,16,17,18,19,20	1	1	XII
E2	2,5,6,7,8,9,10,11,12,13,14,15,16,17,18,19,20	1,2	2	XI
E3	3,4,5,6,7,8,9,10,11,12,13,14,15,16,17,18,19,20	3	3	XII
E4	4,5,6,8,9,10,11,12,13,14,15,16,17,18,19,20	3,4	4	XI
E5	5,8,9,10,11,12,13,14,15,16,17,18,19,20	1,2,3,4,5	5	X
E6	6,8,9,10,11,12,13,14,15,16,17,18,19,20	1,2,3,4,6	6	X
E7	7,8,9,10,11,12,13,14,15,16,17,18,19,20	2,3,7	7	X
E8	8,9,10,11,12,13,14,15,16,17,18,19,20	1,2,3,4,5,6,7,8	8	IX
E9	9,14,17,18,19,20	1,2,3,4,5,6,7,8,9	9	IV
E10	10,11,12,13,14,15,16,17,18,19,20	1,2,3,4,5,6,7,8,10	10	VIII
E11	11,12,13,14,15,16,17,18,19,20	1,2,3,4,5,6,7,8,10,11	11	VII
E12	12,13,14,15,16,17,18,19,20	1,2,3,4,5,6,7,8,10,11,12	12	VI
E13	13,14,16,17,18,19,20	1,2,3,4,5,6,7,8,10,11,12,13	13	V
E14	14,17,20	1,2,3,4,5,6,7,8,9,10,11,12,13,14,15,16	14	III
E15	14,15,16,17,18,19,20	1,2,3,4,5,6,7,8,10,11,12,15	15	V
E16	14,16,17,18,19,20	1,2,3,4,5,6,7,8,10,11,12,13,15,16	16	IV
E17	17	1,2,3,4,5,6,7,8,9,10,11,12,13,14,15,16,17,18,19,20	17	I
E18	17,18,19,20	1,2,3,4,5,6,7,8,9,10,11,12,13,15,16,18,19,20	18,19,20	II
E19	17,18,19,20	1,2,3,4,5,6,7,8,9,10,11,12,13,15,16,18,19,20	18,19,20	II
E20	17,18,19,20	1,2,3,4,5,6,7,8,9,10,11,12,13,14,15,16,18,19,20	18,19,20	II

Source: Author own compilation

Chapter 14
Load Flow Analysis in Smart Grids

Osman Hasan
National University of Sciences and Technology, Pakistan

Awais Mahmood
National University of Sciences and Technology, Pakistan

Syed Rafay Hasan
Tennessee Technological University, USA

ABSTRACT

Load flow analysis is widely used to estimate the flow of various electrical parameters such as the voltage, current, and power in power grids. These estimates allow us to effectively and reliably manage the given grid under random and uncertain conditions. Given the enormous amount of randomness and uncertainties in the factors that affect the smart grids, compared to traditional power grids, a complete and rigorous load flow analysis holds a vital role in ensuring the reliability of this safety-critical domain. In this chapter, the authors describe smart grids in terms of their basic components and then categorize the factors that affect the loads in smart grids. This is followed by a comprehensive survey of various existing load flow analysis techniques (i.e., numerical, computational intelligence, and probabilistic).

INTRODUCTION

With 19320 TW-hr/yr consumption of electrical energy in the entire world nowadays, the traditional unidirectional power transmission grids are struggling to survive as the number of fluctuations, blackouts and outages is tremendously growing since the last decade (Gao et al., 2012). More reliable and safe distribution networks have become a dire requirement due to the safety and financial-critical nature of electricity these days. For example, a blackout per minute across Silicon Valley costs 75 million and 1 million dollars for Sun Microsystems alone. There are numerous environmental concerns with the present-age power generation methods as well since these methods are largely dependent on fossil fuels, which result in global warming and carbon-dioxide emissions. For example, the United States power system

DOI: 10.4018/978-1-5225-7359-3.ch014

alone is responsible for 40 percent of carbon emission nationwide (Hledik, 2009). Thus renewable energy resources, like solar and wind based solutions, are extensively being advocated throughout the world but the traditional grid does not facilitate their integration in the national grids. Moreover, the traditional power grids are not very efficient in terms of distribution loss management as well. For example, about 17 percent of electrical energy generated in the year 2011 by Pakistan was wasted in distribution systems. Similarly, the problem of electricity theft is also a growing concern in traditional grids.

Smart grids can overcome the above mentioned shortcomings by providing an alternative electric power transmission framework that comprises of Intelligence based Electronic Devices (IED) (Momoh, 2012) for detecting and correcting faults, and advanced metering infrastructure (AMI), to facilitate the integration of multiple renewable energy sources. Some of the distinguishing characteristics of smart girds compared to traditional power grids include:

- **Safety and Reliability:** Smart grids can predict unforeseen situations and autonomously react accordingly to prevent them (e.g., isolating the faulty component of the grid from the entire system (Farhangi, 2010)) and hence improve the safety and reliability (Moslehi and Kumar, 2010) of power distribution and save millions of dollars.
- **Cost-Effectiveness:** Smart grids provide real-time tariff information to the consumers so that they can manage their loads to save energy and costs (Li et al., 2010).
- **Efficiency:** Smart grids allow optional usage of the assets to maximize the efficiency of the grid and thus can have a major performance impact. For example, according to the US Department of Energy (DOE), just a 5% increase in grid efficiency can have the same impact as if fuel and greenhouse gas emissions are eliminated from 53 million cars.
- **Security:** Smart grids allow more secure electrical networks, by using tools like smart meters, and thus electricity theft can be minimized (Khurana et al., 2010, Metke and Ekl, 2010).
- **Environmental Friendliness:** Smart grid allows the integration of environmental friendly generation methods and is inline with the recent advancements in renewable energy research (RER) (Ipakchi and Albuyeh, 2009).

Based on above-mentioned capabilities, the National Academy of Engineering listed "electrification as made possible by the grid" as the most significant engineering achievement of the 20th Century.

Due to the inherent randomness of smart grids, including variable loads, peak consumption times and renewable energy sources with generation capacity depending on varying weather conditions, there is a lot of interest in rigorously analyzing the voltages and load profiles for resilient and effective power delivery to the users. Besides providing means for effectively managing the energy distribution, these profiles can be used by the consumers to change their loads by a smart device from anywhere as per their requirements. Load flow analysis (Van Benthem and Doets, 2001) fulfills the above-mentioned requirements and allows us to find the magnitude and phase angle of the voltage and the real and reactive power flowing in each bus of the smart grid and the optimal parameters for various components, like inductors, conductors, transformers, and shunt capacitors. It also provides statistics about the behavior of the system during on-peak and off-peak loads in order to identify and plan the contingencies. Moreover, load flow studies help us in conducting short-circuit fault analysis and in finding the stability and the steady-state operating state of an electric power system by calculating the voltage drop on each feeder, the power flow in all branches and feeder circuits, X/R ratio in line impedances and the voltage at each bus. Finally, load flow studies can determine if system voltages remain within the given specifications and if any of the expensive equipment of the grid is overloaded. The results of load flow analysis are used to make key decisions and ensure a safe and reliable power distribution.

There are various uncertain and random elements associated with the load consumption in smart grids. For example, the usage of consumer appliances depends on weather conditions and the time of the day. The distributed generation and usage of storage cells also plays a key role in varying the electrical demand. Some of the key factors that influence the loads in smart grids and must be taken into consideration for load flow analysis of smart grids include weather conditions, time-of-day, arbitrary disturbances, electricity prices, demand response, storage cells and electric vehicles.

BACKGROUND

There are various uncertain and random elements associated with the load consumption in smart grids. For example, the usage of consumer appliances depends on weather conditions and the time of the day. The distributed generation and usage of storage cells also plays a key role in varying the electrical demand. The smart grid components may fail randomly and either self-repair or need manual repair to restore their operation. Some components may also have back-up protection. Similarly, the influence of electricity prices on the energy demand cannot be neglected as higher prices usually result in the reduction of energy consumption. Moreover, in smart grids, the consumers are more cautious about costs since they can get the real-time tariffs using smart meters. Time-of-Use (TOU) pricing scheme, which offers low off peak rates, encourages consumers to shift their loads to off peak hours. Moreover, electric vehicles (EVs) also greatly influence load profiles since their charging consumes a significant amount of energy and thus is recommended to be done in the off peak times.

The load flow analysis mainly involves studying the behavior of node voltages and the power entering and leaving the nodes in a smart grid while considering the above-mentioned elements of randomness and uncertainty. In this section, we describe some of the key techniques that have been used for load flow analysis.

Numerical Methods

Numerical methods are one of most widely used analysis methods for load flow. Primarily, these methods are used to solve the following nodal equation:

$$I = V.Ybus \tag{1}$$

where I is the N vector of current in each bus, V is the N vector bus voltages and Ybus is the admittance bus matrix. Iterative numerical methods can solve the Equation (1), however, they generate approximate results mainly because the precision of the results is directly proportional to the number of iterations. Some of the frequently used numerical methods for load flow analysis are described in detail below:

Gauss-Seidel Method

The Gauss-Seidel method (Chapra Steven and Canale Raymond, 2008) is a widely used iterative approach for solving linear equations based on an initial guess. In load flow analysis, we develop an admittance matrix using the Kirchhoffs current law and then an iterative method is used to solve the scheduled reactive and real power at each bus, respectively. The final node voltage is determined in an iterative manner by using intermediate values of node voltages obtained after each iteration. This process converges linearly if the initial estimation is close to the unknown value. A variant of modified Gauss-Seidel method has been used for power flow calculations in (Teng, 2002). Gauss-Seidel method shows slower rates of convergence but its main strength is that it does not need to solve a complex matrix system. Gauss-Seidel method is not suitable for radial distribution systems where there are branch connections between a large set of surrounding buses (Momoh, 2012).

Newton-Raphson Method

The Newton-Raphson (NR) method is another iterative method to solve non-linear load flow equations. The sensitivity matrix is determined from inversing the Jacobian matrix, which consists of the injected power equations. The NR method has been used to solve three phase power flow equations in (Le Nguyen, 1997). This method is very useful for large systems. But it is computationally inefficient because it does not take advantage of the radial distribution systems (Momoh, 2012). Moreover, the method also fails if the Jacobian matrix is singular. Finally, in the case of a low X/R ratio value, the NR method becomes ill conditioned (Momoh, 2012). On the other hand, it offers a fast convergence rate.

Fast Decouple Method

The Fast Decouple method is one of the most effective techniques used in power system analysis and design. Just like the Newton-Raphson method, it also utilizes the Jacobian matrix. On contrary, in this technique small angle approximation is used to calculate the relatively smaller elements of the Jacobian matrix. However, this method shows poor convergence for low values of the X/R ratio (Iwamoto and Tamura, 1981). This method has been successfully used for three phase power flow studies in (Zimmerman and Chiang, 1995).

Besides using the above mentioned mainstream numerical methods for load flow analysis, the other significant contributions in this direction include using the interval arithmetic (Wang and Alvarado, 1992), the holomorphic embedded load flow method (Trias, 2012) and an optimal multiplier method for ill conditioned systems (Iwamoto and Tamura, 1981). Other significant methods include Zero-mismatch method, Ward and Hale method, Glimm and Stagg method and Secondary adjustment method for load flow analysis and more details about them are discussed in (Stott, 1974).

Simulation Methods

The main idea behind simulation is to construct a computer-based model of the given system and then analyze the desired load flow properties by observing the behavior of the model under different test cases. A number of dedicated simulations packages for analyzing power distribution systems are available and have been used to analyze the following key aspects (Glover et al., 2011):

- Arc flash hazard and fault analysis,
- Circuit breaker duty,
- Demand management,

- Distribution reliability evaluation,
- Power factor correction,
- Power loss computations,
- Components sizing,
- Voltages/VAR optimization,
- Power quality and reliability,
- Harmonic analysis, and
- Fault detection.

Simulation is a very user-friendly analysis approach since the analysis requires test pattern generation only. On the down side, simulation cannot guarantee absolute correctness of analysis because there is always a possibility that a corner case is missed in test patterns used for analysis. Some of the common load flow software are explained below:

Electrical Transient Analyzer Program

The Electrical Transient Analyzer Program (ETAP) simulator, developed by Operation Technology Inc., is one of the most widely used load flow analysis software. It allows the user to utilize the built-in templates to quickly construct a model of the entire electrical network. The ETAP simulator can be used to analyze integrated power systems (Khan et al., 2009) and has an ability to track up to 10 million load items. In (Sedighizadeh and Rezazadeh, 2008), a genetic algorithm is used for optimal allocation of distributed generation for improved voltage profile and the correctness of results is evaluated by ETAP. The power flow analysis package in ETAP provides both Newton-Raphson and Accelerated Gauss-Seidel method to solve power flow equations. ETAP offers two methods to calculate the X/R ratio. The first method finds the equivalent resistance and reactance of the entire system, to get a single value X/R ratio for a given location. In the second method, individual branch current contributions (each with a separate X/R ratio) are combined into a single X/R ratio. ETAP has also been used to analyze different distribution system models to minimize their power losses (Ramesh et al., 2009). The developers of ETAP claim that ETAP can also serve as a complete smart grid analysis tool.

GridLAB-D

GridLAB-D is a recently developed open-source power system modeling and simulation tool by the Pacific Northwest National Laboratory (PNNL) of the US DOE. It offers distributed energy resource modeling, integration of transmission and distribution systems, SCADA and metering models. An interesting feature is its external links to MySQL, MATLAB, MS-EXCEL and MS-ACCESS (Schneider et al., 2009). This tool offers timing models ranging from a few seconds to decades. GridLAB-D gives a simulation environment that can be incorporated with a variety of data management and analysis tools. It divides the power flow problem in two parts, i.e., 1) transmission and 2) distribution. It uses Gauss-Seidel iterative method to solve power flows at the transmission side while Forward and Backward Sweep (FBS) method to solve power flow problems at the distribution side (Schneider et al., 2009). Instead of the FBS method, the newer versions of GridLAB-D utilize the Gauss Seidel Three Phase Current Injection method due to the inability of FBS to handle networked distributions (Schneider et al., 2009).

Use of Matlab for Load Flow Simulations

Although MATLAB, which is a high level language for doing intense numerical computations and programming, is not developed particularly for load flow analysis but it can be used for that purpose. Given the general-purpose nature of MATLAB, it provides a very flexible environment for load flow analysis. For example, the NR method is programmed to solve a 5-bus system on MATLAB (Mallick and Hota, 2015). The Power System Analysis Toolbox (PSAT), which is an open source MATLAB toolbox, has been used for power flow calculations (Milano, 2005).

Power World Simulator

The Power World simulator is a commercial grade power system analysis and simulation package developed by the Power World Corporation. It is designed to simulate high voltage power system operations on a time frame ranging from several minutes to several days. The software can solve up to 100,000 buses with high efficiency. It extensively uses state-of-the-art graphics for better and easy understanding. The simulator is mainly based on the Newton-Raphson method for iteratively solving non-linear equations for power flows. Load flow analysis by fast decouple method is also an option available in this simulator. It uses three nested loops to solve power flows.

Simulation and testing are the state-of-the-art analysis techniques; however, as we have seen that they also use the numerical methods for their computations and thus cannot guarantee accuracy. Moreover, the main idea behind simulation and testing methods is to approximate a solution to a query by observing a subset of each probable run. Hence, it is possible that a system bug may not be detected during the simulation-based analysis. Moreover, system models, used in simulation cannot capture the true random behaviors, such as frequent changes in renewable energy generation, variations in network configurations and the peak loads, which are very frequently encountered in smart grids. In most cases, simulation based methods rely on pseudo random number generation methods for modeling these elements of randomness and the reliability of the analysis is dependent on the quality of these random number generators.

Computational Intelligence

Computational Intelligence (CI) methods provide a very efficient alternative for verifying and analyzing complex systems that exhibit random behavior. Thus, the grid can be controlled more reliably and more rapidly than humans by multi-variable nonlinear optimal controller based on CI, without requiring a mathematical model of the grid.

CI techniques primarily consist of Artificial Neural Networks (ANNs) (Haykin, 1994), evolutionary computation (Back et al., 1997), non-linear programming (Sasson, 1969) and fuzzy logic systems (Takagi and Sugeno, 1985). ANN based load forecaster is one of the most successful applications of CI for predicting load flows in an uncertain environment. Evolutionary computational techniques tend to solve combinatorial optimization techniques by learning and adapting to new situations and are primarily based on Genetic Algorithm (GA), particle swarm optimization (PSO) and ant colony optimization methods (Vlachogiannis et al., 2005). GAs have been used to analyze the Reactive power (VAR) with real-time operation that contains randomness and uncertainty (Bakirtzis et al., 2002). PSO (del Valle et al., 2008) uses simple mechanism that mimics social behavior of bird flocking and fish schooling to guide the particles search for globally optimal solution. In (Miranda and Saraiva, 1991), fuzzy modeling is used for optimal load flow. Fuzzy logic is also used for developing a unified power flow controller for damping the power system oscillations (Eldamaty et al., 2005). Linear, nonlinear, dynamic and integer programming methods have also been used for load flow analysis (Momoh, 2012). The above-mentioned CI based techniques have also been used to analyze various power systems (Saxena et al., 2010) including maintenance scheduling, long-term system expansion and planning and load forecasting. A very comprehensive overview about CI techniques and the advantages and disadvantages of Genetic Algorithm (GA), Simulated Annealing (SA), Artificial Neural Networks (ANNs), Expert Systems (ES) and few other techniques for analyzing power distribution systems is presented in (Saxena et al., 2010).

CI techniques of linear programming, non-linear programming, quadratic programming and Newton based techniques (Sun et al., 1984) have also been successfully used to solve various Optimal Power Flow (OPF) problems (Habibollahzadeh et al., 1989, Aoki et al., 1987). These techniques have certain drawbacks as mentioned in (Abido, 2002), like, non-linear programming has convergence problems and it is complex, quadratic programming techniques have problems with cost approximation. Newton based techniques may have convergence failure due to inappropriate initial conditions, hence, they are sensitive to initial conditions. In (Abido, 2002), a particle swarm optimization technique is proposed to do OPF.

Probabilistic Load Flow

Given the large number of uncertainties involved in the load flow analysis of smart grids, probabilistic analysis methods have also been used in this domain. Probabilistic Load Flow (PLF) is primarily done by using Monte Carlo (MC) simulations and the convolution method through Fast Fourier Transforms (FFT). These MC simulations are very time consuming since a large number of samples are usually used. An extended form of PLF that supports non-linear load flow equations is presented in (Allan et al., 1976) based on numerical methods. Another technique, known as Stochastic Load Flow (SLF) (Vorsic et al., 1991), is based on the assumption that the states of the system and power flow outputs are normally distributed. SLF deals with short time uncertainties and is only effective for analyzing system operation and thus the reliability of this method has been questioned (Anders, 1989). PLF has been used in (Allan et al., 1974) for analyzing power flows where the nodal loads and generation are defined as random variables and power flow is computed as a probability density function.

Multi-linear simulation algorithm has demonstrated better results than the MC simulation based methods (Da Silva and Arienti, 1990). Similarly, the method of combined cumulants and Gram-Charlier expansion for probabilistic load flow computation (Zhang and Lee, 2004) has reported a significantly improved performance. With this method, the probability density function of the transmission line flows is obtained. It determines the effect of prolonged uncertainty of transmission network. It also provides a new way of calculating probability density function which requires reduced storage, is applicable to larger systems and is faster than Monte Carlo simulations. It is useful in system expansion planning. Moreover, it also ensures better approximation of the cumulative distribution function curve. This method was practically demonstrated on a WSCC (Western Systems Coordinating Council) test system, which consists of 179 buses and 263 lines. The probability of any overloaded line can be easily computed through this method. The probabilistic load flow analysis using an algorithm based on the point estimate method is given in (Su, 2005). It is based on an assumption that the uncertainties of line parameters and injections in the bus can be measured. The method allows any deterministic load-flow program to be used. In order to calculate the statistical moments of load flow solution distribution for a system having m uncertain parameters, 2m calculations of load flow are used and the value of the solution is weighted at 2m locations. These moments are then used in probability density fittings. This method has been tested on several IEEE test systems and it is verified that performance of point estimate method is better than Monte Carlo simulation and it also requires less number of computations. A detailed comparison between the cumulants method and the point estimate method is presented in (Li and Zhang, 2009) with respect to the load flow analysis.

Analyzing random branch outages in load flow is another area where probabilistic load flow is considerably used since conventional load flow methods cannot cater such problems (Hu and Wang, 2006). PLF allows us to solve discrete distribution part of each state and output variable. The effect of branch outage is greater on system state than uncertainties caused by nodal power injections. For simplifying the convolution of random variables, moments and cumulants are used. Branch outages are simulated by injecting the virtual power to the related nodes. The resulting distribution is found by convolving continuous and discrete distributions. The Dynamic Stochastic Optimal Power Flow (DSOPF) (Momoh, 2009), which is based on Adaptive Dynamic Programming (ADP) technique, is another stochastic method for PLF. ADP tends to cope with the complex power system problems, which can be predicted under uncertainty conditions, and it is useful where there is not enough prior knowledge. These tools usually ensure robustness, scalability, stochasticity, predictivity, adaptability and acquisition of instantaneous data.

FUTURE RESEARCH DIRECTIONS

All the above-mentioned techniques are found to be quite scalable and user- friendly but cannot guarantee the accuracy of the analysis results. The main reasons behind the inaccuracies in the result include the usage of computer-arithmetic based models, which contain round-off errors, and the sampling based nature of the analysis, i.e., the models are analyzed for a subset of all possible scenarios due to limited computational resources. Given the safety-critical nature of smart-grids, the accuracy of load flow analysis results is the most desirable feature, since an undetected fault in the smart grid system can have major impact. For example, the analysis inaccuracy limitations have been reported as the main causes behind the 2003 Northeast blackout in the United States and Canada (Poulsen, 2004a), (Poulsen, 2004b) which approximately affected 55 Million people. As a complementary approach, formal methods (Abrial, 2009) which are computer based mathematical analysis tools, can be used to overcome the inaccuracy limitations in the domain of load flow analysis. However, to the best of our knowledge, no prior work regarding the formal load flow analysis exists so far. In order to fill this gap, we recommend to use probabilistic model checking (J. Rutten and Parker, 2004), which is a widely used formal method for analyzing Markovian models, to ensure accurate results of load flow analysis of smart grids.

CONCLUSION

Load flow analysis plays a vital role in safe and effective working of the smart grid system and a number of analysis methods have been used in this domain. This paper presents a brief overview about smart grids and the main factors affecting the loads in smart grids. This information can be utilized to understand the random and uncertain components in smart grids and thus model them appropriately in their load flow analysis, which can be conducted using Numerical methods, Simulation methods, Computational intelligence and probabilistic load flow methods. Each of these analysis methods have their own advantages and disadvantages and they have been highlighted in this chapter.

REFERENCES

Abido, M. (2002). Optimal power flow using particle swarm optimization. *Electrical Power & Energy Systems*, *24*(7), 563–571. doi:10.1016/S0142-0615(01)00067-9

Abrial, J. (2009). Faultless systems: Yes we can! *Computer*, *42*(9), 30–36. doi:10.1109/MC.2009.283

Allan, R., Borkowska, B., & Grigg, C. (1974). Probabilistic analysis of power flows. Electrical Engineers. *Proceedings of the Institution of*, *121*(12), 1551–1556.

Allan, R., Grigg, C., & Al-Shakarchi, M. (1976). Numerical techniques in probabilistic load flow problems. *International Journal for Numerical Methods in Engineering*, *10*(4), 853–860. doi:10.1002/nme.1620100412

Anders, G. J. (1989). *Probability concepts in electric power systems*. Academic Press.

Aoki, K., Nishikori, A., & Yokoyama, R. (1987). Constrained load flow using recursive quadratic programming. Power Systems. *IEEE Transactions on*, *2*(1), 8–16. doi:10.1109/T-ED.1987.22880

Back, T., Hammel, U., & Schwefel, H.-P. (1997). Evolutionary computation: Comments on the history and current state. Evolutionary computation. *IEEE Transactions on*, *1*(1), 3–17.

Bakirtzis, A. G., Biskas, P. N., Zoumas, C. E., & Petridis, V. (2002). Optimal power flow by enhanced genetic algorithm. Power Systems. *IEEE Transactions on*, *17*(2), 229–236.

Benysek, G., Kazmierkowski, M., Popczyk, J., & Strzelecki, R. (2011). Power electronic systems as a crucial part of smart grid infrastructure-a survey. *Technical Sciences*, *59*(4), 455–473.

Chapra Steven, C., & Canale Raymond, P. (2008). *Numerical methods for engineers*. McGraw-Hill Pub.

Da Silva, A. L., & Arienti, V. (1990). Probabilistic load flow by a multilinear simulation algorithm. Generation. *Transmission and Distribution*, *137*(4), 276–282. doi:10.1049/ip-c.1990.0037

del Valle, Y., Venayagamoorthy, G. K., Mohagheghi, S., Hernandez, J.-C., & Harley, R. G. (2008). Particle swarm optimization: Basic concepts, variants and applications in power systems. *Evolutionary Computation*, *12*(2), 171–195. doi:10.1109/TEVC.2007.896686

Eldamaty, A., Faried, S. O., & Aboreshaid, S. (2005). Damping power system oscillations using a fuzzy logic based unified power flow controller. In Electrical and Computer Engineering, (pp. 1950–1953). doi:10.1109/CCECE.2005.1557364

Farhangi, H. (2010). The path of the smart grid. *Power and Energy Magazine, IEEE*, *8*(1), 18–28. doi:10.1109/MPE.2009.934876

Gao, J., Xiao, Y., Liu, J., Liang, W., & Chen, C. (2012). A survey of communication/networking in smart grids. *Future Generation Computer Systems*, *28*(2), 391–404. doi:10.1016/j.future.2011.04.014

Glover, J. D., Sarma, M. S., & Overbye, T. J. (2011). *Power system analysis and design*. CengageBrain.com.

Habibollahzadeh, H., Luo, G.-X., & Semlyen, A. (1989). Hydrothermal optimal power flow based on a combined linear and nonlinear programming methodology. *Power Systems*, *4*(2), 530–537. doi:10.1109/59.193826

Haykin, S. (1994). *Neural networks: a comprehensive foundation*. Prentice Hall.

Hledik, R. (2009). How green is the smart grid? *The Electricity Journal*, *22*(3), 29–41. doi:10.1016/j.tej.2009.03.001

Hu, Z., & Wang, X. (2006). A probabilistic load flow method considering branch outages. *Power Systems*, *21*(2), 507–514. doi:10.1109/TPWRS.2006.873118

Ipakchi, A., & Albuyeh, F. (2009). Grid of the future. *Power and Energy Magazine, IEEE*, *7*(2), 52–62. doi:10.1109/MPE.2008.931384

Iwamoto, S., & Tamura, Y. (1981). A load flow calculation method for ill-conditioned power systems. *Power Apparatus and Systems*, (4), 1736–1743.

Khan, R. A. J., Junaid, M., & Asgher, M. M. (2009). Analyses and monitoring of 132 kv grid using etap software. In *Electrical and Electronics Engineering* (p. I–113). IEEE.

Khurana, H., Hadley, M., Lu, N., & Frincke, D. A. (2010). Smart-grid security issues. *Security & Privacy, IEEE*, *8*(1), 81–85. doi:10.1109/MSP.2010.49

Le Nguyen, H. (1997). Newton-raphson method in complex form power system load flow analysis. *Power Systems*, *12*(3), 1355–1359. doi:10.1109/59.630481

Li, F., Qiao, W., Sun, H., Wan, H., Wang, J., Xia, Y., ... Zhang, P. (2010). Smart transmission grid: Vision and framework. *Smart Grid*, *1*(2), 168–177. doi:10.1109/TSG.2010.2053726

Li, G., & Zhang, X.-P. (2009). Comparison between two probabilistic load flow methods for reliability assessment. In Power & Energy Society General Meeting, (pp. 1-7). IEEE. doi:10.1109/PES.2009.5275534

Mallick, A. K., & Hota, B. S. (2015). *Load flow study in power system*. Retrieved from http://ethesis.nitrkl.ac.in/2464/1/load flow in power system.pdf

Metke, A. R., & Ekl, R. L. (2010). Security technology for smart grid networks. *Smart Grid*, *1*(1), 99–107. doi:10.1109/TSG.2010.2046347

Milano, F. (2005). An open source power system analysis toolbox. *Power Systems*, *20*(3), 1199–1206. doi:10.1109/TPWRS.2005.851911

Miranda, V., & Saraiva, J. (1991). Fuzzy modelling of power system optimal load flow. In *Power Industry Computer Application* (pp. 386–392). IEEE; doi:10.1109/PICA.1991.160606

Momoh, J. (2012). *Smart grid: fundamentals of design and analysis* (Vol. 63). Wiley; doi:10.1002/9781118156117

Momoh, J. A. (2009). Smart grid design for efficient and flexible power networks operation and control. In *Power Systems Conference and Exposition* (pp. 1–8). IEEE; doi:10.1109/PSCE.2009.4840074

Moslehi, K., & Kumar, R. (2010). A reliability perspective of the smart grid. *Smart Grid, 1*(1), 57–64. doi:10.1109/TSG.2010.2046346

Poulsen, K. (2004a). *Software bug contributed to blackout.* Retrieved from http://www.securityfocus.com/news/8016

Poulsen, K. (2004b). *Tracking the blackout bug*. Retrieved from http://www.securityfocus.com/news/8412

Ramesh, L., Chowdhury, S., Chowdhury, S., Natarajan, A., & Gaunt, C. (2009). Minimization of power loss in distribution networks by different techniques. *Energy and Power Engineering, 2*(1).

Rutten, J., Kwiatkowska, M. G. N., & Parker, D. (2004). *Mathematical techniques for analyzing concurrent and probabilistic systems*. American Mathematical Society.

Sasson, A. M. (1969). Nonlinear programming solutions for load-flow, minimum-loss, and economic dispatching problems. *Power Apparatus and Systems*, (4), 399–409.

Saxena, D., Singh, S., & Verma, K. (2010). Application of computational intelligence in emerging power systems. Engineering. *Science and Technology*, *2*(3), 1–7.

Schneider, K. P., Chassin, D., Chen, Y., & Fuller, J. C. (2009). Distribution power flow for smart grid technologies. In Power Systems Conference and Exposition, 2009. PSCE'09. IEEE/PES, (pp. 1-7). IEEE. doi:10.1109/PSCE.2009.4840078

Sedighizadeh, M., & Rezazadeh, A. (2008). Using genetic algorithm for distributed generation allocation to reduce losses and improve voltage profile. World Academy of Science. *Engineering and Technology, 37*, 251–256.

Stott, B. (1974). Review of load-flow calculation methods. *Proceedings of the IEEE, 62*(7), 916–929. doi:10.1109/PROC.1974.9544

Su, C.-L. (2005). Probabilistic load-flow computation using point estimate method. *Power Systems, 20*(4), 1843–1851. doi:10.1109/TPWRS.2005.857921

Sun, D. I., Ashley, B., Brewer, B., Hughes, A., & Tinney, W. F. (1984). Optimal power flow by newton approach. *Power Apparatus and Systems*, (10), 2864–2880.

Takagi, T., & Sugeno, M. (1985). Fuzzy identification of systems and its applications to modeling and control. *Systems, Man and Cybernetics*, (1), 116–132.

Teng, J.-H. (2002). A modified Gauss–Seidel algorithm of three-phase power flow analysis in distribution networks. *Electrical Power & Energy Systems, 24*(2), 97–102.

Trias, A. (2012). The holomorphic embedding load flow method. In *Power and Energy Society General Meeting*, (pp. 1–8). IEEE.

Van Benthem, J., & Doets, K. (2001). Higher-order-logic. In *Handbook of Philosophical Logic* (pp. 189–243). Springer.

Vlachogiannis, J. G., Hatziargyriou, N. D., & Lee, K. Y. (2005). Ant colony system-based algorithm for constrained load flow problem. *Power Systems*, *20*(3), 1241–1249. doi:10.1109/TPWRS.2005.851969

Vorsic, J., Muzek, V., & Skerbinek, G. (1991). Stochastic load flow analysis. In *Electrotechnical Conference*, (pp. 1445–1448). IEEE.

Wang, Z., & Alvarado, F. L. (1992). Interval arithmetic in power flow analysis. *Power Systems*, *7*(3), 1341–1349. doi:10.1109/59.207353

Zhang, P., & Lee, S. T. (2004). Probabilistic load flow computation using the method of combined cumulants and gramcharlier expansion. *Power Systems*, *19*(1), 676–682. doi:10.1109/TPWRS.2003.818743

Zimmerman, R. D., & Chiang, H.-D. (1995). Fast decoupled power flow for unbalanced radial distribution systems. *Power Systems*, *10*(4), 2045–2052. doi:10.1109/59.476074

ADDITIONAL READING

Endrenyi, J., Anders, G., & Leite da Silva, A. (1998). Probabilistic evaluation of the effect of maintenance on reliability. an application. *Power Systems*, *13*(2), 576–583. doi:10.1109/59.667385

Fan, J., & Borlase, S. (2009). The evolution of distribution. *Power and Energy Magazine, IEEE*, *7*(2), 63–68. doi:10.1109/MPE.2008.931392

Miceli, R. (2013). Energy management and smart grids. *Energies*, *6*(4), 2262–2290. doi:10.3390/en6042262

Schneider, J., Gaul, A. J., Neumann, C., Hogräfer, J., Wellßow, W., Schwan, M., & Schnettler, A. (2006). Asset management techniques. *Electrical Power & Energy Systems*, *28*(9), 643–654. doi:10.1016/j.ijepes.2006.03.007

KEY TERMS AND DEFINITIONS

Asset Management Systems: Asset management of smart grids is a core requirement due to the huge investments involved. For example, according to the US Department Of Energy (DOE), around 1.5 trillion dollars have been invested in the US electricity infrastructure so far. Asset management applies to both tangible and intangible assets.

Distribution Management System: The Distribution Management System (DMS) may be regarded as the control center of the smart grid. The DMS mainly uses the fault location, Geographic Information Systems (GIS) and Outage Management System (OMS) to improve the reliability of the smart grid by reducing outages and sustaining the frequency and voltage levels. The most important role of DMS is to check the faults and isolate the faulty part out of the system. The "intelligent nodes" of the DMS can communicate with one-another periodically and if a fault occurs then they work together to reconfigure the system.

Energy Management System: The Energy Management System (EMS) is used for monitoring and controlling the performance of the generation and transmission system. It allows getting real-time updates from power plants about their conditions and generation parameters. The monitor and control functions are implemented through Supervisory Control and Data Acquisition (SCADA).

Renewable Energy Integration: Integrating various renewable energy sources is the most desirable feature of smart grids. However, this component faces various challenges. Advanced energy storage at the transmission, distribution, and residential levels, Static VAR compensators and synchro-phasors within the transmission grid, dynamic pricing demand response, micro grids, virtual power plants, and smart wind and solar technologies are some of the tools for Renewable Energy Integration (REI).

This research was previously published in the Encyclopedia of Information Science and Technology, Fourth Edition edited by Mehdi Khosrow-Pour, pages 3103-3113, copyright year 2018 by Information Science Reference (an imprint of IGI Global).

Chapter 15
Methodology of Climate Change Impact Assessment on Forests

Mostafa Jafari
Regional Institute of Forest and Rangelands (RIFR), Iran

ABSTRACT

Climate change is one of the challenging issues in various countries. Climate change and climate variability and global warming and its effects on natural resources, plants, animals, and on human life are among the subjects that received the attention of scientists and politicians in recent years. Climate change challenges need to be considered in various dimensions. To both understand the present climate and to predict future climate change, it is necessary to have both theory and empirical observation. Any study of climate change involves the construction (or reconstruction) of time series of climate data. How these climate data vary across time provides a measure (either quantitative or qualitative) of climate change. Types of climate data include temperature, precipitation (rainfall), wind, humidity, evapotranspiration, pressure, and solar irradiance. This chapter explores a methodology of measuring climate change's impact on forests.

INTRODUCTION

Climate change is one of the main challenging issues in various countries (Jafari, 2013b) in current century. Climate change and climate variability and Global Warming and its' effects on natural resources, plants, animal and in general on human life are among subjects that received attention of scientists and politicians in recent years. Climate change challenges need to be considered in various dimensions (Jafari, 2013c). To both understand the present climate and to predict future climate change, it is necessary to have both theory and empirical observation. Any study of climate change involves the construction (or reconstruction) of time series of climate data. How these climate data vary across time provides a measure (either quantitative or qualitative) of climate change. Types of climate data include temperature, precipitation (rainfall), wind, humidity, evapotranspiration, pressure and solar irradiance (aric, 2008). Climate change assessments and evaluation should be done by using recorded observation data as well as prepared and provided proxy data (Jafari, 2010). Plant ecophysiological study has very important

DOI: 10.4018/978-1-5225-7359-3.ch015

role to recognize climate changes (Jafari, 2007). Trees and also woods can be used as archive of past events. Climate change will strongly affect water resources, plant communities and wildlife in the arid and semi-arid regions (FAO, 2009). Water, environment humidity and temperature are main factors of plant growth. Majority of plant and forest ecosystems on the earth are formed under these two main factors. Whatever amount of humidity and required water are available and also favorable temperature for plant growth cause plant community reach higher plants and trees and forest ecosystems would develop. In fact plants are important climate indicators. Trees are not an exception. Plants, especially, trees are sensitive to their environmental changes, and tree-ring width is one of the reliable proxies of ambient environmental conditions. Climate and environmental changes affect natural ecosystems as well as planted forests (Kiaee and Jafari, 2014). Investigation of quantity and quality of these growths could help to consider past climatic conditions. Measuring and recording tree rings' widths and its' densities of early woods and late woods can provide valuable data resources to produce time series and consider its correlation with climate factors in the same time periods (Figure 1).

Seasonal changes in temperate climatic region effect on tree rings widths periodically. In spring and summer time plants grow better than unpleasant seasons like fall and winter. The outermost layer of a tree is composed of bark. Bark itself is composed of two tissues: an innermost layer of live phloem, and an outer layer of periderm (the bark 'proper'), which has an outermost layer of waterproofing cork (phellum) which protects the wood to some degree from insects, etc (Figure 2). The cork has its own cambium (phellogen) between the phloem and cork layer. Only the outermost layer of a tree is alive (essentially only the phellogen, phloem, cambium, and maturing xylem of the current year's growth).

Figure 1. Tree ring width and densities, Fagus orientalis (beech tree), Mazandaran province mid-elevation forest (MA II F3)
(Author, 2010)

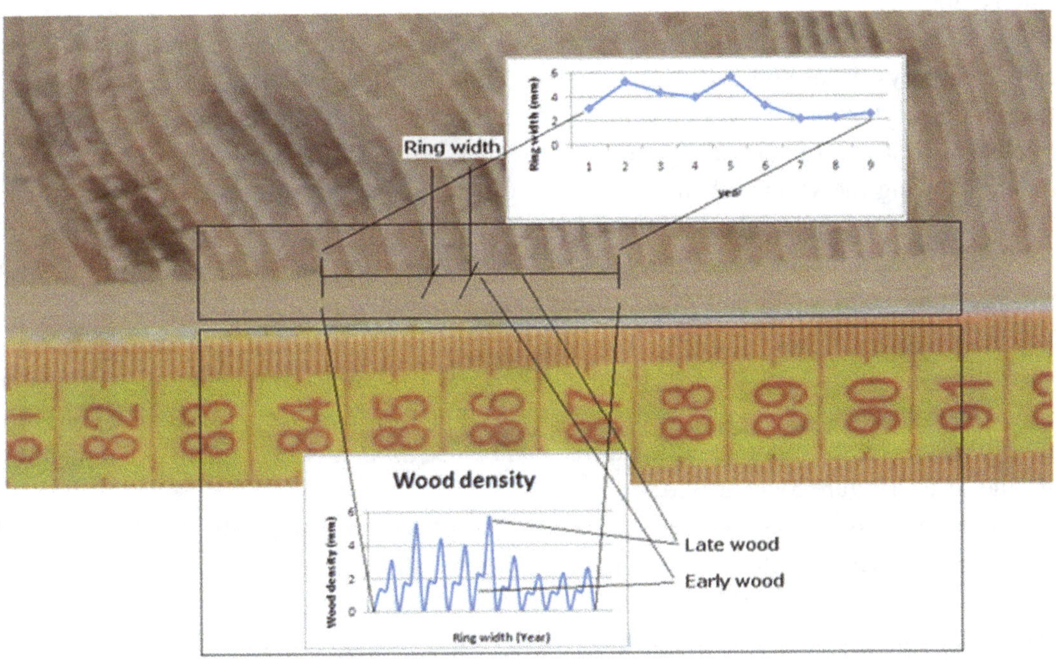

Figure 2. Anatomy of a tree trunk
(Encyclopedia Britannica, 2000)

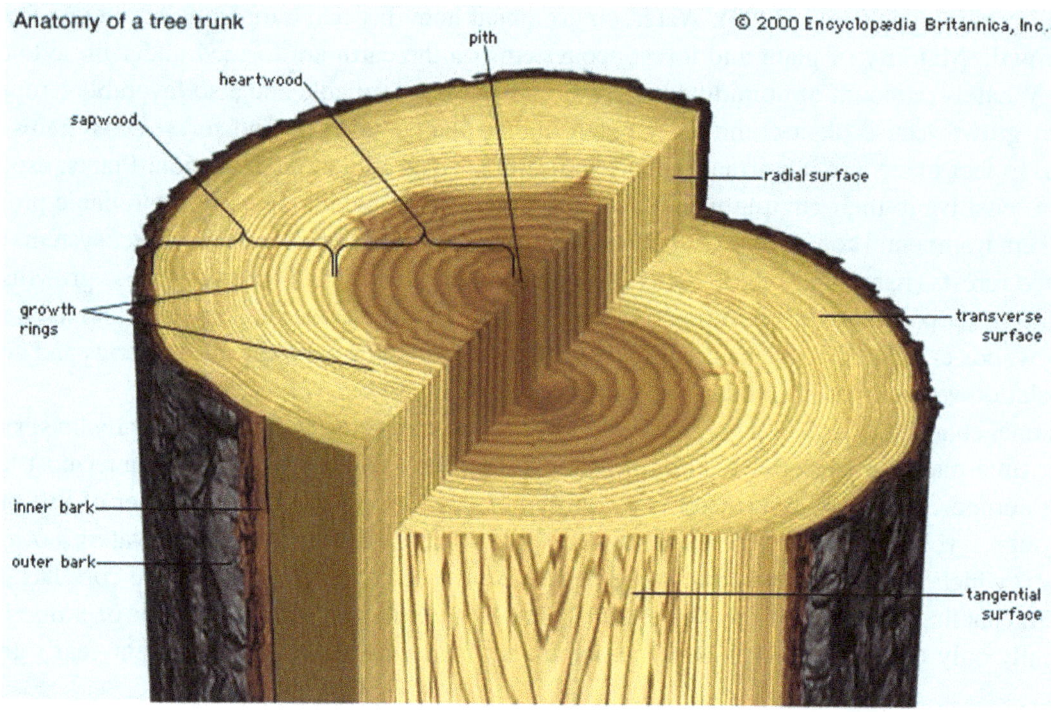

Consequently, the majority of the trunk does not require gaseous exchange. The bark is punctuated by lenticels, a sort of giant stoma, which allows the thin outermost living layers of the trunk to 'breathe' (Anonymous, 2008a),

Growths of the vascular cambium tissue produce wood as secondary xylem production. Sapwood is xylem that conveys water and dissolved minerals from the roots to the rest of the tree. The darker heartwood is older xylem that has been infiltrated by gums and resins and has lost its ability to conduct water. Each growth layer is distinguished by early wood (springwood), composed of large thin-walled cells produced during the spring when water is usually abundant, and the denser latewood (summerwood), and composed of small cells with thick walls. Growth rings vary in width as a result of differing climatic conditions; in temperate climates, a ring is equivalent to one year's growth. Certain conducting cells form rays that carry water and dissolved substances radially across the xylem. Bark comprises the tissues outside the vascular cambium, including secondary phloem (which transports food made in the leaves to the rest of the tree), cork-producing cells (cork cambium), and cork cells. The outer bark, composed of dead tissue, protects the inner region from injury, disease, and desiccation (Encyclopedia Britannica, 2006). A big trunk of a harvested tree can be use as an archive of data and may provide its life long time series (Figure 3)(Jafari, 2010).

Main objectives of dendrochronology are: a) Put the present in perspective of the past, b) Better understand current environmental processes and conditions, and c) Improve understanding of possible environmental issues of the future. To meet these objectives, the exact year of formation of each growth ring must be known: a) Merely counting rings doesn't ensure accurate dating, and b) Crossdating, also known as pattern matching, ensures accurate dating (Sheppard, 2013).

Figure 3. Hyrcanian forest research site, an old trunk of Fagus orientalis L. (Asalem, Guilan province forest)
(Jafari, 2010)

BACKGROUND

Dendrochronology is an accepted and reliable method in considering climate change impact on forest ecosystem through study tree ring widths (Jafari, 2015). The cambium of the trees growing in temperate zones becomes dormant in the falls and reactivates each spring. This leads to annual rings and the vessels produced in the spring are often larger than in the fall; the large vessels allow for rapid sap movement in the spring, whereas the narrow vessels minimize the risk of cavitations under dry conditions in late summer. This leads to a ring-porous pattern in the wood as opposed to the diffuse porous pattern where vessels are more even in size. As trees age the vessels in the center of the stem become air-filled and cease to carry water; they still function for support and storage of waste products, some of which are colored; this is the heartwood in contrast to the sapwood which carries water and is confined to the outer few annual rings (The Ohio State University, 2013).

Recording temperature and using thermometers have only been widely used since around 1850. Thus, the instrumental record for earlier times is quite poor and full of gaps. Essentially nothing is available in the way of quantitative measurements of weather conditions for the time before 1800 A.D. To reconstruct climate change, therefore, we need to use indirect indicators. One source of information is historical records: logs, dairies, lists on when the wine harvest began, reports on when the ice first broke up in a northern river, or when the cherry trees first blossomed. In some cases, such reports go back hundreds of years, although rarely in unbroken sequence. Logs and dairies are treasured finds; they do not exist for most regions of the planet (Anonymous, 2008b).

Climate factors data can be measured by direct observation in different meteorological station (like climatology and synoptic stations) or can be recorded by different instrument in different locations and with different time intervals. This information is more or less confidential for judgments on the past events, and good tools for the future projections.

Applications of Dendrochronology

As definition point of view, the word dendrochronology is composed of: dendro (using trees, or more specifically the growth rings of trees), chrono (time, or more specifically events in past time), and logy (the study of). Applications in dendrochronology include: ecology (insect outbreaks, forest stand structure, past fires) (Jafari, 2012a), climatology (past droughts or cold periods), geology (past earthquakes, volcanic eruptions), and anthropology (past construction, habitation, and abandonment of societies) (Sheppard, 2013). Also some new terms provided (Jafari, 2013a) for related applications such as, dendro-productivity (Jafari & Khoranke, 2013), dendro-genetic (Jafari *et al.,* 2012b), dendro-medical (Jafari, 2014a).

In region where the seasons provide clear seasonal climate difference, trees develop annual rings of different properties depending on weather, rain, temperature, soil pH, plant nutrition, CO_2 concentration, etc. in different years. These variations are used in dendroclimatology to infer past climate variations. Annual rings width of old trees wood sample is valuable data source as a live archive document for past climate changes, if year of growth and cross dating be well recognized. In case of sample from standing live trees, growth year is identifiable. But in case of wood samples, which could be found in archeological sites, recognizing growth year by producing skeleton graph and cross dating is necessary. Wood samples could show the years of past events like fire, drought or flood in growing site of sample tree. Fire-scarred ponderosa pine (*Pinus ponderosa*) from Ashenfelder Basin, Laramie Peak, Wyoming, (Figure 4). Low to moderate intensity fires that burned through a forest may injure or scar surviving trees, leaving a clear record of their passage. (Swetnam and Baisan, 2002)

Figure 4. Fires may injure or scar surviving trees
(Swetnam and Baisan, 2002)

Proxy Data and Climate Change

Climatologists who study past – or paleo – climates (Paleo-climatologists) use the term "proxy" to describe a way that climate change is recorded in nature, within geological materials such as ocean or lake sediments, tree-rings, coral growth-bands, ice-cores, and cave deposits.

For a proxy to be useful, it must first be established that the proxy (i.e. tree-ring width, stable isotope composition of ice, sediment composition) is in fact sensitive to changes in temperature (or some other environmental parameter). This phase of research is known as calibration of the proxy. Perhaps the most frequently used temperature proxy is the relative abundance of microfossils in sediments. That microfossils bear witness to temperature was recognized early in the history of oceanography.

Measuring and recording of tree-ring width is another reliable source of proxy of ambient environmental conditions. When a tree grows at high elevation, near the tree limit, its growth is limited by temperature, and the thickness of its growth rings contains clues about whether the growing season was warm or cold. An equation can then be written relating the changes in ring width to temperature change. Similarly, if the growth is limited by water (say, in a warm semi-arid setting) the ring width can be used to calculate changes in rainfall. Climate proxies have been utilized to provide a semi-quantitative record of average temperature in the Northern Hemisphere back to 1000 A.D (Anonymous, 2008b).

To provide paleo proxy data paleo- climatologists gather proxy data from natural recorders of climate variability such as tree rings, ice cores, fossil pollen, ocean sediments, corals and historical data. By analyzing records taken from these and other proxy sources, scientists can extend our understanding of climate far beyond the 100+ year instrumental record.

Principle sources of the major types of proxy climatic data for palaeoclimatic reconstructions can be categorized as following groups (Jafari, 2010): Glaciological (Ice Cores), Oxygen isotopes, Physical properties, Trace element & micro-particle concentrations

- **Geological:** A. Sediments, 1. Marine (ocean sediment cores), i) Organic sediments (plank-tonic & benthic fossils), Oxygen isotopes, Faunal & floral abundances, Morphological variations, ii) Inorganic sediments: Mineralogical composition & surface texture, Distribution of terrigenous material (provided by river erosion), Ice-rafted debris.
- **Geochemistry:** 2. Terrestrial, Periglacial features, Glacial deposits & erosion features, Glacio-eustatic features (shorelines and sea level changes), Aeolian deposits (sand dunes), Lacustrine deposits/varves (related to the lakes), B. Sedimentary Rocks, Facies analysis, Fossil/microfossil analysis, Mineral analysis Isotope geochemistry.
- **Biological:** Tree rings (width, density, isotope analysis), Pollen (species, abundances), Insects.
- **Historical:** Meteorological records, Para-meteorological records (environmental indicators), Phenological records (biological indicators).

Proxy material differs according to its: a) its spatial coverage; b) the period to which it pertains; and c) its ability to resolve events accurately in time (Bradley, 1985). Some proxy records, for example ocean floor sediments, reveal information about long periods of climatic change and evolution, with a low-frequency resolution. Others, such as tree rings are useful only during the last 10,000 years at most, but offer a high frequency (annual) resolution. The choice of proxy record (as with the choice of instrumental record) very much depends on what physical mechanism is under review. As noted, climate responds to different forcing mechanisms over different time scales, and proxy materials will contain necessary climatic information on these to a greater or lesser extent, depending on the three factors mentioned (aric, 2008).

Natural Archives

Growth conditions can be recorded in tree rings. A wide ring could be define as plenty of warm days and sufficient water, a narrow ring means nasty conditions, either a short growing seasons because summer was late in coming (up on the mountain), or a severe water shortage (in the foothills, in areas where water is limiting). The mixture of conditions recorded (time of snow-melt, intensity of winter rain, temperature in June, etc.) depends on what a given tree cares about in terms of growth. Hence, a tree is a "reporter," and the same is true for all other organisms recording climate change. What a scientist can extract from tree rings depends on how many properties of a ring can be measured (width, density of early wood, density and width of wood grown late in the season), how clever the statistical methods are, and how well the items of interest (say, spring temperature or annual rainfall) are correlated with the properties measured. For instance, special measurements can be made on the isotope chemistry of the wood. This kind of information can yield insights on the composition of the rainfall (from oxygen isotopes) and on the rate of photosynthesis (from carbon isotopes) (Anonymous, 2008b).

While tree growth is influenced by climatic conditions, patterns in tree-ring widths, density, and isotopic composition reflect variations in climate. In temperate regions where there is a distinct growing season, trees generally produce one ring a year, and thus record the climatic conditions of each year. Trees can grow to be hundreds to thousands of years old and can contain annually-resolved records of climate for centuries to millennia.

Tree Rings Measurement Instruments

Outcome from dendrochronological research studies played an important role in the early days of radiocarbon dating. Tree rings provided truly known-age material needed to check the accuracy of the carbon 14 dating method. During the late 1950s, several scientists (notably the Dutchman Hessel de Vries) were able to confirm the discrepancy between radiocarbon ages and calendar ages through results gathered from carbon dating rings of trees. The tree rings were dated through dendrochronology.

Even now, tree rings are still used to calibrate radiocarbon determinations. Libraries of tree rings of different calendar ages are now available to provide records extending back over the last 11,000 years. The trees often used as references are the bristlecone pine (*Pinus aristata*) found in the USA and waterlogged Oak (*Quercus* sp.) in Ireland and Germany. Radiocarbon dating laboratories have been known to use data from other species of trees (BETA, 2013).

Borer core samples submitted to the laboratory are registered, followed by preparation of optimal surfaces for analysis across several growth radii of the tree. Subsequently, tree-ring series are measured manually (Figure 5) and registered and or using specially designed measuring devices (Lintab and Aniol) connected to computers (Figure 6), screens and printers. The soft wares used for data storage, cross-correlation and statistical analyses are CATRAS and ITRDBLIB. Samples subject to wood anatomical determination are analyzed with light microscopy and compared against the laboratory's extensive reference collection of European woody plants (Hammarlund, 2013). Major Equipment in the Bioge-

Figure 5. Measuring disc sample tree ring widths in the Golestan research centre lab. using manual measurements
(Author, 2010)

Figure 6. Measuring core sample tree ring widths in the dendrochronology lab. using computer equipped machine
(Author, 2012)

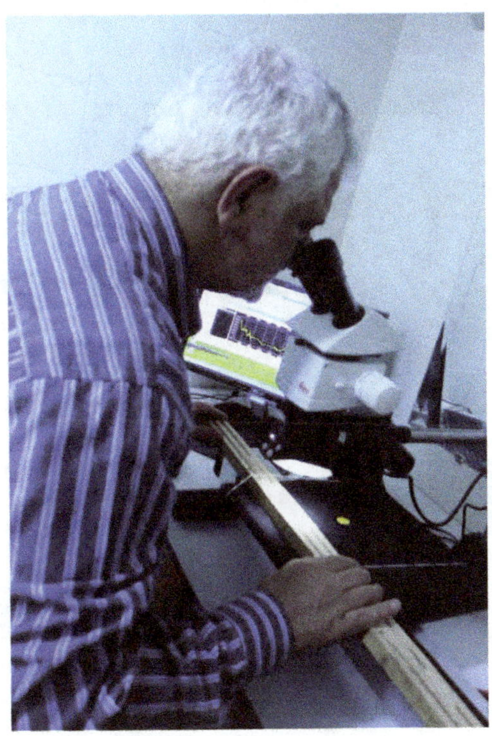

ography/Dendrochronology Laboratory currently houses one Velmex Measuring Machine connected to a Stereozoom Microscope on a boom stand and a microcomputer analysis system. The lab equipment also includes: - incremento borers, - stereozoom microscopes, - belt sanders (4X24"), - desktop and labtop computers, - GPS units, Stihl Chain saw Stihl 046 with a 24" bar, Hand saws, - Hood, - cruiser packs, - Soil Sampling probes, - Munsell Charts, - Measuring tapes, - map tubes, - Paper straws, Poplar core mounts, skeleton plot paper (Indiana State University, 2013).

Discs or Borer Core Samples

The stem cross section is the best way to have a good surface on which to observe tree ring series. These discs can be sometimes obtain in co-operation with foresters when timbering is programmed. In most of the cases, cross dating and then measurements of ring-width as well as densitometry analysis are performed with small cores 4mm in diameter extracted from the tree by an increment borer. In order to avoid dissymmetry in radial tree-growth measurement 2 or 3 cores are extracted on each tree. Consequently, on each sampling site 20 to 45 cores are collected and brought back to the laboratory. Precise cares have to be taken in coring, particularly when densitometry analysis will be performed. The most important is the position of the borer on the trunk. In order to obtain lately an observation surface the most perfectly perpendicular to the long axis of tracheids and fibers, the borer has to be also positioned perpendicularly to trunk axis (TGTC, 2008).

In case which disc of harvested trunk is not available (Figure 7) or it is not possible to take disc, core sampling is an alternative. There are different types of drill instrument for this purpose (Figure 8). Increment borers are instruments that take a small cylindrical core from trees and allow determination if radial growth or age of the tree (Figure 8c). The borer consists of three parts (Figure 8): the case or handle, the borer bit, and the extractor. Increment borers come in various sizes, from 4 inches (101.6

Figure 7. Big trunk disc sample taken during harvesting process
(Author 2010)

Figure 8. Wood sample by increment borer: a) Using borer in the trunk, b) Extracting core sample from borer, c) sample in wooden holder
(Eng Khoshnevis, photos by the author, 2012)

mm) to 30 inches (762 mm) length or more. The ones which are in usual use are in the 8-18 inch range (457.2 -203.2 mm). Smaller borers are used for small trees or where only recent growth is needed from larger trees (Figure 8a, b, and c, Figure 9).

In the field sampling experiment, when the cores remove from the borer, for the safety, it is needed to lay them into an increment core holder that has been pre-glued. Tightly wind the glued cores with cotton string (Figure 10) so as to apply pressure during the drying process (McCarthy, 2008).

Immediate observation of rings on cores extracted from the borer is rarely possible, and moreover rings width measurement quite impossible. Good observation and measurement need a perfect transverse section. After a correct reorientation of the core as the piece of wood was in the trunk, such a section is obtained either by refreshing with a razor blade or polishing the surface selected in order to obtain a plane surface allowing to access to the cell structure (TGTC, 2008).

Measured Data Analysis

A great number of laboratories in different part of the world have been established to study on wood and climate changes. Palaeo-climate changes are on the target of the most of these institutes. The International Tree-Ring Data Bank is maintained by the NOAA Paleoclimatology Program and World Data Center for Paleoclimatology. The Data Bank includes raw ring width or wood density measurements, and site chronologies (growth indices for a site). Tree-ring measurement series from other parameters are welcome as well. Reconstructed climate parameters, including North American Drought, are also avail-

Figure 9. Borer samples in wooden holder
(Author, 2014)

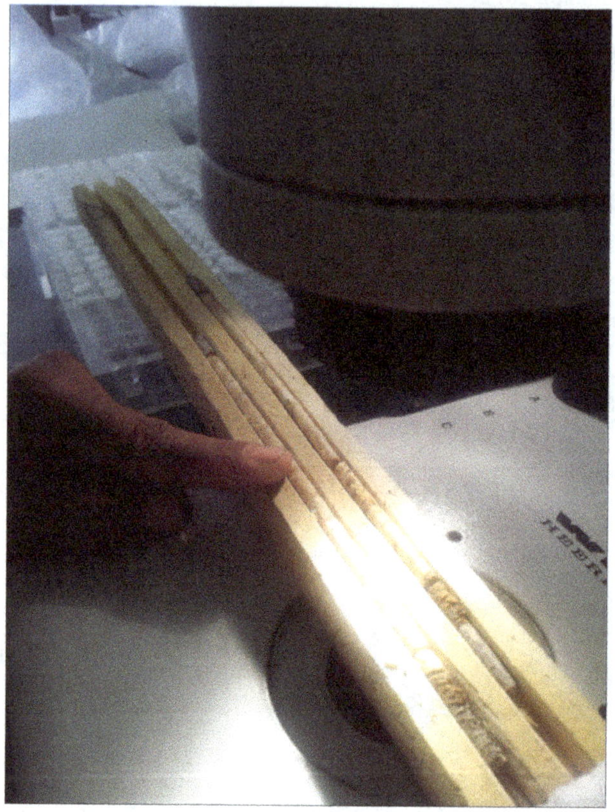

Figure 10. Borer sample and sample holder in the field
(Author, 2012)

able for some areas. Over 2000 sites on six continents are included (WDC, 2008). The objective of the measuring ring widths would be to develop tree-ring records of climate over the past several centuries, to understand inter-annual to century scale variability in climate. This study will improve the capability of understanding environmental variability and key features of the regional environment, e.g. persistence of drought, reliability of stream flow (Hughes & Touchan, 1997). Analyzing data by using software by accuracy of 0.01 mm (Robinson & Evans 1980) or more accurate up to 0.001 mm.

Statistical Analysis of the Climate Factors

Tree- rings can provide continuous yearly paleoclimatic records for regions or periods of time with no instrumental climate data. However, different species respond to different climate parameters with, for example, some sensitive to moisture and others to temperature. For example four common species which grow in Northern Ireland and their suitability for climate reconstruction are beech, oak, ash and Scots pine. Beech and ash are the most sensitive to climate, with tree-ring widths more strongly Influenced by precipitation and soil moisture in early summer than by temperature or sunshine. Oak is also sensitive to summer rainfall, where as Scots pine is sensitive to maximum temperature and the soil temperature. The moisture-related parameters, rainfall and the Palmer Drought Severity Index (PDSI), and to a lesser extent, maximum and mean temperatures, can be reconstructed. Reconstructions of climate parameters with tree-rings as proxies may be relatively stable for some seasons such as May–July. The combinations of species are more successful in reconstructing climate than single species (GarcÍa-Suárez *et al.*, 2009). The development of dendrochronological time series in order to analyze climate-growth relationships usually involves first a rigorous selection of trees and then the computation of the mean tree-growth measurement series. A change in the perspective, passing from an analysis of climate-growth relationships that typically focuses on the mean response of a species to investigating the whole range of individual responses among sample trees (Carrer, 2011).

Crossdating

Primary faze of crossdating work, under a good dissecting microscope, begins by counting backwards from the first known year behind the bark. Using a fine mechanical pencil, place a single dot on each decadal ring (e.g., 2010, 2000, 1990, etc.), place two dots on each 50-year ring (2010, 1960, 1910, etc.), and three dots on each century ring (2010, 1910, 1810, etc.). At this stage, these marks are just temporary year assignments. The actual years will be confirmed after skeleton plotting (McCarthy, 2008).

Using a mm graph paper is first step to draft skeleton plots. The decades are labeled on the x-axis and a vertical line is drawn on a y-scale composed of 10 units. Any ring that is smaller than its neighbor rings (± 3 on either side) gets a line drawn on the paper. If the ring is very small, the line may be 10 units. If the ring is half as small as its predecessor you might code it as a 5, etc. (rings that are coded less than a 5 are rarely useful in crossdating). This is a bit counterintuitive because the longer the line, the smaller the ring. According to the provided skeleton graphs, cross-dating among different samples comparing with control sample would be possible. It is also possible to recognize different years of various samples from different sources for cross-dating.

A more precise method of dating volcanic deposits of recent age is to identify anomalous growth patterns among the annual rings of trees growing at the time the deposits were emplaced. Trees that were injured but not killed by tephra or lahars may show a sequence of narrow rings beginning at the time of impact. “Cross-dating,” the matching of ring-width variation patterns in one tree with corresponding ring patterns in another, should be used to ensure that dating errors are not introduced by missing rings.

Missing rings can often identified by drawing narrow rings in cross-dating progress. Control tree shows 23 annual rings between the 1472 and 1495 narrow rings (Figure 11), while the tree in the We (Mount St. Helens) tephra-fall zone shows only 20 rings. The series of missing and narrow rings starting with the 1482 ring were caused by tree injury during fallout of layer we. Because of possible missing rings, dates of past volcanic events cannot be determined unequivocally by counting back to a series of narrow rings (Brantley *et al.,* 1986).

Since, the same set of environmental factors influence tree growth throughout a region, the patterns of ring characteristics, such as ring widths, are often common from tree to tree. These patterns can be matched between trees in a process called crossdating (Figure 12), which is used to assign exact calendar year dates to each individual ring (The University of Arizona, 2013). The chronology provides two main types of information:

1. The chronology can be used as a tool for dating events that caused tree death or a marked change in the appearance of a ring or set of rings. The death date can be used to date the tree cutting involved in the construction of wooden dwellings. Scars can record the timing of events such as fire, flood, avalanche, or other geomorphologic events, while sequences of suppressed or larger rings record events such as insect infestation, effects of pollution, or changes in forest dynamics.

Figure 11. Tree-ring cross-dating
(Brantley et al., 1986)

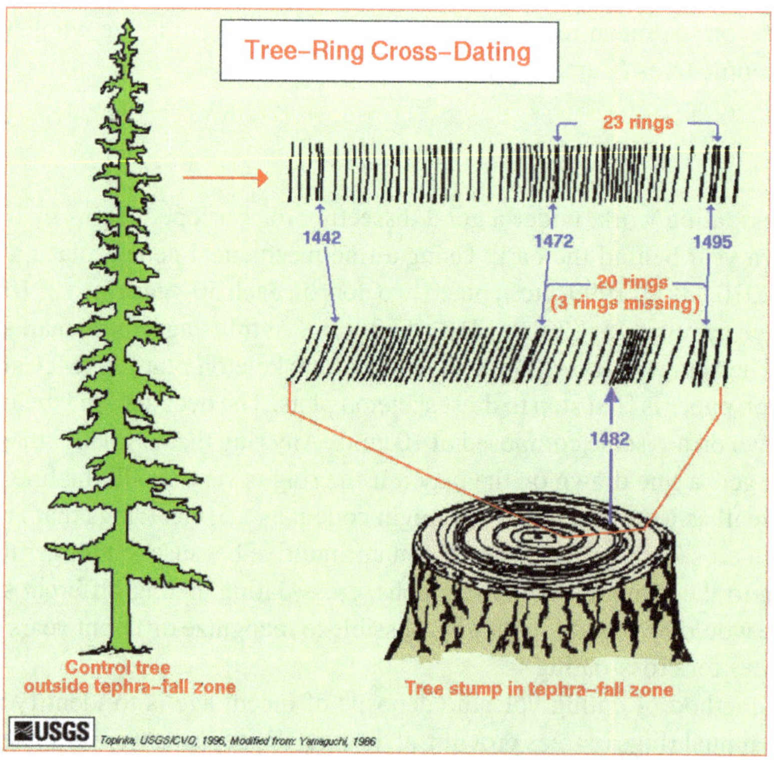

2. The chronology is an average of coherent variations in growth from a number of trees. It enhances the common pattern of variation or "signal" -usually related to climate- while the non-common variance or "noise" is dampened. Chronologies from trees that are sensitive to climate can be used to reconstruct past variations in seasonal temperature, precipitation, drought, stream flow, and other climate-related variables (The University of Arizona, 2013).

The techniques of dendrochronological study were used to date a spruce coffin board from Pukatawagan Bay, Manitoba received from Manitoba Historical Resources. The sample contained 74 annual rings, although the outermost 16 rings were rotten and not measured. The ring width series from the coffin board was matched against records from living spruce growing near South Indian Lake. Crossdating shows that the coffin board was cut in 1878. The board also contains a ring containing poorly developed latewood (a 'light' ring) in 1817 that is found commonly in spruce records across northern Manitoba (Figure 12) (Scott & Nielsen, 2002).

Using different wood materials of different periods, by cross-dating of the samples comparing with control one, it would be possible to record and estimate changes from present time back to the ancient area (Figure 13).

Figure 12. Cross-dating according to the annual ring growth of different samples from live standing trees to old or new wood samples, Illustration showing an example of time series crossdating
(Adapted from Dendrologisches Labor Hamburg, 2015, http://www.scinexx.de/dossier-detail-186-7.html)

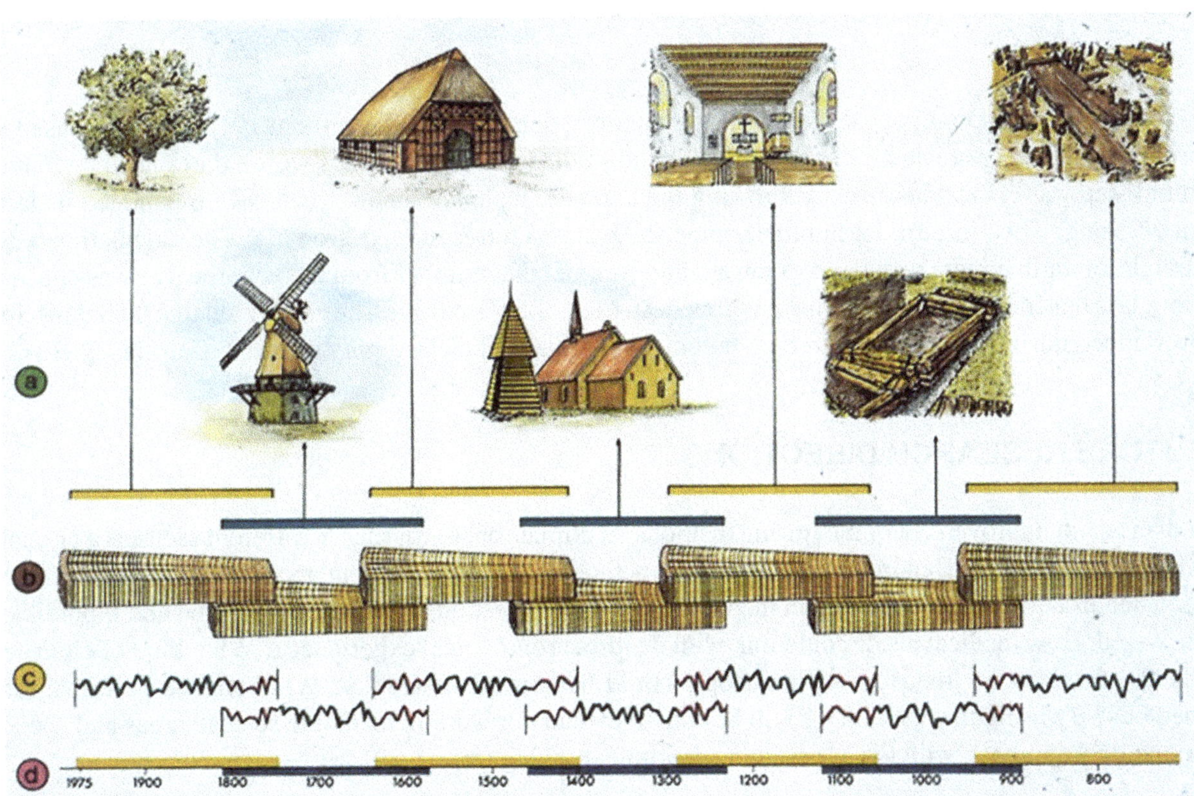

Figure 13. The bridging method, cross-dating of annual rings from present live tree to the past wood samples, for establishing a tree-ring sequence
(Kuniholm, 2001)

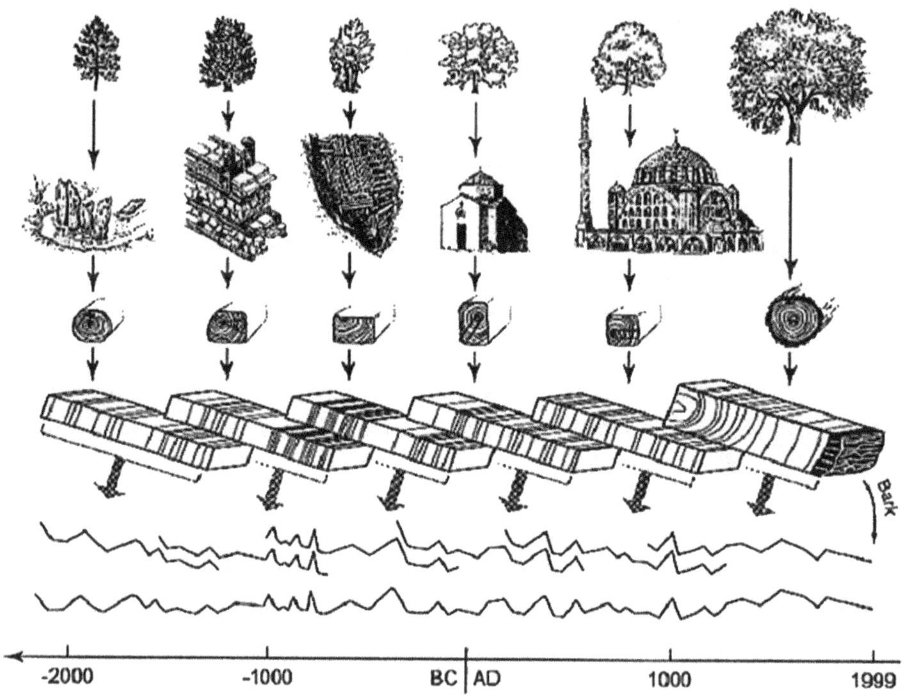

Dr. Andrew E. Douglass, an astronomer, developed dendrochronology about 1913. Douglass used a bridging method to create his chronology. First he studied recently cut trees whose dates he knew. This initial step was critical because by knowing the cut date, Douglass knew when each tree added its last growth ring. This, in turn, let him determine the year each tree started growing. The calculation was straightforward: count the dark rings inward and subtract that number from the year the tree was cut. As Douglass matched and recorded ring patterns from trees of different ages, he confirmed that their patterns overlapped during the years the trees simultaneously lived (The University of North Carolina, 2013).

FUTURE RESEARCH DIRECTIONS

Enhance and improvement of measuring methods on climate change issues in different sector is a crucial and important needs. Dendrochronology study method has several different applications. It is crucial to used dendrochronological method in medical field. Some question need to be considered and if possible answered. How medical needs could link with dendrochronological experiences? What kind of element may be detected on tree rings? How people could benefit of the results? What are the best analysis methods? By using this method we will be able to extend our work in medicine science areas and speed up medical approach with lower cost and economical saving (Jafari, 2014a).

CONCLUSION

The AFOLU sector (AR5, IPCC, WGIII, Chapter 11) is responsible for just under a quarter (~10–12 GtCO_2eq/yr) of anthropogenic GHG emissions mainly from deforestation and agricultural emissions from livestock, soil and nutrient management (robust evidence; high agreement) (Smith *et. al.*, 2014).

Climate change impact will cause changes in biomass production in natural ecosystems. It is a need to consider the vulnerability of Net Primary production (NPP) in forest ecosystem (Jafari, 2014b).

Climate change issue is an important subject in current century. In all possible ways we need to cope with this phenomenon to enhance our understanding knowledge. Dendrochronology as an able and certified study method could be implying in a wide range of applications.

REFERENCES

Anonymous. (2008a). *Plant Growth, in Science*. Retrieved from http://www.steve.gb.com/images/science/wood_layers.jpg

Anonymous. (2008b). *Climate Change, Past and Future, Calspace Courses*. Retrieved from http://earthguide.ucsd.edu/virtualmuseum/climatechange1/cc1syllabus.shtml

aric. (2008). *Global Climate Change Student Guide*. Author.

BETA. (2013). *Beta Analytic, Radiocarbon Dating*. Retrieved from http://www.radiocarbon.com/tree-ring-calibration.htm

Bradley, R. S. (1985). *Quaternary Palaeoclimatology: Methods of Palaeoclimatic Reconstruction*. London: Unwin Hyman.

Brantley, S., Yamaguchi, D., Cameron, K., & Pringle, P. (1986). Tree-Ring Dating of Volcanic Deposits. *Earthquakes and Volcanoes*, *18*(5), 184–194.

Carrer, M. (2011). Individualistic and Time-Varying Tree-Ring Growth to Climate Sensitivity. *PLoS ONE*, *6*(7), e22813. doi:10.1371/journal.pone.0022813 PMID:21829523

FAO. (2009). *Guidelines for good forestry and range practices in arid and semi-arid zones of the Near East*. Cairo: FAO Regional Office for the Near East.

García-Suárez, A.M., Butler, C.J., & Baillie, M.G.L. (2009). Climate signal in tree-ring chronologies in a temperate climate: A multi-species approach. *Dendrochronologia, 27*, 183–198. doi:10.1016/j.dendro.2009.05.003

Hammarlund, D. (2013). *Equipment and sample handling*. Retrieved from http://www.geol.lu.se/dendro/en/equipment.htm

Hughes, M., & Touchan, R. (1997). *Preparation for Collaborative Work on Past Climate in Jordan*. Science Storm, A world of wonder. Retrieved from http://www.sciencestorm.com/

Indiana State University. (2013). *Dendrochronology Lab, Indiana State University*. Retrieved from http://dendrolab.indstate.edu/equip_fac.htm

Jafari, M. (2007). *Review on needfulness for plant ecophysiological study and investigation on climate change's effects on forest, rangeland and desert ecosystems*. Presented at Workshop, Climate Change in South-Eastern European Countries: Causes, Impacts, Solutions, Orangerie, Burggarten, Graz, Austria. Retrieved from http://www.intecol.net/pages/002_personal.php?mode=list&table=blog&tb_kind=Research&id=mjafari&vvt=aa&bidx=27531

Jafari, M. (2010). Climate Change Impacts on Iranian Ecosystems with review on Climate Change Study Methods (Pub. No. 421). RIFR.

Jafari, M. (2012a). A new approach to dendroecological studies: Climate change impact on forest' wood production in Astara (Gilan). *Iranian Journal of Wood and Paper Science Research, 27*(4), 690–706.

Jafari, M. (2013a). *New method on dendrochronology study in forest ecosystems; two new approaches: Dendro-productivity and Dendro-genetic*. Paper ID: 1076-ADA2013. Presented in The 3rd International Conference of Asian Dendrochronology Association (ADA2013), Climate Change and tree Rings, Tehran, Iran. Retrieved from http://ada2013.ut.ac.ir/paper?manu=7148

Jafari, M. (2013b). Integration of forestry and rangeland management institutions in the Islamic Republic of Iran. In *Challenges in adopting an integrated approach to managing forest and rangelands in the Near East Region, Part II, Country studies* (pp. 37–67). Cairo: FAO.

Jafari, M. (2013c, July-December). Challenges in Climate Change and Environmental Crisis: Impacts of Aviation Industry on Human. *Urban and Natural Environments in International Journal of Space Technology Management and Innovation, 3*(2), 24–46.

Jafari, M. (2014a). *Dendro-Medical as New Approach on Dendrochronological Study method*. Presented in the 9th International Conference on Dendrochronology, as Poster No. PO086, Ref. 1590, ThemeZ01, Melbourne, Australia.

Jafari, M. (2014b). Change and vulnerability of Net Primary Production (NPP) in Iranian forest, rangeland and desert ecosystems impacted by climate change. *Iranian Journal of Range and Desert Research, 21*(1), 139-153.

Jafari, M. (2015). Dendrochronology and Climate Change. In *Encyclopedia of Information Science and Technology* (3rd ed., pp. 2917–2930). IGI Global.

Jafari, M. & Khorankeh S. (2013). Impact of climate and environmental changes on forest ecosystem's productivity (case study: Galugah). *Iranian Journal of Forest and Poplar Research, 21*(1), 166-183.

Jafari, M., Maghsudloo, M. K., Khanjani Shiraz, B., Karimidoost, A., & Hemmati, A. (2012b). New approach in dendro-genetic method for identification of beech tree variability. *Iranian Journal of Rangelands and Forests Plant Breeding and Genetic Research, 20*(2), 284–294.

Kiaee, M., & Jafari, M. (2014). Investigation and consideration of forest tree reaction to climate and environmental changes (Case study: Lavizan forest park). *Journal of Plant Research, 27*(1), 130–141.

Kuniholm, P. I. (2001). *Dendrochronology and Other Applications of Tree-ring Studies in Archaeology. In The Handbook of Archaeological Sciences*. London: John Wiley & Sons, Ltd.

McCarthy, B. C. (2008). *Introduction to Dendrochronology, Lab procedures*. Retrieved from http://www.plantbio.ohiou.edu/dendro/

RMTRR. (2008). *Rocky Mountain Tree-Ring Research (RMTRR)*. Retrieved from http://www.rmtrr.org/gallery.html

Robinson, W. J., & Evans, R. (1980). A microcomputer-based tree-ring measuring system. *Tree-Ring Bulletin, 40*, 59–64.

Schweingruber, F. H. (1988). *Tree Rings: Basics and Applications of Dendrochronology*. Dordrecht: Reidel; doi:10.1007/978-94-009-1273-1

Scott, St. G., & Nielsen, E. (2002). Dendrochronological analysis of a coffin board recovered from Pukatawagan, Bay, Manitoba. Terrain Sciences Division, Geological Survey of Canada and Manitoba Geological Survey, Manitoba Industry, Trade and Mines.

Sheppard, P. R. (2013). *Crossdating Tree Rings*. Laboratory of Tree-Ring Research, The University of Arizona. Retrieved from http://www.ltrr.arizona.edu/skeletonplot/applications.htm

Smith, P., Bustamante, M., Ahammad, H., Clark, H., Dong, H., Elsiddig, E. A., ... Tubiello, F. (2014). Agriculture, Forestry and Other Land Use (AFOLU). In *Climate Change 2014: Mitigation of Climate Change. Contribution of Working Group III to the Fifth Assessment Report of the Intergovernmental Panel on Climate Change*. Cambridge, UK: Cambridge University Press.

Swetnam Thomas, W., & Christopher, H. B. (2002). Fire Regimes in Sierran Mixed-Conifer Forests. In Sierra Nature Notes, (vol. 2). Laboratory of Tree-Ring Research, The University of Arizona.

TGTC. (2008). *Tree growth and the tree-site complex*. TGTC.

The Ohio State University. (2013). *Woody plants*. Department of Horticulture & Crop Science, The Ohio State University. Retrieved from http://www.hcs.ohio-state.edu/hcs300/anat2.htm, http://hcs.osu.edu/

The University of Arizona. (2013). *Paleo Slide Set: Tree Rings: Ancient Chronicles of Environmental Change*. Laboratory of Tree-Ring Research, The University of Arizona.

The University of North Carolina. (2013). *Tree-Ring Dating, Lesson 2.4*. Research Laboratories of Archeology. The University of North Carolina at Chapel Hill. Retrieved from http://rla.unc.edu/lessons/Lesson/L204/L204.htm

WDC. (2008). *The World Data Center for Paleoclimatology Mirror Site (WDC)*. Retrieved from http://wdc.obs-mip.fr/treering.html

ADDITIONAL READING

Barnola, J. M., Raynaud, D., Korotkevich, Y. S., & Lorius, C. (1987). Vostok ice core provides 160,000 year record of atmospheric CO_2. *Nature*, *329*(6138), 408–414. doi:10.1038/329408a0

Bergthorsson, P. (1969). An estimate of drift ice and temperature in Iceland in 1000 years. *Jökull*, *19*, 94–101.

Briffa, K. R., & Schweingruber, F. H. (1992). Recent dendroclimatic evidence of northern and central European summer temperatures. In: Climate Since A.D.1500. Routledge, London, pp. 366-392. COHMAP Members, 1988. Climatic change of the last 18,000 years: Observations and model simulations. *Science*, *241*, 1043–1052.

Davis, M. B. (1963). On the theory of pollen analysis. *American Journal of Science*, *261*(10), 899–912. doi:10.2475/ajs.261.10.897

Emmanuel, W. R., Shugart, H. H., & Stevenson, M. P. (1985). Climate change and the broad-scale distribution of terrestrial ecosystem complexes. *Climatic Change*, *7*(1), 29–43. doi:10.1007/BF00139439

Fritts, H. C. (1971). Dendroclimatology and dendroecology. *Quaternary Research*, *1*(4), 419–449. doi:10.1016/0033-5894(71)90057-3

Fritts, H. C. (1976). *Tree rings and climate* (pp. 80–96). London: Academic Press.

Schweingruber, F. H., Fritts, H. C., Bräker, O. U., Drew, L. G., & Schär, E. (1978). The X-ray technique as applied to dendroclimatology. *Tree Ring Bull.*, *38*, 61–91.

KEY TERMS AND DEFINITIONS

AFOLU: Agriculture, Forestry and Other Land Use (AFOLU), main players on emission reduction and mitigation aspect of climate change, chapter 11, WGIII, IPCC AR5 2014.

Climate Change: Global Warming, climate change and climate variability are a definition of deviation of climatic factors from its normal trends mainly impacted by human activities. Global Warming and its' effects on natural resources, plants, animal and in general on human life are among subjects that received attention of scientists and politicians in recent years.

Dendrochronology: Dendrochronology was developed about 1913, and is a (climate change) method to study tree ring widths in terms of time. The word dendrochronology is compose of: dendro (using trees, or more specifically the growth rings of trees), chrono (time, or more specifically events in past time), and logy (the study of). Dendrochronology as an able and certified study method could be implying in a wide range of applications.

Proxy Data: Climate change assessments and evaluation should be done by using recorded observation data as well as prepared and provided proxy data. Paleoclimatologists (climatologists who study past – or paleo – climates) use the term "proxy" to describe a way that climate change is recorded in nature, within geological materials such as ocean or lake sediments, tree-rings, coral growth-bands, ice-cores, and cave deposits.

Tree Rings: Trees growing in temperate climatic region are under seasonal changes. In spring and summer time plants grow better than unpleasant seasons like fall and winter. Each growth layer is distinguished by early wood (springwood), composed of large thin-walled cells produced during the spring when water is usually abundant, and the denser latewood (summerwood), and composed of small cells with thick walls.

This research was previously published in the Encyclopedia of Information Science and Technology, Fourth Edition edited by Mehdi Khosrow-Pour, pages 3114-3130, copyright year 2018 by Information Science Reference (an imprint of IGI Global).

Chapter 16
Model for Assessment of Environmental Responsibility in Health Care Organizations

María Carmen Carnero
University of Castilla-La Mancha, Spain & University of Lisbon, Portugal

ABSTRACT

Sustainability is considered a paradigm for businesses in the 21st century. Despite this, the existing tools for helping to introduce strategies and manage activities to promote sustainable business are few. These deficiencies become more important in healthcare organizations owing to their particular conditions of resource consumption and waste production. It is, therefore, essential to have objective tools to assist in monitoring environmental sustainability in this type of organization. This chapter therefore sets out a multicriteria assessment system constructed by extension to a fuzzy environment of the technique for order preference by similarity to ideal situation (TOPSIS) to assess the environmental responsibility of a healthcare organization. This model allows joint evaluation of a significant number of decision criteria. The aim is to provide a hospital with a model that is easy to apply, with criteria specific to healthcare, and that allows its responsibility with regard to the environment to be monitored over time. The model has been used in a public hospital.

INTRODUCTION

Sustainability is considered a paradigm for businesses in the 21st Century (Garcia et al., 2016). Despite this, the existing tools for helping to introduce strategies and manage activities to promote sustainable business are few (Garcia et al., 2016). These deficiencies become more important in Health Care Organizations owing to its particular conditions of resource consumption and waste production. Health Care Organizations are the only type of company which can generate all the classes of waste, from waste without risk to waste that is potentially infectious, carcinogenic, mutagenic, teratogenic or radioactive. The risk to people and to the environment from this waste is much greater if it is not correctly segregated. It is also vital to carry out action to reduce the consumption of limited natural resources such as water

DOI: 10.4018/978-1-5225-7359-3.ch016

and energy, while increasing the protection and conservation of the environment, including reducing the emission of pollutant gases, protecting biodiversity or considering the role of suppliers in action to prevent or reduce waste.

It is, therefore, essential to have objective tools to assist in monitoring environmental sustainability in this type of organization, taking into account a number of factors. That is, by assessing how improvement actions, within a process of continuous improvement, are contributing to improvements in sustainability. However, it is clear that there is little linkage between sustainability reporting and management control systems (Cintra & Carter, 2012).

Nonetheless, despite its importance, the literature on the development of systems for environmental assessment in Health Care Organizations is very limited.

This Chapter therefore sets out a multicriteria assessment system constructed by extension to a fuzzy environment of the Technique for Order Preference by Similarity to Ideal Situation (TOPSIS), to assess the environmental responsibility of a Health Care Organization. This model allows joint evaluation of a significant number of decision criteria, which include any event that may cause adverse effects on water, ground, seas and rivers, wild species or their habitats; it also considers the existence of possible measures to be carried out in Health Care Organizations to minimize the probability of an event, or to eliminate all risk. However, it should be noted that this model is not intended to perform an environmental audit in the field of health care, as it would need to include economic, technical, legal and other criteria, or a system of environmental impact that would require the assessment of a variety of risks and consequences. The aim is to provide a hospital with a model which is easy to apply, with criteria specific to health care, and which allows its responsibility with regard to the environment to be monitored over time. Following the methodology laid down in Carnero (2015), criteria were used that were assessed depending on the number of admissions or annual services provided, making it possible to compare results over time for a single organization, or between organizations. The model has been used in a Public Hospital.

BACKGROUND

The literature includes a large number of contributions on environmental questions (Aragones-Beltran et al., 2009; Higgs et al., 2008; Hsu & Hu, 2008; Kang et al., 2007; Lamelas et al., 2008; Liang et al. 2006; Madu, Kuei & Madu, 2002; Pilavachi, Chatzipanagi, & Spyropoulou, 2009; Tzeng & Lin, 2005; Tseng, Lin, & Chiu, 2009; Van Calker et al., 2006). However, these are invariably related to manufacturing, transport or energy companies. In the case of service companies and, in particular, in Health Care Organizations, the contributions are practically non-existent (Carnero, 2015).

Health Care Organizations, places dedicated to the improvement and development of preventive measures in health care, with respect to their users, those who live in the area and their workers, should be involved in minimizing their own environmental impact, as there is a strong correlation between the two (Comunidad de Madrid, 2005). In order to improve environmental sustainability of a Health Care Organization, however, it is vital to monitor sustainability over time for decision making and management of activities that constitute an organization's system processes (Salvado et al., 2015).

A system of environmental assessment should combine many factors, which may be technical, social, political, economic and environmental, which often conflict with one another (Lahdelma et al., 2000); it may also be necessary to include a number of individuals or decision groups, with different perspectives or responsibilities within the Health Care Organization; as well as the need to incorporate a great deal of information, quantitative but in many cases qualitative, relating to uncertainties, scenarios, goals, etc. (Munda, 2005). These characteristics make the use of Multi-Criteria Decision Analysis (MCDA) methods highly suitable for supporting decisions about sustainability (Santoyo-Castelazo & Azapagic, 2014). The fact that the model produced, based on mathematical techniques, is objective, also helps to guarantee public acceptance of the solution or result obtained (Huang, Keisler, & Linkov, 2011). MCDA methods then, although not suited to all environmental problems, are very convenient in environmental impact assessment, as they give information in a structured fashion, which can be easily interpreted by the decision makers (Neste & Karjalainen, 2013). All this has led to an increase in the literature applying MCDA in the environmental field over the last two decades (Carnero, 2014).

Because of the pressure brought to bear on companies by different stakeholders and by society to address ecological and social sustainability (Garcia et al., 2016) a variety of research has been carried out applying MCDA methods in this area. Gumus (2009) presents a methodology for selection of the most suitable hazardous waste transportation firms using a fuzzy Analytical Hierarchy Process (AHP). A similar technique is used by Heo et al. (2010) to get the weightings of the criteria to establish ex-ante and ex-post stages of renewable energy dissemination programmes in Korea. Reza et al. (2011) combine morphological analysis and AHP to choose new sustainable products from the earliest stages of conception. The study of Chan et al. (2012) is along similar lines, but using fuzzy AHP. Boran et al. (2012) use intuitionistic fuzzy TOPSIS to assess renewable energy technologies for electricity generation, such as photovoltaic, hydro, wind, and geothermal energy in Turkey. Wang et al. (2012) produce a model for selecting of green initiatives in the fashion industry. Vinodh et al. (2014) describe an assessment model to determine the best method for recycling plastics. Pourebrahim et al. (2014) made a selection of criteria and alternatives for conservation development in a coastal zone. Galvez et al. (2015) propose a model combining Mixed Integer Linear Programming optimization and AHP to assess possible scenarios for the implementation of an anaerobic co-digestion logistics network used to create sustainable energy production processes from biogas. Al Garni et al. (2016) use AHP to assess renewable power generation sources including solar photovoltaic, concentrated solar power, wind energy, biomass, and geothermal, finding that in the case of Saudi Arabia the photovoltaic, followed by concentrated solar power are the first-placed technologies.

The literature which uses MCDA methods for the assessment of environmental sustainability and environmental impact includes the following contributions. Hermann et al. (2007) describe the tool COMPLIMENT, which allows the overall environmental impact of an organization to be found; they combine life cycle assessment, AHP and Environmental Performance Indicators (EPIs). Kaya and Kahraman (2011) use fuzzy AHP to build an environmental impact assessment methodology for urban industrial planning. Viaggi et al. (2011) develop and apply a multicriteria methodology to estimate the environmental effectiveness of European Union agri-environment schemes in Ireland and Emilia-Romagna (Italy). Larimian et al. (2013) use fuzzy AHP to assess environmental sustainability from the point of view of security in different areas in a region of Tehran. Egilmez et al. (2015) assess environmental sustainability in 27 Canadian and US cities using fuzzy multicriteria decision-making models. The research of Zhang et al. (2016) is also related to City sustainability evaluation via MCDA methods, in this case, of 13 cities in China. Khalili and Duecker (2013) describe a methodology for designing a sustainable environmental management system built using ELECTRE. This system serves as a back-up to monitor the efforts made by companies in the area of sustainability, for example through product design, operational development or the modelling of the supply chain. Salvado et al. (2015) use AHP to calculate a sustainability index which allows companies and their supply chains to get information about their own level of economic, social and environmental sustainability. A further review of the literature on environmental questions analysed via multicriteria techniques can be seen in Huang et al. (2011) and Neste and Karjalainen (2013).

However, in the field of environmental assessment of Health Care Organizations, the only contribution is Carnero (2014) which describes a model using a fuzzy AHP together with utility functions. This model is applied to a recently opened public hospital, giving a utility of 0.764 out of 1, and showing how this type of model can be very positive in the process for certification to standard ISO 14001. Carnero (2015) shows another, more advanced, model also using fuzzy AHP, but assessing new criteria, and most are assessed based on number of care services provided annually, which allows results to be compared over time for one Health Care Organization, or between Health Care Organizations.

MODEL FOR ASSESSMENT OF ENVIRONMENTAL RESPONSIBILITY

The development of new information technologies, together with the development and application of new concepts in environmental sustainability, require a constant updating of the assessment models in this area.

The choice of criteria and subcriteria takes account of Rodríguez et al. (2005), García et al. 2010, Mata et al. (2011), Tejedor (2012), Bon-García (2012), Yanguas (2012), Galdakao-Usansolo Hospital (2012), Carnero (2014) and Carnero (2015). The criteria and subcriteria used in the multicriteria model are:

- Annual water consumption (C1).
- Annual energy consumption. Two subcriteria are considered:
    - Annual consumption (MW/h) by the Hospital of electricity, refrigerating energy, thermal energy and natural gas (C2).
    - Consumption of renewable energies (C3).
- Environmental accidents and incidents (C4). Accidents have potentially serious environmental implications, such as uncontrolled spillages into the water supply, fire, x-ray emissions, spillage of dangerous substances on the ground, leaks or spillages of natural gas or diesel, spillage of acetylene or refrigerating gas, leaks of ethylene oxide, etc. Incidents are matters that give rise to internal non-compliance with standard ISO 14001 and require an analysis of the cause and corrective action. Although they are not as serious as environmental accidents, they should be analysed and their causes eliminated as quickly as possible. For example, unsuitable storage of waste or chemicals, complaints from patients or neighbours about noise, etc.
- Biodiversity (C5). An assessment is made of the capacity of the organization to adapt to the rural and forest environment, to take care of endangered species, and also of action taken to continuously improve the environmental impact of the organization.
- Activities to promote and spread the environmental message (C6). This includes annual planning of congresses, celebration, exhibitions, promotion of activities related to environmental prevention and protection, spreading of awareness of environmental aims of the Health Care Organization, etc.
- Training and cooperation on environmental matters (C7). The organization is assessed with respect to training programmes among care and non-care staff in the Health Care Organization, and planning of groups to analyse problems and improvements. This also considers the exis-

tence of surveys and systems for gathering complaints and suggestions, and for dealing with them efficiently.

- Noise inside and outside. This criterion is made up of the subcriteria:
  - Noise inside the Hospital (C8). This should not exceed 30dBA, and optimum noise is between 15 and 25 dBA.
  - Noise near the Hospital (C9). This should not exceed 55dBA, and optimum noise is between 35 and 45dBA.
- Waste production. Assess annual waste production. It is divided into the subcriteria:
  - Group I waste (C10). This is general waste, with no risk, such as edible oils, plastics, paper and cardboard, clothes, glass, etc.
  - Group II waste (C11). Sanitary waste that may be treated as urban waste.
  - Group III waste (C12). Dangerous waste products, including industrial oil, batteries, non-halogenated solvents, chemical waste, radiology liquid, out-of-date or retired medicines, anatomical remains with formaldehyde, cytostatic waste, etc.

Other waste, in other groups, including radioactive waste, is subject to special legislation which must be complied with, and so is not included in the assessment system.

- Green purchasing (C13). Inclusion in conditions for purchasing products and contracting services, of guarantees of respectful treatment of the environment. Assessment of suppliers with regard to certification with standard ISO 14001. Minimizing consumption of paper, cardboard and plastic in transactions and packages, etc.

All the criteria and/or subcriteria are assessed with respect to the number of annual admissions or services provided by the Health Care Organization. This means that the results can be compared over time for a single organization, and comparisons can also be made between Health Care Organizations (Carnero, 2015).

Although different multicriteria techniques can be used to build the model, in this case, applying fuzzy TOPSIS allows simultaneous valuation of a significant number of criteria.

A fuzzy MCDA will be used, as this allows the uncertainty, ambiguous situations or vagueness of the judgements of the decision makers to be included (SeongKon et al., 2011). Furthermore, decision makers usually feel more confident in giving interval judgements rather than fixed value judgments (Isaai et al., 2011).

A triangular fuzzy number $\tilde{a} = \left(l,m,u\right)$ is defined by the membership function $\mu_{\tilde{a}}\left(x\right)$ which satisfies the conditions of normality and convexity. l, m and u are real numbers which satisfy $\leq m \leq u$. The membership function is defined in Equation (1) (Chang, 1996).

$$\mu_{\tilde{a}}\left(x\right) = \begin{cases} \dfrac{x}{m-l} - \dfrac{l}{m-l} & x \in \left[l,m\right] \\ \dfrac{x}{m-u} - \dfrac{u}{m-u} & x \in \left[m,u\right] \\ 0 & otherwise \end{cases} \tag{1}$$

Kaufmann and Gupta (1988) give the main algebraic operations of triangular fuzzy numbers $\tilde{A} = (l,m,u)$ and $\tilde{B} = (l,m,u)$:

$$(l_1,m_1,u_1) \oplus (l_2,m_2,u_2) = (l_1 + l_2, m_1 + m_2, u_1 + u_2) \tag{2}$$

$$(l_1,m_1,u_1) \ominus (l_2,m_2,u_2) = (l_1 - u_2, m_1 - m_2, u_1 - l_2) \tag{3}$$

$$(l_1,m_1,u_1) \otimes (l_2,m_2,u_2) \approx (l_1 l_2, m_1 m_2, u_1 u_2) \tag{4}$$

$$(l_1,m_1,u_1) \oslash (l_2,m_2,u_2) \approx (l_1 / u_2, m_1 / m_2, u_1 / l_2) \tag{5}$$

$$(l_1,m_1,u_1)^{-1} \approx (1/u_1, 1/m_1, 1/l_1) \quad \text{for } l,m,u > 0 \tag{6}$$

$$\lambda \otimes (l_1,m_1,u_1) \approx (\lambda l_1, \lambda m_1, \lambda u_1), \quad \text{»} > 0,\ \text{»} \in \mathrm{R}^+ \tag{7}$$

Fuzzy TOPSIS (FTOPSIS) was proposed by Chen (2000). In a decision problem with criteria $(C_1, C_2, \ldots, C_n)$ and alternatives $(A_1, A_2, \ldots, A_m)$, the best alternative in FTOPSIS is such that should have the shortest distance to a fuzzy positive ideal solution (FPIS) and the farthest distance from a fuzzy negative ideal solution (FNIS). The FPIS is calculated using the best performance values for each criterion and the FNIS looks at the worst performance values.

In FTOPSIS the decision makers use linguistic variables to obtain the weightings of the criteria and the ratings of the alternatives. If there is a decision group made up of K individuals, the fuzzy weight and rating of the *k*th decision maker with respect to the *i*th alternative in the *j*th criterion are respectivamente:

$$\tilde{w}_j^k = (w_{j1}^k, w_{j2}^k, w_{j3}^k) \tag{8}$$

$$\tilde{x}_{ij}^k = (a_{ij}^k, b_{ij}^k, c_{ij}^k) \tag{9}$$

where $i = 1,2,\ldots,m$ and $j = 1,2,\ldots,n$.

The aggregate fuzzy weights $\tilde{w}_{ij}$ of each criterion given by *K* decision makers are calculated using Equation (10).

$$\tilde{w}_j = \frac{1}{K} \otimes \left(\tilde{w}_j^1 \oplus \tilde{w}_j^2 \oplus \ldots \oplus \tilde{w}_j^k \right) \tag{10}$$

While to calculate the aggregate ratings of the alternatives Equation (11) is used (Ouma, Opudo, & Nyambenya, 2015).

$$\tilde{x}_{ij} = \frac{1}{K} \otimes \left(\tilde{x}_{ij}^1 \oplus \tilde{x}_{ij}^2 \oplus \ldots \oplus \tilde{x}_{ij}^k \right) \tag{11}$$

A fuzzy multicriteria group decision-making problem which can be expressed in matrix format is shown in Equation (12) (Chen, 2000).

$$D = \begin{bmatrix} \tilde{x}_{11} & \tilde{x}_{12} & \cdots & \tilde{x}_{1n} \\ \tilde{x}_{21} & \tilde{x}_{22} & \cdots & \tilde{x}_{21} \\ \cdot & \cdot & \cdot & \cdot \\ \cdot & \cdot & \cdot & \cdot \\ \cdot & \cdot & \cdot & \cdot \\ \tilde{x}_{m1} & \tilde{x}_{m2} & \cdots & \tilde{x}_{mn} \end{bmatrix}$$

$$\tilde{W} = \left(\tilde{w}_1, \tilde{w}_2, \ldots, \tilde{w}_n \right) \tag{12}$$

where $\tilde{w}_j$ and $\tilde{x}_{ij}$ are linguistic variables which can be described by triangular fuzzy numbers.

The weightings of the criteria can be calculated by assigning directly the linguistic variables shown in Table 1. The ratings of the alternatives are found using the linguistic variables of Table 2 (Chen, 2000).

The linear scale transformation is used to transform the various criteria scales into a comparable scale. And thus we obtain the normalized fuzzy decision matrix R (Rodrigues, Osiro, & Ribeiro, 2014).

Table 1. Linguistic variables for the weights

Linguistic Variables for the Weights	Fuzzy Number
Very Low (VL)	(0, 0, 0.1)
Low (L)	(0, 0.1, 0.3)
Medium Low (ML)	(0.1, 0.3, 0.5)
Medium (M)	(0.3, 0.5, 0.7)
Medium High (MH)	(0.5, 0.7, 0.9)
High (H)	(0.7, 0.9, 1.0)
Very High (VH)	(0.9, 1.0, 1.0)

(Chen, 2000)

Table 2. Linguistic variables for the ratings

Linguistic Variables for the Ratings	Fuzzy Number
Very Poor (VP)	(0,0,1)
Poor (P)	(0,1,3)
Medium Poor (MP)	(1,3,5)
Fair (F)	(3,5,7)
Medium Good (MG)	(5,7,9)
Good (G)	(7,9,10)
Very Good (VG)	(9,10,10)

(Chen, 2000)

$$\tilde{R} = \left[\tilde{r}_{ij}\right]_{mxn} \quad i = 1,2,\ldots,m; j = 1,2,\ldots,n.. \tag{13}$$

where

$$\tilde{r}_{ij} = \left(\frac{l_{ij}}{u_j^+}, \frac{m_{ij}}{u_j^+}, \frac{u_{ij}}{u_j^+}\right) and\, u_j^+ = max_i u_{ij} \text{ in the case of benefit criteria type}$$

$$\tilde{r}_{ij} = \left(\frac{l_j^-}{u_{ij}}, \frac{l_j^-}{m_{ij}}, \frac{l_j^-}{l_{ij}}\right) and\, l_j^- = max_i l_{ij} \text{ in the case of cost criteria type}$$

Next, the weighted normalized decision matrix, $\tilde{V}$ is calculated, by multiplying the weightings of the criteria $\tilde{w}_j$, by the elements $\tilde{r}_{ij}$ of the normalized fuzzy decision matrix.

$$\tilde{V} = \left[\tilde{v}_{ij}\right]_{mxn} \quad \text{donde}\, \tilde{v}_{ij} = \tilde{x}_{ij} \otimes \tilde{w}_j \tag{14}$$

The distances d_i^+ and d_i^- of each weighted alternative from the FPIS and FNIS are calculated using Equations (15) and (16).

$$d_i^+ = \sum_{j=1}^{n} d_\nu\left(\tilde{\nu}_{ij}, \nu_{ij}^+\right) \tag{15}$$

$$d_i^- = \sum_{j=1}^{n} d_\nu\left(\tilde{\nu}_{ij}, \nu_{ij}^-\right) \tag{16}$$

where $d_\nu\left(\tilde{a}, \tilde{b}\right)$ is the distance measured between the fuzzy numbers $\tilde{a}$ and $\tilde{b}$. This distance is calculated from Equation (17) (Rodrigues, Osiro, & Ribeiro, 2014).

$$d_\nu\left(\tilde{a}, \tilde{b}\right) = \sqrt{\frac{1}{3}\left[\left(l_a - l_b\right)^2 + \left(m_a - m_b\right)^2 + \left(u_a - u_b\right)^2\right]} \tag{17}$$

Finally, the closeness coefficient, CC_i, is calculated for each alternative *i* using Equation (18). This parameter allows the degree of fuzzy satisfaction to be evaluated for each Health Care Organization.

$$CC_i = \frac{d_i^-}{d_i^- + d_i^+} \tag{18}$$

SOLUTIONS AND RECOMMENDATIONS

The model was applied to a Health Care Organization whose mission is working to improve the health of the people it serves, with quality, safety and sustainability. It has 30 medical specialities, 386 beds and 1,744 staff, and provides medical cover to an area with 300,000 inhabitants.

An expert in environmental matters was used to obtain a pairwise comparison matrix for the decision criteria. The decision maker was asked to evaluate the importance of the criteria or subcriteria that could be assessed by applying the fuzzy scale set out in Table 1. The resulting weightings are shown in Table 3. Next, the procedure is applied to evaluate each alternative (different years to be assessed in the Health Care Organization) via the linguistic variables from Table 2. This gives the fuzzy weighted normalized decision matrix shown in Table 4. The distances d_i^+ and d_i^-. of each weighted alternative from the FPIS and FNIS are calculated using Equations (15) and (16), giving the results shown in Table 5.

Table 3. Fuzzy weights

Criteria/Subcriteria	Weightings
C1	(0.900, 1.000, 1.000)
C2	(0.900, 1.000, 1.000)
C3	(0.300, 0.500, 0.700)
C4	(0.900, 1.000, 1.000)
C5	(0.700, 0.900, 1.000)
C6	(0.100, 0.300, 0.500)
C7	(0.300, 0.500, 0.700)
C8	(0.300, 0.500, 0.700)
C9	(0.100, 0.300, 0.500)
C10	(0.100, 0.300, 0.500)
C11	(0.300, 0.500, 0.700)
C12	(0.900, 1.000, 1.000)
C13	(0.300, 0.500, 0.700)

(Created by the author)

Table 4. The fuzzy weighted normalized decision matrix

	C1	C2	C3	C4	C5	C6	C7	C8	C9	C10	C11	C12	C13
Hospital year 1	(0.630, 0.900, 1.000)	(0.630, 0.900, 1.000)	(0.210, 0.450, 0.700)	(0.300, 0.556, 0.778)	(0.350, 0.630, 0.900)	(0.050, 0.210, 0.450)	(0.090, 0.250, 0.490)	(0.129, 0.357, 0.700)	(0.043, 0.214, 0.500)	(0.090, 0.300, 0.500)	(0.270, 0.500, 0.700)	(0.630, 0.900, 1.000)	(0.150, 0.350, 0.630)
Hospital year 2	(0.810, 1.000, 1.000)	(0.810, 1.000, 1.000)	(0.270, 0.500, 0.700)	(0.300, 0.556, 0.778)	(0.490, 0.810, 1.000)	(0.070, 0.270, 0.500)	(0.270, 0.500, 0.700)	(0.129, 0.357, 0.700)	(0.043, 0.214, 0.500)	(0.090, 0.300, 0.500)	(0.270, 0.500, 0.700)	(0.810, 1.000, 1.000)	(0.210, 0.450, 0.700)
Hospital year 3	(0.810, 1.000, 1.000)	(0.810, 1.000, 1.000)	(0.270, 0.500, 0.700)	(0.500, 0.778, 1.000)	(0.350, 0.630, 0.900	(0.090, 0.300, 0.500)	(0.270, 0.500, 0.700)	(0.129, 0.357, 0.700)	(0.043, 0.214, 0.500)	(0.090, 0.300, 0.500)	(0.270, 0.500, 0.700)	(0.810, 1.000, 1.000)	(0.270, 0.500, 0.700)

(Created by the author)

Table 5. The distance measurement

Alternatives	d_i^+	d_i^-
Hospital year 1	7.035	6.984
Hospital year 2	6.223	7.700
Hospital year 3	6.104	7.813

(Created by the author)

The closeness coefficient of each alternative is set out in Table 6. From *CC* the ranking of the three alternatives is Hospital year 3, year 2 and year 1. The best result is obtained by the hospital in the third year assessed. This shows the process of continuous improvement undertaken by the Health Care Organization and allows a global follow-up to be carried out annually.

FUTURE RESEARCH DIRECTIONS

AHP and fuzzy AHP are the most widely-used multicriteria techniques in the literature in general, and specifically in relation to the environment. However, this Chapter uses the fuzzy TOPSIS technique, due to the ease with which it can assess a large number of criteria and alternatives, but it would be appropriate to validate the environmental assessment models using other techniques successfully applied in a significant number of real cases, such as the Measuring Attractiveness by a Categorical Based Evaluation Technique (MACBETH).

Due to the constant evolution of information communication systems and the development of new technologies, with very dramatic effects on improvements in energy efficiency and consumption of natural resources, this model should be periodically reviewed to include new criteria or to up-date their definitions.

CONCLUSION

Health Care Organizations have an essential responsibility to the environment, being significant consumers of natural resources, as well as producers of large quantities of waste, some of which can cause serious risk to people and the environment unless they are properly handled.

Table 6. Closeness coefficient of each year assessed

Alternatives	CC
Hospital year 1	0.498
Hospital year 2	0.553
Hospital year 3	0.561

(Created by the author)

The monitoring of a number of environmental matters is key to assessing continuous improvement in the actions undertaken. However, this monitoring requires objective tools which consider a series of criteria adapted to each organization.

This Chapter therefore presents a multicriteria model for assessing environmental responsibility in Health Care Organizations. The intention is that the model, although based on mathematical tools, should be easy to apply, and should take into account the uncertainties and ambiguities which characterize the real-life decision process. The fuzzy TOPSIS technique was thus used, as it allows a large number of alternatives to be assessed in a simple manner, which means the results obtained can be compared over time.

To test the utility of the model, it was applied in a public Health Care Organization over three consecutive years, showing how the improvement actions undertaken increase the overall utility of the result.

ACKNOWLEDGMENT

This research was supported by the Junta de Comunidades de Castilla-La Mancha and the European Regional Development Fund under Grant number PPII-2014-013-P.

REFERENCES

Al Garni, H., Kassem, A., Awasthi, A., Komljenovic, D., & Al-Haddad, K. (2016). A multicriteria decision making approach for evaluating renewable power generation sources in Saudi Arabia. *Sustainable Energy Technologies and Assessments*, *16*, 137–150. doi:10.1016/j.seta.2016.05.006

Aragonés-Beltran, P., Mendoza-Roca, J. A., Bes-Pia, A., García-Melon, M., & Parra-Ruiz, E. (2009). Application of multicriteria decision analysis to jar-test results for chemicals selection in the physical–chemical treatment of textile wastewater. *Journal of Hazardous Materials*, *164*(1), 288–295. doi:10.1016/j.jhazmat.2008.08.046 PMID:18829168

Bon-García, J. V. (2012). *Declaración ambiental año 2011*. Consorci Hospital General Universitari Valencia.

Boran, F. E., Boran, K., & Menlik, T. (2012). The evaluation of renewable energy technologies for electricity generation in Turkey using intuitionistic fuzzy TOPSIS. *Energy Sources Part B*, *7*(1), 81–90. doi:10.1080/15567240903047483

Carnero, M. C. (2014). *Model for Sustainability in Health Care Organizations*. In J. Wang (Ed.), *Encyclopedia of Business Analytics and Optimization* (Vol. 3, pp. 437–456). Hershey, PA: IGI Global.

Carnero, M. C. (2015). Assessment of Environmental Sustainability in Health Care Organizations. *Sustainability*, *7*(7), 8270–8291. doi:10.3390u7078270

Chan, H. K., Wang, X., & Raffoni, A. (2012). An integrated approach for green design: Life-cycle, fuzzy AHP and environmental management accounting. *Procedia: Social and Behavioral Sciences*, *57*, 236–242.

Chang, D. Y. (1996). Applications of the extent analysis method on fuzzy AHP. *European Journal of Operational Research*, *95*(3), 649–655. doi:10.1016/0377-2217(95)00300-2

Chen, C. T. (2000). Extension of the TOPSIS for group decision –making under fuzzy environment. *Fuzzy Sets and Systems*, *114*(1), 1–9. doi:10.1016/S0165-0114(97)00377-1

Cintra, Y. C., & Carter, D. (2012). Internalising sustainability: Reflections on management control in Brazil. *International Journal of Strategic Management*, *12*(2), 108–125.

Comunidad de Madrid. (2005). *Guía para la implantación de sistemas de gestión medioambiental en centros hospitalarios*. Madrid: Dirección General de Promoción y Disciplina Ambiental.

Egilmez, G., Gumus, S., & Kucukvar, M. (2015). Enviromental sustainability benchmarking of the U.S. and Canada metropoles: An expert judgment-based multi-criteria decision making approach. *Cities (London, England)*, *42*, 31–41. doi:10.1016/j.cities.2014.08.006

Galdakao-Usansolo Hospital. (2012). *Declaración Ambiental. Ejercicio 2012*. Available online: http:// www. hospitalgaldakao-usansolo.com

Galvez, D., Rakotondranaivo, A., Morel, L., Camargo, M., & Fick, M. (2015). Reverse logistics network design for a biogas plant: An approach based on MILP optimization and Analytical Hierarchical Process (AHP). *Journal of Manufacturing Systems*, *37*(3), 616–623. doi:10.1016/j.jmsy.2014.12.005

Garcia, S., Cintra, Y., Torres, R. C. S. R., & Lima, F. G. (2016). Corporate sustainability management: A proposed multi-criteria model to support balanced decision-making. *Journal of Cleaner Production*, *136*, 181–196. doi:10.1016/j.jclepro.2016.01.110

Gumus, A. T. (2009). Evaluation of hazardous waste transportation firms by using a two step fuzzy AHP and TOPSIS methodology. *Expert Systems with Applications, 36*(2), 4067-4074.

Heo, E., Kim, J., & Boo, K. J. (2010). Analysis of the assessment factors for renewable energy dissemination program evaluation using fuzzy AHP. *Renewable & Sustainable Energy Reviews*, *14*(8), 2214–2220. doi:10.1016/j.rser.2010.01.020

Hermann, B. G., Kroeze, C., & Jawjit, W. (2007). Assessing environmental performance by combining life cycle assessment, multi-criteria analysis and environmental performance indicators. *Journal of Cleaner Production*, *15*(18), 1787–1796. doi:10.1016/j.jclepro.2006.04.004

Higgs, G., Berry, R., Kidner, D., & Langford, M. (2008). Using IT approaches to promote public participation in renewable energy planning: Prospects and challenges. *Land Use Policy*, *25*(4), 596–607. doi:10.1016/j.landusepol.2007.12.001

Hsu, C. W., & Hu, A. H. (2008). Green supply chain management in the electronic industry. *International Journal of Environmental Science and Technology*, *5*(2), 205–216. doi:10.1007/BF03326014

Huang, I. B., Keisler, J., & Linkov, I. (2011). Multi-criteria decision analysis in environmental sciences: Ten years of applications and trends. *The Science of the Total Environment*, *409*(19), 3578–3594. doi:10.1016/j.scitotenv.2011.06.022 PMID:21764422

Isaai, M. T., Kanani, A., Tootoonchi, M., & Afzali, H. R. (2011). Intelligent timetable evaluation using fuzzy AHP. *Expert Systems with Applications*, *38*(4), 3718–3723. doi:10.1016/j.eswa.2010.09.030

Kang, H. Y., Ogunseitan, O., Shapiro, A. A., & Schoenung, J. M. (2007). A comparative hierarchical decision framework on toxics use reduction effectiveness for electronic and electrical industries. *Environmental Science & Technology*, *41*(2), 373–379. doi:10.1021/es060739u PMID:17310694

Kaufmann, A., & Gupta, M. M. (1988). *Fuzzy Mathematical Models in Engineering and Management Science*. Amsterdam: North Holland.

Kaya, T., & Kahraman, C. (2011). An integrated fuzzy AHP–ELECTRE methodology for environmental impact assessment. *Expert Systems with Applications*, *38*(7), 8553–8562. doi:10.1016/j.eswa.2011.01.057

Khalili, N. R., & Duecker, S. (2013). Application of multi-criteria decision analysis in design of sustainable environmental management system framework. *Journal of Cleaner Production*, *47*, 188–198. doi:10.1016/j.jclepro.2012.10.044

Lahdelma, R., Salminen, P., & Hokkanen, J. (2002). Locating a waste treatment facility by using stochastic multicriteria acceptability analysis with ordinal criteria. *European Journal of Operational Research*, *142*(2), 345–356. doi:10.1016/S0377-2217(01)00303-4

Lamelas, M. T., Marinoni, O., Hoppe, A., & de la Riva, J. (2008). Suitability analysis for sand and gravel extraction site location in the context of a sustainable development in the surroundings of Zaragoza (Spain). *Environmental Geology*, *55*(8), 1673–1686. doi:10.100700254-007-1116-9

Larimian, T., Zarabadi, Z. S. S., & Sadeghi, A. (2013). Developing a fuzzy AHP model to evaluate environmental sustainability from the perspective of Secured by Design scheme-A case study. *Sustainable Cities and Society*, *7*, 25–36. doi:10.1016/j.scs.2012.10.001

Lee, S. K., Mogi, G., Lee, S. K., & Kim, J. W. (2011). Prioritizing the weights of hydrogen energy technologies in the sector of the hydrogen economy by using a fuzzy AHP approach. *International Journal of Hydrogen Energy*, *36*(2), 1897–1902. doi:10.1016/j.ijhydene.2010.01.035

Liang, Z. H., Yang, K., Sun, Y. W., Yuan, J. H., Zhang, H. W., & Zhang, Z. Z. (2006). Decision support for choice optimal power generation projects: Fuzzy comprehensive evaluation model based on the electricity market. *Energy Policy*, *34*(17), 3359–3364. doi:10.1016/j.enpol.2005.06.021

Madu, C. N., Kuei, C., & Madu, I. E. (2002). A hierarchic metric approach for integration of green issues in manufacturing: A paper recycling application. *Journal of Environmental Management*, *64*(3), 261–272. doi:10.1006/jema.2001.0498 PMID:12040959

Mata, F., García, J. A., Sierra, S., Giordani, V., Gutiérrez, F. J., Márquez, M., ... Cerrato, R. J. (2011). *EMAS 2009. Declaración ambiental EP Hospital Costa del Sol*. Consejería de Salud de la Junta de Andalucía.

Munda, G. (2005). Measuring sustainability: A multi-criterion framework. *Environment, Development and Sustainability*, *2*(1), 117–134. doi:10.100710668-003-4713-0

Neste, J., & Karjalainen T. P. (2013). *A literature review – The use of multi-criteria decision analysis in Environmental Impact Assessment, Annex 7.2.5.* Inception report IMPERIA EU LIFE11 ENV/FI/905 Action: B1B Report on the use of MCDA in EIA and SEA, 25.4.2013

Ouma, Y. O., Opudo, J., & Nyambenya, S. (2015). Comparison of Fuzzy AHP and Fuzzy TOPSIS for Road Pavement Maintenance Prioritization: Methodological Exposition and Case Study. *Advances in Civil Engineering*, Article ID 140189. .10.1155/2015/140189

Pilavachi, P. A., Chatzipanagi, A. I., & Spyropoulou, A. I. (2009). Evaluation of hydrogen production methods using the analytic hierarchy process. *International Journal of Hydrogen Energy, 34*(13), 5294–5303. doi:10.1016/j.ijhydene.2009.04.026

Pourebrahim, S., Hadipour, M., Mokhtar, M. B., & Taghavi, S. (2014). Application of VIKOR and fuzzy AHP for conservation priority assessment in coastal areas: Case of Khuzestan district, Iran. *Ocean and Coastal Management, 98*, 20–26. doi:10.1016/j.ocecoaman.2014.05.009

Reza, B., Sadiq, R., & Hewage, K. (2011). Sustainability assessment of flooring systems in the city of Tehran: An AHP-based life cycle analysis. *Construction & Building Materials, 25*(4), 2053–2066. doi:10.1016/j.conbuildmat.2010.11.041

Rodrigues, F., Osiro, L., & Ribeiro, L. C. (2014). A comparison between Fuzzy AHP and Fuzzy TOPSIS methods to supplier selection. *Applied Soft Computing, 21*, 194–209. doi:10.1016/j.asoc.2014.03.014

Rodríguez, E., Carmona, M., & Soto, J. (2005). *Memoria sostenibilidad del hospital clínico San Carlos.* Madrid, Spain: Hospital Clínico San Carlos.

Salvado, M. F., Azevedo, S. G., Matias, J. C. O., & Ferreira, L. M. (2015). Proposal of a Sustainability Index for the Automotive Industry. *Sustainability, 7*(2), 2113–2144. doi:10.3390u7022113

Santoyo-Castelazo, E., & Azapagic, A. (2014). Sustainability assessment of energy systems: Integrating environmental, economic and social aspects. *Journal of Cleaner Production, 80*, 1–20. doi:10.1016/j.jclepro.2014.05.061

Tejedor, J. (n.d.). *Declaración Ambiental Enero de 2011-Diciembre de 2011, hm Hospitales.* Available online: http://www.hmhospitales.com/grupohm/Paginas/HMHospitales.aspx

Tseng, M. L., Lin, Y. H., & Chiu, A. S. F. (2009). Fuzzy AHP-based study of cleaner production implementation in Taiwan PWB manufacturer. *Journal of Cleaner Production, 17*(14), 1249–1256. doi:10.1016/j.jclepro.2009.03.022

Tzeng, G. H., Lin, C. W., & Opricovic, S. (2005). Multi-criteria analysis of alternative-fuel buses for public transportation. *Energy Policy, 33*(11), 1373–1383. doi:10.1016/j.enpol.2003.12.014

Van Calker, K. J., Berentsen, P. B. M., Romero, C., Giesen, G. W. J., & Huirne, R. B. M. (2006). Development and application of a multi-attribute sustainability function for Dutch dairy farming systems. *Ecological Economics, 57*(4), 640–658. doi:10.1016/j.ecolecon.2005.05.016

Viaggi, D., Finn, J. A., Kurz, I., & Bartolini, F. (2011). Multicriteria analysis for environmental assessment of agri-environment schemes: How to use partial information from Mid-Term Evaluations? *Agricultural Economics Research, 12*(1), 5–21.

Vinodh, S., Prasanna, M., & Prakash, N. H. (2014). Integrated fuzzy AHP–TOPSIS for selecting the best plastic recycling method: A case study. *Applied Mathematical Modelling*, *38*(19-20), 4662–4672. doi:10.1016/j.apm.2014.03.007

Wang, X., Chan, H. K., Yee, R. W. Y., & Diaz-Rainey, I. (2012). A two-stage fuzzy AHP model for risk assessment of implementing green initiatives in the fashion supply chain. *International Journal of Production Economics*, *135*(2), 595–606. doi:10.1016/j.ijpe.2011.03.021

Yanguas, G. (2012). *Perfil ambiental de España. Informe basado en indicadores*. Ministerio de Agricultura, Alimentación y Medio Ambiental Español. Retrieved may 6, 2013 from http://www.magrama.gob.es/es/

Zhang, L., Xu, Y., Yeh, C. H., Liu, Y., & Zhou, D. (2016). City sustainability evaluation using multi-criteria decision making with objective weights of interdependent criteria. *Journal of Cleaner Production*, *131*(10), 491–499. doi:10.1016/j.jclepro.2016.04.153

ADDITIONAL READING

Griffiths, J. (2006). Environmental sustainability in the national health service in England. *Public Health*, *120*(7), 609–612. doi:10.1016/j.puhe.2006.04.005 PMID:16750230

Herva, M., & Roca, E. (2013). Review of combined approaches and multi-criteria analysis for corporate environmental evaluation. *Journal of Cleaner Production*, *39*, 355–371. doi:10.1016/j.jclepro.2012.07.058

Janssen, R. (2001). On the use of multi-criteria analysis in environmental impact assessment in The Netherlands. *Journal of multi-criteria decision analysis - Optimization. Learning and Decision Support*, *10*(2), 101–109.

Joshi, H., & Latif, M. A. (2004). Environmental impact assessment of water resource projects with matrix and multi-criteria decision-making methods: A case study. *Journal of Environmental Systems*, *3*(3), 207–231. doi:10.2190/E7BK-WAHF-QHF5-V6P6

Mendoza, G. A., & Prabhu, R. (2003). Qualitative multi-criteria approaches to assessing indicators of sustainable forest resource management. *Forest Ecology and Management*, *172*(1-3), 329–343. doi:10.1016/S0378-1127(02)00044-0

Munda, G., Nijkamp, P., & Rietveld, P. (1994). Qualitative multicriteria evaluation for environmental management. *Ecological Economics*, *10*(2), 97–112. doi:10.1016/0921-8009(94)90002-7

Schaltegger, S., Lüdeke-Freund, F., & Hansen, E. (2012). Business Cases for Sustainability. The Role of Business Model Innovation for Corporate Sustainability. *International Journal of Innovation and Sustainable Development*, *6*(2), 95–119. doi:10.1504/IJISD.2012.046944

Uysal, F., & Tosun, Ö. (2012). Fuzzy TOPSIS-based computerized maintenance management system selection. *Journal of Manufacturing Technology Management*, *23*(2), 212–228. doi:10.1108/17410381211202205

World Commission on Environment and Development. (1987). *Our common future*. Oxford: Oxford University Press.

KEY TERMS AND DEFINITIONS

Multi-Criteria Decision Analysis: It is a part of operations research that use multiple criteria in decision-making processes providing acceptable compromise solutions when criteria are in conflict. There is a relevant quantity of tools belonging to this category, for example Analytic Hierarchy Process (AHP), Analytic Network Process (ANP), Measuring Attractiveness by a Categorical Based Evaluation Technique (MACBETH), ELimination Et Choix Traduisant la REalité (ELECTRE), Multi-Attribute Utility Theory (MAUT), The Preference Ranking Organization METHod for Enrichment of Evaluations (PROMETHEE), etc. The MCDA techniques allow to construct objective models to improve understanding of underlying decision processes in the systemic processes.

TOPSIS: The Technique for Order of Preference by Similarity to Ideal Solution (TOPSIS) is a multi-criteria decision-making approach created by Hwang and Yoon in 1981. It is a compensatory aggregation method based on the concept that the best alternative should have the shortest geometric distance to a positive ideal solution (PIS) and the geometric farthest distance from a negative ideal solution (NIS).

This research was previously published in the Encyclopedia of Information Science and Technology, Fourth Edition edited by Mehdi Khosrow-Pour, pages 3131-3143, copyright year 2018 by Information Science Reference (an imprint of IGI Global).

Chapter 17
Potential Benefits and Current Limits in the Development of Demand Response

Clementina Bruno
University of Piemonte Orientale, Italy

ABSTRACT

This chapter, after defining demand response (DR) and its potential benefits, illustrates a set of challenges to DR development. A brief review of recent contributions on DR is provided, illustrating that such challenges can come from different sources. Regulatory, technical, or socio-economic challenges are considered and discussed. Finally, inter-disciplinary research is suggested as solution to overcome challenges, and some examples of future research directions with respect to economics and social science are provided.

INTRODUCTION

Once upon a time, for many families, electricity was a somehow magic and mysterious stuff allowing houses lighting and appliances operation, whose secrets began just behind the switch or the socket. Other, more informed, users knew that it came from generation plants and "travelled" along a grid towards houses or firms. Nowadays, the role of end users has changed a lot. They have a broader knowledge of the electric system, and a certain awareness of being part of it, in some cases not simply as consumption units. Distributed generation (such as residential photovoltaic production) and demand response mechanisms have transformed (residential, industrial or commercial) users in an active part of the electricity supply chain, so that they are often defined as "prosumer" (Crispim et al., 2014).

In particular, Demand Response (DR) is attracting increasing attention from regulators, policy makers and system operators due to its large potential in supporting and, in some cases, substituting generation in providing flexibility to the system. This corresponds, on the academic side, to an exponential growth of scientific production, with focus on the technical or on the socio-economic features of the issue. This chapter will provide a review of some recent contributions on this topic. Far from being exhaustive of

DOI: 10.4018/978-1-5225-7359-3.ch017

the extremely wide related literature, the aim of this chapter is to provide a general presentation of the issue, briefly discussing the main benefits related to DR, as well as the most relevant regulatory, technological and socio-economic challenges that can slow down or hinder its development. Therefore, this work will provide an analysis of the impact and issues related to DR from a socio-economic perspective. Moreover, it will also briefly consider the role of technology (especially information and communication technology) in supporting DR implementation and, more in general, the evolution towards "smart" systems. The rest of the chapter is organized as follows. The next section defines DR and illustrates the most relevant benefits of its development. Subsequently, challenges to DR development are discussed and some recommendations are provided. Future research directions and conclusions close the work.

BACKGROUND

The literature provides a wide set of definitions of DR. Quite common across these definitions is the focus on end-users and on the modification in their electricity utilization patterns (see, for instance, the list provided in Eid et al., 2016). For example, in the FERC (Federal Energy Regulatory Commission) website[1] DR is defined as

Changes in electric usage by demand-side resources from their normal consumption patterns in response to changes in the price of electricity over time, or to incentive payments designed to induce lower electricity use at times of high wholesale market prices or when system reliability is jeopardized.

There are several typologies of DR mechanisms, which can be classified following different criteria (Vardakas et al., 2015). Here we propose the most common classification.

- Time-based retail rates (Cappers et al., 2012), also called rate-based or price-based programs (Siano, 2014), or implicit DR (SEDC - Smart Energy Demand Coalition, 2015), provide incentive to end-users to modify their consumption as response to price variations. Price fluctuations are designed to reflect the dynamics of the wholesale market price or the grid tariff, and ultimately, of the cost of the electric service. Prices can be predetermined but different for given time periods or move dynamically depending on the system and market contingencies.
- Incentive-based retail programs (Cappers et. al, 2012), also defined as event-based programs (Siano, 2014), reliability-based (Shen et al., 2014) or explicit DR schemes (SEDC, 2015) reward consumers through a payment or a bill credit for a reduction in their consumption. Such mechanisms are activated by the entity managing DR services (users can contract directly with the utility or with an aggregator) in response to particular events affecting the electric system, e.g. network congestion[2].

Examples of price-based DR programs are:

- Time of Use tariffs, where prices are different but fixed for given time periods (e.g. times of the day or days of the week).
- Critical Peak Pricing, that applies particularly high prices for a limited period (few hours) in response to critical technical or economic/market events.

- Critical Peak Rebate, where consumers are recognized a bill rebate if they reduce their consumption below a pre-specified baseline during critical hours.
- Real Time Pricing, where prices vary dynamically (e.g. every hour) following the wholesale market price and/or the actual generation costs.

Common forms of incentive-based mechanisms, instead, are:

- Direct Load Control, where the utility has the opportunity to manage directly some consumer's equipment (e.g. air conditioning or heating).
- Interruptible/Curtailable programs, where (usually large) users accept that (a part of) their load can be disconnected, in some cases even without notification.
- Emergency DR programs, which provide end-users a compensation to reduce their loads when the system reliability is endangered.
- Ancillary Service Programs, where consumers provide "reserves" by committing to reduce their load in case of necessity.

For further examples or deeper descriptions of DR programs, see, among the others, Cappers et al. (2012), Darby and McKenna (2012), Shen et al. (2014), Siano (2014), Hu et al. (2015), Vardakas et al. (2015). Figure 1 reports some examples of DR programs.

Figure 1. Examples of DR programs

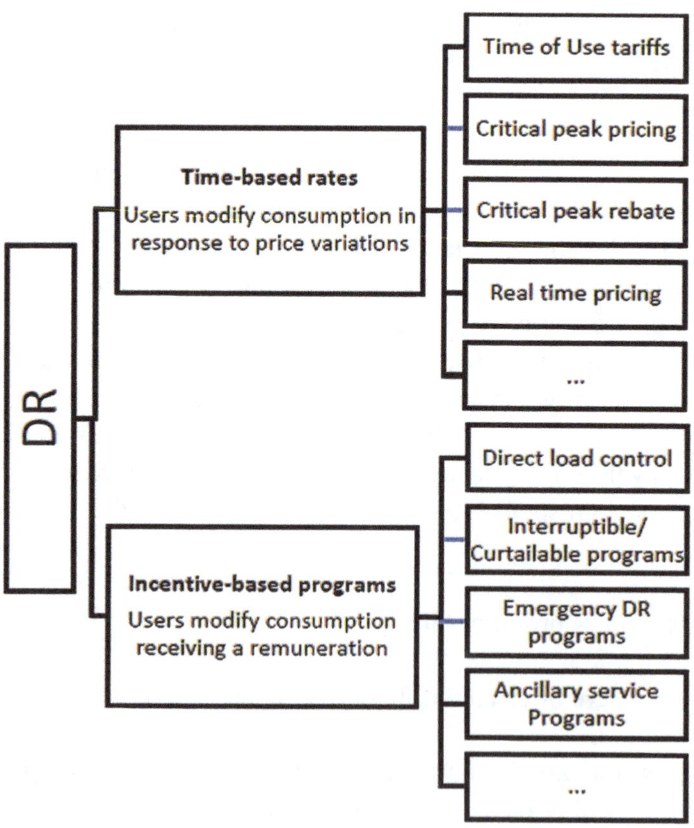

While some mechanisms are well suited even in "traditional" electric systems (e.g. Interruptible/Curtailable programs), other ones present important technological requirements, and can develop their full potential when implemented in smart grid contexts (e.g. Real Time Pricing).

Following Siano (2014; p.462), a smart grid (SG) is "an electric grid able to deliver electricity in a controlled, smart way from points of generation to consumers that are considered as an integral part of the SG since they can modify their purchasing patterns and behavior according to the received information, incentives and disincentives". This definition highlights some relevant peculiarities of SGs with respect to traditional grid structures. First, energy is delivered from generation "points", rather than generation "plants" (thus the definition encompasses also distributed generation). Second, consumers are active subjects, because they provide services to the electric system, by modifying their behaviour. Third, such modifications are driven by (system or economic) information that consumers receive; however, we must notice that also the information flow to the utility is crucial, since it allows the application of the correct incentive or price schemes. This information exchange relies on the implementation of advanced technologies such as those embedded in smart meters, and is not possible with traditional equipment. Therefore, in SGs, both information and energy can flow from the utility to the users and vice versa, as represented in Figure 2. Finally, the delivery of power is "controlled" and "smart", suggesting continuous adjustments of demand and supply to ensure high efficiency in the whole service provision.

DR (especially incentive-based programs) has led to the emergence of a new subject in the electricity value chain, i.e. the aggregator. The aggregator is a market intermediary that "aggregates" the DR capacity from consumers that do not have sufficient dimension, information, knowledge, technology, reliability or availability to participate directly in electricity markets. Usually, the aggregator is an electricity supplier or an independent intermediary (SEDC, 2015; Eid et al., 2015).

The potential positive effects of DR on the electric system and, more generally, on users and on the community as a whole are broadly recognised at the political level. For instance, in the European context, the Electricity Directive (2009) and the Energy Efficiency Directive (2012) recognise the opportunities related to DR development and encourage Member States to foster this form of flexibility and to remove potential barriers (e.g. regulatory barriers) to the effective participation of users in the electricity market.

The advantages deriving from the implementation of DR programs are numerous.

Firstly, DR provides system flexibility that can be effectively employed to contrast the intermittency and limited predictability of some Renewable Energy Sources (RES), such as wind and solar. This is especially the case of Real Time Pricing, within the price-based mechanisms, and of several incentive-based programs, such as Direct Load Control (Cappers et al., 2012). Secondly, DR improves the general system reliability, e.g. by reducing demand in case of outages, thus supporting the system recovery, as highlighted by Siano (2014)[3].Thirdly, DR programs help flattening consumers' demand, for instance by leading them to reduce consumption or to shift it to off-peak periods. In the short term, this reduces the

Figure 2. Energy and information flows in smart grids

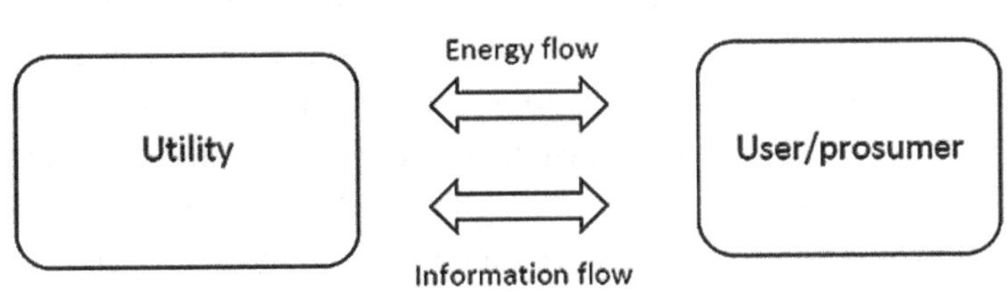

global generation costs, by limiting the use of expensive peak generation technologies. Moreover, in the long term, this reduces or defers new investments in generation or grid capacity. These cost savings can turn into benefits for consumers. In fact, users participating to DR programs may enjoy lower bills or other compensations for their flexibility. Lower tariffs induced by lower wholesale price, however, will benefit all users, even non-participant ones. Fourthly, non-activation of peak units reduces the related emissions. Fifthly, the "distributed" nature of DR resources is also able to reduce the network losses associated to transmission and distribution. Finally, a more elastic demand limits the possibility of generators to exert market power. (Batlle and Rodilla, 2009; Darby and McKenna, 2012; Siano, 2014; Shen et al., 2014; Gils, 2014; Hu et al., 2015; Eid et al., 2016).

Several scientific works support the desirability of DR programs, from different perspectives. For instance, Feuerriegel and Neumann (2016), relying on a mathematical model and German data, find that, when DR is employed for load shifting, it can generate savings in the order of 2.83%. Brouwer et al. (2016) employ a simulation tool to evaluate the economic impact of RES penetration for the year 2050. The results show that system costs would increase with RES penetration, due to investments and RES intermittency, but this effect could be counteracted by some options, including DR development. Dupont et al. (2014) run a simulation based on Belgian data and find that DR reduces system costs and emissions while improving reliability. Schleich et al. (2013) analyse econometrically the results of a trial in Austria and show that receiving feedback on consumption can induce households to reduce electricity use by 4.5% on average. Conversely, Torriti (2012) finds that Time-of-Use tariffs in Northern Italy generate lower bills, but higher average consumption. DR effectiveness for load shifting is supported, but limited to morning peaks. Finally, Bradley et al. (2013) rely on an extensive review of the literature and find support to the economic sustainability of DR in UK markets.

Notwithstanding the broad evidence in favour of DR, its development is just in a starting phase and proceeds slowly, with limited consumers' involvement (Hu et al., 2015). This likely depends on the complexity of adopting and managing DR programs, related to a set of existing challenges hindering DR development. In the following section, some relevant challenges will be discussed.

REGULATORY, TECHNOLOGICAL, AND SOCIO-ECONOMIC CHALLENGES TO DR DEVELOPMENT

Regulatory Challenges: DR Potential and Current Regulation in Europe

In order to understand the potential limits to DR development, we provide an example referred to the European case. SEDC (2015) reports an evaluation of the regulatory framework supporting explicit DR in 2015[4] for 16 European Countries. Regulation is evaluated with respect to four main "areas": "consumer access and aggregation", "program requirements", "measurement and verification", "finance and penalties". Each country receives a score for each area and an overall score resulting from their sum. Higher values correspond to environments more favourable to DR development. Results show that some states have reached an advanced level in the regulation aimed at promoting DR and consumers empowerment, whereas other ones still present relevant inconsistencies between this target and their actual regulatory framework. These differences across countries are highlighted by the diversified grades reported in the last column of table 1, which range from 16 for France and Switzerland (a star indicates existing standardized arrangements involving aggregators) to 3 or 2 for Italy and Spain, respectively.

It is interesting to compare this evaluation with the results reported in Gils (2014). The author provides an assessment of the theoretical DR potential in Europe and some North-African Countries ("theoretical" indicates that the potential includes all equipment available for DR purposes, independently from possible constraints due to technological, economic, legal issues, etc.). Results are detailed by consumer type and by country, and demand flexibility potential is evaluated either in terms of load reduction (due to load shedding or shifting by delaying consumption) and load increase (related to load shifting by anticipating consumption). Focussing only on the 16 Countries analysed in SEDC (2015), we provide, in the second and third columns of table 1, rankings constructed on the basis of the potential load increase or reduction (originally expressed in MW by Gils, 2014). We can see that these two rankings are quite similar, and their comparison with the last column allows some reasoning. France and UK, that present very high DR potentials and favourable regulatory frameworks, seem to be isolated cases. Several other countries with noticeable potential (Germany, Italy, Spain) present very low SEDC grades. Conversely, countries like Switzerland or Ireland, whose regulatory frameworks are supportive to DR, can exploit only limited potential. Keeping in mind the limitations related to the fact that SECD (2015) focuses only on incentive-based DR, nevertheless we can conclude that there is an inconsistency between the regulatory evolution and the amount of DR resources that could be exploited. In fact, several states are neglecting the opportunity offered by DR, or, at least, are accumulating an important delay in fostering its development. This happens especially for countries where DR benefits could be large.

In SEDC (2015), the main regulatory limits in promoting DR seem to be related to the area "Consumers participation and aggregation", for several reasons. For instance, in a number of states, DR resources continue to be excluded from some markets, and generation is still the favourite source of flexibility.

Table 1. DR potential and actual regulation in Europe (our elaboration from SEDC, 2015, and Gils, 2014)

Country	Ranking Based on Potential Load Reduction (Gils, 2014)	Ranking Based on Potential Load Increase (Gils, 2014)	SEDC (2015) Grade
Austria	13	12	10
Belgium	10	11	12
Denmark	14	14	8
Finland	9	10	12
France	2	1	*16
Germany	1	2	6
Ireland	15	15	12
Italy	3	4	3
Netherlands	8	9	10
Norway	11	8	10
Poland	6	6	4
Slovenia	16	16	6
Spain	4	5	2
Sweden	7	7	10
Switzerland	12	13	*16
UK	5	3	12

In some cases, markets are open to DR, but not to aggregated load, thus limiting the participation of medium or small users. Moreover, the possibility of end-users of freely choosing their DR services provider is often limited.

Other limitations appear also with respect to other "areas". For example, excessively high capacity thresholds to access the market, or other unnecessarily binding requirements (probably a result of older market rules designed for large generators only) in terms of DR events duration or frequency can limit consumers' participation. The fairness and the transparency of the methodologies for measuring the actual consumption reduction constitute another issue, together with the procedures to determine the standard consumption "baseline" (i.e. the starting point for evaluating reductions). Moreover, in some cases, participation to DR programs is subject to the provision of relevant bank guarantees, which can constitute a further barrier.

Some of these points are highlighted also in ENTSO-E (European Network of Transmission System Operators for Electricity, 2015), that also underlines, in a discussion including price- and incentive-based programs, the "lack of effectiveness of price signals" (p. 5), as well as the limited price variability in some retail markets, especially for medium-small users. This limits both the ability of prices to reflect actual energy costs and the related consumers' responsiveness. Moreover, to ensure small consumers' participation to DR programs, it is necessary that "price signals remains understandable and manageable" (p. 6).

Technological Challenges: Technology Requirements for Effectively Exploiting DR Potential

In order to be effective in providing DR resources, consumers need to interact actively with suppliers or DR services providers. Moreover, they need to be able to react promptly to price or event-driven signals requiring DR activation. This is often not feasible through manual intervention, especially when signals are frequent or provided with very short notice (Cappers et al., 2012).

To deal with this issue, an extended deployment of smart technologies appear necessary, as well as adequate communication and standardization. Below, these concepts are briefly introduced (see Siano, 2014; Shen et al., 2014; ENTSO-E, 2015; Hu et al., 2015; Vardakas et al., 2015).

Among smart technologies, a key role is played by Advance Metering Infrastructures (AMI), systems able to measure electricity usage, to save and analyse the related data, to receive information from devices and to exchange information with utilities. Important components of AMI are smart meters, electronic devices able to measure users' consumption at fixed time intervals (e.g. one hour) and communicate data to suppliers.

In addition, automation technology allows consumers to program "smart" appliances (such as specifically designed washing machines or refrigerators) or control devices (e.g. "smart thermostats") to automatically respond to event or price signals. For instance, "smart thermostats" can adjust the temperature of rooms in response to electricity price variations. Home Area Networks (HAN) can be employed for connecting all these home devices among each other and with the utilities.

Finally, it is relevant to underline the key role of communication. For an effective use of the communication infrastructure, integration among networks should be promoted, and this would require the adoption of open communication standards. Concerning communication systems (Siano, 2014), they can be one-way (information and DR signals flow from the utility to the end-users), or more expensive but more effective two-way systems, that, additionally, allow utilities to receive feedback in relation to consumers response (e.g. smart meters). They can be wireless (e.g. relying on cellular networks) or wired.

The latter can rely, for instance, on power line communications (PLC, either broadband or narrow band), suitable at local level and not requiring new infrastructure building, or on optic fiber communications, for longer distances. The key role of communication highlights the importance of security (Vardakas et al., 2015), either in terms of protection of consumers' sensitive data and privacy, or in terms of preservation of system and market information from external inference or attacks that can create damages or inconvenience.

Economic and Social Challenges

From the purely economic perspective, two major issues, somehow interrelated, concern the provision of sufficient economic incentives to participate to DR programs (ENTSO-E, 2015) and the distribution of investment costs related to enabling technologies. With respect to the latter point, Eid et al. (2016) underlines the problem of correctly splitting DR cost and benefits along the whole electricity supply chain (e.g. among network operators, retailer and consumers), in such a way to create positive business cases for all the involved parties. Failing in providing proper mechanisms to reach this goal generates the risk that none of the parties would chose to make the "first move" (p. 22). The authors report an example related to investments in smart meters. Moreover, they state that, as the (environmental) benefits of such programs would affect the whole society, this should be the case also for costs. Also Shen et al. (2014), with a similar line of reasoning, suggest that incentives to DR participation and investments could effectively receive public support, in consideration of their positive impact on the society. Hu et al. (2015) states that utilities do not have sufficient incentives to support DR in absence of subsidies. ENTSO-E (2015), however, stresses that general subsidization of DR programs must be minimal and avoid market distortions, since DR "must achieve its full economic potential in fair competition with other sources" (p.8).

Turning, now, to the former point, as highlighted by Cappers et al. (2012), "the opportunities created by the DR service providers must generate sufficient value to customers or else they will eschew these offerings" (p.426). In this perspective, with specific regards to residential users, Darby and McKenna (2012) underlines the key role of rates. Cappers et al. (2012) stress also that a sufficient remuneration is necessary for DR service providers as well, which will be in charge of marketing and managing customers' involvement in DR programs. These points highlight the crucial role played by the design of DR mechanisms: advanced optimization methods provide relevant support in this sense. Vardakas et al. (2015) report a broad survey and discussion of optimization models for DR, highlighting the plurality of available options. For instance, the objective function(s) of the optimization problem may target the minimization of electricity costs or of total consumption, or the maximization of social welfare, or combine two of these objectives. Moreover, based on the form of the objective functions or of the constraints, the optimization problem can be linear or non-linear, while the variable characteristics can lead to integer or mixed-integer optimization problems; each type of problem requires an appropriate technique to be solved. Different sources of complexity can arise in dealing with such models, e.g. problems with no feasible solutions or excessive computational times. Additionally, game theory provides valuable theoretical support to model the interactions among electric system actors (users or utilities). For a deep discussion and examples of optimization algorithms in the context of DR and SGs, see Vardakas et al. (2015) and the references therein.

Moreover, there are some social and cognitive issues potentially affecting users' willingness to be involved in DR, and this holds especially for residential users. In fact, apart from the gratification that can derive from the idea of having contributed to improve the system reliability (Siano, 2014), users could be affected by some kinds of inconvenience. First, they will need to change some consumption habits: while some activities can be more easily shifted to a different moment in time, other activities, such as cooking (and eating) are less likely to be suitable for variations (Darby and McKenna, 2012). In addition, some consumers could have a limited willingness to accept high levels of automation and external control of their activities by a "Big Brother" (Cappers et al., 2012, p.425). Darby and McKenna (2012) suggest that home automation could effectively involve heating and cooling, as well as electric vehicles, while consumption related to other activities can be managed manually. Moreover, the same contribution highlights that consumers used to flat rates are not likely to switch easily to variable tariffs, and that a gradual passage to Time of Use pricing and subsequently to the more effective, but more complex, Real Time Pricing could represent a feasible path.

Many of the mentioned authors, however, agree on the key role of consumers' education and information in order to overcome most of the social issues described above, as long as the DR services providers will be able to achieve a high level of trust from customers and to provide an adequate protection of their rights and privacy.

SOLUTION AND RECOMMENDATIONS

Throughout this chapter, we have drawn a picture of DR as a promising resource developing in the electric system that, however, still presents many unknown features, which make difficult evaluating its potential, thus slowing down or limiting its development.

As in any other case where something "new" is developing on a broad (in this case global) scale, also with respect to DR it is important that all the involved actors (Governments, regulators, utilities, users, etc.) can make their choices on an informed basis. Therefore, as highlighted in Hu et al. (2015), it is crucial that further studies are promoted and developed.

In my opinion, such studies could focus on several fields and adopt different methodologies (examples can range from game-theoretical studies aimed at identifying the best market design to field trials evaluating the practical consequences of specific projects). A multi-disciplinary approach involving different entities, professional figures and expertise (e.g. regulators and utilities, practitioners and academics) seems the most promising strategy, given the multiplicity of perspectives that need to be considered to effectively deal with such a broad issue.

FUTURE RESEARCH DIRECTIONS

Future research focused on DR can take manifold directions. With reference to the economic and social sciences fields, some (not exhaustive) examples could be the following.

- Direct evaluations of consumers' preferences and willingness to accept contracts including different kinds of DR programs could be performed by adopting surveys approaches relying on contingent valuation or conjoint analysis. These approaches can be implemented describing hypothetical scenarios to generic consumers, or directed to subjects involved in actual trials or pilot studies. These methods can be designed also to infer the correct level of monetary incentive to be provided.

- DR affects system operations. Efficiency analysis techniques (parametric and non-parametric) can be effectively employed to evaluate the impact of DR on the performance of firms operating in different phases of the supply chain (generation or network management).
- A more general approach based on cost-benefit analysis, adopting a macro-economic perspective, appears crucial for evaluating specific programs before their implementation. Such an approach should be able to incorporate also non-monetary elements, such as environmental benefits related to reduced emissions.

CONCLUSION

Interest in DR as a source of flexibility is increasing from several direction (policy makers, regulators, utilities, academics and users). The potential benefits related to DR development are numerous and extremely interesting. Nevertheless, the implementation of such programs is still in a starting phase and is advancing in a relatively slow way. This is due to the complexity of the management of this new resource. This chapter has analysed the main regulatory, technological and socio-economic challenges to DR development. As a conclusion, the level of knowledge of the issue needs to be expanded in order to support informed decision-making, and multi-disciplinary approaches can constitute, as this article has discussed, an effective strategy given the multiplicity of perspectives to be considered in dealing with DR issues.

ACKNOWLEDGMENT

This research is supported by the Associazione per lo sviluppo dell'Università nel territorio novarese.

REFERENCES

Batlle, C., & Rodilla, P. (2009). Electricity demand response tools: Current status and outstanding issues. *European Review of Energy Markets*, *3*, 1–27.

Bradley, P., Leach, M., & Torriti, J. (2013). A review of the costs and benefits of demand response for electricity in UK. *Energy Policy*, *52*, 312–327. doi:10.1016/j.enpol.2012.09.039

Brouwer, A. S., Van den Broek, M., Zappa, W., Turkenburg, W. C., & Faaij, A. (2016). Least-cost options for integrating intermittent renewables in low-carbon power systems. *Applied Energy*, *161*, 48–74. doi:10.1016/j.apenergy.2015.09.090

Cappers, P., Mills, A., Goldman, C., Wiser, R., & Eto, J. H. (2012). An assessment of the role mass market demand response could play in contributing to the management of variable generation integration issues. *Energy Policy*, *48*, 420–429. doi:10.1016/j.enpol.2012.05.040

Crispim, J., Braz, J., Castro, R., & Esteves, J. (2014). Smart Grids in the EU with smart regulation: Experiences from the UK, Italy and Portugal. *Utilities Policy*, *31*, 85–93. doi:10.1016/j.jup.2014.09.006

Darby, S. J., & McKenna, E. (2012). Social implication of residential demand response in cool temperate climates. *Energy Policy*, *49*, 759–769. doi:10.1016/j.enpol.2012.07.026

Directive 2009/72/Ec Of The European Parliament And Of The Council of 13 July 2009 concerning common rules for the internal market in electricity. (n.d.). Retrieved July 20th, 2016 at: http://eur-lex.europa.eu/legal-content/EN/TXT/PDF/?uri=CELEX:32009L0072&from=EN

Directive 2012/27/Eu Of The European Parliament And Of The Council Of 25 October 2012 On Energy Efficiency. (n.d.). Retrieved July 20th, 2016 At: http://eur-lex.europa.eu/legal-content/EN/TXT/PDF/?uri=CELEX:32012L0027&from=EN

Dupont, B., Dietrich, K., De Jonghe, C., Ramos, A., & Belmans, R. (2014). Impact of residential demand response on power system operation: A Belgian case study. *Applied Energy*, *122*, 1–10. doi:10.1016/j.apenergy.2014.02.022

Eid, C., Codani, P., Chen, Y., Perez, Y., & Hakvoort, R. (2015). *Aggregation of demand side flexibility in a smart grid: A review of European market design*. EEM15 – 12th International Conference on the European Energy Market, Lisboa,, Portugal.. 10.1109/EEM.2015.7216712

Eid, C., Koliou, E., Valles, M., Reneses, J., & Hakvoort, R. (2016). Time-based pricing and electricity demand response: Existing barriers and next steps. *Utilities Policy*, *40*, 15–25. doi:10.1016/j.jup.2016.04.001

European Network of Transmission System Operators for Electricity – ENTSO-E. (2015). *Market design for demand side response*. Policy paper, November 2015. Retrieved September 12th, 2016 at: https://www.entsoe.eu/Documents/Publications/Position%20papers%20and%20reports/entsoe_pp_dsr_web.pdf

Federal Energy Regulatory Commission (FERC). Retrieved from http://www.ferc.gov/industries/electric/indus-act/demand-response/dem-res-adv-metering.asp

Feuerriegel, S., & Neumann, D. (2016). Integration scenarios of Demand Response into electricity markets: Load shifting, financial savings and policy implications. *Energy Policy*, *96*, 231–240. doi:10.1016/j.enpol.2016.05.050

Gils, H. C. (2014). Assessment of the theoretical demand response potential in Europe. *Energy*, *67*, 1–18. doi:10.1016/j.energy.2014.02.019

Hu, Z., Kim, J., Wang, J., & Byrne, J. (2015). Review of dynamic pricing programs in the U.S. and Europe: Status quo and policy recommendations. *Renewable & Sustainable Energy Reviews*, *42*, 743–751. doi:10.1016/j.rser.2014.10.078

Schleich, J., Klobasa, M., Gölz, S., & Brunner, M. (2013). Effects of feedback on residential electricity demand – Findings from a field trial in Austria. *Energy Policy*, *61*, 1097–1106. doi:10.1016/j.enpol.2013.05.012

Shen, B., Ghatikar, G., Lei, Z., Li, J., Wikler, G., & Martin, P. (2014). The role of regulatory reforms, market changes, and technology development to make demand response a viable resource in meeting energy challenges. *Applied Energy*, *130*, 814–823. doi:10.1016/j.apenergy.2013.12.069

Siano, P. (2014). Demand response and smart grids – A survey. *Renewable & Sustainable Energy Reviews*, *30*, 461–478. doi:10.1016/j.rser.2013.10.022

Smart Energy Demand Coalition. (2015). *Mapping Demand Response in Europe Today 2015*. Retrieved July 20th, 2016 at: http://www.smartenergydemand.eu/wp-content/uploads/2015/09/Mapping-Demand-Response-in-Europe-Today-2015.pdf

Torriti, J. (2012). Price-based demand side management: Assessing the impacts of time-of-use tariffs on residential electricity demand and peak shifting in Northern Italy. *Energy*, *44*(1), 576–583. doi:10.1016/j.energy.2012.05.043

Vardakas, J. S., Zorba, N., & Verikoukis, C. V. (2015). A survey on demand response programs in smart grids: Pricing methods and optimization algorithms. *IEEE Communications Surveys and Tutorials*, *17*(1), 152–178. doi:10.1109/COMST.2014.2341586

ADDITIONAL READING

Aalami, H. A., Moghaddam, M. P., & Yousefi, G. R. (2010). Demand response modelling considering interruptible/curtailable loads and capacity market programs. *Applied Energy*, *87*(1), 243–250. doi:10.1016/j.apenergy.2009.05.041

Agrell, P. J., Bogetoft, P., & Mikkers, M. (2013). Smart-grid investments, regulation and organization. *Energy Policy*, *52*, 656–666. doi:10.1016/j.enpol.2012.10.026

Albadi, M. H., & El-Saadany, E. F. (2008). A summary of demand response in electricity markets. *Electric Power Systems Research*, *78*(11), 1898–1996. doi:10.1016/j.epsr.2008.04.002

Brouwer, A. S., Van den Broek, M., Seebregts, A., & Faaij, A. (2015). Operational flexibility and economics of power plants in future low-carbon power systems. *Applied Energy*, *156*, 107–128. doi:10.1016/j.apenergy.2015.06.065

Earle, R., Kahn, E. P., & Macan, E. (2009). Measuring the capacity impacts of demand response. *The Electricity Journal*, *22*(6), 47–58. doi:10.1016/j.tej.2009.05.014

Feuerriegel, S., & Neumann, D. (2014). Measuring the financial impact of demand response for electricity retailers. *Energy Policy*, *65*, 359–368. doi:10.1016/j.enpol.2013.10.012

Katz, J. (2014). Linking meters and markets: Roles and incentives to support a flexible demand side. *Utilities Policy*, *31*, 74–84. doi:10.1016/j.jup.2014.08.003

Kondziella, H., & Bruckner, T. (2016). Flexibility requirements of renewable energy based electricity systems – a review of research results and methodologies. *Renewable & Sustainable Energy Reviews*, *53*, 10–22. doi:10.1016/j.rser.2015.07.199

Lang, C., & Okwelum, E. (2015). The mitigating effect of strategic behavior on the net benefits of a direct load control program. *Energy Economics*, *49*, 141–148. doi:10.1016/j.eneco.2015.01.025

Rochlin, C. (2009). The alchemy of demand response: Turning demand into supply. *The Electricity Journal*, *22*(9), 10–25. doi:10.1016/j.tej.2009.09.004

Rodrigues, R., & Linares, P. (2014). Electricity load level detail in computational general equilibrium – Part I – Data and calibration. *Energy Economics*, *46*, 258–266. doi:10.1016/j.eneco.2014.09.016

Rodrigues, R., & Linares, P. (2015). Electricity load level detail in computational general equilibrium – Part II – welfare impacts of a demand response program. *Energy Economics*, *47*, 52–67. doi:10.1016/j.eneco.2014.10.015

Ruester, S., Schwenen, S., Batlle, C., & Pérez-Arriaga, I. (2014). From distribution networks to smart distribution systems: Rethinking the regulation of European electricity DSOs. *Utilities Policy*, *31*, 229–237. doi:10.1016/j.jup.2014.03.007

Wang, Q., Zhang, C., Ding, Y., Xydis, G., Wang, J., & Østergaard, J. (2015). Review of real-time electricity markets for integrating Distributed Energy Resources and Demand Response. *Applied Energy*, *138*, 695–706. doi:10.1016/j.apenergy.2014.10.048

KEY TERMS AND DEFINITIONS

Demand Response (DR): Source of flexibility for the electricity system relying on modifications in users' consumption.

Enabling Technologies: With respect to DR, this term indicates those technologies (e.g. equipment, appliances) that allow an effective implementation of DR programs. Examples could be smart meters, allowing communication between users and utilities, or home appliances that automatically react to price or non-price signals by modifying electricity consumption.

Incentive-Based DR: DR mechanisms that rely on consumption modifications induced by non-price signals, for which the involved users receive compensations.

Off-Peak: Points of lower demand (with respect to the peak-load) in a certain time period.

Peak-Load: Within the electric system, it is the point of highest demand in a certain time period. We can have daily peaks, seasonal peaks, etc.

Peak-Shifting: Refers to the possibility of moving demand from peak to off-peak times. This is one of the desirable effects of DR.

Price-Based DR: DR mechanisms that rely on consumption modifications induced by price variations.

ENDNOTES

1 http://www.ferc.gov/industries/electric/indus-act/demand-response/dem-res-adv-metering.asp

2 Siano (2014) identifies as a separated category demand reduction bids, where participant send their bids (offer a demand reductions and request a price) to the utility or the aggregator.

3 The authors refers to DR benefits in a SG context, but most of the proposed lines of reasoning are likely to apply in a more general framework.

4 As mentioned in the report, explicit DR has limited requirements in terms of public investment in technology, therefore regulation plays the more critical role in making DR programs available and suitable for end-users.

This research was previously published in the Encyclopedia of Information Science and Technology, Fourth Edition edited by Mehdi Khosrow-Pour, pages 3144-3155, copyright year 2018 by Information Science Reference (an imprint of IGI Global).

Chapter 18
Waste Gas End-of-Pipe Treatment Techniques in Italian IPPC Chemical Plants

Gaetano Battistella
ISPRA, Italy

Giuseppe Di Marco
ISPRA, Italy

Carlo Carlucci
ISPRA, Italy

Raffaella Manuzzi
ISPRA, Italy

Federica Bonaiuti
ISPRA, Italy

Celine Ndong
ISPRA, Italy

ABSTRACT

Due to more stringent emission regulations, very efficient emission control technologies are required adopting national IPPC permits based on BAT conclusions. Some abatement techniques are operating inside IPPC plants based in Italy. This chapter includes the results of a screening of Italian IPPC plants, trying to highlight operating conditions of abatement devices and possible already existing improvements for several compounds removal. The abatement techniques analyzed operate mainly on VOC content reduction or on inorganic compounds abatement. ISPRA experience has allowed to analyze different operative conditions, related with abatement techniques and their application in IPPC permitted plants. The results of this analysis suggest a possible reconsideration and new assessment for some end-of-pipe devices in order to find other, better-defined operational contexts, different from Italian provisions and an evaluation of current operational performances of the devices, in order to improve their environmental conditions consistently with BAT application.

DOI: 10.4018/978-1-5225-7359-3.ch018

INTRODUCTION

Due to more stringent emission regulations, very efficient new advanced emission control technologies are required adopting National IPPC (Integrated Pollution Prevent and Control) Permits (below AIA) based on Best Available Technologies (below BAT) Conclusions.

Some of these techniques are operating inside Chemical Plants and Refineries based in Italy, such as Oxidation, Adsorption and Absorption devices. Other techniques (i.e. the ones that are new advanced technologies still in research or in demonstration state), are not subject of this Paper, based on describing running situation inside operating IPPC Chemical Plants and Refineries licensed in Italy at National Level.

This paper includes, but are not limited to, the results of a screening of Italian Chemical IPPC Industries and Refineries up to day, trying to highlight operating conditions and possible already existing improvements for removal of:

- VOC and other cancer causing and toxic substances;
- Dust, Mercury and heavy metals;
- NOx and Nitrogen compounds;
- SOx and Sulphur compounds;
- Chlorides and Fluoride compounds.

The abatement techniques analyzed in this work operate mainly on VOC content reduction, through the use of Oxidizing devices or on inorganic compounds abatement (in addition to VOC), through the use of Absorption or Adsorption devices.

Superior Environmental Protection and Research Institute (below ISPRA) experience, mainly developed as Technical Support to Italian Minister of Environment, Land and Sea (below IMELS), has allowed to analyze different operative conditions, related with abatement techniques and their application in IPPC permitted plants.

Many pollutants emitted from IPPC plants (according to Environmental Permits limit values) have been identified and charted a profile of possible application for abatement techniques in these plants in their different IPPC categories.

The results of this analysis allow to suggest a possible reconsideration and, also, new assessment for some end-of-pipe devices, in order to find other better defined operational contexts, different from actually Italian provisions and, also, an evaluation of current operational performances of the devices, in order to improve their environmental conditions, consistently with BAT application.

BACKGROUND

In Italy, IPPC Permit is an authorization released for environmental protection purposes, in order to prevent and control pollution ‘at the source’ by means of an integrated authorization, allowing operation of IPPC industrial activities with specified production’s characteristics and dimensions, at both national and regional levels (Battistella, 2013).

The list of the categories of these specific industrial activities is regulated by the Italian Legislative Decree n. 59/2005 and s.m.i. (Italian Legislative Decree n. 152/2006 and s.m.i.) that adopts and endorses the Directive n. 96/61/EC and s.m.i. (Directive 2008/1/EC and s.m.i.) concerning integrated pollution prevention and control (actually recast in the Directive 2010/75/EU).

IPPC permits – by law definition - plan and perform an integrated prevention and control set in the exact point of pollution ('a la source'), e.g. pollutants are identified, declared, controlled, detected and monitored in the admission/emission points of the IPPC industrial activities, as well as during all activities of industrial plants' operation (Battistella & Di Marco, 2013a; Battistella & Di Marco, 2013b).

This means authorization of plants' operation controlling natural resources' usage, as well as emissions and discharges in the environment inside predefined limit values with prescriptions, adoption of predefined monitoring framework, as self-controls on selected parameters, frequencies and methodologies, with a periodic reporting and planned inspections based on the effective environmental risk (Battistella & Di Marco, 2013a; Battistella & Di Marco, 2013b).

In Italy, AIAs are released by the Competent Authority, as

- By IMELS for national strategic plants;
- By other Authority designed by Region or autonomous Province for others.

In order to accomplish IPPC permits operative performances in terms of assigned limit values, among other provisions, also waste gas end-of-pipe treatment devices are adopted - as well as installed and operated - in order to abate or at least decrease pollutants' contents (even often dangerous substances and compounds) in *waste-gases* before their emission into open air.

Adopted techniques must be considered equal to, or committed to become under IPPC permits period of duration (more or less 10 years), as Best Available Techniques and their operating performances are described in details into *Reference Documents on Best Available Techniques (BRefs[1])*.

BAT CONCLUSIONS APPLICATION IN ITALIAN IPPC PERMITTED PLANTS

The Italian National Environmental Regulatory Framework for IPPC Installations

As regards to Italian Regulations, Attachment X to Part II of Legislative Decree n. 152/06 [9] defines the list of main pollutants to be monitored, if pertinent, with reference to air emissions for IPPC Plants under operational licensee.

Attachments to Part V of Legislative Decree n. 152/06 list minimum emissions values (relevance thresholds) and maximum emissions values (limit values) and related prescriptions (continuous monitoring, process parameters to monitor, measurements' norms, continuous monitoring control procedures, conformity criteria for measurements, etc.) for pollutant substances emitted in atmosphere - as conveyed emissions - at the stacks and specific indications for some substances (as SOx, NOx, CO, Dusts, VOC, etc.) and for some types of plants.

It is relevant to highlight that Legislative Decree n. 152/06 *does not establish specific limit values for emissions of pollutants from end of pipe treatment devices,* as well *as does not even describe specific abatement techniques.*

This is why applied emission limit values are strictly connected with specific pollutants and plant units from where they are generated, as mentioned in Suppl. Annex I at V Part of Italian Legislative Decree n.152/06.

Table 1 shows main pollutants related with operative experience in Italian IPPC plants.

In III Part of mentioned Annex I, emission levels for some specific type of production such as Non ferrous metal, Sulphuric acid, Chlorine, Fertilizers, polymers are described, while in IV Part emission levels for Refineries are described.

For VOC in Annex III at V Part of Italian Legislative Decree n.152/06, emission levels for particular type of compound are described as shown in the following Table 2.

Table 1. Main pollutants related with operative experience in Italian IPPC plants

Pollutant	Class Identity (Annex I, V Part of Italian Legislative Decree n.152/06)	Relevance Threshold (g/h)	Limit Value (mg/Nm^3)
Point 1.1. Cancer-causing, reproduction toxic or mutagen compounds (Table A1)			
Benzene	class III	25	5
PHA (summation)	class I	0,5	0,1
Organic chlorated compounds	class III	25	5
Point 1.2. Compounds that can be especially toxic and be accumulated (Table A2)			
Dioxins and Furans	class I	0,02	0,01
PCB	class II	0,5	0,5
Point 2. Inorganic compounds that mainly come up in dust form (Table B)			
Hg	class I	1	0,2
Point 3. Inorganic compounds that mainly come up in vapour or gas form (Table C)			
SO_x	class V	5000	500
NO_x	class V	5000	500
Chlorine	class II	50	5
Chlorine inorganic compounds (as HCl)	class III	300	30
Fluorine and its compound (as HF)	class II	50	5
H_2S	class II	50	5
NH_3	class IV	2.000	250
Point 4. Organic Compounds that come up in dust, vapour or gas form (Table D)			
Organic compounds (included Chlor Aromatics, Halogenic hydrocarbons and Nitro aromatics compounds)	It depends on the specific compound		
5. Total dust			
Total Dust	-	> 5000	50
		>1000 <5000	150

Table 2. Emission levels for VOC

Pollutant	Identity	Relevance Threshold (g/h)	Limit Value (mg/Nm³)
VOC	Cancer-causing, reproduction toxic or mutagen compounds	10	2
	Alogenated	100	20

Screening of IPPC Permitted Chemical Plants and Refineries in Italy

In Italy, at December 2014, are operating 175 activities so called "strategic" (IPPC permits released at national level) and between these there are: 16 Crude Oil Refineries and 46 large Chemical Plants and about 5.560 Plants have received IPPC Permits, while almost 200 are concluding their procedures (no one with critical conditions to be resolved for EU obligations) in terms of first release (Superior Environmental Protection and Research Institute [ISPRA], 2015).

Figure 1 shows the IPPC Permits situation at the beginning of year 2013 (ISPRA, 2015).

In the Italian procedure to release IPPC permits, ISPRA is on charge of performing the requested comparisons of environmental parameters with the applicable 'Best Available Techniques' - contained inside available published BRef documents – in order to evaluate and to propose to the 'Technical Investigating Group' appropriate limit values, prescriptions and recommendations to be adopted in the 'Technical Advice', to be approved (Battistella, 2013; Battistella & Di Marco, 2013a; Battistella & Di Marco, 2013b).

In this way, also a virtual process is started, in order to compare industrial activities and plants' performances in term of 'resource efficiency', trying to optimize the analyzed processes and to reduce their emissions limits.

Figure 1. IPPC Permits situation at the beginning of year 2013

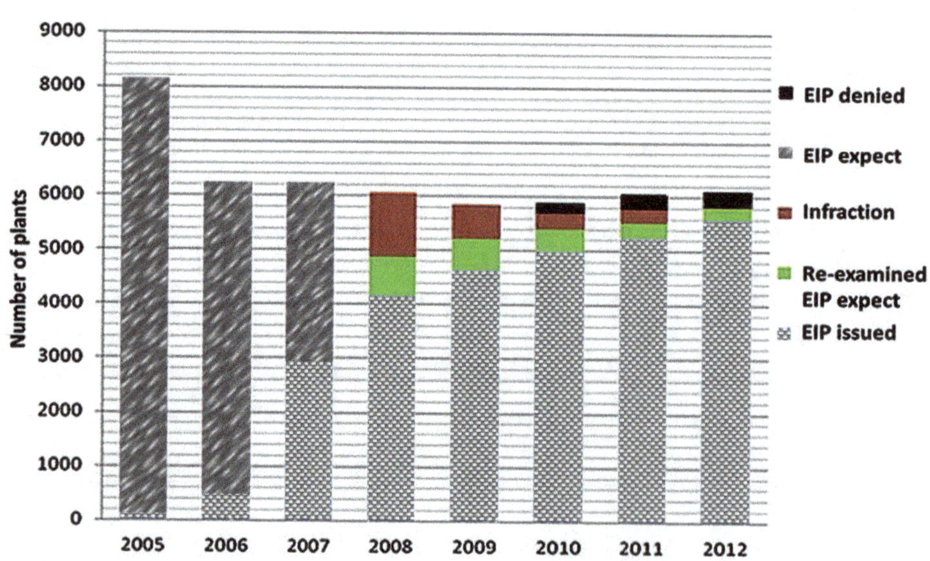

The complete IPPC Permits release procedure is shown in the following Figure 2 (Battistella, 2013).

In Italy, since its start in 2009, this procedure allowed the results shown in the following figures for abatement of main environmental pollutants (air emissions) and at December 2014, every IPPC Plant have been licensed with IPPC Permits, in terms of first release, concluding authorization procedures (ISPRA, 2015).

1. *For 16 Italian Refineries,* the result in pollution abatement for macro pollutants, such as SOx, NOx, CO and PTS, is showed in Figure 3.

Figure 2. IPPC Permits release procedure

Figure 3. Pollution abatement for macro pollutants in Italian refineries

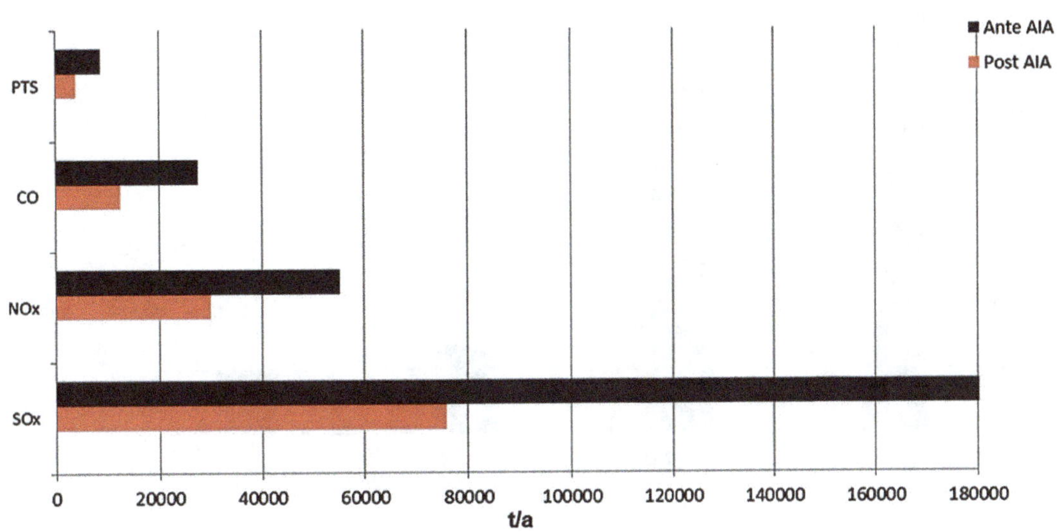

2. *For 46 Italian Chemical Plants,* the result in pollution abatement for macro pollutants, such as SOx, NOx, CO and PTS, is showed in Figure 4.

BAT Conclusions Application in IPPC Permitted Plants

In Italy, recent regulatory innovations introduced by Legislative Decree n. 46/2014– adopting in Italy the European Commission 'Industrial Emissions Directive' - led to the imposition, during operation of IPPC installations as specified in *AIA* Permits, of further specific indications/prescriptions as defined in new BAT for each IPPC sector, with revised emission limit values for pollutants released into the environment, as reported in "BAT Conclusions" of new reviewed BRefs (Battistella, Di Marco, Bonaiuti, Carlucci, 2014).

This new approach applies to all IPPC installations subject to *AIA* Permits (first release) at national level, for which new legislation requires a review and an updating of existing authorizations, through new procedural "review/renewal" mechanisms, now closely linked to the adoption of new technologies as identified and published into new Brefs documents as 'BAT Conclusions'(Battistella et al., 2014).

In numerous 'vertical' BRefs, or rather specific Reference Documents for each type of production plants, end-of-pipe gas treatments are considered as a BAT for VOC and inorganic compounds abatement.

Almost all of these documents send back to 'horizontal' *Reference Document on Best Available Techniques in Common Waste Water and Waste Gas Treatment / Management Systems in the Chemical Sector - February 2003*, for achievable emission limits and operative conditions of end of pipe devices.

Design standards are strictly related to properties of gas flow that define operative conditions, achievable performance and cross-media effects.

In Table 3 are described operative conditions and performance and emission levels achievable by implementation of BAT.

Figure 4. Pollution abatement for macro pollutants in Italian chemical plants

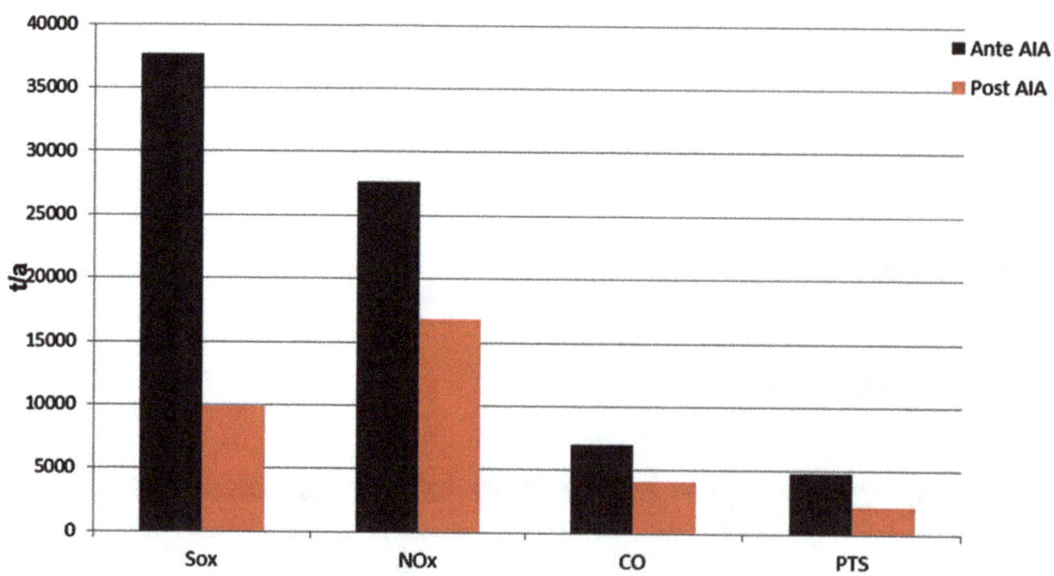

Table 3. Operative conditions and performance and emission levels achievable by implementation of BAT

	Thermal oxidation	Catalytic oxidation	Wet scrubbing (Absorption)	Adsorption
Description[2]	Oxidation with oxygen (air) by heating a gas above its auto-ignition point	Oxidation with oxygen (air) using catalyst to lower auto-ignition point	Mass transfer from gaseous phase into liquid phase	Mass transfer from gaseous phase to solid surface
Application	Emissions from all VOC sources most suitable for autothermal VOC concentrations and final treatment for hazardous substances	Same application as thermal incineration, contaminants restricted to non-poisonous ones. It can be with or without heat recovery	Control of VOC, inorganic compounds and also of dust, depending on variant and scrubbing liquid (water, acidic and alkaline solution). By desorption it's also possible recovery of substances.	Removal of VOC, odorous substances, dioxins, mercury After regeneration it's also possible recovery of substances.
Cross-media effects	CO and NO_x in flue gas Cl- and S-content require flue gas treatment Dioxins can represent a problem if not operated at optimized conditions	Very low NO_x content in flue gas (about 15 mg/Nm^3) Cl- and S- content require flue gas treatment Dioxins are usually not a problem	Waste water to be treated Energy consumption and emissions during regeneration	Waste water generated during regeneration Disposal of adsorbent
Achievable performance [% pollutant removal]	VOC 95-99	VOC 90-99 CO >98 Odour 80-95	VOC 50-99 Inorganic compounds 90-99 SO_2 80-99	VOC 80-95 Odour 80-95 H_2S 80-95
Achievable emission levels [mg/Nm^3]	TOC 1-4	-	HF <1 HCl <10 (<50 with water) SO_2 < 40	Hg < 0,05 Dioxins <0,1 ng/Nm^3 TEQ

The following figures show the state-of-the-art of BAT application for waste-gas cleaning in Italian *AIA* IPPC plants (Chemical plants and Refineries)[3].

Going towards a complete BAT application for waste-gas cleaning operations, Italian *AIA* IPPC chemical plants and Refineries actually apply mainly Absorption and Adsorption techniques (respectively 54% and 30% of total *AIA* IPPC Chemical plants and Refineries adopting BAT[4] as shown in Figure 5).

Absorption devices are mostly installed in IPPC Chemical Plants due to the properties of emitted pollutants (mainly inorganic ones), whereas Adsorption technique in mostly adopted in Refineries because of the mainly presence of organic emitted pollutants (see Figure 6).

Figure 5. IPPC Chemical plants and Refineries installing abatement devices

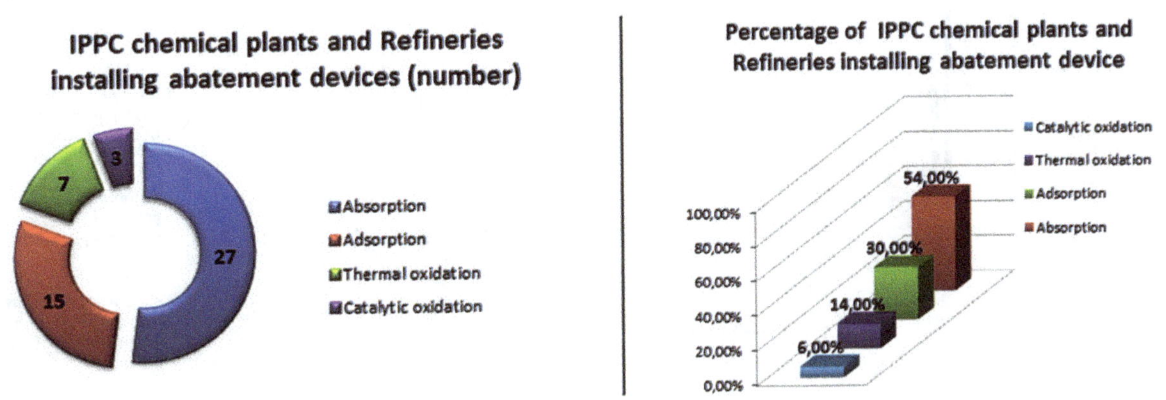

Figure 6. Percentage of Italian IPPC plants using 'sorbents technologies'

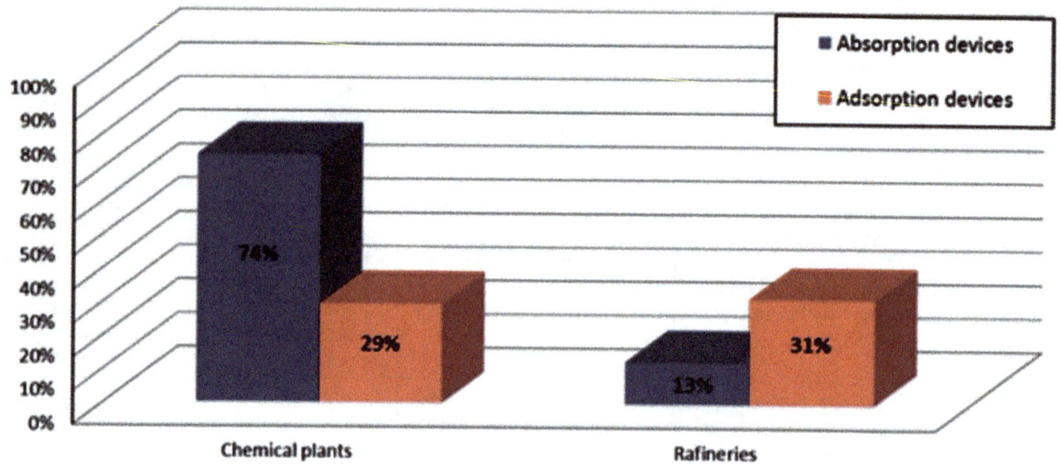

Following graphics (see Figures 7, 8, 9, 10) represent operating data concerning waste gas end-of-pipe treatment techniques as run in licensed Italian Chemical plants and Refineries[5] in order to represent operative conditions and emissions scenarios related with application of BAT in Italian IPPC Plants.

The upper value for each board of graphics represent the *maximum IPPC Permits limit values* for each IPPC Category of activities.

Figure 7. Operating emission data from thermal oxidizers in IPPC permitted plants

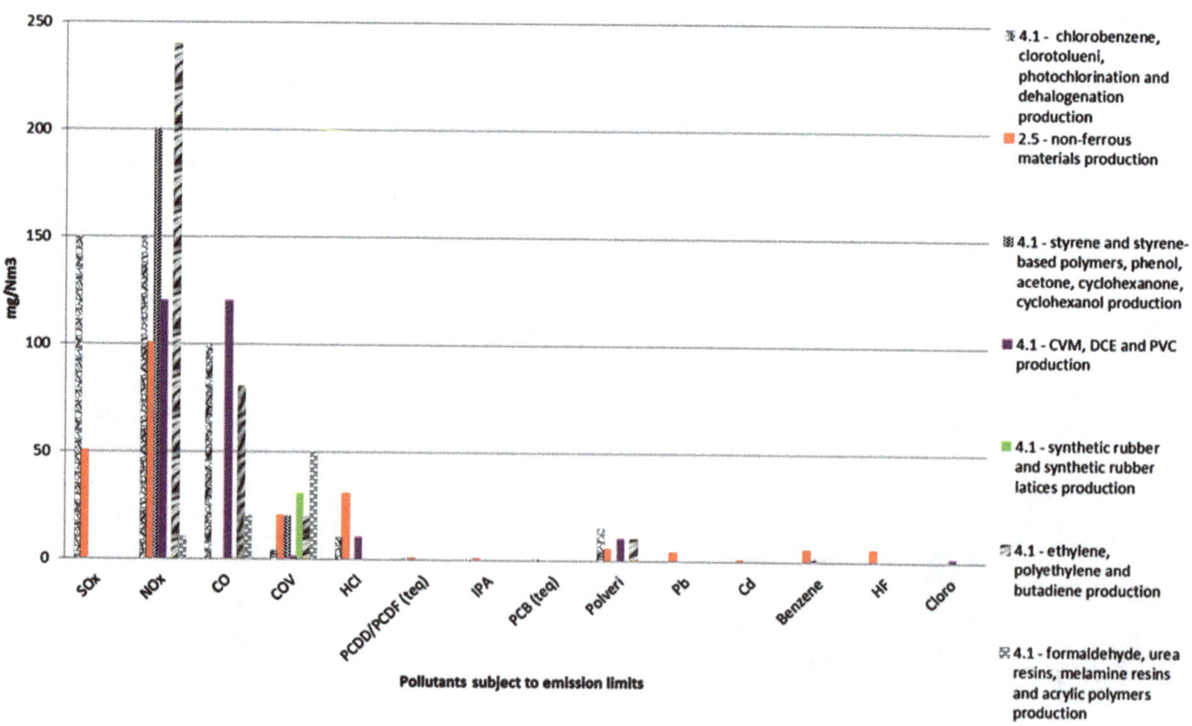

Figure 8. Operating emission data from catalytic oxidizers in IPPC permitted plants

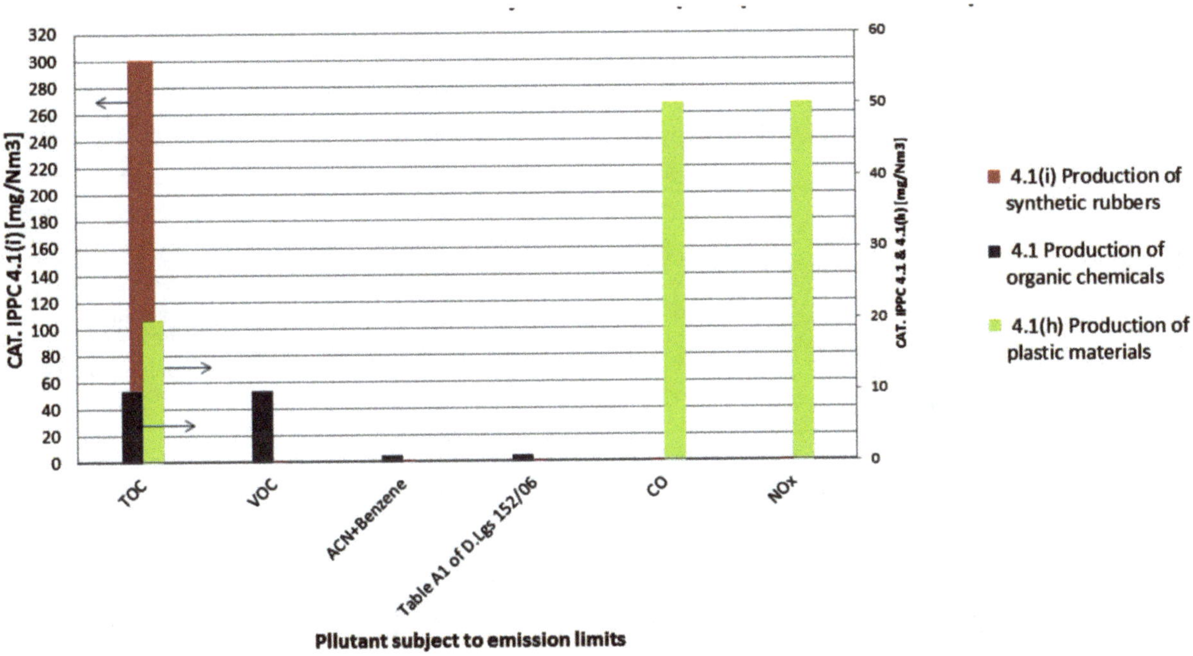

Figure 9. Operating emission data from wet scrubbing devices in IPPC permitted plants

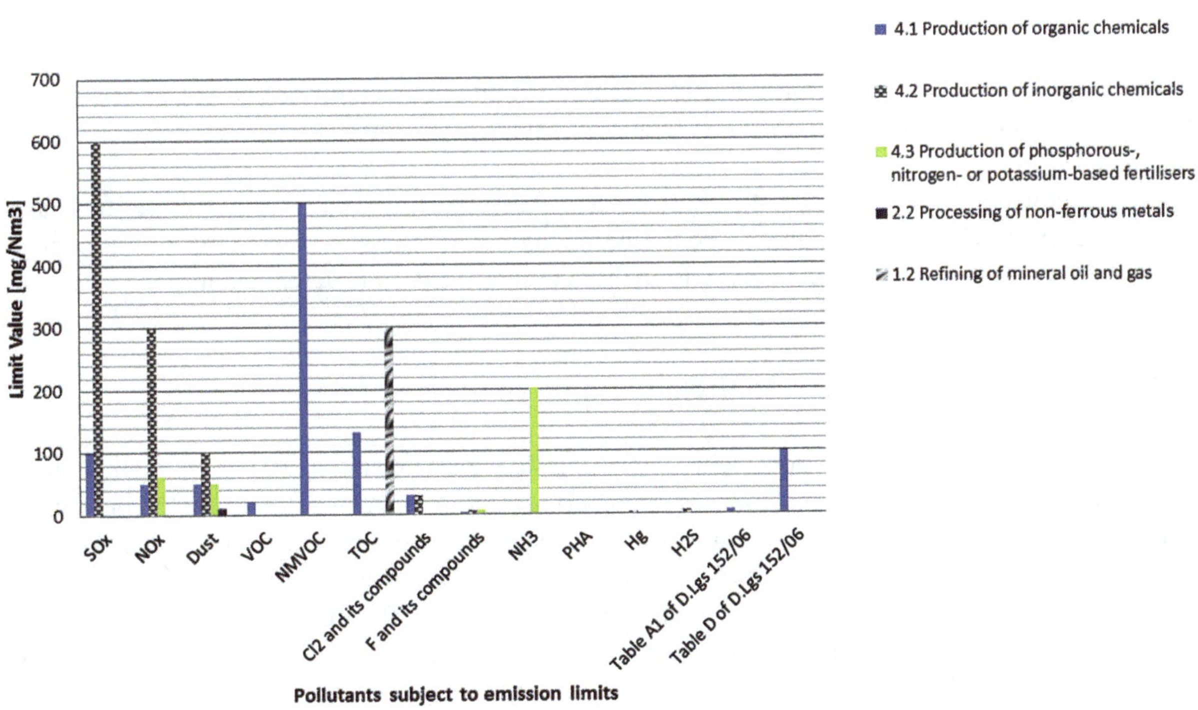

Figure 10. Operating emission data from adsorption devices in IPPC permitted plants

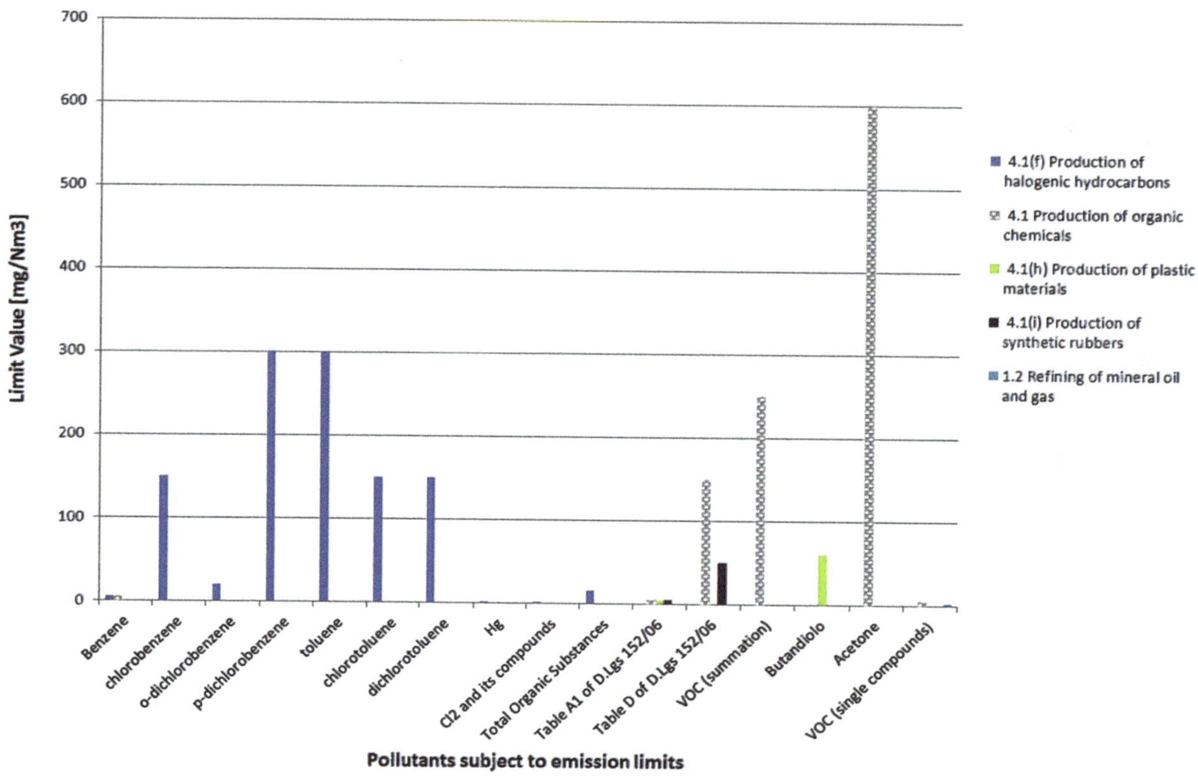

The graphics well represents that VOC (as organic volatile compounds or equivalent carbon) are the widespread pollutants and the choice of treatment device is strictly connected with the amount of pollutants fed in it.

In the following tables (see Tables 4, 5, 6,7) are described operating data in 'first' *AIA* permits partitioned for each IPPC activity: in some cases *AIA* permit emission limit values are described as a range, in order to represent different limit values within these ranges fixed for different plants.

FUTURE RESEARCH DIRECTIONS

Recent regulatory innovations introduced by Italian Legislative Decree n. 46/2014 [9] led to the imposition, during operation of IPPC installations as specified in AIA Permits, of further specific indications as defined in new BAT for each IPPC sector, with revised emission limit values for pollutant released into the environment, as reported in "BAT Conclusions" of new reviewed BRefs.

This new approach applies to all IPPC installations subject to AIA Permits (first released) at national level, for which the new legislation requires an updating of existing authorizations, through new procedural "review/renewal" mechanisms, now closely linked to the identification of new technologies and the publication of BAT Conclusions.

Table 4. AIA permits emission limits range for thermal oxidizers in IPPC permitted plants

THERMAL OXIDATION OPERATING DATA IN *AIA* PERMITS		
IPPC ACTIVITIES	**Pollutants**	***AIA* Permits Emission Limits Range (mg/Nm³)**
4.1 Production of organic chemicals + 4.2 Production of inorganic chemicals	SOx	150
	NOx	150
	CO	100
	VOC	4
	HCl	10
	PCDD/PCDF (as TEQ)	0,1
	PHA	0,1
	PCB (as TEQ)	0,1
	Dust	15
2.5.a Production of non-ferrous crude metals from ore, concentrates or secondary raw materials by metallurgical, chemical or electrolytic processes + 4.2 Production of inorganic chemicals	SOx	50
	NOx	100
	TOC	20
	Dust	5
	Pb	3,5
	Cd	0,2
	Benzene	5
	PCDD/PCDF	0,0005
	PHA	0,1
	HF	5
	HCl	30
4.1 Production of organic chemicals	NOx	10 - 240
	VOC	1-30
	CO	20 - 120
	Dust	10
	TOC	50
	Benzene	1
	PCDD/PCDF (come teq)	0,1 ng/Nm³
	Chlorine	1
	CVM + DCE	0,2 ppm v/v
	HCl	10

In such a way, a second phase of IPPC permits is starting, after the expiring first phase, taking into account 'BAT Conclusions' issued in sectorial BRefs updating and upgraded and also other environmental aspects, as for example renewed attention towards territory inside which IPPC installations stands, based on a new evaluation of site conditions in terms of production/release of relevant hazardous substances, possibility of soil and groundwater contamination, soil vulnerabilities and all available operational data coming from annual reporting.

Table 5. AIA permits emission limits range for catalytic oxidizers in IPPC permitted plants

CATALYTIC OXIDATION OPERATING DATA IN *AIA* PERMITS		
IPPC ACTIVITIES	**Pollutants**	***AIA* Permits Emission Limits Range (mg/Nm³)**
4.1 Production of organic chemicals	TOC	10
	VOC	10
	ACN+Benzene	1
	Table A1 of D.Lgs 152/06	1
4.1(i) Production of synthetic rubbers	TOC	300
4.1(h) Production of plastic materials	TOC	20
	CO	50
	NOx	50

For IPPC installations permitted at national level, these new regulations are going to finalize already existing IPPC permits by means of a 'review and renew' mechanism, giving also new opportunities for research, and focused mainly on these new aspects:

1. More tight respect – by now mandatory – towards the application of new technologies, to reach new and more stringent limit values, as well as synthetically defined inside 'BAT Conclusions' inside new revised BRefs;
2. Greater attention – to be verified from time to time – towards territorial environmental aspects, up to now considered only marginally, as managed only inside the respect of the specific applicable standard;
3. A precise reconsideration of the implementation of adequate control and monitoring systems inside IPPC operating installations, finalized also to validate planned controls.

On account of the advantages reachable with a large use of these abatement devices, it's desirable a faster and faster development in application and correct use to achieve BAT performance and emission levels as soon as possible in a more wide part of IPPC plants.

Considering Italian Regulations, it can be suitable the definition of a specific contest for these kind of devices due to their advantages and peculiarity.

CONCLUSION

The release of IPPC operational permits has given place to hundreds of different scenarios of best available techniques (BAT) implementation in these industrial sites, where environmental pollution is going now under a planned control in order:

1. To enforce BAT application and implementation inside new and already existing IPPC Plants;

Table 6. AIA permits emission limits range for wet scrubbing devices in IPPC permitted plants

ABSORPTION (WET SCRUBBING) OPERATING DATA IN *AIA* PERMITS		
IPPC ACTIVITIES	**Pollutants**	***AIA* Permits Emission Limits Range (mg/Nm^3)**
4.1 Production of organic chemicals	SOx	100
	NOx	50
	Dust	5-50
	VOC	20
	NMVOC	500
	TOC	10-130
	Cl_2 and its compounds	8-30
	F and its compounds	1
	Table A1 of D.Lgs 152/06	5
	Table D of D.Lgs 152/06	5-100
4.2 Production of inorganic chemicals	SOx	75-600
	NOx	100-300
	Dust	5-100
	Cl_2 and its compounds	1-30
	F and its compounds	5
	Hg	0,05
	H_2S	5
4.3 Production of phosphorous-, nitrogen- or potassium-based fertilisers	NOx	50-60
	Dust	20-50
	F and its compounds	6
	NH_3	10-200
2.2 Processing of non-ferrous metals	Dust	10
1.2 Refining of mineral oil and gas	TOC	300
	H_2S	'Bubble' application
	SOx	'Bubble' application

2. To strengthen IPPC Permits via harmonization of integrated management of air, water and soil pollutants emissions an emissions to avoid different approaches towards pollution impacts management could allow pollution transfers from one environmental matrix to another.
3. To perform a properly planned and adequate monitoring and data reporting activities and, if needed, also periodic inspections based on the effective environmental risk found.

BAT application, in waste gas cleaning operations, is normally adopted in Italian IPPC chemical plants and Refineries, nevertheless mainly *Absorption systems and Adsorption systems are widely used, as well as respectively 54% and 30% of total Italian IPPC Chemical plants and Refineries* are installing these types of devices:

Table 7. AIA permits emission limits range for adsorption devices in IPPC permitted plants

ADSORPTION OPERATING DATA IN *AIA* PERMITS		
IPPC ACTIVITIES	**Pollutants**	***AIA* Permits Emission Limits Range (mg/Nm^3)**
4.1(f) Production of halogenic hydrocarbons	Benzene	5
	chlorobenzene	150
	o-dichlorobenzene	20
	p-dichlorobenzene	150-300
	toluene	300
	chlorotoluene	150
	dichlorotoluene	150
	Hg	0,05
	Cl_2 and its compounds	1
	Total Organic Substances	15
4.1 Production of organic chemicals	Benzene	5
	Table A1 of D.Lgs 152/06	1-5
	Table D of D.Lgs 152/06	150
	VOC (summation)	250
	Acetone	600
	VOC (single compounds)	5
4.1(h) Production of plastic materials	Table A1 of D.Lgs 152/06	4
	Butandiolo	60
4.1(i) Production of synthetic rubbers	Table A1 of D.Lgs 152/06	5
	Table D of D.Lgs 152/06	50
1.2 Refining of mineral oil and gas	VOC	2-20

1. *Absorption devices* are mostly installed in Italian IPPC Chemical Plants (74% absorption vs 29% adsorption);
2. *Adsorption techniques* are mostly adopted in Refineries (31% adsorption vs 13% absorption) due to peculiarity of pollutants emitted (mainly inorganic ones in Chemical plants and organic ones in Refineries).

Considering all installed devices:

1. Absorption devices (such as wet scrubbers) are most diffused with an incident percentage of 55,9% relative to total number of devices installed and they are mainly used for abatement of inorganic substances (see Figure 11).
2. *Adsorption and Absorption devices* are still mostly diffused for abatement of VOC, respectively 43.28% and 31.34% of total devices analyzed in this work (see Figure 12).

Figure 11. Percentage of installed 'end of pipes abatement devices'

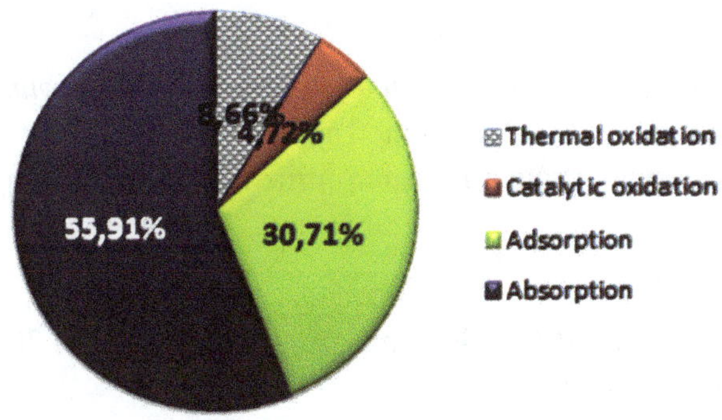

Figure 12. Percentage of installed 'end of pipes devices' for abatement of VOC

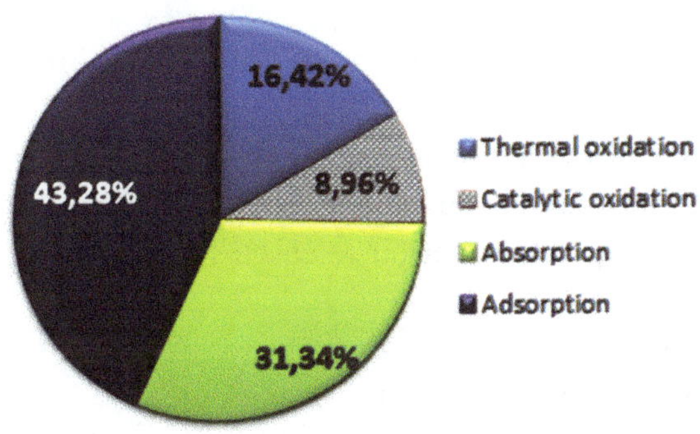

For these devices operative management is relatively simple and cross media effects are at minimum compared with Thermal and Catalytic Oxidation devices (Battistella et al., 2014).

A correct use and management of waste gas cleaning devices can bring:

- An abatement of organic substances (cancer causing ones too), generally coming from vents, storage tanks and handling of liquid substances containing VOC, with a high efficiency of VOC removal and recovery in case of adsorption devices;
- A strong reduction of inorganic pollutants in the emissions from production processes with simple maintenance and minimum cross media effects (in case of Adsorption and Absorption);
- The recovery of combustion heat and generation of steam (in case of Thermal Oxidation);
- A high thermal efficiency using recuperative and regenerative oxidation (in case of Thermal and Catalytic Oxidation).

REFERENCES

Battistella, G. (2013, October). *L'innovazione tecnologica per le attività di licensing, monitoraggio e controllo degli insediamenti IPPC soggetti ad AIA nazionale*. Paper presented at Veneto Region Educational Course 'Emissione negli impianti di processo', Mestre, Italy.

Battistella, G., & Di Marco G. (2013). *Air emissions measurements in environmental monitoring of IPPC plants*. IGI Global.

Battistella, G., & Di Marco, G. (2013, June). *Some Open Issues in Environmental Monitoring of IPPC Plants Air Emissions Measurements*. Paper presented at 4th Imeko TC19 Symposium on Environmental Instrumentation and Measurements Protecting Environment, Climate Changes and Pollution Control, Lecce, Italy.

Battistella, G., Di Marco, G., Bonaiuti, F., & Carlucci, C. (2014, March). *La Ossidazione Termica come migliore tecnica disponibile per gli Off-Gas*. Paper presented at the AIDII 20th Conference of Industrial Hygiene 'Le giornate di Corvara', Corvara, Italy.

KEY TERMS AND DEFINITIONS

Absorption: Mass transfer between a soluble gas and a solvent – often water – in contact with each other.

Adsorption: Heterogeneous reaction in which gas molecules are retained on a solid surface (adsorbent) that prefers specific compounds to others and thus removes them from effluent streams.

AIA: (Italian IPPC Permit): Operating permit released for environmental protection purposes, in order to prevent and control pollution.

BAT: Best Available Technologies.

BREF: Reference Document on Best Available Technologies.

Catalytic Oxidation: Oxidation process of combustible gases and odorants in a waste gas stream, by heating a mixture of contaminants with air or oxygen above its auto-ignition point. The gas, after passing through the flame area, passes through a catalyst bed. The catalyst has the effect of increasing the oxidation reaction rate, enabling conversion at lower reaction temperatures than in thermal oxidation units.

IMELS: Italian Minister of Environment, Land and Sea.

IPPC: Integrated Pollution Prevent and Control.

ISPRA: Superior Environmental Protection and Research Institute.

PTS: Total Suspended Particulate, the atmospheric particulate matter.

Thermal Oxidation: Oxidation process of combustible gases and odorants in a waste gas stream, by heating a mixture of contaminants with air or oxygen above its auto-ignition point in a furnace and maintaining it at high temperature for sufficient time to complete combustion to carbon dioxide and water.

TOC: Total Organic Carbon, the amount of carbon found in an organic compound.

VOC: Volatile Organic Compound, organic chemicals that have a high vapor pressure at ordinary room temperature. Their high vapor pressure results from a low boiling point, which causes large numbers of molecules to evaporate from the liquid form.

ENDNOTES

1 Available in the web portal at the URL http://eippcb.jrc.ec.europa.eu/reference.

2 For further technical descriptions see Section 3 of Reference Document on Best Available Techniques in Common Waste Water and Waste Gas Treatment / Management Systems in the Chemical Sector - February 2003.

3 It were analyzed only data concerning 'first' AIA Permits. This scenario could change because of *AIA* Permits provisions or modifications.

4 Some *AIA* IPPC plants adopt more than one different technique considered as BAT.

5 It were analyzed only data concerning first *AIA* Permits. The scenario could be changed because of *AIA* Permits provisions or modifications.

This research was previously published in the Encyclopedia of Information Science and Technology, Fourth Edition edited by Mehdi Khosrow-Pour, pages 3156-3171, copyright year 2018 by Information Science Reference (an imprint of IGI Global).

Section 6

Geographic Information Systems

Chapter 19
Application of Geospatial Mashups in Web GIS for Tourism Development

Somnath Chaudhuri
Maldives National University, Maldives

Nilanjan Ray
Adamas University, India

ABSTRACT

This chapter examines current development in Web GIS with the implementation of geospatial mashup technologies, such as Google Map, in the context of map mashups and presents a classification of map Mashups and their application in tourism management and promotion. On the Web GIS context, mashup is the process of merging multiple sources of data, both spatial and non-spatial, into a single integrated spatial display. It is about extracting spatial data from a non-spatial source and combining with other spatial data and finally displaying it on a map. This chapter demonstrates that geospatial mashup has great potential to facilitate and widen the rapid development of the future web mapping technology in Web GIS in tourism development. It also highlights on the basic architecture and working principles of map mashups in the context of tourism management. The final section of this chapter emphasizes some issues and limitations inherent to the current mashup technologies like privacy protection, copyright issues, etc., which need to be worked out before its wider adoption.

INTRODUCTION

Tourism is defined by the World Tourism Organisation *(NSCB, 2004)*, as the act travelling for the purpose of recreation and the provision of services for this act. Currently, tourism industry is one of the fastest growing industries all over the world. This smokeless industry is basically a kind of service industry, as it renders service to the tourists and all other supporting industries related to tourism like, hotel industry, transport industry etc. This business involves many socio-economic activities like promotion and advertising tourist spots and destinations, providing effective transport facility, fooding-lodging,

DOI: 10.4018/978-1-5225-7359-3.ch019

entertainment etc. At the same time when the tourism industry is flourishing it helps in socio-economic development of those tourist destinations. It also helps in strengthening the economical status of the country by earning foreign currencies without exporting national wealth. So, it is obvious that if this industry becomes more effective and efficient, it will definitely be the major source of revenue and will take a leading role in the overall economic development of the nation. Information and Communication Technology (ICT) can lead tourism to emerge as a new mantra for alternative economic development (Buhalis, 1998). Information Technology breaks the geographical boundaries so it is shared to the global audiences. Information Technology integrates between tourism product and requirement of the tourists. Due to changes in tourists or visitors behavior, the tourism market is becoming more segmented with each potential tourist belonging to a number of market segments (Cheng et al., 2002). Tourist operators need to be aware of these changes and be equipped to respond or better still, take a proactive approach. Technological revolution during 1990s brought with it new opportunities and challenges for the tourism industries. Technology has become fundamental to the ability of the global tourism industry to operate effectively and competitively. Information technology is being rapidly diffused throughout the tourism industry and that no player will escape from information technologies impacts. These technological innovations started in the 1970s when the main airlines set up CRSs (Computerized Reservation Systems), with the strategic aim of building a global distribution network for their products. Connecting travel agencies to the CRSs set off a process of distribution automation involving an ever-increasing number of tour operators, carriers, and car hire firms, individual hotels, hotel chains, and other hospitality firms. Geographical Information System, an ICT tool has been extensively used for tourism promotion and management. It was in use for GIS data design and collection, database design management and application of tourism analysis and problem solving. Currently, Internet has become the inseparable part of the Information and communication Technology. Online technologies within the tourism industry have significantly impacted on communications, transactions and relationships between the various industry operators and with the customer, as well as between regulators and operators. The Internet provides many advantages in the tourism industry (Ray et al., 2014). The GIS technicians and researchers started research on how to share the GIS features online, rather than using it as a standalone system. In the year 1993, the Web GIS started evolving rapidly. The online static maps slowly changed to interactive dynamic maps over the World Wide Web. This is the first step of Web GIS. The greatest advantage was to get rid of traditional desktop GIS, its installation and data sharing hazard. Today's web user can create content on the web both collaboratively and individually, allowing for a personalized web experience through wikis, blogs, podcasts, photo sharing, and other technologies. GIS and mapping applications have both benefited from and contributed to these trends, collectively called "Web 2.0" (Pierce et al., 2009). This provides the concept of Geospatial Mashups, especially Map Mashups. The latest trend in the field of geospatial science and technology in Web GIS, is *Geospatial Mashups.* Integration of multiple data layers from multiple sources, is one of the most common and effective functional requirements of Web GIS applications. On the Web GIS context, a Mashup is the process of merging multiple sources of data, both spatial and non-spatial, into a single integrated spatial display. It is about extracting spatial data from a non-spatial source and combining with other spatial data and finally displaying it on a map. This research paper discusses the basic architecture of the Geospatial Mashups in Web GIS and its application in visual impact analysis and strategic management in tourism.

Objectives of the Study

This present study seeks to the application of Web-Based Geographical Information System (GIS), an ICT tool for tourism management and promotion particularly through internet, with a future plan to develop this type of promotion by implementing GIS tools for tourism. In the context of tourism management this present study penetrates the usage of Geospatial Mashups, a spatial technical tool of Web GIS. It disseminates maximum level of information for tourism promotion in a collaborative manner. This paper examines current development in Web GIS with the implementation of Geospatial Mashup technologies, such as Google Map in the context of map Mashups, and presents a classification of map Mashups and their application in tourism management and promotion. Geospatial Mashup has great potential to facilitate and widen the rapid development of the future web mapping technology in Web GIS in the context of tourism development.

Background and Literature Survey

Geographic Information System is one of the most popular ICT tools for capturing, storing, retrieving, manipulating, mapping and analyzing spatial and non-spatial geographical data in the digital format. GIS is the information system that provides functions including visual 3D presentations about any geographical locations, advanced analysis of digital geospatial information by processing them in an integrated manner.

GIS technology integrates common database operations, such as storage of data, retrieval through query and converting those data to information through statistical analysis (see Table 1). GIS manages region-based information and provides tools for depictation and analysis of various statistics, including population density, economic development, transport facility, types of vegetation etc. GIS helps us to store the detailed information of any region in the databases and maps to create dynamic displays. Additionally, it provides tools to convert and display raw data in the form of 3D maps, run any query, and overlay those databases in ways which is not possible with traditional spreadsheets. These special-efficient as well as effective abilities distinguish GIS from other information systems, and make it one of the most effective ICT tool to a wide range of public and private enterprises for elucidation of events, predicting outcomes, and scheduling strategies. The United States Geological Survey (USGS) defined "GIS as a computer hardware and software system designed to collect, manage, analyze and display geographically (spatially) referenced data".

If we want to define GIS covering all its features then it can be categorised as shown in Table 1.

Basically, it can be stated that GIS can use any information that includes location or region. Now, the location of a particular region can be expressed in various manner. It can be done through address, pin code, latitude-longitude etc. GIS can store, compare and analyse different types of information. The system can include information about the land, such as the location of water bodies, different varieties of vegetation, different kinds of soil, contours, transport facilities like roads, railway tracks, subways etc. It can include data about the density of population, economic standard, education level etc. GIS can also include information about the different buildings in the region like sites of factories, farms, and schools, roads, and electric power lines. Now days the GPS provide the satellite images of the shopping malls, cinema halls, hotels, restaurants, banks and even small retail stores also.

Table 1. Features and analytical functions of GIS

Features of GIS	Related Definition	Related Analytical Functions
A Process	A system for capturing, storing, checking, manipulating, analyzing and displaying data which are spatially referenced to the earth (DoE, 197: 132)	Presentation and thematic mapping Data Query Spatial Query Database
A Toolbox	Containing tools for collecting, storing, retrieving, transforming and displaying spatial data (Burrough, 1986: 6)	Integration Route finding
A Database	Of spatially referenced entities (Smith et al., 1987)	Point in polygon analysis Overlays
An Application	Cadastral information system, marketing information system, planning information system, etc.	Buffering Visualization and 3D modelling
A Decision Support System	Integrating spatial data within a problem solving environment (Cowem, 1988)	

Source: Bahaire and White (1999, p. 161)

Now the question is, how this vast information is stored in this system. Actually, GIS can store data, not the information. Data is stored in the system and it is retrieved and analysed to convert it into relevant information. Data in different forms and formats can be entered into GIS. Putting information into GIS is called data capture. GIS being a digital system can capture the digital data directly. Data that are already in digital form, such as images taken by satellites and most tables, can simply be uploaded into GIS. An example of this kind is data collected by satellites that show land use—the location of farms, towns, or forests. GIS can also include data in table form, such as population information. But any other non-digital or analog data should be digitised and then punched into GIS database; viz.

Maps must be scanned or converted into digital data. While punching the data in the database, GIS must make the information from all the various maps and resources align, so that they fit together. This is one of the most vital steps of data storage, this is necessary as the maps collected from different sources may be of different scales. A scale is the virtual relationship between the distance on a map and the real distance on the Earth. GIS integrates all the information from different sources in such a way that it will have all the same scale and no ambiguity during analysis. GIS technology allows all these different types of information, no matter their source or original format, to be overlaid on top of one another on a single map. This map or sometimes statistical reports or virtual models are generated as outputs just on a single click of the mouse of the users' computer. The complete process of GIS function is depicted in Figure 1.

From the basic features we can redefine GIS as:

- Geographical
    - It contains data and concepts which are concerned with spatial distributions and inventory mapping
- Information
    - It implies some notion of storage, retrieval and analyses of data, usually as an aid to decision-making.
- System
    - It involves the sequence of inputs, processes and outputs (Chaudhuri et al., 2015).

Figure 1. Functionalities of GIS

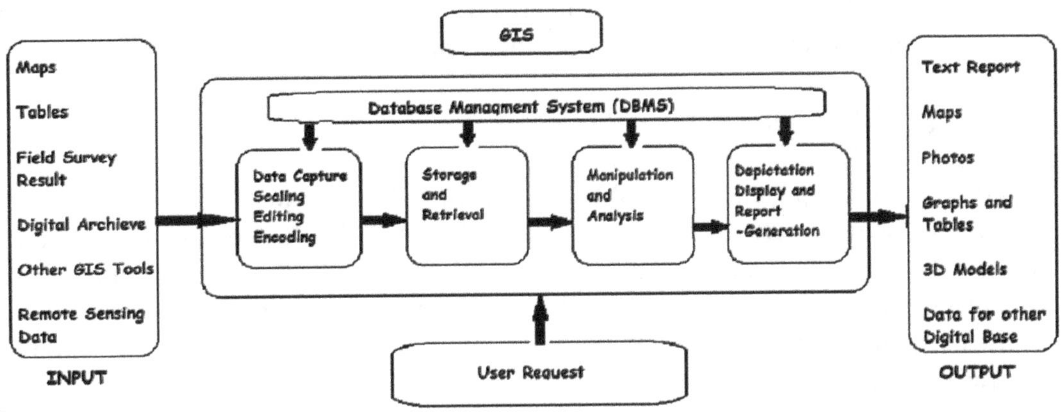

Tourism Classification and Problems

Tourism is viewed as travel for recreation, pleasure or instruction, often in organized groups. It could be visits to attractions, city breaks, and trips for business meetings, sports events, concerts or visits to friends and relatives. Tourism is becoming an integral part of our lives and culture (Balogun et al., 2010). According to Ayeni (2004) tourism can be categorized as follows:

- Cultural Tourism
    - It includes religious and national festivals, art galleries, historical monuments, arts and crafts, museum etc.
- Ecological Tourism
    - It includes geomorphological, geophysical and geological objects like waterfalls, mountains, sea-beach, desert, national parks etc.
- Modern Tourism
    - It includes modern notable engineering structures or architecture, amusement parks, travel and accommodation facilities.

The tourists get information about the popular tourist sites and then travel in search of cultural, ecological or modern tourism aspect. There are many locations or even countries which are sated with numerous ecologically or culturally attractive and significant tourist sites. But many of those locations are barely explored. Fajuyigbe and Balogun (2007) itemized those problems as follows:

- It is a time consuming, hectic and expensive work to updating existing graphical tourist guides and maps.
- Lack of digital online information for tourism amenities and destinations. There is hardly any comprehensive information over the internet.
- Inadequate technology-driven approach for sustainable tourism
- Lack of motivation for strategic marketing techniques and management

All the tourists are not tech-savvy to install a standalone GIS application in their individual computer system and do research about the tourist spots. Rather, it will be much simple if they can collect information over the internet. That leads towards the transition from GIS to Web GIS.

Tourism Using Web-Enabled GIS

Internet has become an inseparable part of the Information and communication Technology. The GIS technicians and researchers started research on how to share the GIS features online, rather than using it as a standalone system. In the year 1993, "Web 2.0", the Web GIS started evolving rapidly. The online static maps slowly changed to interactive dynamic maps over the World Wide Web and Internet. This is the first step of Web GIS. The greatest advantage was to get rid of traditional desktop GIS, its installation and data sharing hazard. Web GIS is any Geographic Information System that uses Web technologies. In a narrower definition, Web GIS is any GIS that uses Web technology to communicate between two components (Fu et al., 2011). Traditionally, geographic information systems were realized as monolithic and platform-dependent applications (Wong et al., 2002). The major concern is to use GIS as a tool for communication between different interest groups like planners, decision makers, and the public. The rapid growth of the internet provides highly customized, accessible and interactive source of public information and is changing the ways that people capture and manipulate spatial information (Balogun et al., 2010). In 2004, Information Technology Planning Management and Development of Goa, India was developed by Dr. P. K. Pandley and Ruma Chakraborty to advice the Government in developing and managing tourism in Goa (Pandey et al., 2004). Ghana Geographical Information System (GGIS) was developed by Longmatey et al. (2002). It provides GIS in tourism management, marketing and promotion in Ghana. It is one of the principal GIT tool in Ghana tourism market. Fajuyigbe et al. (2007) developed a web-enabled GIS for the management of tourism in Oyo state, Nigeria. The basic architecture of this application has two sections,

- Server-end and
- Client-end.

The client can access the geospatial information over the internet and the server will process and customize the query for execution. But it doesn't incorporate the Virtual tourism features. An implementation of Web-enabled GIS as a potential tool in sustainable tourism in western Nigeria was done by Fajuyigbe and Balogun (2007). This was an advanced version of Fajuyigbe et al. (2007)'s system. It was used as a potential tool for projecting data and analysis of critical information for the efficient management and promotion of tourism in most of the states in western Nigeria. Having reviewed the above cited works the requirement for integration of multiple datasets from different independent sources in Web GIS applications, has become prominent and inevitable.

Web GIS Architecture

The "Web GIS" became a synonym for Web information systems that provides a functionality of geographic information systems on the Web through HTTP and HTML (Shanzhen et al., 2001).

The simplest architecture of Web GIS should have at least two components, a Web-application Server and one or more Clients. Client can be as simple as a Web browser or a mobile client. The GIS service providing server can be connected by the client using Internet (HTTP). This dynamic connection is interactive in nature. The basic architecture is represented as in Figure 2.

The simple client side architecture allows the users to carry out some data manipulation, data customization and analysis locally on their own machines. The greatest advantage of this model is its low overhead on the server and high interactivity at the client side. Most of the computing tasks are performed by the client end, as a result, less workload and computation on the server side. With direct access to the database, the capability if client computer can be maximized to match the traditional client/server application and reserves the platform-dependent characteristics at client side concurrently (Ben et al., 1999). On the other side the server side architecture allows clients to submit query for data and analysis to a server. The server executes the query, retrieve the result from the database and process the result. The spatial and tabular data remain on the server. This architecture provides cross-platform capability at the client side. No special plug-in or application software is required for this interactive full-duplex communication. But though the architecture is providing a two-way communication without installing any third party application software, it has a severe drawback in terms of its speed. It is slow because each time a request is made, a new HTML document containing a graph, or map has to be generated and transferred to the client end. The next section will explain how Web GIS works as a web application over the internet. The working principal of Web GIS is basically a query request and response cycle. Suppose the client is accessing the GIS application over the internet using any web browser or from a mobile application. The client is searching for a specific location. The client posts the query in the application or in the browser interface. The Web Server will receive the query. In the next step the request will be forwarded from the Web Server to the GIS Server. The query will be executed by the GIS Server and required data will be retrieved from the GIS database. The result can be a simple map or, a satellite image or, a Mashup map, which depends upon the request from the client. The result is then forwarded to that Web Server from where the GIS Server has received the query and finally forwarded to the client. The browser interface or the mobile application will display the result. Based upon the data processing in the Web GIS application the complete workflow has been categorized into three sections, namely (Figure 3):

- Presentation tier,
- Logical tier and
- Data tier.

Figure 2. Basic Architecture of Web GIS

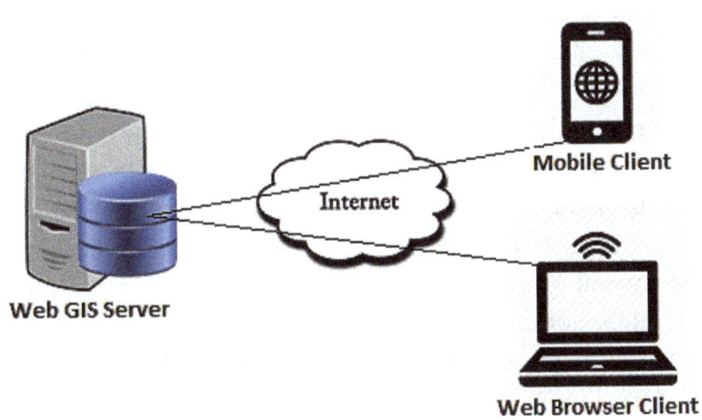

Figure 3. Web GIS Workflow

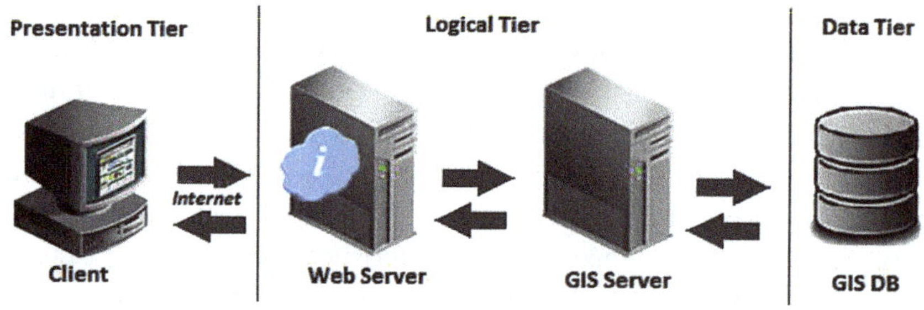

This complete Web GIS workflow is depicted in Figure 3.

The Presentation tier is the Web browser interface or the mobile application interface from where the client can send a request. Using the same interface the client can view the result. The web server will display the customized result in the presentation tier in the form of graph or, chart or, may be tabular data. Web server forwards the request query to the GIS server in the Logical tier. The key processing of the request and the query analysis is performed in this tier. SQL query is executed by the GIS server and it retrieves the data from the GIS database. After the query is executed the result is forwarded from the GIS server to the Web server. The principal role of the Data tier is to maintain normalised data for effective query processing.

Geospatial Mashups

The three tier Web GIS system is definitely an easier and user-friendly application to get access to geo-spatial data over the internet just on a click. But the effectiveness and reliability of a Web GIS system is determined by the standard of the GIS server it uses. The quality and standard of the Web GIS application is dependent on the functionality, scalability and ability to be customized of the GIS server. The GIS database is the backend support for any Web GIS application. The quality of the data in the GIS database, determines the standard of the result generated by the Web GIS application. Professional applications definitely need the up-to-date, reliable and high-quality data set. Slowly, it was becoming obvious that to maintain a quality and up-to-date Web GIS database it needs a collaborative contribution. There are clients who are accessing the system for information, but at the same time there might be some other end-users who can update the system with recent information about their locality. The first type of clients are just sending query in the system and accessing the information. They don't have any idea what is going on at the backend of the system, they don't need to know about the backend processes. But the Web GIS system administrator slowly realised the need of another group of authenticated end-users who can perform some geo-spatial processing tasks. Web GIS initiated from a basic assumption that the global geo information cannot be captured, saved and customized under a single organizational unit. The second group of authenticated end users can provide the up-to-date information and can be accessed globally. This concept of collaborative contribution for the system leads to the new avenue of Geospatial science and technology called Geospatial Mashups. It is the modern technology of providing web mapping or GIS services globally. Integration of multiple data layers or datasets, often from multiple sources, is one of the most common functional requirements of GIS applications (Fu et al., 2011).

The term "Mashup" is derived from pop music, where the musicians create new songs by remixing multiple sound tracks together. Today's web user can create content on the web both collaboratively and individually, allowing for a personalized web experience through wikis, blogs, podcasts, photo sharing, and other technologies. GIS and mapping applications have both benefited from and contributed to these trends, collectively called "Web 2.0" (Pierce et al., 2009). According to OGC (2002), Web Feature Service is one of the most important web data service. It allows a client to retrieve heterogeneous geospatial data from multiple geospatial data servers. Using HTTP as the distributed computing platform it allows fundamental database Query operations like, Select, Update, Insert, Delete etc. The counterpart Web Map Service is defined by OGC (2004) as the service that is capable of creating and projecting maps that come concurrently from different heterogeneous sources in any of the following standard image format, SVG, PNG, GIF, and JPEG. WMS consists of three basic operations, namely,

- *GetCapabilities*,
- *GetMap* and
- *GetFeatureInfo*.

These three operations lead to a new feature of remixing data in the web called as the Mashup feature of the Web GIS. In Web GIS, geomashup is being used. A geomashup is a Mashup where at least one of the contents or functions is georeferenced. Geospatial Mashups integrate multiple data sources based on common geographic locations. Mashups carry the overlay functions. In an overlay function, information from different heterogeneous sources are merged together and represented as a single informative Mashup layer. It includes topological overlay which restructures into a single vector dataset and a graphic overlay which superimpose images or maps on top of other. Most of today's online Mashups are graphic overlays. The simple architecture of Web GIS consists of a server end and client or browser end. The Mashup operation can be implemented in both ends. If the Mashup operation is performed in the server end, it is called a Server-side Mashup and if on the client end it is called as Browser-side Mashup. In the Server-side Mashup architecture, the Mashup web server sends requests to different web services, receives the responses and merges the results. Figure 4 depicts the Server-side Mashup architecture.

In the Browser-side Mashup architecture, the web browser sends requests to different services, receives the responses and displays the composite result. Figure 5 depicts the Server-side Mashup architecture.

Both the architecture has some advantages and disadvantages. The advantage of Server-side Mashups is that the server has more powerful hardware and software than the browser, but these Mashups also require specialized programming tools and complex-server side programming, which includes more manual work during development and deployment. Because of this complex programming and tedious work, the original Server-side Mashups were mainly limited to professional programmers. But most of the today's Mashups are Browser-side. It uses mainly lightweight programming; as a result it has become easy to develop and professional GIS companies have adopted this approach as an easy and fast way to develop Web applications. ArcGIS by ESRI is one such example.

This recent trend of Geospatial Mashup has significant impact on Web GIS. The next section will highlight on the design pattern of Mashups in Web GIS. The common design pattern consists of three components: Basemaps, Operational Layers, and Tools.

Figure 4. Server-side Mashup Architecture

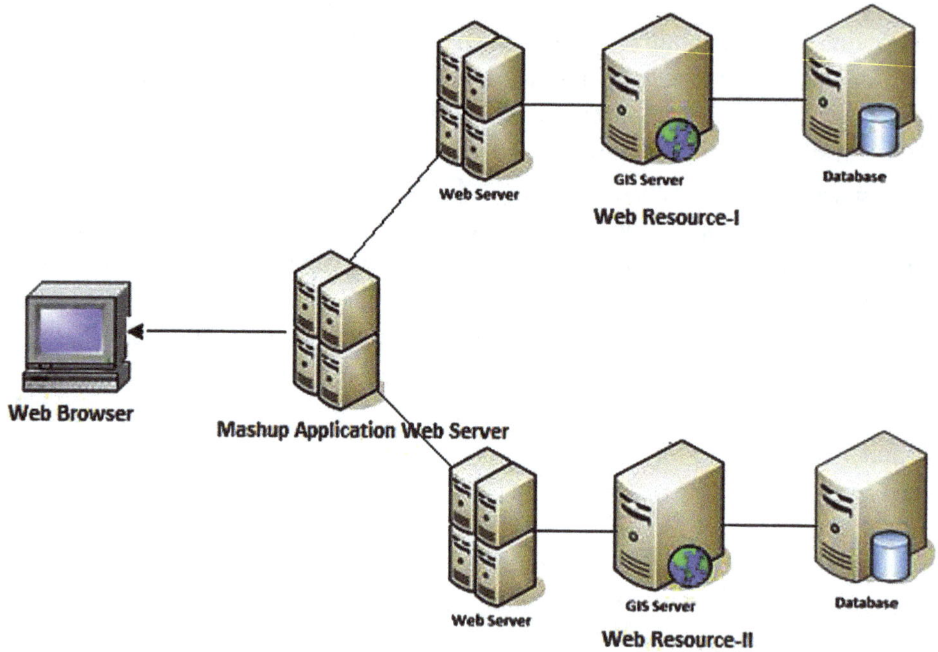

Figure 5. Browser-side Mashup Architecture

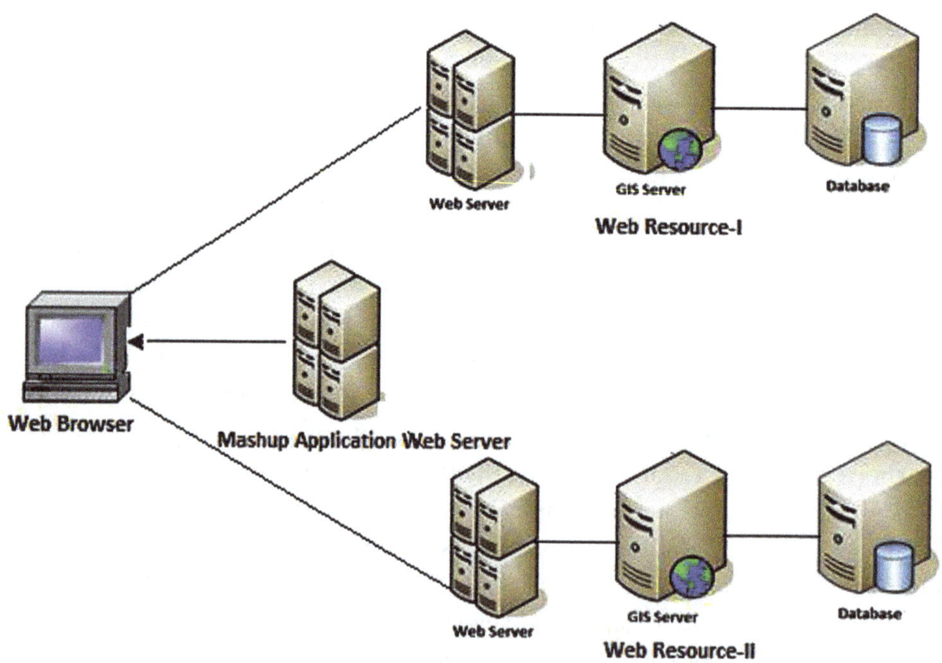

- Basemaps
  - It provides the initial framework for any geomashup. It is the frame of reference like, the aerial imagery maps often from the web services of Google, Yahoo, ArcGIS Online etc.
- Operational Layers
  - It is also known as the thematic layers, usually embedded on the top of a basemaps, like Café, building names, street names, bus stops etc.
- Tools
  - It can execute logical and analytical functions.

Mashups can potentially combine any type of contents and functions over the web, regardless of whether a formal programming interface is available. The vast majority of contents over the web are HTML pages and photos that don't have formal APIs, but they contain a huge amount of valuable information that can be scraped and geospatially tagged or referenced and then reused to build new and value added applications. Virtually, the whole web can be remixed, which opens the door to unlimited value-added Mashup applications (Fu et al., 2011).

Application of Geospatial Mashup in Tourism

The ability to examine data in its spatial context is extremely important. The geocoding process takes a simple street address and matches it to a database containing address ranges for every link in the street network. In this way customer lists may be converted into simple dot maps, or coupled with census data and displayed as maps of market penetration (Goodchild et al., 1984). This research paper concentrates mainly on the application of Geospatial Mashups in tourism sector. With respect to the tourism industry the map Mashup can be classified into four broad sections:

- Informative Mashup
- Participatory Mashup
- Collaborative Mashup
- Enterprise Mashup

Check the Google map illustrated in Figure 6.

The initial map was developed without the detail information about the stadium, the tourist lodge, the hotels, the schools, colleges, banks and the historical and religious architectures like Nandalal temple, Chinnamasta mandir etc. Initial aerial imagery map was only with the land structure and the roads, that is the Basemaps. Other than the road and basic framework the basemaps include the water-bodies like the Lal Bandh or other basemaps' objects mention in blue in the above Google Mashup map. Next in the Operational Layers other end users have implemented Participatory or Collaborative Mashups. Some end users are providing the information of the historical place of interest like the temples. Some other client has supplied the information about hotels, restaurants and banks. Few others may have contributed in providing information about the local school, colleges and health centers. There is high probability of receiving the data from different clients in heterogeneous form. When all these heterogeneous data is compiled and remixed using WFS and WMS service, it results into this detailed map, containing every in and out about the location. Thus, this Mashup service is providing detailed information about a location in a collaborative manner. The basic information are required by a potential tourist to visit any tourist

Figure 6. Mashup in Web GIS
Source: www.google.com/maps/

location. Initially, the tourists have to depend on the information provided by the travel agents about a location. The agents used to provide the information over telephonic conversation or sharing documents over email. The information may not be up-to-date and the tourists have to solely trust the agents. The tourists could have viewed the aerial imagery maps over the internet, but those maps don't provide detail information about the road, hotels, place of interest, transport facilities available etc. But now with the detailed geospatial Mashup services over any Web GIS applications, the tourists can collect every in and out detail information about their place of choice. They have to just browse the internet, run the online Web GIS application and run a query. They will receive detailed information about the location. Another greatest advantage in this Participatory or Collaborative Mashups is that, the globally accessible geomashups are always up-to-date; they contain the most recent information about the location. Even the tourists while visiting those tourist sites if they found something is not matching with the Mashup information over the internet, they can customize those published information with the updated one. This will be beneficial for other potential tourists. These are the benefits of updating information of those locations which are already established as tourist spots. Now, if some entrepreneur or travel agency wants to project a new location as a tourist spot, they can perform initial research work to analyse whether their local area can be a potential tourist location or not. First and foremost they have to find out whether that particular location has any Cultural, Ecological or Modern tourism component or not. If it is there they can update that information using the Mashup option in Web GIS application. Next, they have to update other facilities like, transport, hotels, café, health service etc. even internet facilities available or not. That means, the locations which can be potential tourist spots can slowly come into limelight just by some simple initiatives taken by locals. One such online application provided by Google is http://www.editor.giscloud.com/ where the end users can update and create a map online contributing detailed information about their locality or any other place of interest. They don't need any Government intervention. In this way the Goespatial Mashups are making hundreds of new Cultural, Ecological or Modern tourist locations popular. This is effecting overall economic development of the location through tourism industry. On the other hand, in today's competitive marketplace, leading companies are analyzing and listening what customers really care about. Traditional forms of marketing are no longer as effective as they used to be in the past. This is where Web GIS comes in and allowing the companies to visualize where potential customers are located by analyzing demographic, psychographic, purchasing, and spending characteristics for accurate customer segmentation and helping the companies to find more like them (GIS for Marketing: Where Strategy Meets Opportunity, 2010). The tourism sector is implementing the Web GIS Mashup application as a promotional and marketing tool.

Challenges

Mashups provide different opportunities for sharing data collaboratively and using information in new applications and systems. It has become an integral design pattern for Web GIS applications. As more and more functions and applications become available over the Web and more policies and technologies favour the use of Mashups, this branch of geospatial science and technology will grow exponentially in future and will contribute in tourism industry. Like all other technologies Mashup also has few challenges and demerits. New research works are going on to address the following drawbacks of Mashups:

- Quality and Reliability of Information
    - Using the Participatory and Collaborative Mashups anyone can publish contents over the internet. No one is authenticating the genuineness of the uploaded information. Quality can vary enormously, and if questionable sources are implicated in a chain of services, the results of uncertainty or error can be propagated over the internet. This can produce misleading and even wrong information (Goodchild et al., 2010).
- Copyright and Ownership
    - While the spirit of open exchange is still the main characteristics of the web, many websites post terms of use and retain their copyright. Copyright and terms of use can get complex when a Mashup uses multiple web resources, or even more complex when a Mashup uses information generated by another Mashup. It is important for Mashup developers to comply with provider's terms of use to avert future disputes and to understand the impacts of copyright on their application (Fu et al., 2011).
- Security
    - Enterprise Mashups can involve confidential information sharing over the internet, making security a key consideration.
- Technical Challenges
    - Technical challenges in dealing with map Mashups are mainly related to technical incompatibility, data integration and quality assurance, interfaces and functionality, level of programming skills required, enterprise integration, etc (Li et al., 2008).

CONCLUSION

Impact of ICT in tourism is very vital. For instance, a guide requires historical and cultural information about the destinations, tour agency requires information on transformation, reservations, accommodation in Hotel, and for marketing, a tourist requires information from Cell Phone Application (SMS), Internet, Wireless (Wi-Fi), VOIP, GPS, GIS, Digital Radio, On demand application and so on. Role of Information Technology in tourism industry cannot be underestimated because it has vital contributions. Once the tourism information or data is in the system (computer) this may be transmitted to the proposed tourists through different channels viz. FAX, GPS, Website etc. One of these informative IT tools is Geospatial Information Technology (GIT). GIS is an information system that provides functions including visual presentation about the tourism destination, advanced analyses, etc., of digital geospatial information by processing it in an integrated manner on electronic maps of the tourist site. The benefits of Web GIS, especially Geospatial Mashups far outweigh its limitations and challenges. Web GIS disseminates maximum level of information to the potential tourists, at the same time allows the travel agents or organizations to promote an unknown location having adequate features to be a popular historical, ecological or modern tourist spot. The potential tourists can analyze demographic, economic, psychographic, market characteristics of the location just on a click. Thus, providing support for strategic planning, Collaborative information sharing, online marketing techniques, improving efficiency in tourism management and attaining competitive locational and business advantage for the tourism agencies. The future research scope of this present paper depicts the application of Web GIS and Geospatial Mashups and how this modern GIT system can be used well as an analytical and strategic tool in tourism management and promotion.

Followings are the domain of future research in Web GIS and Geospatial Mashups in context to tourism business.

1. Web GIS Applications in tourism resource inventories
2. Web GIS application to location suitability under conflicting demands
3. Monitor and control tourism activities using Web GIS
4. Application of Web GIS in visual impact analysis in tourism research
5. Application of Web GIS community involvement and participation in tourism research
6. Web GIS uses as a decision supporting tools in tourism research
7. Application of Geospatial Mashups in tourism promotion
8. Limitations in Geospatial Mashups and their solution

REFERENCES

Ayeni, O. O., Saka, D. N., & Ikwuemesi, G. (2004). Developing a Multimedia GIS Database for Tourism Industry in Nigeria. *ISPRS 2004: 20th International Society for Photogrammetry and Remote Sensing (ISPRS) Congress: Geo-Imagery Bridging Continents.*

Balogun, V. F., Thompson, A. F., & Sarumi, O. A. (2010). A 3D Geo-Spatial Virtual Reality System for Virtual Tourism. *The Pacific Journal of Science and Technology*, *11*(2), 601–609.

Buhalis, D. (1998). Strategic use of information technologies in the tourism industry. *Tourism Management, 19*(5), 409–421. doi:10.1016/S0261-5177(98)00038-7

Butler, R. (1993). Alternative Tourism: The Thin Edge of the Wedge. In V. Smith & W. Eadington (Eds.), *Tourism Alternatives* (pp. 31–46). Chichester, UK: Wiley.

Chaudhuri, S., & Ray, N. (2015). Application of Web-Based Geographical Information System (GIS). In Tourism Development, Encyclopedia of Information Science and Technology. IGI Global.

Cheng, C., & Piccoli, G. (2002). Web-Based Training in the Hospitality Industry: A Conceptual Definition, Taxonomy and Preliminary Investigation. *International Journal of Hospitality Information Technology, 2*(2), 19–33. doi:10.3727/153373402803617737

Davin, K. (1997). Effects of Computer Support on Group Decisions. *Journal of Hospitality & Tourism Research (Washington, D.C.), 21*(2), 44–57. doi:10.1177/109634809702100204

Fajuyigbe, O., & Balogun, V. F. (2007). Web-Enabled GIS: A Potential Tool for Sustaining Tourism in (Western) Nigeria, *Nigeria. Computers & Society*, (June): 15–22.

Fajuyigbe, O., Balogun, V. F., & Odembe, O. M. (2007). Web-Based Geographical Information System (GIS) for Tourism in Oyo State, Nigeria. *Information Technology Journal, 6*(5), 613–622. doi:10.3923/itj.2007.613.622

Fesenmaier, D. R., Vogt, C. A., & Stewart, W. P. (1993). Investigating the influence of welcome center information on travel behavior. *Journal of Travel Research, 31*(3), 47–52. doi:10.1177/004728759303100308

GIS for Marketing: Where Strategy Meets Opportunity. (2010). Retrieved on June 20, 2015, from http://www.esri.com/marketing

Goodchild, M. F. (1984). ILACS: A Location-Allocation Model for Retail Site Selection. *Journal of Retailing, 60*, 84–100.

ICT & Tourism USAID from the American People. (n.d.). *International Journal of Knowledge and Learning, 3*(6), 628-648. doi:10.1504/IJKL.2007.016837

Bahaire, T., & Elliott-White, M. (1999). The Aplication of Geographical Information Systems GIS in Sustainable Tourism Planning: A Review. International *Journal of Sustainable Tourism, 7*(2), 159–174.

Jafar Jafari. (2003). Encyclopedia of Tourism (p. 308). Academic Press.

Li, S., & Gong, J. (2008). MASHUP: A New Way of Providing Web Mapping/GIS Services, *The International Archives of the Photogrammetry. Remote Sensing and Spatial Information Sciences, 37*(B4), 639–648.

Longmatey, D., Samuel, A., & Prah, B. K. (2002). *Management and Promotion of Tourism in Ghana: A GIS Approach from African Tourism Websites*. Retrieved on May 20, 2015. http://www.africantourism.com/ghana.htm

Pandey, O. K., & Chakraborty, R. (2007). *GIS: Information technology for planning management and development of GOA tourism master plan in next millennium*. Retrieved on May 20, 2015. http://www.gisdevelopment.net/application

Pierce, T., Dobson, G., & Phillips, M. (2009). GIS Mashups on the Web: Using Google, ESRI, and Open Source to Develop Weather and Climate Applications for the Web. *Geospatial Infrastructure Solutions Conference*, Tampa FL

Pinde, F., & Sun, J. (2011). *Web GIS, Principles and Applications.* ESRI Press.

Rahaman, M. (2010). Application of GIS in EcoTourism Developemnt: A Case Study in Sundarbans, Bangladesh. *International Journal of Entrepreneurial Behaviour And Research, 10*(4),247-59.

Ray, N., Das, D. K., Ghosh, S., & Banerjee, A. (2011). A Study on Application of Information Technology on Tourism Development, IEEE Proceeding Catalog No: CFP1141H-ART ISBN978-1-4244-6579- China

Ray, N., Das, D. K., & Chaudhuri, S. (2014). Tourism Promotion Through Web. In Strategic Marketing in Fragile Economic Condition. IGI Global.

Shanker, S. (2008). *ICT and Tourism: Challenges and Opportunities, Conference on Tourism in India – Challenges Ahead.* IIMK.

Shanzhen, Y., Lizhu, Z., Chunxiao, X., Qilun, L., & Yong, Z. (2001). Semantic and Interoperable WebGIS. *Proceedings of the Second International Conference on Web Information Systems Engineering,* 2, 42-47. 10.1109/WISE.2001.996702

Sigala, M. (2008). *Web 2.0, social marketing strategies and distribution channels for city destinations.* Academic Press.

Wong, S. H., Swartz, S. L., & Sarkar, D. (2002, April). A Middleware Architecture for Open and Interoperable GISs. *IEEE MultiMedia, 9*(2), 62–76. doi:10.1109/93.998065

ADDITIONAL READING

Buhalis, D. (1998). Strategic use of information technologies in the tourism industry. *Tourism Management, 19*(5), 409–421. doi:10.1016/S0261-5177(98)00038-7

Chervenak, L. (1993). Hotel Technology at the start of the New Millennium. *Hospitality Research Journal., 17*(1), 113–120. doi:10.1177/109634809301700110

Cho, Y., & Fesenmaier, D. R. (2001). A new paradigm for tourism and electronic commerce: Experience marketing using the virtual tour. In E. Laws & D. Buhalis (Eds.), *tourism distribution channels: practices, issues and transformation* (pp. 351–370). Thomson.

Collins, C., Buhalis, D., & Peters, M. (2003). Enhancing SMTEs Business Performance Through the Internet and U-learning Platforms. *Education + Training, 45*(9), 483–494. doi:10.1108/00400910310508874

Fesenmaier, D. R. (1994). Traveler use of visitor information centers: Implications for development in Illinois. *Journal of Travel Research, 33*(1), 44–50. doi:10.1177/004728759403300106

Klein, S. (2002). Web impact on the distribution structure for flight tickets. In K. W. Wöber, A. J. Frew, & M. Hitz (Eds.), *Information and Communication Technologies in Tourism 2002* (pp. 219–228). New York: Springer-Verlag Wien. doi:10.1007/978-3-7091-6132-6_22

O'Brien, D., & Fitzgerald, B. (2006). Mashups, remixes and copyright law. *Internet Law Bulletin, 9*(2), 17–19.

Werthner, H., & Klein, S. (1999). ICT and the changing landscape of global tourism Distribution. *Electronic Markets, 9*(4), 256–262. doi:10.1080/101967899358941

KEY TERMS AND DEFINITIONS

CRS: Computer Reservation Systems is a computerized system used to store and retrieves information and carry out transactions related to the information or booking of airline ticket, railway ticket, hotel booking etc. CRS has made the online booking and collecting information much easier, faster and effective.

Geospatial Mashups: Mashup is picking data or functionality from different sources (both homogeneous and heterogeneous) on the web and combining them within the browser to present to the user. On the Web GIS context, a Mashup is the process of merging multiple sources of data, both spatial and non-spatial, into a single integrated spatial display. It is about extracting spatial data from a non-spatial source and combining with other spatial data and finally displaying it on a map.

GIS: Geographic Information System is one of the most popular tools of GIT. It is the modern technology for capturing, storing, retrieving, manipulating, mapping and analyzing spatial and non-spatial geographical data in the digital format. GIS is the information system that provides functions including visual 3D presentations about any geographical locations, advanced analysis etc. of digital geospatial information by processing them in an integrated manner.

GIT: Geospatial Information Technology is a collection information communication tools used to capture, store, transform, manipulate, analyze, and produce information related to the surface of the Earth. This data may exist as maps, graphs, 3D virtual models, tables, and/or lists. Example: GIS, GPS etc.

GPS: Global Positioning System is one of the most popular tools of GIT. GPS is a radio navigation system that allows land, sea, and airborne users to determine their exact location, velocity, and time 24 hours a day, in all weather conditions, anywhere in the world. The GPS is being used in science to provide data that has never been available before in the quantity and degree of accuracy that the GPS makes possible.

ICT: Information and Communication Technologies, it refers to technologies that provide information through telecommunications. ICT and Information Technology (IT) are more or less similar, but ICT focuses primarily on communication technologies. This includes the Internet, wireless networks, cell phones, satellite systems and other communication mediums.

Tourism: Tourism is a social, cultural and economic phenomenon related to the movement of people to places outside their usual place of residence. Tourism comprises the activities of persons travelling to and staying in places outside their usual environment for not more than one consecutive year for leisure, business and other purposes not related to the exercise of an activity remunerated from within the place visited.

Tourist: Someone who travels for pleasure and recreation and exchange of culture.

VR: Technology (VR): Virtual Reality Technology is a computer-simulated environment that can simulate virtual reality, i.e. an artificial environment created by software and presented to the user in such a way that the user accepts it as a real environment. In VR, highly visual 3D environments are created by using CAD software and graphics hardware. VR has its application in different sectors viz. Military training, Architecture, Game, Medical Procedures etc.

VOIP: Voice over Internet Protocol is a collection of technology for the full duplex real-time transmission of voice communications and multimedia sessions over the internet. VoIP offers a substantial cost savings over traditional long distance telephone calls.

This research was previously published in the Encyclopedia of Information Science and Technology, Fourth Edition edited by Mehdi Khosrow-Pour, pages 3403-3418, copyright year 2018 by Information Science Reference (an imprint of IGI Global).

Chapter 20
Archaeological GIS for Land Use in South Etruria Urban Revolution in IX–VIII Centuries B.C.

Giuliano Pelfer
University of Florence, Italy

ABSTRACT

A GIS is developed for analysis of formation, growth, and collapse of past societies. The Urban Revolution in the Mediterranean area between 9th and 13th centuries B.C. pushed by development of Mediterranean market area is a good case study. The process leads to the growth of urban centers and of population. In Ancient Etruria the changes produced the abandonment of older sites and the origin of proto-urban centers on hill plateaus. The GIS is developed for explaining the factors affecting the urbanization process in Tarquinia. Settlement strategy was linked to available resources in the territory. The research focused on land use for ancient agriculture by reconstructing features of land use. The results show that the selection of plateau allowed a better resources management, required by population growth as well by the Mediterranean market demand. The changes explain the transition from a subsistence economy to a production of agrarian surplus in cereals. A new organization based on the private household property of agrarian plots could satisfy a better resource allocation.

INTRODUCTION

Urban Revolution in Ancient Etruria Between IX-VIII Centuries B.C. Origin of the Etruria Protocities

Archaeological discussion about protourban centers growth in middle Tyrrhenian and in Ancient Etruria regions outlined the original and revolutionary features of such process, called Villanovan Revolution for resuming the deep transformations occurring in a few decades.

DOI: 10.4018/978-1-5225-7359-3.ch020

New largest protourban settlements grew through the concentration, fusion and absorption of the earlier scattered sites on the plateaux of South Etruria, extended for many hectares and well defended (Bartoloni, 2002; Bietti Sestieri, 1996; Carandini, 2003; Mandolesi, 1999; Pacciarelli, 2000; Peroni, 1994, 1996).

Geographic position of plateaux and agrarian quality of adjacent soils were the key factors for settlement position selection. Tens of scattered older villages were abandoned with simultaneous people transfer, called synoecism, to the plateaux from the late final Bronze age to the first Iron age. The process features, the plateaux selected and their area extension would prove existence of some urban planning. New settlements extension, until to about 200 hectares, would testify the great dimension of the new communities born from synoecism (Peroni, 1969, 1994) and a gradual reduction of settlements number with significant increase of their population size and surface extension (Carandini, 2003; Pacciarelli, 2000).

Analysis Objectives

Although many aspects of protourban centers growth, like the foundation of Rome and its organization in *curiae* are investigated, relevant criticism remains about the same processes in South Etruria. In fact, some issues about the strong changes in settlement patterns observed in this period have to be yet clarified and deepened (see Background section - Bartoloni, 2002; Pacciarelli, 2000; Peroni, 1994).

The present research has the goal to explain the main factors that favored such historical process as well the features of the settlement strategies.

Use of GIS is the most powerful technology introduced to archaeology since the introduction of carbon 14 dating and seems the most suitable for specific information.

Most widespread use of this technology is for Cultural Resource Management, data visualization, excavations but also for prediction of archaeological site locations.

This paper focuses on the use of GIS for archaeological predictive modeling of ancient land use, by critically applying this new technology and exploring its theoretical and analytical implications. Archaeological data are point like geographical and temporal data with more or less large uncertainty. Work of the archaeologists is to connect all the point like data related to a geographic region and to an historical period using a qualitative and quantitative based narration that fit in the best way with available data and with the theory and model.

An archaeological GIS is designed with the aim to record the existing and reconstructed data on a database, to visualize data by thematic maps and to use such data for advanced statistical and spatial analysis. Archaeological GIS is realized installing GRASS GIS on OS Linux with interface to PostgreSQL database, with its extension PostGIS for geographic information and to R package for statistical and geostatistical analysis.

BACKGROUND

Archaeological Theories on Protourban Centers Genesis

Archaeologists suggested two approaches to protourban centers origin: the first approach underlined a continuity between protourban centers and earlier settlements through gradual development of settled areas, between the late final Bronze age and the beginning phases of first Iron age (Pacciarelli, 2000); the second approach suggested an ungradual transition from the scattered villages on the territory to the protourban centers with an extension 30 times larger than the villages of preurban phase (Peroni, 1994, 1996). Surface surveys of last fifty years showed an uniform distribution of the villanovan pottery remains, attesting a protourban phase after 1,000 B.C., with occupation of large areas, because of transformations in social and economical structures (Pacciarelli, 2000).

Villanovan Urban Revolution During X-IX Centuries B.C.

Studies of Peroni and Rittatore Vonwiller in the 1960s (as cited in Pacciarelli, 2000, pp. 11-12) outlined the close relationship of Urban Revolution in South Etruria with social, political and economic transformations starting from 1,000 B.C. Changes concerned the techniques of agrarian production, the work organization and the social relationships.

A new economic organization of the agrarian property is based on the private ownership of land by division into lots for households (Mandolesi, 1999; Pacciarelli, 2000; Peroni, 1994).

In this context the nuclear family emerged as the basic social cell of the protourban centers, replacing the older multi familiar clans.

Social hierarchies reflected kinship and rank relationships, as shown from the funerary data since the X century B.C. (Bartoloni, 2002, 2003; Iaia, 1999; Pacciarelli, 2000; Peroni, 1996).

Urban Revolution developed between XII-VIII centuries B.C. in the Mediterranean area (Moscati, 2001) and along Mediterranean sea coasts, favored by the emergence of a global Mediterranean market area. New situation produced huge changes: villages in a few tenths of years disappeared and protourban centers grew.

Collapses of the great eastern territorial empires and of the earlier palatial states, also due to the invasions by the People of the Sea, promoted the propulsive role of Phoenicians in the new Mediterranean global market between XII-VIII centuries B.C.

Such role of Phoenicians has been recognized in the last decades (Braudel, 2002; Giardino, 1998; Guidi, 1998; Liverani, 1988; Moscati, 2001; Pacciarelli, 2000; Trigger, Kemp, O'Connor, & Lloyd, 2000). Because of Phoenicians maritime activities, autonomous urban centers developed in the Etruria, middle Tyrrhenian and in central Aegean areas.

Case Study: Tarquinia

The Tarquinia settlement is an important case study on the genesis of protourban centers (see Figure 1.)

The surface surveys testified the uniform distribution of the villanovan pottery in all area, showing a full plateaux extension of the protourban center. Data show that the main center covered about 120 hectares on main plateaux of La Civita, including the oldest final Bronze age La Castellina and Cretoncini sites (Di Gennaro, 1986; Hencken, 1968; Mandolesi, 1999; Pacciarelli, 2000; Peroni, 1996).

The protourban center geographical structure is organized in three settlement subsystems: La Civita and the Calvario, Infernaccio, Acquetta villages on Monterozzi hill, their corresponding cemeteries, and the satellite coastal settlements of Fontanile delle Serpi and le Saline. The oldest settlement system developed on La Civita and on Castellina and Cretoncini sites with their cemeteries (Bietti Sestieri, 1996; Hencken, 1968; Iaia, 1999; Linington, 1982; Mandolesi, 1999; Pacciarelli, 2000; Peroni, 1996).

Funerary data and cemeteries distribution fit well with all the approaches about the protourban genesis and for this reason such information cannot be used for supporting neither theory (Carandini, 2003; Mandolesi, 1999).

Figure 1. The Tarquinia protourban center between final Bronze age and first Iron age

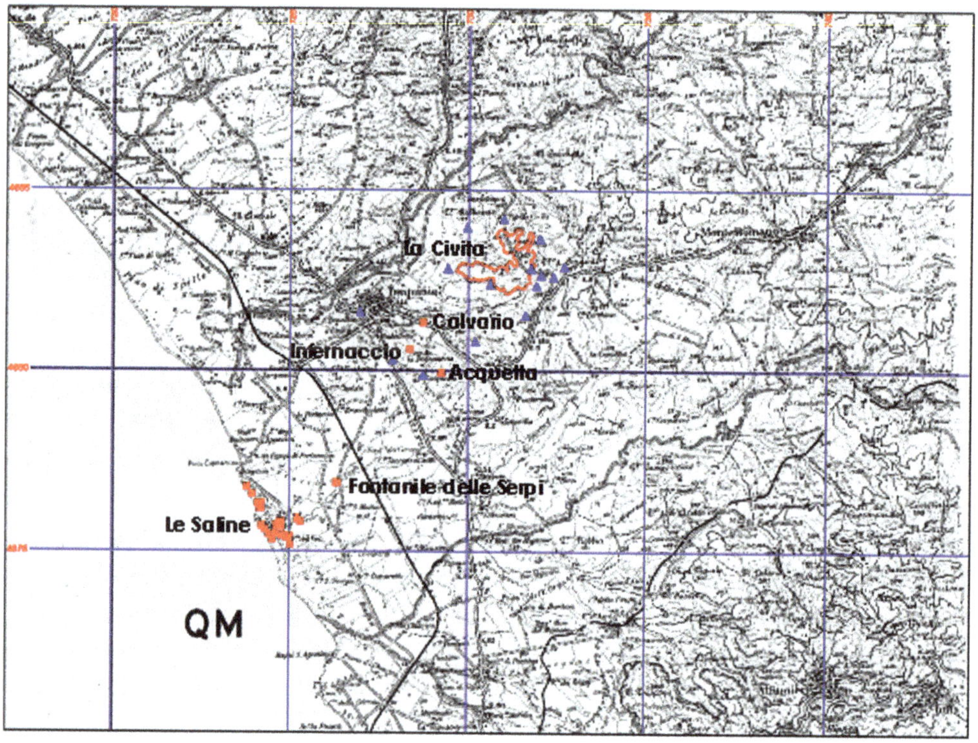

The social evolution of the protourban center, linked to familiar subgroups, likewise to the organization of the future *curiae* and *gentes* in the archaic Rome (Giardino, 1998; Mandolesi, 1999), extended the settlement system to the Monterozzi hill as shown by the cemetery at Le Rose near to the Calvario village (Carandini, 2003; Moscati, 2001; Pacciarelli, 2000). Monterozzi hill becomes important for its favorable position for coastal plain visibility and control.

MAIN FOCUS

Agrarian Production in Final Bronze Age

Ancient sources and archaeological data recognized key economic role of agriculture and sheep farming in the Etruria villanovan communities (Colonna, 2000; Pallottino, 1937; Sassatelli, 2009). The increasing demand of cereals characterized exchanges intensification during Urban Revolution. From late Bronze age, the agrarian activities prevailed on other ones, as proved by small amount of wild paleozoological remains. Changes in agrarian activities are due to major economic and social transformations as well to larger diffusion of bronze working tools rather than to great innovations in the farming technology (Ampolo, 1980; Peroni, 1994; Sassatelli, 2009). In fact, heavy soils plowing required a larger work and time amount supplied by motivated people and by new work organization, before the iron plowshare developed (Forni, 2002; Peroni, 1994; Van Joolen, 2003).

The Main Crops

Ancient sources testify that selected cereals like the hulled wheats as *triticum monococcum* and *triticum dicoccum* (spelt) and the naked wheats as *triticum aestivum* and *triticum compactum*, were used in agriculture until to the VIII century B.C. (Bartoloni, 2002; Bietti Sestieri, 2002; Sassatelli, 2009). Sources clarify which were the more diffused crops and the suitable requirements for soil features (Columella, II, 9, 5-9; Varro, I, 9; Cato, 6,1).

All the variables linked to cultivated crops and agrarian techniques are used for understanding and modeling the productive process. The grazing and the breeding were also important activities in villanovan economy (Bedini 1997; Sassatelli, 2009).

This is testified by paleozoological remains in many protohistorical sites.

The breeding could be practiced both around plateaux and on coastal plain, due to proximity of rivers and lagoons as water resources satisfying animals daily needs.

Bovines were utilized for agrarian work and for increasing land productivity by their manure, but not for food, because religious tradition forbidden to eat their meats (Ampolo, 1980).

Instead, sheep and goats grazed on free terrains as well in the coastal plain marshy areas and were used in particular for milk and the secondary products, as wool and skins.

Territory Control and Viewshed

An optimal position respect to available resources and to communication ways could also explain the settlements location choice. The resources availability in surrounding territory was crucial factor for protourban centers growth, like its control by the main settlement.

The viewshed analysis can outline these factors and is also useful for defining the cultural and cognitive landscape (Van Leusen, 2002).

Issues and Problems

Archaeological land evaluation estimates ancient land uses of the studied territories.

Soil scientists developed land evaluation methods for estimating alternative soil uses.

The outcome is a limited number of alternative land uses for the different situations, according to geomorphological framework that allows to classify the landscape in different Landscape Systems, identified as homogeneous areas for their physical, geomorphological and ecological features.

The main goal of such classification is to distinguish such territorial areas with their specific features for favoring the agrarian soils classification with the suitability categories for different crops.

These methods are based on reorganization and interpretation of inventories of soil, vegetation, climate and other aspects of land, for comparing alternative land uses, and their requirements with land resources.

This approach, used in physical geography, has then to be replaced by models built on protohistorical context based on ethnographic, archaeological and historical sources.

SOLUTIONS AND RECOMMENDATIONS

Methodology

Tables and maps for the ancient land use are created by many and complex steps.

All the thematic maps needed for land use analysis are defined.

The list includes the Digital Elevation Model (DEM), the topographic maps, the geological and hydrological maps, the land cover maps, the soils maps, the land use maps and any other useful map. Data available were generally not digitized and georeferenced.

Available maps displaying the present or the historical situation used different datums and coordinate systems. Any map had to be represented by using the same datum and the same coordinate system by suitable mathematical transformations.

Data finally have been stored in PostgreSQL database with PostGIS extension.

The same group of thematic maps, as in defined list, was built following an historical order, that starts from the present and includes historical periods for which data useful for analysis, although not complete, were available. Any group of thematic maps is characterized by the parameter of time indicating the historical period referred to the group of maps itself.

A time series of thematic maps has been built to infer the information needed for ancient soils maps. The maps information from historical sources and from archaeological remains, going back in time, is incomplete or missing.

GIS based model for reconstructing past soils maps and for viewshed analysis has been built. The model used a geostatistical approach to the analysis of the time series of thematic maps, to predict the ancient soils maps: an approach similar in some extent to the one used to predict the future soils maps (Rossiter, 1994; Van Joolen, 2003).

The evaluation of ancient soils suitability to main used crops required further information on ancient crops, on used techniques, on work organization due to the large scale changes in agrarian activities. For establishing the soils suitability to the main used crops, the physiographic and pedological features of Land Use Requirements (LUR), as drain categories, typology, depth, slope, and so forth, have to be related with the needed Land Use Qualities (LUQ), as moisture, workability, nutrients, and so forth, that represent the soil capacity to satisfy the proper requirements. This is shown in Table 1.

The latter allow to define the specific Land Use Types (LUT).

From the previous tables other ones can be extracted for the specific land use where the needed requirements correspond to the pertaining features according to the general table itself.

LUR have to be measured and related to the LUQ, classified in Severity Levels and Degrees of Limitation, that are the categories of LUR for any LUQ.

From this information and from the land features, three main suitability categories for Severity Levels and Degrees of Limitations for any specific land use, are defined: suitable, partially suitable, not suitable (Cremaschi, 2000).

For such information look at Table 2 and Table 3.

Territory maps for any soil feature, as slope, soil typology, available moisture, texture, and so forth, can be built by GIS. The maps obtained with this method are superimposed.

The suitability category of a soil is defined by its more unfavorable feature, whatever is the assigned value to the other features.

Main crops suitability are shown in Table 4 and Table 5.

Table 1. Relationships between LUQ and LUR for soils crop suitability

Land Use Qualities	Land Use Requirements
Moisture	Drain categories
	Soil suitable depth
	Field capacity
Nutrients	Soil typology
	Soil thickness
Workability	Slope
	Stones or rocks presence
	Poor drainage
	Soil texture
Erosion risk	
Rooting conditions	Stages
	Soils structure and texture

Table 2. LUT for spelt

Land Use Type (LUT)		
Spelt cultivation on medium-sized plots		
Land Use Quality (LUQ)	Land Use Requirements (LUR)	Suitability categories
Moisture	Soil depth	
	• Less than 30 cm • More than 30 cm	• Not suitable = 3 • Suitable = 1
	Drain categories	
	• Poorly drained • Drained • Very drained	• Suitable = 1 • Suitable = 1 • Not suitable = 3
Nutrients	Soil typologies	
	• Luvisois,Vertisols,Fluvisols • Arenosols,Regosols,Planosols, Leptosols • All other types	• Suitable = 1 • Not suitable = 3 • Suitable = 1
	Lime content in the soil	
	• Very calcareous • Calcareous • Non-calcareous	• Not suitable = 3 • Partially suitable = 2 • Suitable = 1
Workability	Soil texture	
	• Sandy soils • Loamy soils • Clayey soils	• Not suitable = 3 • Suitable = 1 • Suitable = 1
	Slope class	
	• 0-13% • 13-55% • >55%	• Suitable = 1 • Partially suitable = 2 • Not suitable = 3
	Stoniness class	
	• 0-3% • 3-90% • >90%	• Suitable = 1 • Partially suitable = 2 • Not suitable = 3
	Rockiness class	
	• 0-10% • 10-50% • 50-100%	• Suitable = 1 • Partially suitable = 2 • Not suitable = 3

Table 3. LUT for other cereals

Land Use Type (LUT)		
Other cereals cultivation on medium-sized plots		
Land Use Quality (LUQ)	Land Use Requirements (LUR)	Suitability categories
Moisture	Soil depth	
	• Less than 30 cm • More than 30 cm	• Not suitable = 3 • Suitable = 1
	Drain categories	
	• Poorly drained • Drained • Very drained	• Not suitable = 3 • Partially suitable = 2 • Suitable = 1
Nutrients	Soil typologies	
	• Luvisols,Vertisols,Fluvisols • Arenosols,Regosols,Planosols, Leptosols • All other types	• Suitable = 1 • Not suitable = 3 • Partially suitable = 2
	Lime content in the soil	
	• Very calcareous • Calcareous • Non-calcareous	• Partially suitable = 2 • Suitable = 1 • Not suitable= 3
Workability	Soil texture	
	• Sandy soils • Loamy soils • Clayey soils	• Not suitable = 3 • Partially suitable = 2 • Suitable = 1
	Slope class	
	• 0-13% • 13-55% • >55%	• Suitable = 1 • Partially suitable = 2 • Not suitable = 3

Table 4. Land suitability for spelt

Spelt	Suitable	Partially suitable	Not suitable
Soil requirements	At least 30 cm thickness		< 30 cm thickness
	Clayey soils, loamy clayey soils, loamy soils	Loamy soils	Sandy soils
	Moderately drained (wet) or drained	Marshy	Very drained (dry)
	Firmly structured		Loose
	Marginally fertile or fertile	Fertile	Very fertile
	Non-calcareous	Calcareous	Very calcareous
		Medium salinity	
Cultivation requirements	Rotation system of one to two years		
	No manuring		
	No irrigation		

Data Analysis and Results for Land Use

The analysis provides an archaeological evaluation of land use to define a potential adaptability of soils of Tarquinia area to various types of land uses and to the recognized major cereal crops, like spelt or cereals that demand requirements other than those of spelt.

Data available, reconstructed or estimated by environmental, historical and archaeological information are used. Land requirements needed to cultivate specific crops are calculated by some predictive models developed to infer the ancient soils adaptability.

Models used a set of synchronic thematic maps from present to past for any single chronological step. Ancient sources, the studies carried out so far and any added element are used to complement the analysis on soils and their requirements, up to build tables and maps on LUR, LUQ, LUT (Bedini, 1997; Boerma, 1989; Cremaschi, 1992; Hunt, Malone, Sevink, & Stoddart 1990; Kamermans, Loving, & Voorrips 1985; Mallegni, Rottoli, & Bruni 2001; Peroni, 1994; Van Joolen, 2003; Van Leusen, 2002; Renfrew & Bahn 1995; Rottoli, 1997).

The soils requirements suited to investigated crops are shown in the maps, in different colors. The analysis shows that the plateaux of La Civita are placed at the center of a large area, including the coastal plain, the river valleys and the surrounding land, suitable without restrictions or partially suitable to the spelt specific culture as the coastal plain and the river valleys, while is partially suitable for other cereals as the belt surrounding the plains and the hills of the interior geographic system, including the Monti della Tolfa. Figure 2 and Figure 3 show the results.

Data Analysis and Results for Viewshed

The visibility analysis established which parts of the territory are visible and from which points, and which positions were strategic for territory control with its territorial hierarchies and its cultural and cognitive landscapes (Van Leusen, 2002; Wheatley, 1995).

The identification of the points used for visibility analysis into settlements is the first important step. In fact, settlements and urban centers were not surrounded by wall with watchtowers. The intervisibility

Table 5. Land suitability for other cereals

Other cereals	Suitable	Partially suitable	Not suitable
Soil requirements	Deep soils		< 30 cm thickness
	Clayey soils	Loamy soils	Sandy soils
	Drained or very drained	Drained	Poorly drained
	Firmly structured		Loose
	Fertile or very fertile		Unfertile
	Calcareous (Marls)	Calcareous	Non-calcareous
	Plains	Hill slopes	
	Relatively warm		Cold
	Open and elevated position		
Cultivation requirements	No clear indications		

Figure 2. Map of soils suitable, partially suitable and not suitable for spelt cultivation

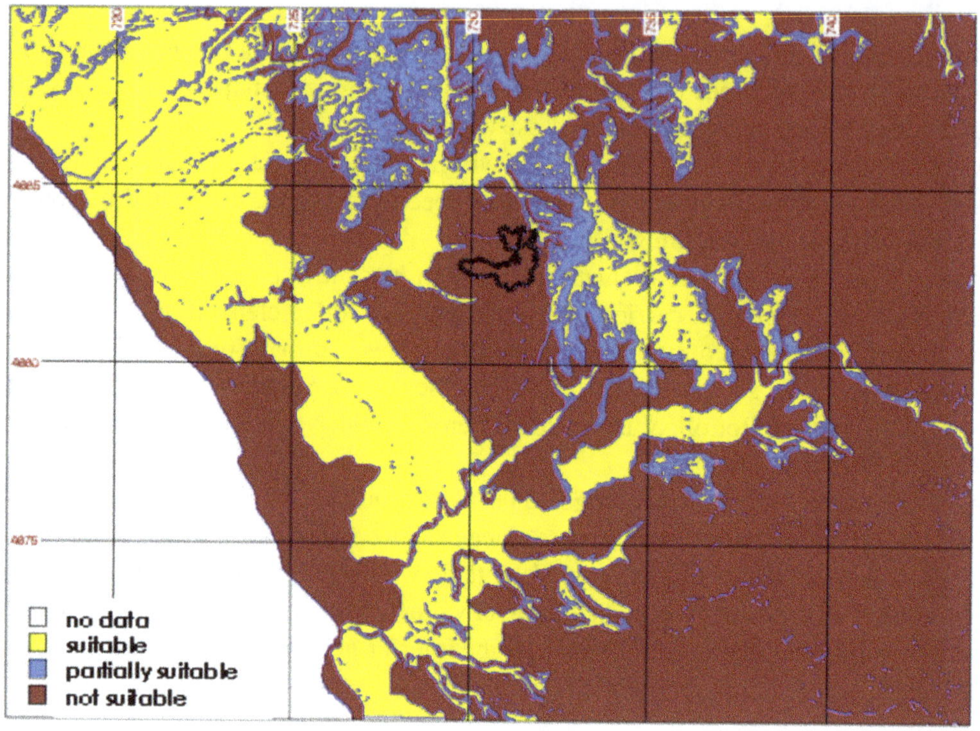

Figure 3. Map of soils suitable, partially suitable and not suitable for other cereals cultivation

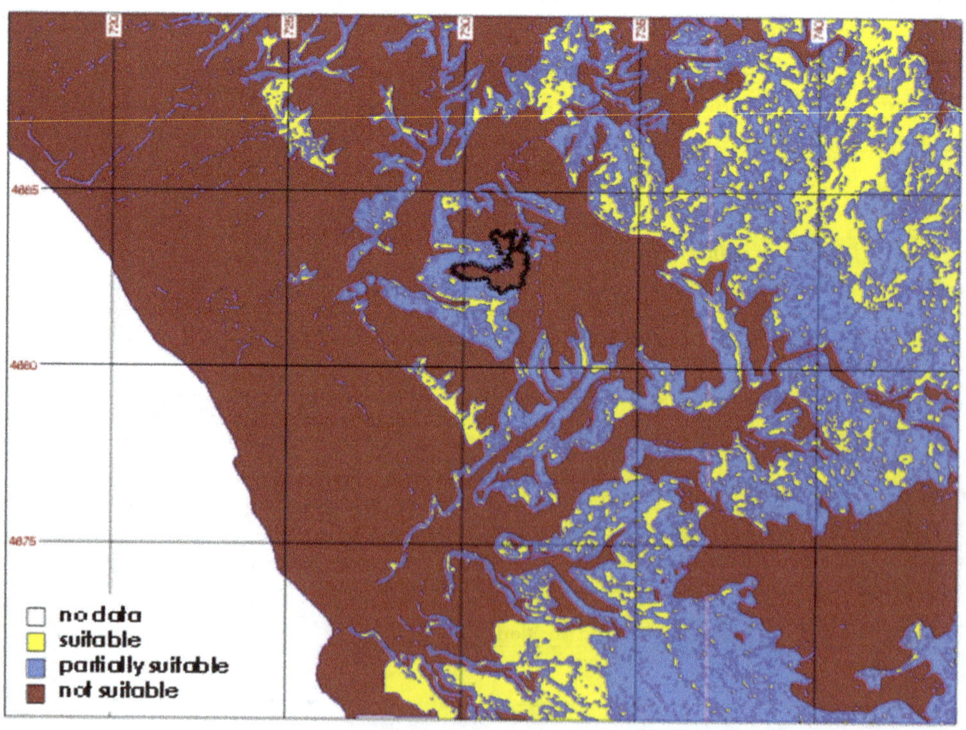

between the three Monterozzi villages and the Pian della Regina on La Civita, emerged as cumulative viewshed of the territory visible from these settlements, can explain the settlements location on the plateau highest point. Also viewshed analysis confirmed central position of the protourban center located on the main plateaux of La Civita and demonstrated that from first Iron age settlements on Monterozzi hill it was possible to control all the coastal plain, the Mignone and Marta river valleys, the La Civita plateau (Mandolesi, 1999).

The viewshed analysis demonstrated also that the settlement location on La Civita plateaux is further justified by its visual control on important areas like Marta valley, coastal plain, roads, Saline harbour (see Figure 4).

FUTURE RESEARCH DIRECTIONS

Archaeological GIS and Multi Agent Based Models (MABM) From Models to Archaeological Data

Although GIS based model used for reconstruction of ancient land use integrate well in a spatiotemporal uniform framework the available data, its approach based on analysis and interpretation on main important patterns for going back to the protourban processes, is static.

Integration of MABM simulations with archaeological GIS can allow to overcome GIS model limitations to better explain the dynamics of protourban processes.

Figure 4. Cumulative viewshed between La Civita, the coastal plain and the villages on the Monterozzi hill

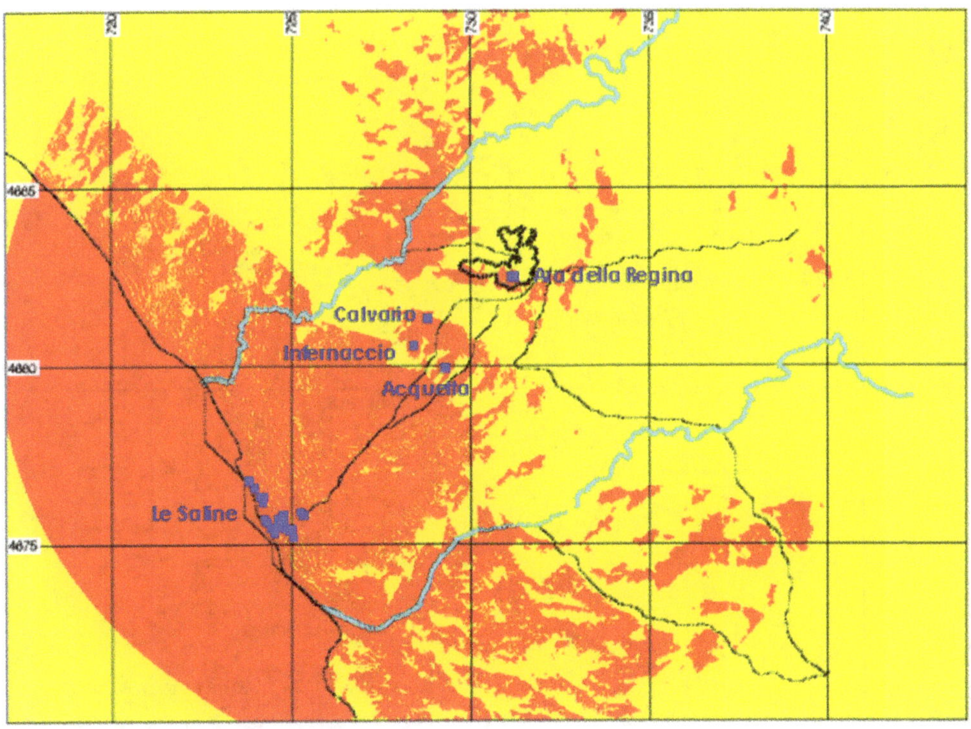

So achievement of an investigation level more precise using computer simulations of dynamical evolution of the archaeological landscape is possible.

Tarquinia is an emblematic case study for South Etruria protocities origin, due to complexity of the settlement pattern starting from medium Bronze age, when small settlements coexist with large ones located on plateaux (Mandolesi, 1999) (See Figure 5).

The Multi Agent Based Models – MABM – analysis seems the best way for analyzing Tarquinia protocity dynamics and development between X-VIII centuries B.C., for discriminating between different hypotheses.

MABM simulations are necessary if the analysis purposes are a better interpretation of protourban processes dynamics, including paleoagriculture and paleoproduction (Kohler, 2005). MABM simulations start from archaeological and anthropological theories and models and its results are compared with archaeological records in an approach top to bottom.

The archaeological GIS integrated with MABM develops into a testbed laboratory for validation of models on social and cultural processes dynamics (Parisi, 2001).

CONCLUSION

Land Use Results Discussion

Also heavy clay soils of the Marta and Mignone rivers valleys, partially suitable for cereals, became cultivable only after deforestation, drainage and tillage.

Figure 5. Preurban settlements of the older Bronze ages in the Tarquinia territory

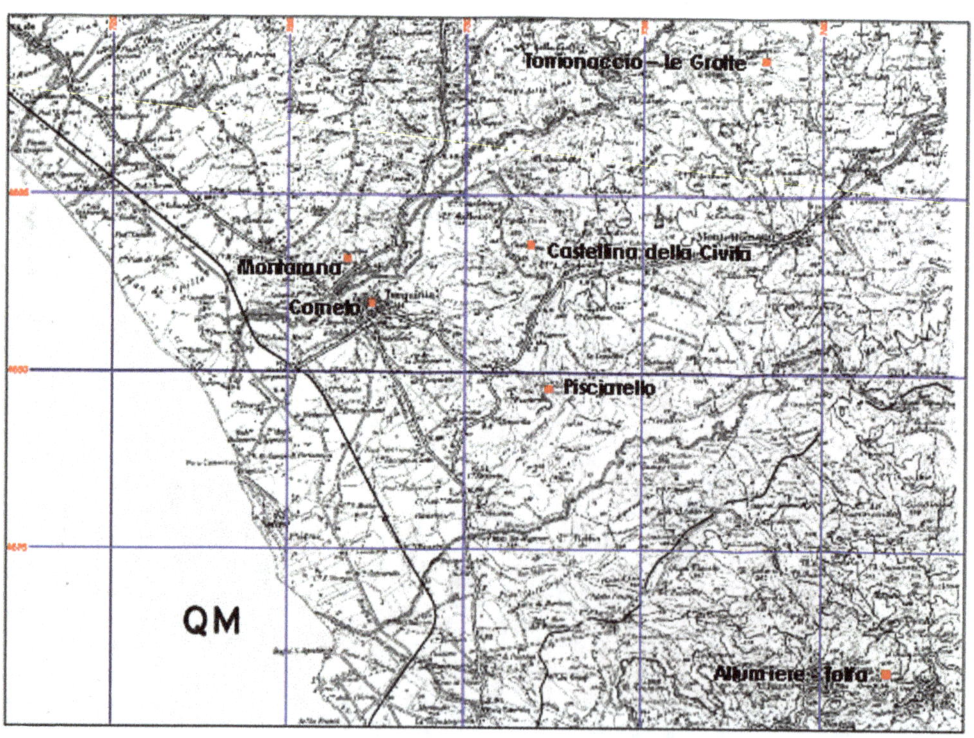

Agricultural productivity was improved by intensive use of available technologies.

Agricultural practices and production increased for incentives caused by endogenous and exogenous factors that affected paleoproductivity indexes of various wheats (Ampolo, 1980).

These factors were a consequence of protourban center development.

In fact, the hypothesis that the people displacement on the plateaux, from the final Bronze age, was due to the availability of large adjacent amount of land suitable to cereal cultivations, is confirmed. Other factors affecting the settlements choice were favorable geographical and environmental conditions, due to large homogeneous districts of plateaux, to position near to sea and to two rivers, accessible but also well protected, and to well accessible woods and pastures.

For instance, the sites settled from the final Bronze age on the coastal plain like Fontanile delle Serpi in central position of land suitable to spelt cultivation and to grazing, meet similar criteria.

Viewshed Results Discussion

Viewshed analysis is very important for archaeological interpretation of protourban process and is relevant for many discussed aspects.

Viewshed analysis showed an easy control of the territory by the protourban center, but also by the settlements on Monterozzi hill. Control was very important during the deforestation of the coastal plain and river valleys. In fact, the agrarian practices of igniculture were used for obtaining soils suitable for agrarian activities in large areas. The control of fire diffusion during these practices was certainly very important to avoid dangerous consequences.

Other coastal and landing sites could play similar control of coast.

Visibility analysis helps to predict new sites not yet discovered.

In first Iron age, Saline settlement was the coastal plain control site, as before were the Monterozzi villages. Throughout Iron age, the Saline settlement controlled waterfront and Monterozzi villages will be abandoned. When Tarquinia later developed into City State, Monterozzi hill became, with Arcatelle necropolis, the Tarquinia larger burial area, integrating La Civita main town with Ara della Regina temple and coastal Gravisca and Le Saline settlements.

Overall Discussion

The major reasons for populating plateaux were the need of land accessible and suitable to cultivate cereals and of areas for building households. In late final Bronze age motivations for making cultivable the fertile lands represented by the wetlands of the coastal plain and river valleys can be explained by social and economical changes.

In fact, the Fontanile delle Serpi site of final Bronze age, located in the alluvial coastal plain on the edge of an ancient lagoon, was created in relation with intensive spelt cultivation on the coastal alluvial soils, and the Saline first Iron age was important site for the exploitation of the coastal plain resources. Agrarian surplus production is linked with growing demand due to increasing exchanges and to demographic development in the villanovan Tarquinia between X- VIII centuries B.C. The subsistence economy of early villages located on the hills and river terraces, based on collective land ownership, was replaced by agricultural soils intensive exploitation to a greater extent, connected with new land ownership by households from final Bronze age (Bartoloni, 2002).

The protourban centers growth is a dynamic process carried forward from transformations in social structure and in production organization, compared with preurban clans organization. Villanovan society organized in households landowners reacted to market expansion in Mediterranean area. Demand of commodities like cereals and wheat increased for demography growth and for development of Mediterranean global market.

The settlement evolution shows the transition from resources management in catchment area plenty for the subsistence preurban villages to resources growth linked to the increasing scale of economic activities for emerging protourban centers.

The problem of protourban centers is how to produce a larger amount of agrarian products on a wider territory with more productive methods. For such reason new social organization linked to intensive agrarian production is based on private agrarian ownership. This new structure made possible a more effective work organization and a more rational resource management by deployment and concentration of the productive forces and ways of production on larger scale.

This pattern is reflected from the household clusters divided in *curiae*, according to the future model of the archaic Rome, using cooperation and coordination for managing a growing economy (Carandini, 2003).

In this context, plateaux work as possible territorial attractors: that is, places where households converged to favor common activities in agrarian works as tillage with igniculture and economical exchanges, both for external and internal markets in protourban centers, where material and human resources were concentrated.

With the synoecism, whatever bonds, internal and external to the villages, could favor people concentration on same area, for supplying new economical requests and transformations. Demographic growth was an important factor for the production increase but also is the consequence of such increase: so that demographic growth has to be intended as cause and effect of the villanovan society changes.

These conclusions show the relevance of the GIS and of its results on ancient land uses, for selecting between the theories on the origin of the protourban centers (Bartoloni, 2003).

REFERENCES

Ampolo, C. (1980). La formazione della città nel Lazio. Le condizioni materiali della produzione. Agricoltura e paesaggio agrario. *Dialoghi di Archeologia, 1*(2), 15–46.

Bartoloni, G. (2002). *La civiltà villanoviana all'inizio della storia etrusca*. Roma, Italy: Carocci.

Bartoloni, G. (2003). *Le società dell'Italia primitiva. Lo studio delle necropoli e la nascita delle aristocrazie*. Roma, Italy: Carocci.

Bedini, E. (1997). I resti faunistici. In M. Bonghi Jovino & C. Chiaramonte Treré (Eds.), *Tarquinia. Testimonianze archeologiche e ricostruzione storica: scavi sistematici dell'abitato. Campagne 1982-1988* (Vol. 1-2, pp. 103–141). Roma, Italy: Tarchna.

Bietti Sestieri, A. M. (1996). *Protostoria. Teoria e pratica.* Roma, Italy: La Nuova Italia Scientifica.

Bietti Sestieri, A. M. (2002). L'agricoltura in Italia nell'età dei Metalli. In G. Forni & A. Marcone (Eds.), *Storia dell'agricoltura italiana. L'età antica. La preistoria* (Vol. 1-2, pp. 205–217). Firenze, Italy: Polistampa.

Boerma, J. A. K. (1989). Land evaluation in prehistoric perspective: some observations. In Haex, O.C.M., Curvers, H.H., & Akkermans, P.M.M.G. (Eds). To the Euphrates and beyond (pp. 17-28). Rotterdam, The Netherlands: CRC Press.

Braudel, F. (2002). *Il Mediterraneo. Lo spazio la storia gli uomini le tradizioni*. Milano, Italy: Bompiani.

Carandini, A. (2003). *La nascita di Roma. Dei, Lari, eroi e uomini all'alba di una civiltà* (Vol. 1-2). Torino, Italy: Einaudi.

Colonna, G. (2000). I caratteri originali della civiltà etrusca. In M. Torelli (Ed.), *Gli Etruschi, Catalogo della mostra, Venezia, Palazzo Grassi* (pp. 25–41). Milano, Italy: Bompiani.

Cremaschi, M. (1992). Insediamenti e sussistenza nella media età del Bronzo, Italia settentrionale. *Rassegna di Archeologia, 10*, 145–188.

Cremaschi, M. (2000). *Manuale di geoarcheologia*. Roma-Bari, Italy: Laterza.

Di Gennaro, F. (1986). *Forme d'insediamento fra Tevere e Fiora dal Bronzo Finale al principio dell'età del Ferro*. Firenze, Italy: Istituto di Studi Etruschi e Italici.

Forni, G. (2002). L'agricoltura: coltivazione ed allevamento. Genesi, evoluzione, contesto. In G. Forni & A. Marcone (Eds.), Storia dell'agricoltura italiana. L'età antica. La preistoria (pp. 7-157). Firenze, Italy: Polistampa.

Giardino, C. (1998). *I metalli nel mondo antico. Introduzione all'archeometallurgia*. Roma-Bari, Italy: Laterza.

Guidi, A. (1998). The emergence of the state in Central and Northern Italy. *Acta Archaeologica, 69*, 139–161.

Hencken, H. (1968). *Tarquinia, Villanovans and Early Etruscans. American School of Prehistoric Research Bulletin, 23.*

Hunt, C., Malone, C., Sevink, J., & Stoddart, S. (1990). Environment, soils and early agriculture in Apennine Central Italy. *World Archaeology, 22*(1), 34–44. doi:10.1080/00438243.1990.9980127

Iaia, C. (1999). *Simbolismo funerario e ideologia alle origini di una civiltà urbana. Forme rituali nelle sepolture "villanoviane" a Tarquinia e Vulci, e nel loro entroterra. Grandi contesti e problemi della protostoria italiana*. Firenze, Italy: All'Insegna del Giglio.

Kamermans, H., Loving, S. H., & Voorrips, A. (1985). Changing patterns of prehistoric landuse in the Agro Pontino. In C. Malone & S. Stoddart (Eds.), Papers in Italian Archaeology, IV, Part I, The Human Landscape (pp. 53-68). Oxford, UK: BAR International Series 243.

Kohler, T. A., Gumerman, G. J., & Reynolds, R. G. (2005). Simulating Ancient Societies. *Scientific American, 293*(1), 77–84. doi:10.1038cientificamerican0705-76 PMID:16008305

Linington, R. E. (1982). Il villaggio protostorico nella località Calvario sui Monterozzi a Tarquinia. In Studi in onore di Ferrante Rittatore Vonwiller (pp. 245-256). Como, Italy: Società archeologica comense.

Liverani, M. (1988). *Antico Oriente. Storia società economia.* Roma-Bari, Italy: Laterza.

Mallegni, F., Rottoli, M., & Bruni, S. (2001). I dati delle discipline consociate. In A.M. Moretti Sgubini (Ed.), Tarquinia etrusca. Una nuova storia, Catalogo della mostra. Tarquinia, Museo Archeologico Nazionale (pp. 59-61). Roma, Italy: L'Erma di Bretschneider.

Mandolesi, A. (1999). *La "prima" Tarquinia. L'insediamento protostorico sulla Civita e nel territorio circostante. Grandi contesti e problemi della Protostoria italiana.* Firenze, Italy: All'Insegna del Giglio.

Moscati, S. (2001). *Civiltà del mare. I fondamenti della storia mediterranea.* Napoli, Italy: Liguori.

Pacciarelli, M. (2000). *Dal villaggio alla città. La svolta protourbana del 1000 a.C. nell'Italia tirrenica. Grandi contesti e problemi della Protostoria Italiana.* Firenze, Italy: All'Insegna del Giglio.

Pallottino, M. (1937). Tarquinia. *Monumenti Antichi della R. Accademia Nazionale dei Lincei.* 36.

Parisi, D. (2001). *Simulazioni. La realtà rifatta nel computer.* Bologna, Italy: Il Mulino.

Peroni, R. (1969). Per uno studio dell'economia di scambio in Italia nel quadro dell'ambiente culturale dei secoli intorno al 1000 a.C. *La Parola del Passato, 125,* 134–160.

Peroni, R. (1994). *Introduzione alla protostoria italiana.* Roma-Bari, Italy: Laterza.

Peroni, R. (1996). *L'Italia alle soglie della storia.* Roma-Bari, Italy: Laterza.

Renfrew, C., & Bahn, P. (1995). *Archeologia. Teorie, Metodi, Pratica.* Bologna, Italy: Zanichelli.

Rossiter, D. G. (1994). *Land Evaluation.* Cornell University, College of Agriculture & Life Sciences, Department of Soil, Crop & Atmosferic Sciences. Retrieved August 1994, from http://www.css.cornell.edu/faculty/dgr2/docs/Scas494/s494ch1.pdf

Rottoli, M. (1997). I resti vegetali. In M. Bonghi Jovino & C. Chiaramonte Treré (Eds.), *Tarquinia. Testimonianze archeologiche e ricostruzione storica: scavi sistematici dell'abitato. Campagne 1982-1988* (Vol. 1-2, pp. 92–96). Roma, Italy: Tarchna.

Sassatelli, G. (2009). Gli Etruschi e la loro economia. In V. Castronovo (Ed.), *Storia dell'economia mondiale. Dall'antichità al Medioevo. Dal Neolitico agli albori del I millennio* (Vol. 1-2, pp. 99–116). Roma-Bari, Italy: Laterza.

Trigger, B. G., Kemp, B. J., O'Connor, D., & Lloyd, A. B. (2000). *Storia sociale dell'antico Egitto.* Roma-Bari, Italy: Laterza.

Van Joolen, E. (2003). *Archaeological land evaluation. A reconstruction of the suitability of ancient landscapes for various land uses in Italy focused on the first millennium B.C.* (Doctoral dissertation). Retrieved from http://dissertations.ub.rug.nl/faculties/arts/2003/e.van.joolen/

Van Leusen, M. (2002). *Pattern to process. Methodological investigations into the formation and interpretation of spatial patterns in archaeological landscapes* (Doctoral dissertation). Retrieved from http://dissertations.ub.rug.nl/faculties/arts/2002/p.m.van.leusen

Wheatley, D. (1995). Cumulative viewshed analysis: a GIS-based method for investigating intervisibility, and its archaeological application. In G. Lock & Z. Stancic (Eds.), *Archaeology and Geographical Information Systems: A European Perspective* (pp. 171–186). London: Taylor & Francis.

ADDITIONAL READING

Allen, M. S., Green, S. W., & Zubrow, E. B. W. (Eds.). (1990). *Interpreting Space: GIS and Archaeology*. London, England: Taylor & Francis.

Bivand, R., Pebesma, E., & Gómez-Rubio, V. (2008). *Applied Spatial Data Analysis with R*. New York, NY: Springer.

Christiansen, J. H., & Altaweel, M. R. (2006) Understanding Ancient Societies: A New Approach Using Agent-Based Holistic Modeling. *Structure and Dynamics: eJournal of Anthropological and Related Sciences Vol. 1, No. 2, Article 7.* Retrieved from http://repositories.cdlib.org/imbs/socdyn/sdeas/vol1/iss2/art7

Conolly, J. (2006). *Geographical Information Systems in Archaeology*. Cambridge, NY: Cambridge University Press. doi:10.1017/CBO9780511807459

Harvey, F. (2016). *A Primer of GIS. Fundamental Geographic and Cartographic Concepts* (2nd ed.). New York, NY: Guilford Press.

Langran, G. (1993). *Time in Geographic Information Systems*. Washington, DC: Taylor & Francis.

Wurzer, G., Kowarik, K., & Reschreiter, H. (Eds.). (2014). *Agent-based Modeling and Simulation in Archaeology*. Charm, Switzerland: Springer.

KEY TERMS AND DEFINITIONS

Historical Ecology: Ecology of the historical natural environment.

Landscape Archaeology: Archaeology of ancient landscapes.

Multi-Agent Based Models: Simulation with multi-agent based models for the analysis of ancient social systems dynamics.

Paleoagriculture: Ancient agrarian techniques and crops.

Protourban Center: Urban settlement before the City State.

Synoecism: Collective population transfer from one site to another.

Territory Control: Ways and geographical places for territory control.

Viewshed: Territory area visible by selected geographical points.

This research was previously published in the Encyclopedia of Information Science and Technology, Fourth Edition edited by Mehdi Khosrow-Pour, pages 3419-3433, copyright year 2018 by Information Science Reference (an imprint of IGI Global).

Chapter 21
Exploring Tourism Cluster in the Peripheral Mountain Area Based on GIS Mapping

Ya-Hui Hsueh
National Taichung University of Education, Taiwan

Huey-Wen Chuang
National Taichung University of Education, Taiwan

Wan-Chiang Hsieh
National Taichung Girls' Senior High School, Taiwan

ABSTRACT

This chapter locates a set of points of tourist spots distributed on a peripheral mountain area by GIS mapping, illustrates that accessibility and neighbor to community are the influencing factors of tourism cluster in a peripheral mountain area, and further analyzes the benefits of tourism cluster of establishing cooperation network by tourism associations. This research investigates the linkages and local impacts between tourism and agriculture of a well-known organic agricultural area of fruits in Taiwan. This research aims to demonstrate the increased tourism development due to the tourist attractions of organic agriculture and religious spots in a peripheral region through a tourism clustering process. By collecting a set of points of tourist spots specified to GIS slope raster and point density surface, tourism-agriculture linkages and their local impacts are demonstrated.

INTRODUCTION

Increasingly, developing tourism and making linkages with other economic sectors are strategically as a tool for regional economic development. In this article we will briefly resume and discuss the main results in the field of linkage between tourism and agriculture. In particular, we will survey the features of research area field data based on extensions of the linkage between tourism and agriculture on the several small agricultural settlements in Taiwan, which are the most relevant for mainstream application

DOI: 10.4018/978-1-5225-7359-3.ch021

development. Finally, we will survey the currently available implementations of tourist spots by specified their locations on environmental conditions. Tourism in peripheral area is a relatively vulnerable to environment because of limited resource and disaster possibility. However, tourism provides alternative employment choice and greater economic inflexibility, so in peripheral area tourism development is still welcomed based on economic benefits. Clarifying the locational feature of the tourist spots is to realize the condition of tourism cluster, especially focus on the influencing factors of terrain conditions and accessibility. This research aims to explore the tourist attractions of a peripheral mountain area in the central Taiwan, analyze the important influencing locational factors of tourism cluster based on GPS (Global Positioning System) data of a set of tourist spots by processing GIS (Ggeographical Information System or Geographic Information System) mapping, and finally clarify what impacts the tourism cluster to the local environment by DTM (Digital Terrain Model) data.

BACKGROUND

The potential for creating synergistic relationships between tourism and agriculture has been widely recognized by development planners, policy makers. Commonly, economic leakage is the main factor for why the linkage of tourism and agriculture to promote local economic development in peripheral regions. Food is an essential component of tourism and also represents a significant part of tourism expenditure, and creating and strengthening the linkages between tourism and local food production sectors can provide a proximate market. The concept of "farm-to-fork" demonstrates the linkages between the dimensions of sustainable agriculture, sustainable cuisine and tourism by increasing demand for local products. Farm-to-fork concept also can lead to a range of related direct and indirect tourism activities such as food festivals, farm visits, factory tours and souvenir food merchandise, thus further enhancing the benefits to the local (Berno, 2011).

Torres (2002) explores the linkages between tourism and agriculture in the Yucatan Peninsula, observes that the principal force driving hotel purchasing hotel food differences by tourist nationality and type of tourist based on tourist food consumption and preferences. So, in the Yucatan Peninsula Mexican foods, tropical fruits and organic produce are identified as this study area significant potential tourist food for linking tourism and local agriculture. Seaton (1999) examines tourism attraction in a peripheral region, identified critical success factors to small scale sustainable development, indicated that peripheral areas, distance from core areas with sparse populations and low GDP economic structure can motivate visitors to through some kind of special attraction such as book town for retailing. Gardiner and Scott (2014) investigates successful tourism cluster on the Gold Coast, Australia to develop the youth tourism market, through joint promotion and product development as an attractive destination. They propose that niche tourism clusters are often used to improve competitiveness to achieve economic advantages, through use of concepts of strategic alliances and networks. Cluster is therefore a fundamental factor to creating successful tourism industry within a destination zone.

Food supply chains of tourism accommodation providers in the coastal region of KwaZulu-Natal, South Africa, make the pro-poor tourism build of linkages between tourism and agriculture as a whole, and that revealing significant implications of tourist food consumption on destinations. Most research examining tourism and agriculture linkages has focused on hotel food procurement patterns while failing to address the main driving force of hotel purchasing tourist food based on tourist consumption and preferences (Pillay & Rogerson, 2013). The role of tasting room in the direct marketing of southwest

Michigan wines educates visitors about Michigan wines, a way to differentiate the Michigan wines, and offers spectacular views of fields, orchards is varied according to the different locations of the wineries. Commonly, these wineries all provide supporting services of cellar door sales and direct shipping for promoting local wine (Che & Wargenau, 2011).

The linkages between tourism and related sectors vary widely from nation to nation, region to region. There are numerous successful examples of strategic alliances between tourism and the agricultural related sectors in Taiwan, especially the Jhong-Liao Township, in Nantou County of central Taiwan. Jhong-Liao Township composes of several small agricultural settlements with a population of about 16000, is located on the elevation from 200 meters to 1264 meters. Jhong-Liao is not the most famous destination in Taiwan, but it is a popular mystery destination for its distant from urban area-Taichung City. During Ching Dynasty it was a stopping place of wildness space along a transport route from costal areas to mountainous areas of central Taiwan rather than a tourist spot. During Japanese Occupation Period (1895-1945) it was a high productive place of agricultural activity, planted in plenty of banana, so the reputation for Jhong-Liao is also called "Banana Mountain". 20 years ago, a visitor to the Jhong-Liao seemed to be a lonely trip into an empty mountain area, until 1999 there were almost no tourist spots in Jhong-Liao. It's famous for as a tourist attraction was traced to 1999, when an earthquake disaster dramatically destroyed here, an atmosphere of "Ghost Town" spread out in Taiwan. Since 1999 many tourists visited here to see the destroyed landscapes after earthquake, and the number of tourist spots in Jhong-Liao has increased to 70, so now it is transformed to one of most popular destination of Nantou County.

In recent years, the linkage between tourism and agriculture for environmental sustainability has been discussed as the development of alternative tourist food, which was perceived to be traditional and local agrarian products, appeals to the visitor's desire for authenticity within the holiday experience (Hall, 2005; Michael, 2007; Sims, 2009). Thus, tasting local food performs as an authentic tourism experience that transforms traditional agriculture products to create cuisine culture of the destination. The significant implications of tourist food consumption on destinations have received a growing research interest recently (Telfer & Wall, 2000; Torres, 2002, 2003; Torres & Momsen, 2004, 2011; Pillay & Rogerson, 2013), and this means the linkage between tourism and agriculture has play important role in tourism development because tourism seems to help farmers to overcome economic growth constraints, upgrade and face new competitions in distant markets. Exploring the tourist consumption preferences is critical to the analysis of linkages between tourism and agriculture for creating either backward linkages to local agriculture or increased economic growth. Tourist consumption and demand for food vary greatly according to a number of factors, including tourist preference and type of tourist. Local food may be particularly popular with tourists because of its consideration of "iconic" products that capture the "typical" nature of a particular place. Local food is popular because it is associated with a host of value, and souvenir purchase is making local food as a symbol of place and culture (Sims, 2009).

GIS MAPPING WAY

The concept of tourism cluster has attracted attentions during the past decades, both as descriptive of an increasingly important strategy and an effective tool to enhance economic development in peripheral regions. This research used empirical data to explore the linkages of tourism, agriculture and religion for different implications both in terms of relevant evidence and the scope for promotional policies. In order to realize how many tourist spots clustering in this study area, this research locates tourist spots based on

GPS data to collect data, and then specified all the points to raster layer of TWD 97 coordination system on GIS processing system. With elevation analysis, slope analysis, point density analysis, buffer analysis and overlay analysis of GIS mapping way, this research aims to distinguish the significant association between the core clustering area of tourist spots and the influential factors such as environmental ones and accessibility. Finally, we conduct in-depth interviews of the holders of organic farmers, agricultural officers in government and the travelers visiting to Jhong-Liao, and defined there are 72 the total number of tourist spots, including:

- 38 religious temples,
- 14 organic leisure farms,
- 7 artificial landscapes,
- 6 natural sceneries,
- 5 mountain tracks, and
- 2 historical heritages.

Terrain Analysis of GIS Mapping

Hsueh (2013a) used terrain analysis of GIS mapping to specify a set of points based on the accommodations of a mountain area to compare the different percentages of accommodations on the different terrain conditions. Hsueh (2013b) also used terrain analysis of GIS mapping to explore the significant clustering area and locations of mushroom huts based on different terrain conditions in an agricultural mountain area. Hsueh (2015) analyzed the neighborhood of clustering area of a set of points of tea firms located on a plat form terrain by using terrain analysis of GIS mapping combined of related layers to process multiple buffer analysis to define the significant core area of tea firms clustering. This research aims to process terrain analysis to display the elevation and slope conditions of tourism cluster of tourist spots in the Jhong-Liao area. Jhong-Liao is a productive area of organic agricultural products in the central Taiwan. There are 14 organic farms in Jhong-Liao, thus Jhong-Liao has the advantage of tourist foods for its available for a lot of local organic foods, the most popular and favorite foods in Taiwan. As the representative agricultural products of oranges, longyans, bananas, it is obvious that Jhong-Liao has a lot of organic agricultural products, which have converted to tourist foods to meet tourist consumption and preferences recently.

With many farmers switching to organic products of fruits, consolidation occurred in the organic agriculture as the number of Farm Alliance of Jhong-Liao increases gradually. It was not until 1999 that the He-Xing Organic Cultural Association created the "Original and Organic Cuisine" (OOC) whose purpose is to start to control the agricultural productive environment as well as ecology conservation and making health preserving cuisine as tourist foods based on local products. The linkages between tourism and agriculture become more apparent when we analyze what kinds of tourist foods and experiences tourists are seeking during their holiday. By telling the story of "vegetable dyes" in Jhong-Liao, it is possible to use the long scarf DIY activity of making tourist's souvenir for authenticity to encourage the tourism development based on local agricultural products that will boost environmental sustainability and economic benefits. The experience of vegetable dyes with a particular focus on tourist attraction in Jhong-Liao is obtained from different local plant sources—roots, berries, bark, and leaves. Typically, the dye material is put in a pot of water and then the textiles tied with special techniques to be dyed are added to the pot, which is heated and stirred until the color is transferred. Without use of chemicals, the color of the vegetable dye is brilliant, permanent and with creative diversified local patterns, which is due to the different local plant sources such as:

- Betel nuts, Litchi, Longyan, and Tungoil Tree creating into light red shades;
- Cape Jasmine, Common Garcinia, kunyit, and onion into yellow shades;
- Chinese ixora, Subcostate Crape Myrtle, Kassod tree into green shades;
- Litchi, Longyan, rhus chinensis, and onion into brown shades.

By the interactions of cultivating technology of organic agricultural products through those professional associations' promotions, the tourist food reputation of Jhong-Liao thus enhanced. With their members concentrated in the same area, those professional associations are easy to share and offer professional knowledge to help the farmers in Jhong-Liao to establishment organic brand and organic certification. The tourism cluster of tourist spots strengthened the ties between tourism and agriculture providing for high quality and productive network of tourist food. There are a lot of agriculture related and supporting leisure farms co-located within the Jhong-Liao area to take an advantage based on close working relationships, and the cluster of leisure farms offers an empirical research of how tourism related sectors create competitive advantages. Organic farm holders, for instance, interact regularly with agricultural associations on tourist foods processing technologies and learn about how to promote tourist foods to visitors.

The notion of a new form of niche tourism has gone further than ever before. No longer is improved health on holiday or an escape from the routine work and the movement to a place with a cleaner climate, the rise of niche tourism has become the central theme of tourism in an active rather than a passive sense. The tourism literatures suggest there are different types in tourist attractions based on various characteristics of tourist spots. The important characteristics identified in the literature are tourist origin (Telfer & Wall, 1996; Telfer, 2001) and type of tourist (Sharkey & Momsen, 1995; Momsen, 1998; Telfer & Wall, 1996). Types of tourists range from the highly organized conservative mass tourist to the more flexible independent tourist. Most visitors who have the travel motivations to Jhong-Liao is triggered by religion, namely these visitors are flexible independent pilgrims by religion tourist attractions.

These diversity characteristics of the tourist spots locations enable comparisons based on elevation and slope conditions whether 'suitable' or 'unsuitable' tourist spots. Table 1 and Figure 1 show that all the total 72 tourist spots are near to rivers and the popular elevation of tourist spots is located on an elevation < 300 meter accounting for 66% (48 tourist spots), 32% (23 tourist spots), while tourist spots on the area proximity to elevation > 600 meter is almost none. There are 27 tourist spots of 37% with a slope < 5%, 28 tourist spots of 39% with a slope 5%~15%, and 17 tourist spots of 24% with a slope 15%~30%, respectively (Table 2, Figure 2). Obviously, a flat slope is suitable for the management of tourist spots, beneficial to reach through an unfrequented mountain track for travelers, especially in a

Table 1. Number of tourist spots on different elevation condition

Elevation (Meters)	< 300	300 ~ 600	600 ~ 900	900 ~ 1200	1200 ~ 1500
Number of tourist spots	48	23	1	0	0
(Percentage)	66%	32%	2%	0%	0%

Figure 1. Spatial distribution of tourist spots -based on elevation condition

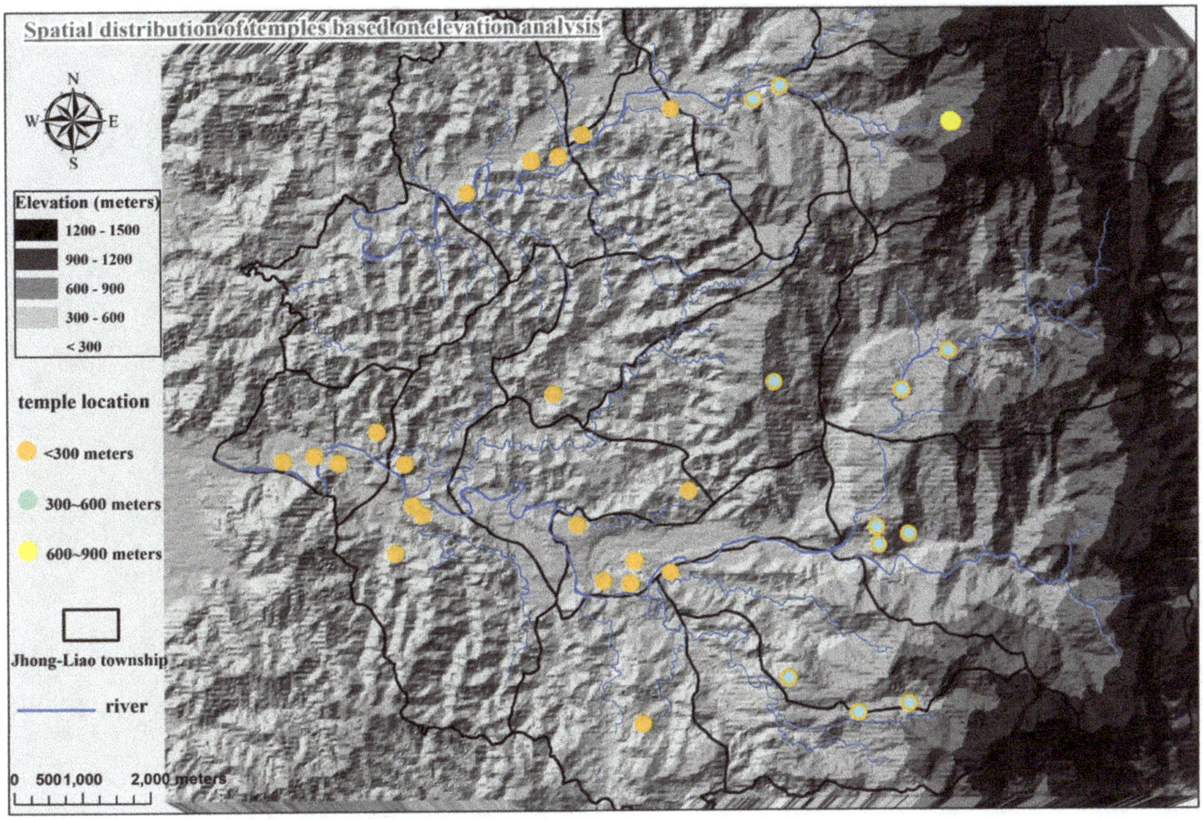

Table 2. Number of tourist spots on different slope condition

Slope	< 5%	5% ~ 15%	15% ~ 30%	30% ~ 40%	40% ~ 55%
Number of tourist spots	27	28	17	0	0
(Percentage)	37%	39%	24%	0%	0%

peripheral mountain area. Based on GIS mapping, we can realize terrain for tourist spots location is not only the main influencing factor in a peripheral mountain area, but also access to waters is becoming the important factor.

A cluster is a geographically proximate group of interconnected companies and associated institutions in a particular field, linked by commonalities and complementarities (Porter, 1990, p. 199). The local associations playing important role on Jhong-Liao's tourism cluster, such as Long-Yan-Lin Welfare Association, promotes the tourism image of North Jhong-Liao through the marketing network agriculture production through "organic brand" of the leisure farms; Farm Alliance of Jhong-Liao establishes banana market through the cooperation of banana farmers to promote Jhong-Liao's banana brand. Although the linkages of tourism and agriculture can result in higher levels of economic growth instead of becoming

Figure 2. Spatial distribution of tourist spots - based on slope condition

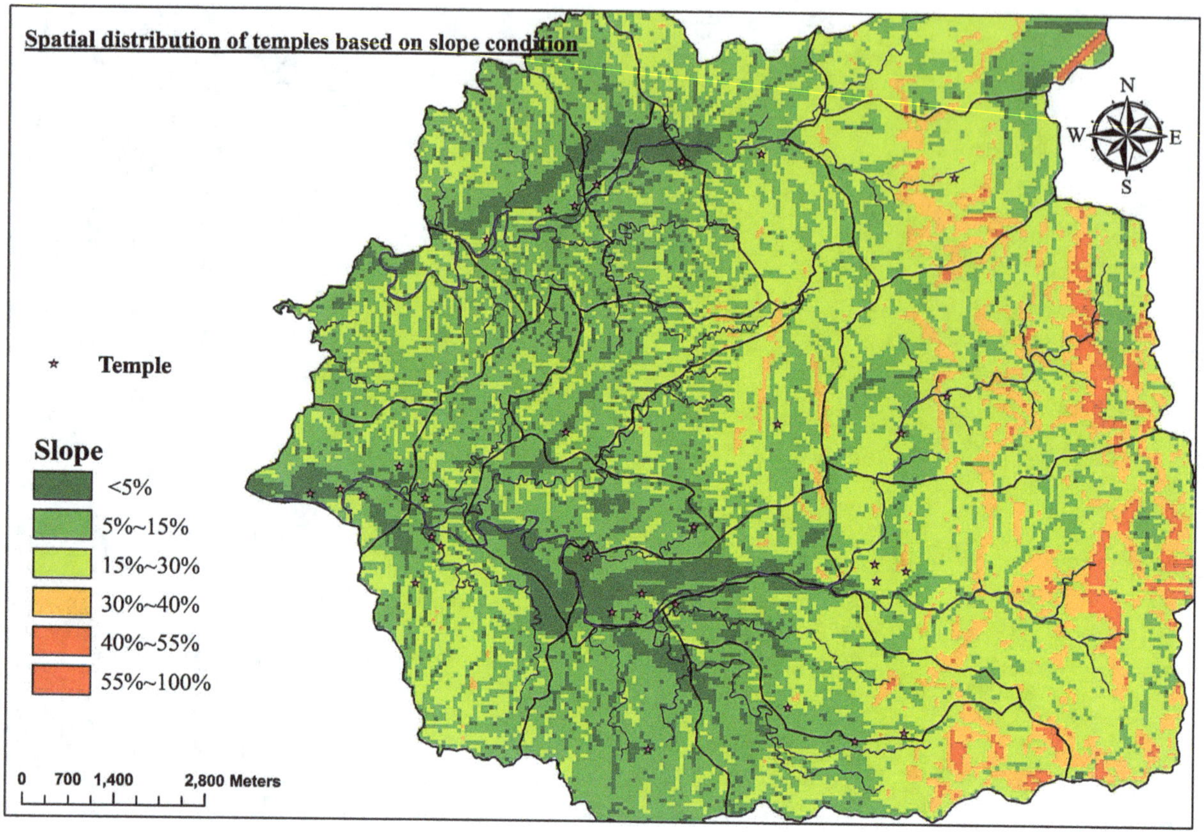

poor agrarian producers, it also contributes significantly to environmental impacts. There are 18 tourist spots of 25% with a slope < 5%, 31 tourist spots of 43% with a slope 5%~15%, and 23 tourist spots of 32% with a slope 15%~30%, respectively (Table 3, Figure 3).

Apparently, several tourist spots located on unsuitable terrain condition because of its easy suffered natural disasters if there is heavy storm rain or typhoon. In Jhong-Liao, the high density clustering areas of tourist spots is apparent on the location with steep slope of slope > 5%, where is inclined to have debris flow disaster and slope land slide or collapse. Environmental control and resources conservation are essential missions for Jhong-Liao sustainable development on the tourism clustering process. The high-density clustering areas of tourist spots is apparent on the location with steep slope of slope > 5%, where is inclined to have debris flow disaster and slope land slide or collapse. Thus, slope factor may be an essential influencing factor for tourism development in peripheral mountain area.

Table 3. Number of tourist spots based on slope condition

Slope	< 5%	5% ~ 15%	15% ~ 30%	30% ~ 40%	40% ~ 55%
Number of tourist spots	18	31	23	0	0
(Percentage)	25%	**43%**	**32%**	0%	0%

Figure 3a. Spatial distributions of the tourist spots based on slope condition

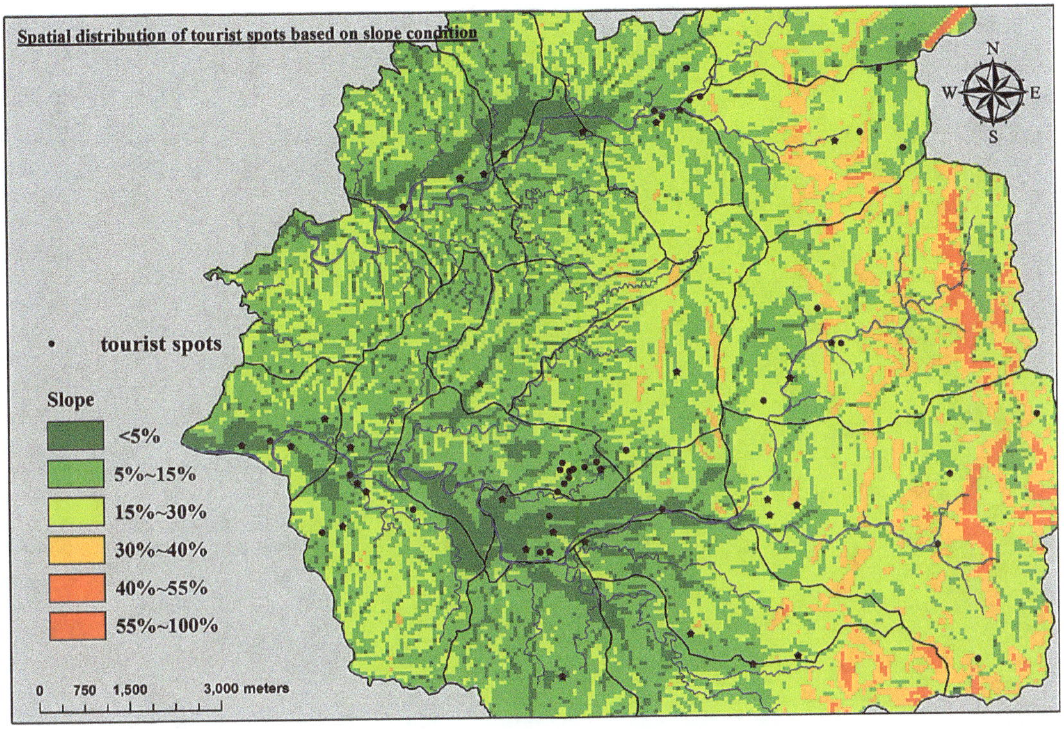

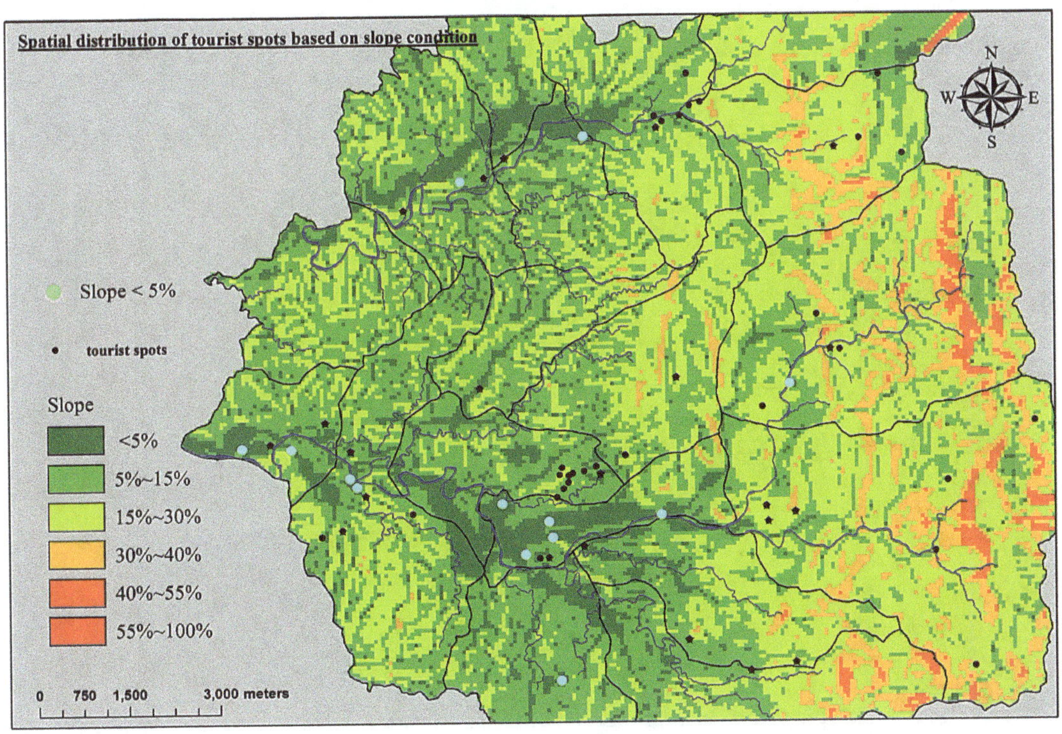

Figure 3b. Spatial distributions of the tourist spots based on slope condition

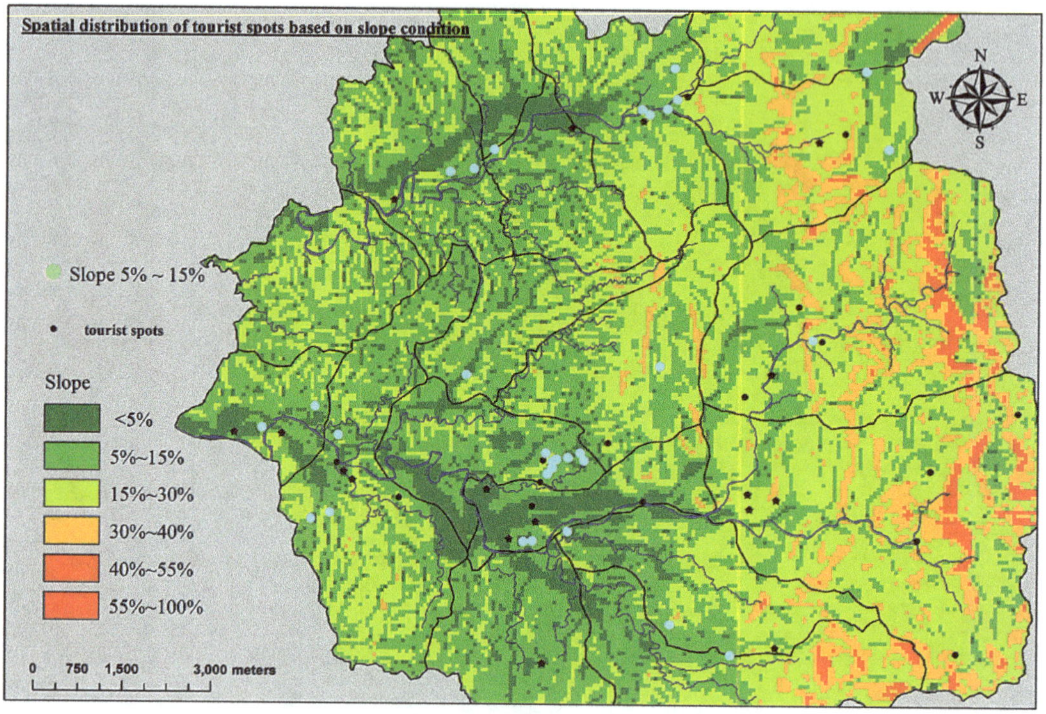

F

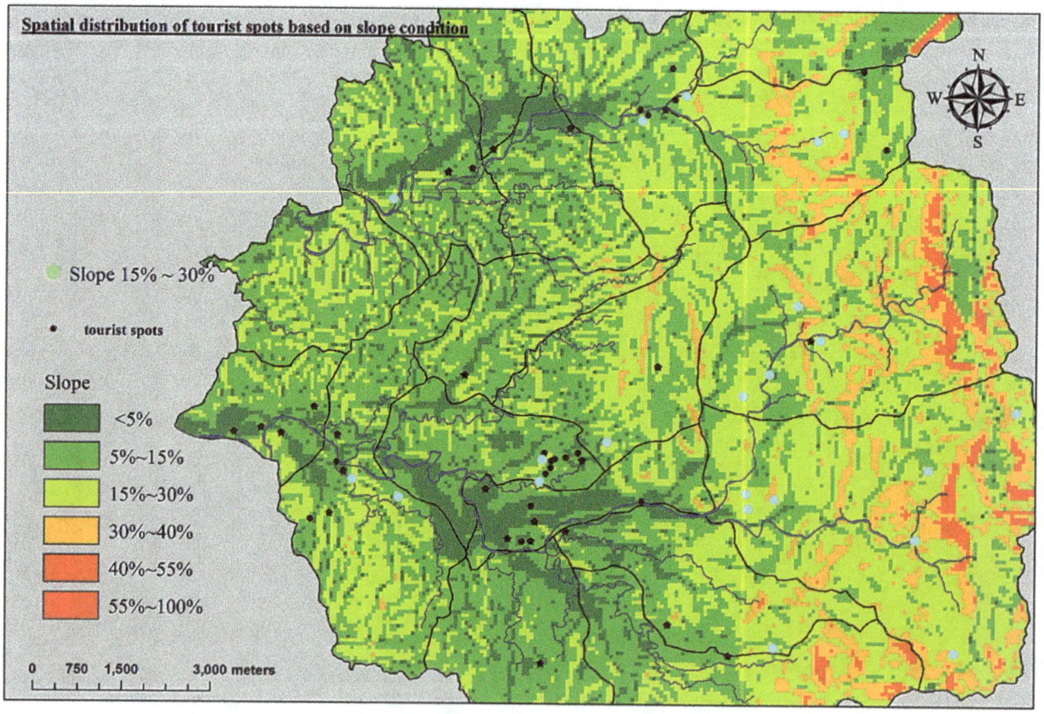

Point Density Analysis of GIS Mapping

He and Gebhardt (2012) explore the spatial clustering of creative industries based on point density analysis of GIS mapping by locating the sites of creative industry parks to visualize the core area of creative industries cluster in Shanghai. To explore the tourism cluster of Jhong-Liao, this research also used point density analysis of GIS mapping to locate a set of points of the total 72 tourist spots, specified all the points of identified tourist spots' locations of TWD 97 coordination system. The obtained density surface of tourist spots distribution of Jhong-Liao by GIS mapping is essential to define the concentration center of tourism cluster. The calculated method of point density value for comparing the extents of clustering of a raster surface is dependent on the estimating cell size and searching radius. On a point density surface, individual cell value is calculated by the specific point that falls within the searching radius. This research used 300 meter x 300 meter of cell size and 500 meters of searching radius to process the quantity of points per cell of each searching radius. By using point density analysis, we calculated the density value to explore the spatial concentration trends of tourist spots, and define the most concentrated zones of point locations of tourist spots, namely to distinguish the significant clusters of spatial distribution of tourist spots. With the point density surface of locations of the tourist spots, many layers such as DTM terrain raster, the transport raster, and the settlement raster can be overlaid together, and then the influencing factors of suitable locations accounting for tourism cluster render to be demonstrated.

In peripheral mountain area continuous monitoring on negative effects of tourism should be carry out and vulnerable sites should either be protected or visitors should be guided away from them. Micro-clusters are visible in some forms of tourism, where groups of tourist spots co-located and interact to deliver particular kinds of outputs, revealing that cluster is not confined solely to metropolitan areas but occurs also in peripheral mountain area. Hsueh (2012) used Porter's concept of cluster co-location for the example of tourism cluster by GIS mapping to locate a set of points of tourist spots along a route, clarified the core-periphery relationship among the tourist spots due to accessibility, and illustrated the core area of tourism clustering due to the increasing tourist attraction of destinations drawing more and more visitors, while the relative peripheral area is undergoing marginalized through the spatial competition process. The conception of tourism cluster creating tourist attraction is also demonstrated by the antiques industry clustered in a mountain area as a trip generator and supporting sustainable form of economic growth (Michael, 2002).

Based on the point density analysis (searching radius 500 meter, cell value 300x300 meters), obvious cluster cores of tourist spots are identified (Figure 4). The areas with the most high point density value of between 10.18591615~12.73239517 are located on the west northern part and the west southern part of Jhong-Liao. Figure 4 also shows that accessibility is also an influential factor on the locations of tourist spots clustering. Apparently, the west southern part of Jhong-Liao, with the highest point density value, is not only due to the geographical proximity to urban area-Nantou City for providing more tourism related services but also convenient to leave Jhong-Liao through Route 139 to back home. Obviously, the representative tourism cluster districts, in the west southern part and west northern part of Jhong-Liao, agglomerate along the main linkages on the Route 139 and Route 22 reveals that accessibility still has a decisive role in terms of location selection.

Geographical proximity characterizes tourism cluster as collaborative network with strong competitiveness and significant opportunities to economic development, so the collaborative networks were especially important for Jhong-Liao's tourism development. Establishing a stable network of producers and generating external economies are also the premises of the local associations for promoting the tourism development for sustainability. Why tourist spots cluster along the main route – Route 22 and Route 139 of Jhong-Liao? Along a route is suitable for tourist spots because of visitors to access a tourist spot and the related supporting services of community, especially in a peripheral mountain area. Apparently, tourist spots also clustered on the neighborhood of settlements because settlement serve as the function of community by providing related supporting services, especially in a peripheral mountain area with confined activity space. There is considerable evidence that it is efficient for tourist spots next to community to access tourism services based on point density analysis by GIS mapping. So, access to community may be the basic factor for Jhong-Liao's tourism cluster. There are several tourism related networks in the community between tourism associations and agriculture activities. The professional tourism associations tourism of Jhong-Liao were established from 1999, such as Long-Yan-Lin Welfare Association, Dab-Hand Vegetable Dye Workshop, Xi-Di-Yao Agriculture Learning Garden, He-Xing Organic Cultural Association, Kan-Ding Cinnamon, Making Wine Specialized Garden, and Long-An Medicinal Botanical Garden. One of the reasons for those professional associations playing important role on the process of tourism cluster is their creative knowledge on shaping tourist attractions and related support service networks between tourism and agriculture for segment market of niche tourism. For example, there are common facilitates for vegetable dying, fruit-picking tools shared by the small workshops or leisure farms or education providing by professional associations, which also offers agriculture information to members.

Figure 4. Point density analysis of tourist spots

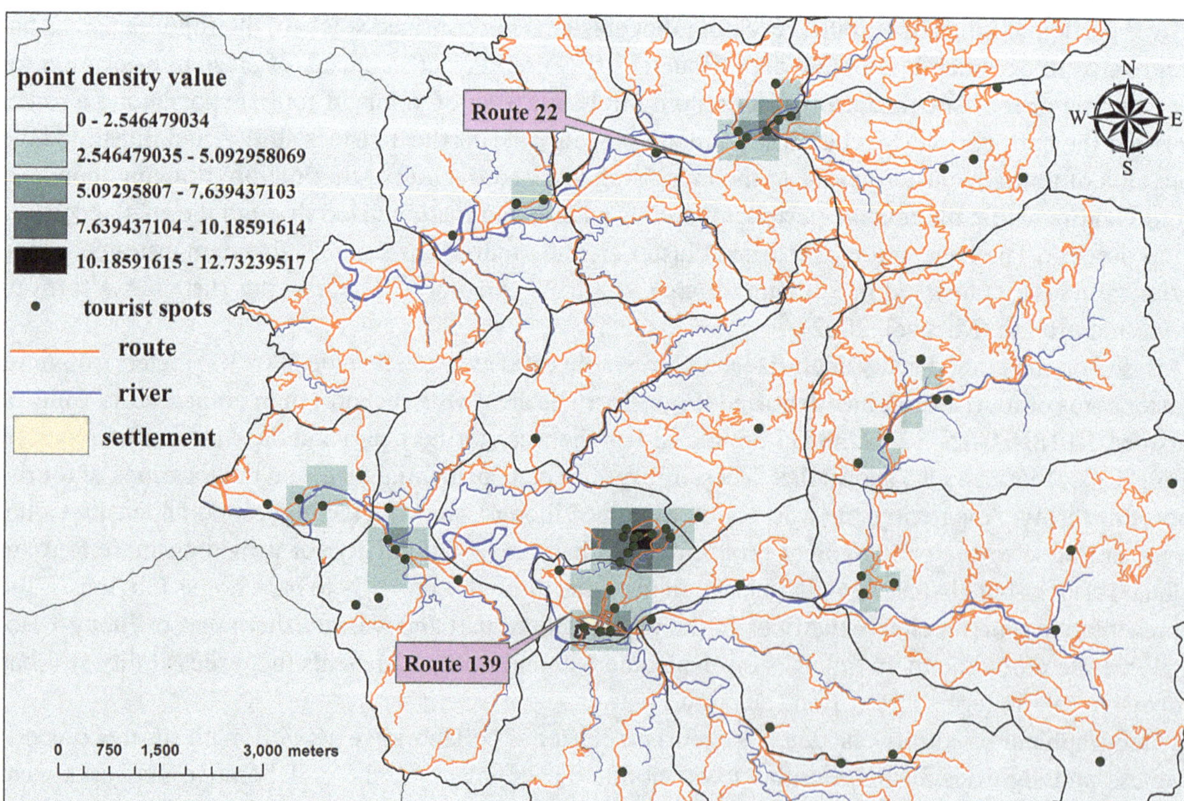

FUTURE RESEARCH DIRECTIONS

Future research should follow the direction of a location extension of spatial data base and analysis way in GIS mapping with advanced aspects in the linkage of tourism and agriculture implementations, including, for instance, which location is suitable for developing tourism based on agriculture products. Implementation of the tourist foods and their marketing integration with government sectors have not performed yet, so further pulling forces for the tourist foods marketing and development efforts in this direction by GIS mapping are required in the next further research.

CONCLUSION

In this article, we reviewed some of the most prominent theoretical results and practical achievements of research in the GIS mapping field. Although the ways of GIS mapping applied to the linkage between tourism and agriculture have been studied, we focused on the terrain analysis including elevation condition and slope condition and point density analysis, discussed their main features and surveyed their emerging implementations on tourist spots in the study area. Based on GIS mapping way, we realize the promoting of tourist attractions is not only by tourism cluster of tourist spots but also by linkages between tourism and related and supporting sectors such as agriculture. Linkage between tourism and agriculture makes a destination easier to shape its destination image, establish tourism activity network, and segment market. Most related literature reviews about tourism cluster focus on cluster benefits as the key elements of analyzing optimal location, but this research focuses on environmental impacts of tourism cluster for small scale tourist spots in a peripheral mountain area, reassesses the role of micro tourism cluster, like the leisure farms distinguished by organic agriculture and religious tourist spots, as tourist attractions of peripheral tourism. Instead of focusing on the benefits or patterns of tourism cluster, this research explores the environmental impacts of tourism cluster in a fragile mountain area, examines terrain factors of elevation and slope by GIS mapping way to realize how tourism cluster is influenced and further emphasize tourism cluster promoting peripheral area economic.

REFERENCES

Berno, T. (2011). Sustainability on a Plate: Linking Agriculture and Food in the Fiji Islands Tourism Industry. In R. Torres, & J. Momsen (Eds.), *Tourism and Agriculture: New Geographies of Production and Rural Restructuring*, (pp. 87-103). London: Routledge.

Botterill, D., Owen, R., Emmanuel, L. N., Foster, N. T., Gale, T., Nelson, C., & Selby, M. (2002). Perceptions from the Periphery: The Experience of Wales. In F. Brown & D. Hall (Eds.), *Tourism in the Peripheral Areas,* (pp. 7-38). Clevedon: Channel View Publications.

Buhalis, D. (1999). Tourism on the Greek Islands: Issues of Peripherality, Competitiveness and Development. *International Journal of Tourism Research*, *1*(5), 341–358. doi:10.1002/(SICI)1522-1970(199909/10)1:5<341::AID-JTR201>3.0.CO;2-0

Chaperon, S., & Bramwell, B. (2013). Dependency and Agency in Peripheral Tourism Development. *Annals of Tourism Research*, *40*(1), 132–154. doi:10.1016/j.annals.2012.08.003

Che, D., & Wargenau, A. (2011). Visiting Winery Tasting Rooms: Venues for Education, Differentiation and Direct Marketing. In R. Torres, & J. Momsen (Eds.), *Tourism and Agriculture: New Geographies of Production and Rural Restructuring,* (pp. 192-204). London: Routledge.

Connell, J. (2006). Medical Tourism: Sea, Sun, Sand and Surgery. *Tourism Management*, *27*(6), 1093–1100. doi:10.1016/j.tourman.2005.11.005

Eastman, C. M. (1995). Tourism in Kenya and the Marginalization of Swahili. *Annals of Tourism Research*, *22*(1), 172–185. doi:10.1016/0160-7383(94)00069-5

Gardiner, S., & Scott, N. (2014). Successful Tourism Clusters: Passion in Paradise. *Annals of Tourism Research*, *46*, 163–184. doi:10.1016/j.annals.2014.01.004

Getz, D. (1992). Tourism Planning and Destination Life Cycle. *Annals of Tourism Research*, *19*(4), 752–770. doi:10.1016/0160-7383(92)90065-W

Hall, C. M. (2005). Rural Wine and Food Tourism Cluster Network Development. In D. Hall, I. Kirkpatrick & M. Mitchell (Eds.), Rural Tourism and Sustainable Business, (pp. 149-164). Clevendon: Channel View Publications.

He, J. L., & Gebhardt, H. (2012). Socio-Culture Embeddedness of Spatial Agglomeration of Creative Industries in Shanghai. *Sociology Study*, *2*(8), 605–614.

Hsueh, Y. H. (2012). The Influence of Transportation Revolution to Local Tourism Development-Example for Puli Area of Taiwan. *Sociology Study*, *2*(8), 615–625.

Hsueh, Y. H. (2013a). Exploring the Clustering Location of Accommodation Units through the Tourism Development in the Cing-Jing Area of Taiwan. *International Journal of Basic & Applied Sciences*, *13*(4), 34–43.

Hsueh, Y. H. (2013b). Mushroom huts diffusion analysis based on GPS data. In Y. H. Hsueh (Ed.), Global positioning systems: Signal structure, applications and sources of error and biases, (pp. 115-124). New York: Nova Science Publishers, Inc.

Hsueh, Y. H., & Chen, C. H. (2015). Clustering of Tea Processing Industry Based on GIS Analysis. *International Journal of Agricultural Sciences and Natural Resources*, *2*(3), 58–66.

Marshall, J. (2001). Women and Strangers: Issues of Marginalization in Seasonal Tourism. *Tourism Geographies*, *3*(2), 165–186. doi:10.1080/14616680010030266

Michael, E. (2002). Antiques and Tourism in Australia. *Tourism Management*, *23*(2), 117–125. doi:10.1016/S0261-5177(01)00053-X

Michael, E. J. (2007). Micro-Clusters: Antiques, Retailing and Business Practice. In E. J. Michael (Ed.), Micro-Clusters and Networks: The Growth of Tourism, (pp. 63-78). London: Elsevier Publications.

Momsen, J. (1998). Caribbean Tourism and Agriculture: New Linkages in the Global Era? In T. Klak (Ed.), *Globalization and Neoliberalism: The Caribbean Context*. Oxford, UK: Rowman & Littleman Publishers, Inc.

Pillay, M., & Rogerson, C. M. (2013). Agriculture-Tourism Linkages and Pro-Poor Impacts: The Accommodation Sector of Urban Coastal KwaZulu-Natal, South Africa. *Applied Geography (Sevenoaks, England)*, *36*, 49–58. doi:10.1016/j.apgeog.2012.06.005

Prideaux, B. (2002). Building Visitor Attractions in Peripheral Areas: Can Uniqueness Overcome Isolation to Produce Viability? *International Journal of Tourism Research*, *4*(5), 379–389. doi:10.1002/jtr.387

Seaton, A. V. (1999). Book Towns as Tourism Developments in Peripheral Areas. *International Journal of Tourism Research*, *1*(5), 389–399. doi:10.1002/(SICI)1522-1970(199909/10)1:5<389::AID-JTR204>3.0.CO;2-0

Sharkey, D. A., & Momsen, J. (1995). Tourism in Dominica, West Indies: Problems and Prospects. *Caribbean Geography*, *6*(1), 40–51.

Sims, R. (2009). Food, Place and Authenticity: Local Food and the Sustainable Tourism Experience. *Journal of Sustainable Tourism*, *17*(3), 321–336. doi:10.1080/09669580802359293

Sindiga, I. (1996). International Tourism in Kenya and the Marginalization of the Swahili. *Tourism Management*, *17*(6), 425–432. doi:10.1016/0261-5177(96)00051-9

Telfer, D. J. (2001). Strategic Alliances along the Niagara Wine Route. *Tourism Management*, *22*(1), 21–30. doi:10.1016/S0261-5177(00)00033-9

Telfer, D. J., & Wall, G. (1996). Linkages between Tourism and Food Production. *Annals of Tourism Research*, *23*(3), 635–653. doi:10.1016/0160-7383(95)00087-9

Telfer, D. J., & Wall, G. (2000). Strengthening Backward Economic Linkages: Local Food Purchasing by Three Indonesian Hotels. *Tourism Geographies*, *2*(4), 421–447. doi:10.1080/146166800750035521

Torres, R. (2003). Linkages between Tourism and Agriculture in Mexico. *Annals of Tourism Research*, *3*(3), 546–566. doi:10.1016/S0160-7383(02)00103-2

Torres, R., & Momsen, J. H. (2004). Challenges and Potential for Linking Tourism and Agriculture to Achieve Pro-Poor Tourism Objectives. *Progress in Development Studies*, *4*(4), 294–318. doi:10.1191/1464993404ps092oa

Weaver, D. B. (1998). Peripheries of the Periphery Tourism in Tobago and Barbuda. *Annals of Tourism Research*, *25*(2), 292–313. doi:10.1016/S0160-7383(97)00094-7

ADDITIONAL READING

Carvera, S., Comberb, A., McMorranc, R., & Nuttera, S. (2012). A GIS Model for Mapping Spatial Patterns and Distribution of Wild Land in Scotland. *Landscape and Urban Planning*, *104*(3-4), 395–409. doi:10.1016/j.landurbplan.2011.11.016

Gregoretti, C., Degetto, M., & Boreggio, M. (2016). GIS-Based Cell Model for Simulating Debris Flow Runout on a Fan. *Journal of Hydrology (Amsterdam)*, *534*, 326–340. doi:10.1016/j.jhydrol.2015.12.054

Haworth, B., Bruce, E., & Iveson, K. (2013). Spatio-Temporal Analysis of Graffiti Occurrence in an Inner-City Urban Environment. *Applied Geography (Sevenoaks, England)*, *38*, 53–63. doi:10.1016/j.apgeog.2012.10.002

Kumarn, S., & Bansal, V. K. (2016). A GIS-Based Methodology for Safe Site Selection of a Building in a Hilly Region. *Frontiers of Architectural Research*, *5*(1), 39–51. doi:10.1016/j.foar.2016.01.001

Lacher, R. G., & Nepal, S. K. (2010). From Leakages to Linkages: Local-Level Strategies for Capturing Tourism Revenue in Northern Thailand. *Tourism Geographies*, *12*(1), 77–99. doi:10.1080/14616680903493654

Lee, S. H., Choi, J. Y., Yoo, S. H., & Oh, Y. G. (2013). Evaluating Spatial Centrality for Integrated Tourism Management in Rural Areas Using GIS and Network Analysis. *Tourism Management*, *34*, 14–24. doi:10.1016/j.tourman.2012.03.005

Scaini, C., Felpetob, A., Martí, J., & Carniel, R. (2014). A GIS-Based Methodology for the Estimation of Potential Volcanic Damage and Its Application to Tenerife Island, Spain. *Journal of Volcanology and Geothermal Research*, *278–279*, 40–58. doi:10.1016/j.jvolgeores.2014.04.005

Taramelli, A., Valentini, E., & Sterlacchini, S. (2015). A GIS-Based Approach for Hurricane Hazard and Vulnerability Assessment in the Cayman Islands. *Ocean and Coastal Management*, *108*, 116–130. doi:10.1016/j.ocecoaman.2014.07.021

Zhai, S., Xu, X., Yang, L., Zhou, M., Zhang, L., & Qiu, B. (2015). Mapping the Popularity of Urban Restaurants Using Social Media Data. *Applied Geography (Sevenoaks, England)*, *63*, 113–120. doi:10.1016/j.apgeog.2015.06.006

KEY TERMS AND DEFINITIONS

DTM: A digital representation of ground-surface topography or terrain represents the bare ground surface without any objects like plants and buildings. DTM is a topographic model of the bare earth-terrain relief that can be manipulated by computer programs. Vegetation, buildings and other man-made artificial features are removed digitally-leaving the underlying terrain. The data files contain the spatial elevation data of the terrain in a digital format which usually presented as a rectangular grid, and can further be transformed to visualize slope, aspect and viewshed conditions of the terrain.

Elevation Analysis: An analysis way of terrain based on GIS mapping. Elevation analysis can perform certain analytical tasks quickly and easily, without having to collect or update an authoritative set of base data. Elevation analysis is essential for many GIS applications example for natural resource management, conservation, agriculture, transportation, risk management.

GIS: A system designed to capture, store, manipulate, analyze, manage, and present all types of spatial or geographical data. GIS tools allow users to create interactive queries, analyze spatial information, edit data in maps, and present the results of all these operations. GIS can specify locations or extents in the coordinate system recorded as dates/times of occurrence, and x, y, and z coordinates representing, longitude, latitude, and elevation by using location as the key index variable.

GIS Mapping: A method of digital mapping and a technology that offers different tools in which we produce and use the maps to manage our communities and industries. GIS creates intelligent super maps through which sophisticated planning and analysis can be performed based on the transformation of spatial locations of real-world features and visualize the spatial relationships among them.

GPS: A satellite-based navigation system made up of a network of 24 satellites placed into orbit by the U.S. Department of Defense. GPS was originally intended for military applications, but in the 1980s, the government made the system available for civilian use. GPS works in any weather conditions, anywhere in the world, anytime in a day.

Point Density Analysis: A tool of GIS mapping that can calculate the density of point features around each output raster cell. A neighborhood is defined around each raster cell center, and the number of points that fall within the neighborhood is totaled and divided by the area of the neighborhood. Possible uses include finding density of disease cases, crime cases, or firms clustering. For example, one address might actually represent a crime case, or some crime cases might be weighted more heavily than others in determining crime rates. Increasing the radius will not greatly change the calculated density values. Although more points will fall inside the larger neighborhood, this number will be divided by a larger area when calculating density. The main effect of a larger radius is that density is calculated considering a larger number of points, which can be outside from the raster cell.

Slope Analysis: An analysis way of terrain based on GIS mapping. Slope analysis calculates the maximum rate of change between each cell and its neighbors, and the maximum change in elevation over the distance between the cell and its eight neighbors. Every cell in the output raster has a slope value. The lower the slope value, the flatter the terrain, while the higher the slope value, the steeper the terrain. In GIS mapping, slope analysis is very important for suitability analysis and predicting potential hazards. Analyzing the terrain slope of a given location plays an important role in the research of flood disaster, site planning, and conservation.

This research was previously published in the Encyclopedia of Information Science and Technology, Fourth Edition edited by Mehdi Khosrow-Pour, pages 3434-3447, copyright year 2018 by Information Science Reference (an imprint of IGI Global).

Chapter 22
Geographic Information System (GIS) Modeling Analysis and the Effects of Spatial Distribution and Environmental Factors on Breast Cancer Incidence

Akram Gasmelseed
University of Science and Technology, Sudan

Ali H. Alharbi
Qassim University, Saudi Arabia

ABSTRACT

Breast cancer is a public health problem among women in the United States. According to the American Cancer Society, about 40,290 women in the United States are expected to die in 2015 from breast cancer. This chapter contains an analysis of the breast cancer distribution in the United States by comparing the spatial distribution of breast cancer cases against physical environmental factors using geographic information system (GIS). Results show that the exposure to some spatial environmental factors seems likely to have a major impact on the overall trends in breast cancer rates. Moreover, the possibility to develop cancer is existing naturally through the environment factors.

INTRODUCTION

Breast cancer is a major health issue in all countries affecting thousands of women (Tazzite et al., 2013; Dube & Gupta, 2015). So far its cause(s) are unknown and the national and international strategies to reduce its morbidity and mortality levels are based on early detection of cancer through screening and treatment according to clinical guidelines. Thus, knowledge of which women are at risk and why they are at risk is therefore essential component of disease prevention and screening. Researchers from

DOI: 10.4018/978-1-5225-7359-3.ch022

the International Agency for Research on Cancer (IARC) and the World Health Organization (WHO) reports that globally breast cancer might contribute to the greatest burden on women's health when compared to other cancer sites (World Health Organization, n.d.). In 2015, an estimated 231,840 new cases of invasive breast cancer are expected to be diagnosed in women in the United States, along with 60,290 new cases of non-invasive (in situ) breast cancer (Siegel et al., 2015). However, all locations are not equal for breast cancer risk and thus support a major role of the geography in breast carcinogenesis (Akram & Nanna, 2003).

The purpose of this work is to provide a more detailed analysis of the breast cancer distribution in the United States by comparing the spatial distribution of breast cancer cases against physical environmental factors using Geographic Information System (GIS) (Figure 1). Further, it gives background information to the GIS and its applications in health-related research.

BACKGROUND

Breast Cancer Facts/Spatial-Based Patterns

Previous reports have shown that the Northeast United States has a 16% higher breast cancer mortality rate than the rest of the country (Kulldorff et al., 1997). The probability of breast cancer risk is not equal for all locations which indicate that geography plays a very important role in the etiology of breast cancer.

Figure 1. Medical-based GIS

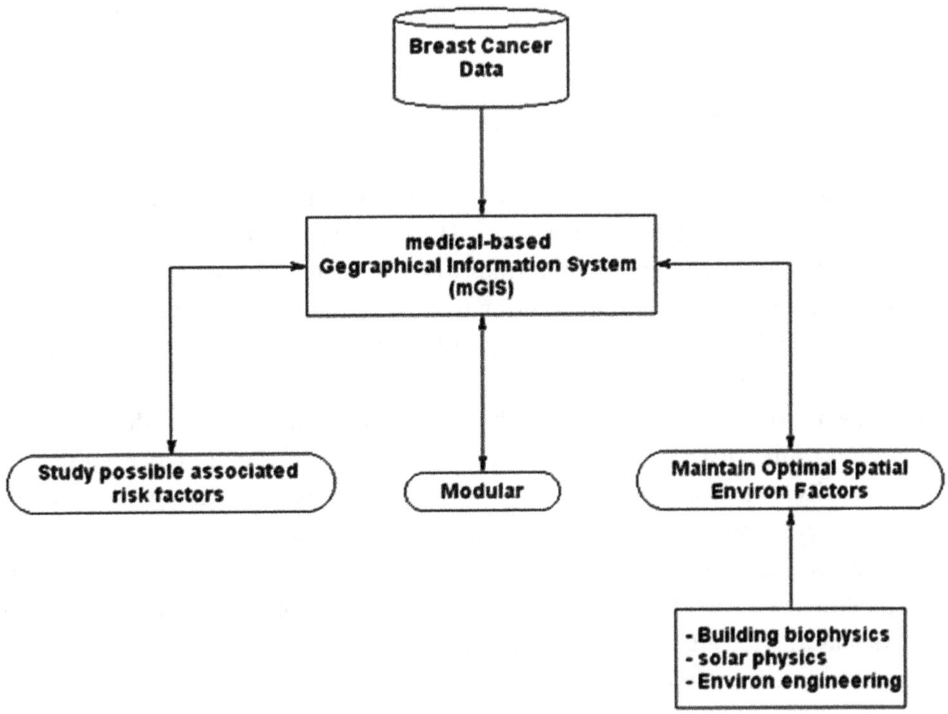

There are geographic patterns of high cases of breast cancer, and the analysis of these patterns is very important in the formulation of hypotheses about risks and focus investment more effectively in research and intervention on the most significant areas (Laden et al., 1997).

In general, breast cancer incidence rates have continued to increase since 1980, although the rate of increase slowed down in the 1990s, compared to the 1980s (American Cancer Society, 2004). Furthermore, during the more recent time period, breast cancer incidence rates have increased only in those aged 50 and over. The mortality rates declined by around 1.4% per year during 1989-1995 and by 3.2% afterwards, with the largest decreases in younger women in both whites and African Americans. These decreases are probably due to the result of both earlier detection and improved treatment. Clearly, the ultimate cause of breast cancer is unknown (Roche, 1998), but several risk factors appear to play a role.

Previous literatures have shown that breast cancer mortality and incidence rates vary geographically according to the different regions of the United States (Devesa et al., 1999; Joseph et al., 2004). Generally, the disease is most common in North America and Western Europe, account for about one in four female cancers in these regions, while in the Far East (China and Japan) it is very much rare (Le et al., 2002). Furthermore, the disease rates among Asian-Americans are lower than those of U.S. whites but considerably higher than rates prevailing in Asia. Thus, it is suspected that migration to the US brings about a change in endocrine function among Asian women, although reasons for this change remain obscure (Wu et al., 1996). In fact, the geographic patterns of cancer around the world and within countries have provided important clues to the causes of cancer (Robert, 1996). The highest incidence rates of all are found in Hawaii, where a rate of 93.9 per 100,000 female populations has been reported, and in US white women. The incidence rises with age from about age 30. Moreover, there are ethnic variations, such as a high incidence in Israeli Jews compared with non-Jews in Israel. It is more common in single women, in higher social classes, and in urban rather than rural areas. On the other hand, about 1% of cases occur in males. Mortality has increased less rapidly than incidence, but breast cancer is most common cause of cancer death.

What Is Geographic Information System (GIS)?

Maps are well-known to be one of the most used visual tools in our life today. They have very long tradition and they have been used from ancient time. They are very useful tools to visually show a place and its features. However, maps, at least in their traditional form, have limited capabilities in the amount and type of information they can depict. In addition to this, the capability of human processing of huge information is limited. Gathering information from different sources, integrate, analysis, and interpret them is a complex and time-consuming task.

Data visualization is one of the areas that have been improvement due to the adoption of new technology and visualization frameworks and systems. Advancement in information technologies has affected each aspect of our daily life. Data integration, processing and visualization are no exception. Today, intelligent computerized tools are available to help users enter and process information effectively and efficiently. The capability of computer systems in processing and make sense of data is importing dramatically in the last few years. Data visualization is the branch of information processing that deals with finding ways and methods to present data using various visualization techniques in order to make it easy for the users to interpret and make sense of the data.

One the general and most comprehensive definitions of GIS is the one provided by ESRI (Environmental System Research Institute) (GIS dictionary, ESRI, n.d.), in which GIS is described as:

An integrated collection of computer software and data used to view and manage information about geographic places, analyze spatial relationships, and model spatial processes. A GIS provides a framework for gathering and organizing spatial data and related information so that it can be displayed and analyzed.

This definition stresses the fact that GIS is more than just the technology behind this system, but it is actually a comprehensive framework to support decision making by the integration of spatial data and other factors using visualization techniques.

(GIS) can be defined as:

A collection of science and technology tools used to manage geographic relationships and integrate information. GIS helps us analyze spatially-referenced data and make well-informed decisions based on the association between the data and the geography.

Applications of GIS in Health Science

Applications of GIS in health science have increased dramatically in the last few years with many innovative approaches have emerged. Understating the relationship between health and contextual factors, such as sociodemographic, environmental and political variables, can be complex. GIS can facilitate the integration and presentation of these contextual variables to gain insight into their influence on health. This section reviews the research related to the use of GIS to improve health care.

A previous study conducted a review that focuses on the applications of GIS in public health and health promotion (Nykiforuk & Flaman, 2011). The final review was composed of 621 journal articles and book chapters reporting health-related applications of GIS. According to the review, GIS applications in public health can be classified into four main themes: disease surveillance, risk analysis, health access and planning, and community health profiling. Moreover, some other reviewed literature on the applications of GIS in health science research (Lyseen, et al., 2014), and present a conceptual framework that can be used to better categorize and understand research in this area, and thus put emphasis on areas that need further research. The sample size of the study consists of a total of 865 articles pulled out from the research databases Scopus and Web of Science. By applying ground-theory data analysis method, the study presents a framework that comprises four conceptual domains to represent GIS applications in health science. These domains are:

- Spatial analysis of diseases,
- Spatial analysis of health service planning,
- Public health, and
- Health technology.

Disease Surveillance

One of the most well-known applications of using GIS in public research is for disease surveillance, which is the integration and tracking of data related to the incidence, prevalence, and the spread of the disease in the community. Epidemiology seek to answer basic questions such as when and where are cases of an infectious disease currently occupying, can we predict the next occurrence of the disease, and how can we prevent the spread of the disease? GIS can help increase the speed and occurrence of finding answers to epidemiology these questions by integrating spatial information on the disease with other types of geographic information to track and map the spread of the disease (Eisen & Eisen, 2014).

Research in this area can be classified into two interrelated categories: disease mapping and disease modelling. Disease modeling is the extension of GIS capabilities to support the decision making process. It can be applied to the data to (a) predict the outbreak of a disease, (b) identify Planning of Healthcare Services Literature reported different ways of applying GIS technology to support decision making in the planning and distribution of healthcare service.

Strategic planning is required to achieve optimal and demand-based distribution of healthcare centers. An example, GIS was used to assess hospital distributions in Seoul, the capital of South Korea (Lee & Moon, 2014). The study proved that GIS can be used to identify and examine the factors that influence health service distribution.

In developing countries, planning and delivery of healthcare services can be critical. Brijnath and De Souza (Brijnath & De Souza, 2012) discussed different ways in which GIS can be utilized to enhance health service planning and delivery for infectious diseases in low-income countries. according to the paper GIS be used to:

1. Get insight into the optimized locations for distribution healthcare centers,
2. Provide real-time surveillance system for the control of diseases,
3. Serve as an evidence-based accountability tool between funding agencies and healthcare provides, and finally
4. Transform data into accessible format to make it easy for stakeholders to make informed decisions.

METHODOLOGY

In this research we obtained mortality estimates (Figure 2) in 50 states of the United States from the National Center of Health Statistics (NCHS) for the years 1994-98 (National Cancer Institute, n.d.). We excluded those states for which the corresponding data was not available.

The environmental factors used in the analysis were limited to seven spatial environmental factors collected from NSTATE, LLC (United States Geography, n.d.). In this work Regression analysis have been used to detect and evaluate the statistical significance between the mortality rates and the spatial environmental factors.

RESULTS

Preliminary results show an inverse relationship between the death rates and mean elevation ($T = -6.25$; $p = 0.000$) and positive relationship with the average temperature ($T = 3.07$; $p = 0.004$). The high F-value 21.71 and low p-value indicate there is a statistically significant difference between mortality rates and the selected spatial environmental factors. The differences are certified as statistically significant and are

Figure 2. Female breast cancer mortality (1994-98) rate by state

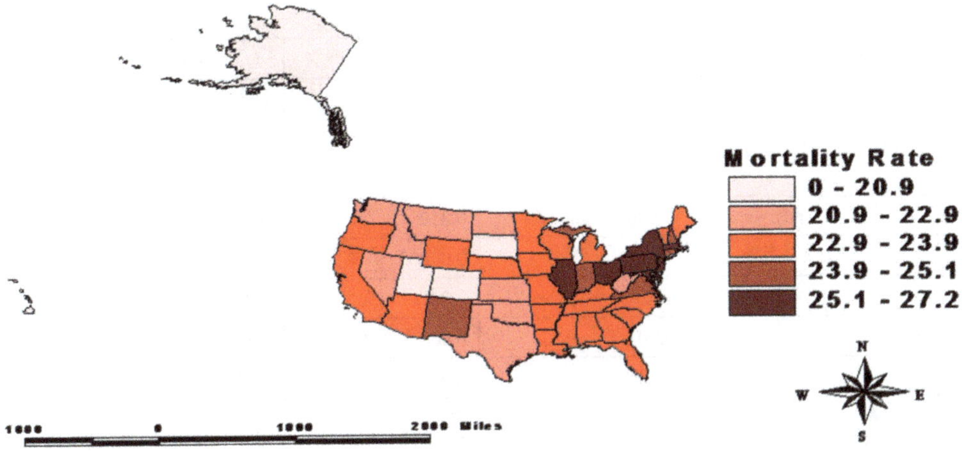

probably not due to sampling variability only. All the relevant finding related to the explanatory power of the independent variables, in the goodness of fit of all possible models can be enhanced by looking at the best-fit models with the significant (at the 5% level, $p \leq 0.05$) explanatory underlying variables, for each of the mortality/incidence pattern groups according to the spatial environmental factors (Table 1 and Table 2).

Table 1. Stepwise multiple regression, explanatory variables best-fit model according to mortality rates, 1991-95, 1992-96, 1993-97 and 1994-98.

Step	1	2	3	4
Constant	21.14	21.10	21.39	25.27
Ave. Temp	0.048	0.069	0.067	0.069
T-Value	3.42	6.20	6.09	6.27
P-Value	0.001	0.000	0.000	0.000
Elevation		-0.00080	-0.00075	-0.00067
T-Value		-11.24	-10.29	-8.14
P-Value		0.000	0.000	0.000
Land Area			-0.00000	-0.00000
T-Value			-2.53	-3.00
P-Value			0.012	0.003
Low Ave. Temp				-0.046
T-Value				-2.06
P-Value				0.041
S	2.13	1.65	1.63	1.61
R-Sq	5.81	43.54	45.40	46.61
R-Sq (adj)	5.32	42.95	44.53	45.47

Table 2. Stepwise multiple regression, explanatory variables best-fit model ($p \leq 0.05$) according to incidence rates, 1990-94, 1991-95, 1993-97 and 1994-98.

Step	1	2	3	4
Constant	105.2	182.8	185.2	182.7
Ave. Temp	0.026	0.055	0.130	0.171
T-Value	0.40	1.00	2.34	3.07
P-Value	0.690	0.317	0.020	0.003
Elevation	-0.00128	0.00033	0.00050	0.00073
T-Value	-3.29	0.84	1.34	1.98
P-Value	0.001	0.402	0.183	0.050
Low Ave. Temp		-0.93	-1.04	-1.05
T-Value		-7.86	-8.94	-9.30
P-Value		0.000	0.000	0.000
Water/Land Ratio			13.7	27.2
T-Value			3.98	4.88
P-Value			0.000	0.000
Water Area				-0.00014
T-Value				-3.03
P-Value				0.003
S	8.75	7.37	7.03	6.84
R-Sq	6.79	34.25	40.64	44.15
R-Sq (adj)	5.53	32.92	39.02	42.24

DISCUSSION

Effect of Average Temperature on Breast Cancer

Temperatures cause a variety of physiological changes, e.g. of blood composition, blood pressure (Stout & Crawford, 1991) and circadian rhythms (Reiter, 1991), which in turn are believed to contribute to an increased mortality/incidence of breast cancer (Kliukiene et al., 2001). Furthermore, published laboratory studies have reported that melatonin might have cancer-preventive activities (Coleman & Reiter, 1992). Moreover, melatonin has been observed to exert potent inhibition on cancer growth by acting as natural anti-angiogenic molecule, with a following opposition or angiogenesis-dependent cancer proliferation (Lissoni et al., 2001). This has been demonstrated in certain human breast cancer cell lines such as MCF-7, with additional in-vivo effects on breast oncogenesis in various rat models. Melatonin (N-acetyl-5-methoxytryptamine) is a hormone secreted by the pineal gland of vertebrates, which controls several physiological functions associated with circadian rhythm (Becker-Andre et al., 1994). Temperature is a major regulator of circadian rhythms. Circadian rhythms control many physiological activities. Synchronization of biological clocks to environmental time is adaptive and important for physiological homeostasis and for the proper timing of species-specific behaviors. A number of investigations have been conducted to examine the effects of temperature and light on melatonin rhythms (Underwood & Calaban, 1987). Based on our analysis, we proposed a second order equation ($y = a + bx + cx^2$) to describe the correlation between the average temperature and mortality rates as this model seems to be the one that best describes the relationship between the two parameters. Hence, based on this model, the estimated temperature (Te) values at which the mortality rates are low is in the range $49 < T_e < 11°C$.

To further examine the idea that temperature can influence breast cancer rates, we tested the relationship (if any) between mortality rates and geographic coordinates: latitude and longitude.

Clearly, the principal controls of temperature variations are latitude and elevation (altitude). The latitude for a given point on the earth has an effect on temperature; the farther away from the Equator (0°) a point is the cooler average temperature that point is likely to have. We obtained worldwide breast cancer mortality statistics from WHO Cancer Mortality Database, WHO/IARC (WHO Cancer Mortality Database, n.d.). Then, we investigated the possible correlation between the geographic coordinates of the countries and the age-specific rates (ASR) of breast mortality. The result revealed that there is existing negative relationship between latitude and mortality rates, i.e. by increasing latitude the death rate increases. Variations in latitude lead to variations in the temperature and sunlight intensity and hence for disturbance of melatonin rhythms. A previous report, investigators showed evidence for a link between cosmic ray intensity and cancer due to variations in the latitude (Juckett & Rosenberg, 1997). Moreover, it was demonstrated that light-induced melatonin suppression in humans is sensitive to short wavelength light (420 – 480 nm; l_{max} @ 460 nm) (Skene, 2003).

Jet Lag and Circadian Rhythm

Investigators reported that "Jet Lag" is caused from air travel through changing time zones, and hence leading to disturbance of melatonin rhythms (Samel & Wegmann, 1997). The main but not only cause of jet lag is crossing time zones. Usually going east is worse than going west (Delagrange & Guardiola-Lemaitre, 1997). Finally, the obtained negative relationship between mortality data and elevation is clear in the sense that reduced temperature occur at higher elevation where there is less water vapor to trap and hold heat (Figure 3).

Further, we found a large, significant relationship between women at ages 30-74 and the geographic latitude. For women aged 65 years and older, breast cancer mortality is 26% higher in New England than in the south, while incidence is only 3% higher. Mortality rates from breast cancer are approximately 25% higher for women in the northeastern United States than for women in the South or West. This can be depicted by the correlation with the geographic longitude, see Figure 3. Investigators reported that "Jet Lag" is caused from air travel through changing time zones, and hence leading to disturbance of melatonin rhythms.

The exposure to some spatial environmental factors seems likely to have a major impact on the overall trends in breast cancer rates. The possibility to develop cancer is existing naturally through the environment factors. Unbalanced in these nature factors and/or other existing risk factors, e.g. socioeconomic conditions will increase the chance for cancer development (Figure 4).

Figure 3. Influence of "jet lag" on breast cancer development

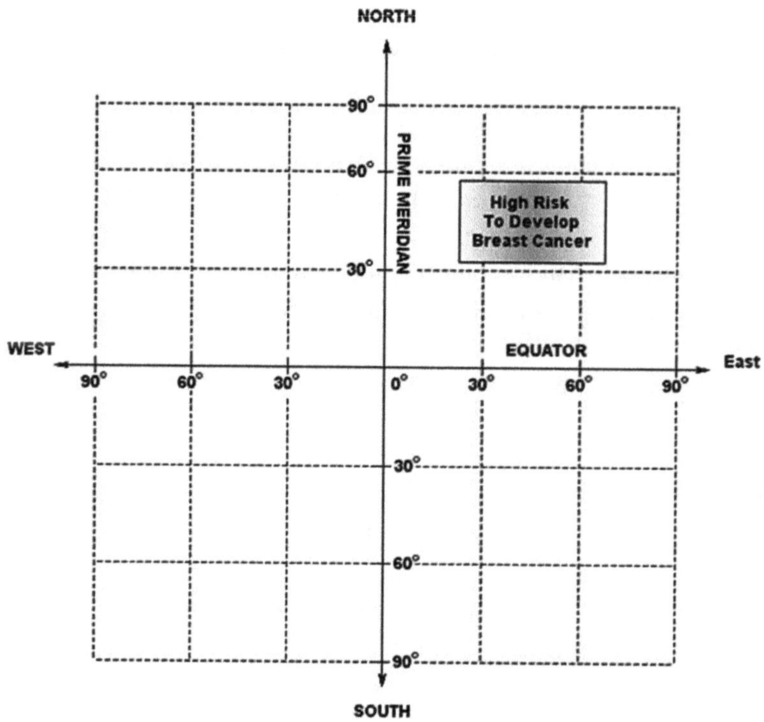

Figure 4. Role of spatial environmental factors in breast cancer development

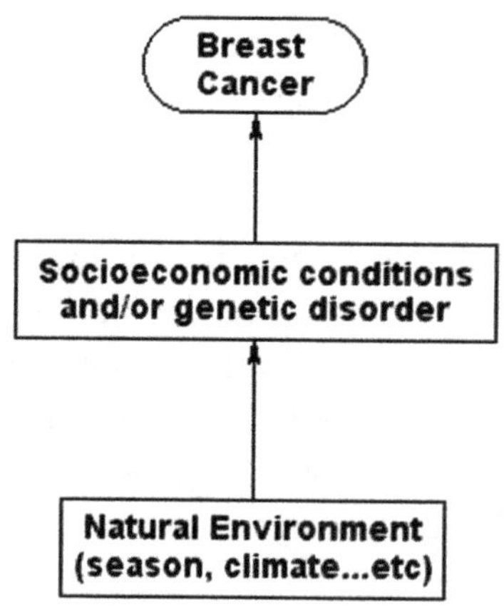

FUTURE RESEARCH DIRECTIONS

The results presented in this work need to be elaborated based on the latest advancement in information technology. Hence, for future research, our goal is to incorporate cutting-edge information technology services to increase our understanding of variations in the breast cancer disease and geographical factors, to describe the disease burden, and to obtain guidance on the action to be taken.

Cloud computing is one of the cutting-edge technologies that may be feasible in the future. Thus, the integration of GIS with cloud computing technology can provide new insights into breast cancer research by providing shared information through internet.

Moreover, adopting such approach could maintain a reliable data as well as device and location independency.

CONCLUSION

We have combined data and techniques from the areas of environmental physics and cancer epidemiology to explore the hypothesis that spatial environmental factors can induce changes in melatonin rhythms, which can lead to increases in breast cancer rates. The results are interpreted as evidence that temperatures can modulate cancer via changes induced in melatonin rhythms.

Results show that the exposure to some spatial environmental factors seems likely to have a major impact on the overall trends in breast cancer rates. Moreover, the possibility to develop cancer is existing naturally through the environment factors.

Furthermore, this study provides valuable reference information for clinicians and health administrators, as well as baseline for more detailed studies of patient survival for individual environmental/geographical sites. The earlier the cancer is detected, the smaller the chance that it already has spread beyond the limited primary focus. This implies that the extend of intervention needed is lesser and the prognosis improved.

Finally, map-based exploration of georeferenced health statistics will lead to a better understanding of health/environment interaction. By utilizing geographical information system technology we can enhance our ability to manage, estimate and predict breast cancer distribution.

REFERENCES

Akram, A., & Nanna, S. (2003). *Medical-based geographical information system (mGIS): diagnosis versus healing*. Global Health Network Supercourse Project. Retrieved October 2, 2015, from http://supercourse.isciii.es/lecture/lec10671/index.htm

American Cancer Society. (2004). *Breast Cancer Facts and Figures 2003-2004*. Atlanta, GA: American Cancer Society. Retrieved July 28, 2005, from https://www.cancer.org/acs/groups/content/@nho/documents/document/caff2003brfpwsecuredpdf.pdf

Becker-Andre, M., Wiesenberg, I., Schaeren-Wiemers, N., Andre, E., Missbach, M., Saurat, J. H., & Carlberg, C. (1994). Pineal gland hormone melatonin binds and activates an orphan of the nuclear receptor superfamily. *The Journal of Biological Chemistry*, *269*(46), 28531–28534. PMID:7961794

Brijnath, B., & De Souza, D. K. (2012). Four Ways Geographic Information Systems Can Help to Enhance Health Service Planning and Delivery for Infectious Diseases in Low-Income Countries. *Journal of Health Care for the Poor and Underserved*, *23*(4), 1410–1420. doi:10.1353/hpu.2012.0146 PMID:23698658

Cancer (Fact sheet N 297). World Health Organization. (n.d.). Retrieved from http://www.who.int/mediacentre/factsheets/fs297/

Coleman, M. P., & Reiter, R. J. (1992). Breast cancer, blindness and melatonin. *European Journal of Cancer, 28*(2-3), 501–503. doi:10.1016/S0959-8049(05)80087-5 PMID:1591073

Delagrange, P., & Guardiola-Lemaitre, B. (1997). Melatonin, its receptors, and relationships with biological rhythm disorders. *Clinical Neuropharmacology, 20*(6), 482–510. doi:10.1097/00002826-199712000-00002 PMID:9403224

Devesa, S. S., Grauman, D. J., Blot, W. J., Pennello, G. A., Hoover, R. N., & Fraumeni, J. F. (1999). *Atlas of cancer mortality in the United States, 1950-94*. Bethesda, MD: National Institutes of Health, National Cancer Institute.

DictionaryG. I. S. ESRI. (n.d.). Retrieved from http://support.esri.com/

Dubey, A. K., Gupta, U., & Jain, S. (2015). Breast cancer statistics and prediction methodology: A systematic review and analysis. *Asian Pacific Journal of Cancer Prevention, 16*(10), 4237–4245. doi:10.7314/APJCP.2015.16.10.4237 PMID:26028079

Eisen, R. J., & Eisen, L. (2014). Use of Geographic Information Systems in Infectious Disease Surveillance. In N. M. M'ikanatha & J. K. Iskander (Eds.), *Concepts and Methods in Infectious Disease Surveillance*. Oxford, UK: John Wiley & Sons, Ltd. doi:10.1002/9781118928646.ch20

Joseph, S. T., DeChello, L. M., Kulldorff, M., Gregorio, D. I., Gershman, S., & Mroszczyk, M. (2004). The geographic distribution of breast cancer incidence in Massachusetts 1988 to 1997, adjusted for covariates. *International Journal of Health Geographics, 3*(1), 17. doi:10.1186/1476-072X-3-17 PMID:15291960

Juckett, D. A., & Rosenberg, B. (1997). Time series analysis supporting the hypothesis that enhanced cosmic radiation during germ cell formation can increase breast cancer mortality in germ cell cohorts. *International Journal of Biometeorology, 40*(4), 206–221. doi:10.1007004840050043 PMID:9225597

Kliukiene, J., Tynes, T., & Andersen, A. (2001). Risk of breast cancer among Norwegian women with visual impairment. *British Journal of Cancer, 84*(3), 397–399. doi:10.1054/bjoc.2000.1617 PMID:11161406

Kulldorff, M., Feuer, E. J., Miller, B. A., & Freedman, L. S. (1997). Breast cancer clusters in the northeast United States: A geographic analysis. *American Journal of Epidemiology, 146*(2), 161–170. doi:10.1093/oxfordjournals.aje.a009247 PMID:9230778

Laden, F., Spiegelman, D., Neas, L. M., Colditz, G. A., Hankinson, S. E., Manson, J. E., ... Hunter, D. J. (1997). Geographic variation in breast cancer incidence rates in a cohort of U.S. women. *Journal of the National Cancer Institute, 89*(18), 1373–1378. doi:10.1093/jnci/89.18.1373 PMID:9308708

Le, G. M., Gomez, S. L., Clarke, C. A., Glaser, S. L., & West, D. W. (2002). Cancer incidence patterns among Vietnamese in the United States and Ha Noi, Vietnam. *International Journal of Cancer, 102*(4), 412–417. doi:10.1002/ijc.10725 PMID:12402312

Lee, K.-S., & Moon, K.-J. (2014). Hospital distribution in a metropolitan city: Assessment by a geographical information system grid modelling approach. *Geospatial Health, 8*(2), 537–544. doi:10.4081/gh.2014.43 PMID:24893031

Lissoni, P., Rovelli, F., Malugani, F., Bucovec, R., Conti, A., & Maestroni, G. J. (2001). Anti-angiogenic activity of melatonin in advanced cancer patients. *Neuroendocrinology Letters*, *22*(1), 45–47. PMID:11335879

Lyseen, A. K., Nohr, C., Sorensen, E. M., Gudes, O., Geraghty, E. M., Shaw, N. T., & Bivona-Tellez, C. (2014). A Review and framework for categorizing current research and development in health related geographical information systems (GIS) studies. *Yearbook of Medical Informatics*, *9*(1), 110–124. doi:10.15265/IY-2014-0008 PMID:25123730

National Cancer Institute. (n.d.). *SEER*Stat version 5.3.1*. Bethesda, MD: National Cancer Institute. Retrieved from http://www.seer.cancer.gov/seerstat

Nykiforuk, C., & Flaman, L. (2011). Geographic information systems (GIS) for health promotion and public health: A review. *Health Promotion Practice*, *12*(1), 63–73. doi:10.1177/1524839909334624 PMID:19546198

Reiter, R. J. (1991). Pineal melatonin: Cell biology of its synthesis and its physiological interactions. *Endocrine Reviews*, *12*(2), 151–180. doi:10.1210/edrv-12-2-151 PMID:1649044

Robert, N. P. (1996). *Cancer wars: How politics shapes what we know and don't know about cancer*. New York: Basic Books.

Roche, L. M. (1998). *Breast cancer in New Jersey, 1979-1995*. New Jersey Dept. of Health and Senior Services, Cancer Surveillance Program, Office of Cancer Epidemiology.

Samel, A., & Wegmann, H. M. (1997). Bright light: A countermeasure for jet lag? *Chronobiology International*, *14*(2), 173–183. doi:10.3109/07420529709001154 PMID:9095377

Siegel, R. L., Miller, K. D., & Jemal, A. (2015). Cancer statistics, 2015. *CA: a Cancer Journal for Clinicians*, *65*(1), 5–29. doi:10.3322/caac.21254 PMID:25559415

Skene, D. J. (2003). Optimization of light and melatonin to phase-shift human circadian rhythms. *Journal of Neuroendocrinology*, *15*(4), 438–441. doi:10.1046/j.1365-2826.2003.01006.x PMID:12622847

Stout, R. W., & Crawford, V. (1991). Seasonal variations in fibrinogen concentration among elderly people. *Lancet*, *338*(8758), 9–13. doi:10.1016/0140-6736(91)90004-9 PMID:1676131

Tazzite, A., Jouhadi, H., Saiss, K., Benider, A., & Nadifi, S. (2013). Relationship between family history of breast cancer and clinicopathological features in Moroccan patients. *Ethiopian Journal of Health Sciences*, *23*(2), 150–157. PMID:23950631

Underwood, H., & Calaban, M. (1987). Pineal melatonin rhythms in the lizard Anolis carolinensis: I. Response to light and temperature cycles. *Journal of Biological Rhythms*, *2*(3), 179–193. doi:10.1177/074873048700200302 PMID:2979659

United States Geography. (n.d.). Retrieved from http://www.netstate.com

WHO Cancer Mortality Database. (n.d.). World Health Organization. Retrieved from http://www-dep.iarc.fr/WHOdb/ WHOdb.htm

Wu, A. H., Ziegler, R. G., Horn-Ross, P. L., Nomura, A. M., West, D. W., Kolonel, L. N., ... Pike, M. C. (1996). Tofu and risk of breast cancer in Asian-Americans. *Cancer Epidemiology, Biomarkers & Prevention*, *5*(11), 901–906. PMID:8922298

ADDITIONAL READING

Barrett, R. K., & Takahashi, J. S. (1995). Temperature compensation and temperature entrainment of the chick pineal cell circadian clock. *The Journal of Neuroscience*, *15*(8), 5681–5692. PMID:7643210

Barrett, R. K., & Takahashi, J. S. (1997). Lability of circadian pacemaker amplitude in chick pineal cells: A temperature-dependent process. *Journal of Biological Rhythms*, *12*(4), 309–318. doi:10.1177/074873049701200403 PMID:9438879

Boulos, Z., Macchi, M. M., Sturchler, M. P., Stewart, K. T., Brainard, G. C., Suhner, A., ... Steffen, R. (2002). Light visor treatment for jet lag after westward travel across six time zones. *Aviation, Space, and Environmental Medicine*, *73*(10), 953–963. PMID:12398256

Choi, W. S., Parker, B. A., Pierce, J. P., & Greenberg, E. R. (1996). Regional differences in the incidence and treatment of carcinoma in situ of the breast. *Cancer Epidemiology, Biomarkers & Prevention*, *5*(4), 317–320. PMID:8722225

Garcia, E., Hurley, S., Nelson, D. O., Hertz, A., & Reynolds, P. (2015). Hazardous air pollutants and breast cancer risk in California teachers: A cohort study. *Environmental Health*, *14*(1), 14. doi:10.1186/1476-069X-14-14 PMID:25636809

Garfinkel, L., Boring, C. C., & Heath, C. W. Jr. (1994). Changing trends: An overview of breast cancer incidence and mortality. *Cancer*, *74*(1Suppl), 222–227. doi:10.1002/cncr.2820741304 PMID:8004590

Keegan, T. H., Chang, E. T., John, E. M., Horn-Ross, P. L., Wrensch, M. R., Glaser, S. L., & Clarke, C. A. (2007). Recent changes in breast cancer incidence and risk factor prevalence in San Francisco Bay area and California women: 1988 to 2004. *Breast Cancer Research*, *9*(5), R62. doi:10.1186/bcr1768 PMID:20210979

Lacey, J. V., Devesa, S. S., & Brinton, L. A. (2002). Recent trends in breast cancer incidence and mortality. *Environmental and Molecular Mutagenesis*, *39*(2-3), 82–88. doi:10.1002/em.10062 PMID:11921173

Medina-Leendertz, S., Valero, N., Chacin-Bonilla, L., Anez, F., Giraldoth, D., Arias, J., ... Bonilla, E. (2001). High intensity light increases olfactory bulb melatonin in Venezuelan equine encephalitis virus infection. *Neurochemical Research*, *26*(3), 231–234. doi:10.1023/A:1010964500370 PMID:11495546

Moyer, R. W., Firth, B. T., & Kennaway, D. J. (1995). Effect of constant temperatures, darkness and light on the secretion of melatonin by pineal explants and retinas in the gecko Christinus marmoratus. *Brain Research*, *675*(1-2), 345–348. doi:10.1016/0006-8993(95)00073-Y PMID:7796151

Nair, I., & Morgan, M. G. (1990). Electromagnetic fields: The jurys still out. 1. Biological effects. *IEEE Spectrum*, *27*(8), 23–27. doi:10.1109/6.58428

Rea, M. S., Bullough, J. D., & Figueiro, M. G. (2002). Phototransduction for human melatonin suppression. *Journal of Pineal Research*, *32*(4), 209–213. doi:10.1034/j.1600-079X.2002.01881.x PMID:11982788

Reiter, R. J., & Richardson, B. A. (1992). Magnetic field effects on pineal indoleamine metabolism and possible biological consequences. *The FASEB Journal*, *6*(6), 2283–2287. PMID:1544540

Reynolds, P., Hurley, S., Goldberg, D. E., Anton-Culver, H., Bernstein, L., Deapen, D., ... Ziogas, A. (2004). Regional variations in breast cancer among california teachers. *Epidemiology (Cambridge, Mass.)*, *15*(6), 746–754. doi:10.1097/01.ede.0000134863.45834.50 PMID:15475725

Sturgeon, S. R., Schairer, C., Gail, M., McAdams, M., Brinton, L. A., & Hoover, R. N. (1995). Geographic variation in mortality from breast cancer among white women in the United States. *Journal of the National Cancer*, *87*(24), 1846–1853. doi:10.1093/jnci/87.24.1846 PMID:7494228

Tosini, G., & Fukuhara, C. (2003). Photic and circadian regulation of retinal melatonin in mammals. *Journal of Neuroendocrinology*, *15*(4), 364–369. doi:10.1046/j.1365-2826.2003.00973.x PMID:12622835

Wang, R. J. (1975). Lethal effect of daylight fluorescent light on human cells in tissue-culture medium. *Photochemistry and Photobiology*, *21*(5), 373–375. doi:10.1111/j.1751-1097.1975.tb06688.x PMID:1208665

Wright, K. P. Jr, Hughes, R. J., Kronauer, R. E., Dijk, D. J., & Czeisler, C. A. (2001). Intrinsic near-24-h pacemaker period determines limits of circadian entrainment to a weak synchronizer in humans. *Proceedings of the National Academy of Sciences of the United States of America*, *98*(24), 14027–14032. doi:10.1073/pnas.201530198 PMID:11717461

KEY TERMS AND DEFINITIONS

Breast Cancer: It is tumor growth that starts in the cells of the breast. The disease occurs almost entirely in women, and it can affect men as well.

Circadian Rhythm: A description of physiological and behavioral patterns altered within 24 hours.

Geographic Information System (GIS): It is defined as computer- established information system that is particularly designed to view and manage information about geographic places, analyze spatial relationships, and model spatial processes.

Incidence and Mortality Rates: Are defined as how many people get and die from breast cancer respectively.

Jet Lag: It is a medical condition which results as consequence of alterations to circadian rhythms due to long hours of travel.

Medical-Based Geographic Information System (mGIS): It is defined as GIS applications and the related medical data for monitoring and analyzing medical risk factors associated with the spatial environmental factors.

Regression Analysis: It is a statistical tool for estimating the relationships between variables.

This research was previously published in the Encyclopedia of Information Science and Technology, Fourth Edition edited by Mehdi Khosrow-Pour, pages 3448-3459, copyright year 2018 by Information Science Reference (an imprint of IGI Global).

Chapter 23
Geographic Information Systems

Paula Remoaldo
University of Minho, Portugal

Vitor P. Ribeiro
University of Minho, Portugal

Hélder Silva Lopes
University of Minho, Portugal

Sara Catarina Gomes Silva
University of Minho, Portugal

ABSTRACT

The emergence of technological systems with computerized cartographic application allowed the resolution of certain military problems between the 1950s and 1960s. The first GIS was created in Canada Geographic Information Systems in 1964 for Tomlinson. At this time, GIS was in a consolidation phase in multiple areas and for various purposes. The geographical science growth with the development of GIS were in connection to the subject of the new geography, justified by the application of the methods of quantitative analysis in various spatial nature of research. In this context, the GIS feature a number of operational advantages and have allowed the proliferation of new fields of endeavor in open access systems across multiple forms of acquisition, management, interpretation, and spatial information analysis. The main goal of this chapter is to underwrite the concept of GIS, as well as distinguish the diversity of applications from the past until the present, and to identify new paths to accommodate recent scientific approaches with extensive range of application possibilities.

INTRODUCTION

One of the main challenges of the 21st century are caused by the large amount of geospatial information through a GIS. Throughout time there have been many attempts to define Geographic Information Systems (GIS). Yet there is no consensus on define it and restrict it to one definition is limited. In the acronym - Geographic Information Systems - the geographic refers to the Earth's surface and near-surface, therefore, all human production and activity, and non-human are possible spatialization in GIS.

DOI: 10.4018/978-1-5225-7359-3.ch023

GIS is recognized as an analytical and decision-making tool with many uses in different fields. Likewise it is used in many industries plus commercial, education or government. It is powerful for

- Land administration,
- Statistical mapping,
- Transport,
- Network and environment management,
- Remote sensing images,
- Water/waste management,
- Maintenance and management of public lighting,
- Regional and urban planning,
- Tourism planning,
- Healthcare planning, and in
- Crime and security management.

In broad terms GIS is a special class of information systems that keep track not only of events, activities, and things, but also their location. Computerization has opened a vast new potential in the way people communicate, analyze our surroundings and take decisions. The available data represent layers of the real world that can be stored, processed and presented later to answer future needs (Bernhardsen, 2002).

In the process of acquisition, processing and spatial representation there is the involvement of a multiplicity of inputs and outputs that can be managed on databases, which invariably seek analytical and graphical spatial embodiments. In the graphical display, vector or raster elements can be chosen, depending on the degree of specificity of the database and the type of expected results.

These databases can be collected at different scales and using a plurality of data types, including population census, aerial photography or satellite imagery. It allows to address multiple operating phases of the planning management process in a multiscale perspective with the challenge to meet more effective and efficient solutions. Due to this, nowadays it is frequently used as a spatial decision support system (SDSS) (Crossland, 2005).

Well-designed GIS should be able to provide a good computer system, because traditional GIS are intended to users operating on local servers. Traditionally GIS includes hardware and software. The hardware are the physical parts of the computer itself and associated peripherals (e.g., plotters and printers); and the software is interoperable, supporting the many data formats (in the infrastructure life cycle) and implementation may be custom-designed for an organization.

Even so a GIS can have two types of groups typically called as "GIS carries" and "GIS users", which are respectively responsible for the management and analysis. The heart of GIS technology is the ability to conduct spatial analysis, overlay data and integrate other solution and systems. Geoprocessing operations facilitate to link or merge data, spatial characteristics of data; search for particular characteristics or features in an area, update data quickly and cheaply and model data assess alternatives (maps, graphs, address lists, reports and summary statistics) tailored to meet particular needs.

Nonetheless GIS feature a number of operational advantages and have allowed the proliferation of new fields of endeavor in open access systems across multiple forms of acquisition, management, interpretation and spatial information analysis. This can be seen in the first item of the present paper where the background and GIS starting point is explored. The main goal of this paper is to underwrite the concept of GIS evolution and to identify new paths to accommodate recent scientific approaches with extensive range of application possibilities.

THEORETICAL BACKGROUND: THE STARTING POINT

GIS is the advent of a new stage of cartography. The evolution of this type of system is relatively recent, between the 50s and 60s of the XX century, but knowledge and technology have grown rapidly recently. The emergence of technological systems with computerized cartographic application arose from the need of the resolution of certain military problems and public administration domains. Many contributors and diverse influences concerning concepts and principles, data and issues of spatial infrastructure, software vendors, application areas, allowed a cohesive growth (Figure 1). GIS organizational structure is as diverse well as the multitude of roots from which it originated multitude of proprietary and public domain GIS software packages (Hendriks, 2005). Nowadays, applicability of this type of systems widened for commercial, non-profit and academic areas.

The mid 1960s witnessed the initial development of GIS in combining spatially referenced data, spatial data models and visualization. The actual roots of GIS are complex and difficult to determine (Miller & Goodchild, 2015).

Most authorities cite the Canada Geographic Information System (CGIS), designed around 1964, with a project led by Tomlinson (Bruno & Giannikos, 2015; Mordechai et al., 2008; Tomlinson, 1967). The objective was to obtain means for summaries and tabulations of areas of land from the Canada Land

Figure 1. Timeline of major GIS events, Source: Authors' own elaboration

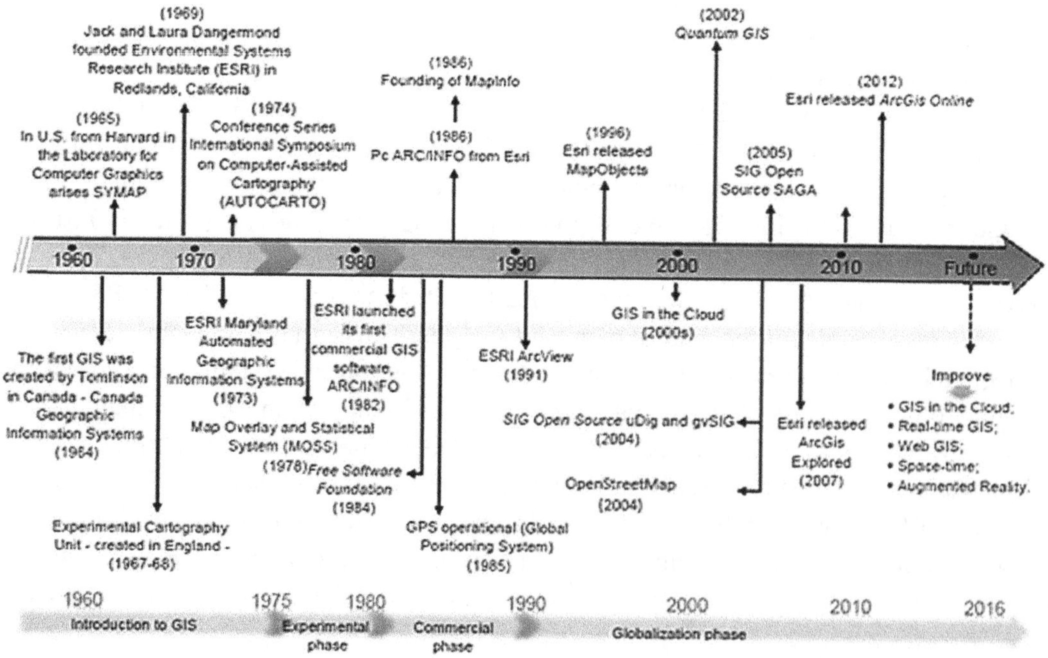

Inventory. For the registration of this lands, was made a massive federal-provincial effort to assess the utilization and potential of the Canadian land base. CGIS arose from the need to answer the challenges to measure accurately the areas of irregular geographic patches of homogeneous utilization and to overlay/ compare different themes (Goodchild, 2006).

The period of the 1970s was characterized by rapid evolution, and ability of computer mapping automatic using data format and the solution of a wide range of technical issues. In the 1980s, democratization of access to computer allowed expand use of GIS. These innovations led to the first commercial viability of GIS, started to become popular as a standard computer application in government departments, universities, and private corporations. Accordingly, the ability to select, sort, extract, classify and display geographic data on the basis of complex topological and statistical criteria was available to users (Goodchild, 2006; Pourabbas, 2014).

The 1990s saw map analysis and modeling advances in GIS, and these systems became real management information tools as computing power increased. During this decade, the Open GIS Consortium (OGC), aimed at developing publicly available geoprocessing specific actions, was founded. OGC is an international industry consortium, including government agencies and universities.

In 2000, the advent of web 2.0 and, more recently, the Web 3.0 (the semantic web) allow the Open Source GIS growth. Open source is a software that allows code source access to be open and free to distribute and modify. Nowadays it is possible for programmers to add new functions very rapidly and cheaply due to recent GIS advances. Mobile and internet devices, cloud computing, NoSQL databases, Semantic Web, Web services offer new ways of accessing, analyzing, and elaborating geospatial information in both real-world and virtual spaces, both for open source and commercial GIS (Pourabbas, 2014).

Despite the progress made during the twenty-first century, there are numerous challenges still lay ahead for geospatial sciences in various fields as shown in Table 1 (Yue et al., 2013).

DATA ANALYSIS AND TREATMENT SYSTEMS

Common citizens are constantly interacting with spatial dimension. Thereby, a GIS can be defined as a computer system operated by people, which comprises different aspects to be efficiently operated. Firstly, a GIS have the hardware, software and data components. Those multidimensional components can be articulated to give us the basis to develop spatial analysis. However human interaction is crucial to develop a conceptual model approach, plan, operate and analyze the information. Nowadays a GIS

Table 1. The challenges for geospatial sciences for types of data intensity

Intensity	Description
Data	Collect of a multitude of data from space by day and accumulation at a similarly high rate.
Processing	Intensive modes of processing information in different spatiotemporal spectra.
Competition	Action of a multitude of end users accessing parallel to information (reception of a large number of users simultaneously are possible, because of development of the several services (*e.g.*, Google Maps and Bing Maps).
Spatiotemporal	Set of spatial and temporal dimension. It can be distinguished into two types of information – dynamic or static.

Source: (Adapted from Yue et al., 2013.)

can be very helpful to different enterprises size, organizations and persons where geographic patterns can be modelled and predicted. Spatial/Geographical data is representing real world through layers or objects where spatial position is crucial. Typically geographical data have descriptive or spatial information (Faiz & Krichen, 2012). It can be used to represent discrete data, typically through a vector-based representation (points, line polygons), or as a continuous data through a cell-based or raster mode that uses a matrix representation.

The most interesting part of a GIS project is the Spatial analysis (Heywood et al., 2011). It is related with the capability to visual analysis of maps and imagery, computational analysis of geographic patterns, finding optimum routes, site selection, and advanced predictive modeling (ESRI, 2013:6). The world is complex, but the exponential growth of technologies, such as Global Position Systems (GPS), real time-sensors or GIS made possible there simplification (ESRI, 2013). GIS are efficient tools for recording, exploiting, analyzing and displaying geographical data that can be applied in:

- Transportation,
- Health, environmental,
- Urbanism,
- Political activities,
- Water/waste management,
- Geomarketing,
- Security,
- Tourism,
- Viticulture/enology,
- Education or
- Crime.

This broad kind of applications covering private and public sectors are growing exponentially. Open-source and proprietary software development have been contributing to this development due to its recent growth and attention.

Aside the raster and vector in spatial modelling more recently three and four-dimensional data is being investigated (Lin et al., 2013). Virtual Geographic Environments (VGEs) interest are growing faster in last years. It is characterized as being a bridge between the three scientific requirements of Geographic Information Science: multi-dimensional visualization, dynamic phenomenon simulation, and public participation (Lin et al., 2013).

GIS TRENDS

Currently, the key trends that face in GIS concerns are geospatial web, GIS in the cloud, space-time GIS, augmented reality and real-time GIS. Web-based in GIS combines information systems and geographic web technology (Chakraborty et al., 2015). Web GIS is a paradigm shift from a model based on national governments as major players in the production of geospatial data to one based on the collaboration between citizens and private institutions (Goodchild et al., 2007; Grossner et al., 2008). The enhanced participation of different actors in the generation of geospatial data, it is increasingly difficult to distinguish the producers and users (Budhathoki & Nedovic-Budic, 2008), mainly due to the free software and open source tools availability (Crampton, 2009). Due to the opening of the GIS world to a multitude of actors involved in the mapping data and Web GIS tools, refers to ads to meet end-user needs (Elwood, 2009). GIS based on the web are accessible not just from a computer but also on different multiple devices, including laptops, smartphones or tablets (Chakraborty, 2015).

The opening of the source code and the use of free software aims to contribute to greater openness to collective voluntary participation in the use, study and modification of the software (Chakraborty et al., 2015). In the beginning of the 21st century started an innovative and leading geolocation-based service of crowdsourcing at a massive scale named as Open Street Map project (OSM). This project improves Volunteered Geographical Information' and aims to create a free digital map of the world. Those collaborative platforms are empowering citizens to create a global patchwork of geographic information (Goodchild, 2007; Mordechai Haklay, 2010). The international non for profit Open Geospatial Consortium was founded in 1994. It is an international voluntary organization that led the standards development process for geospatial and location services (Haklay et al., 2008). Among the three most relevant standards OGC include: Web Feature Service (WFS), Web Map Service (WMS) and Web Coverage Service (WCS) (Giuliani et al., 2016; Parker & Dominguez, 2015).

This new forms of utilisation of GIS on the web environment using a distributed and asynchronous requires a client-server architecture (C/S). This is characterized by a client request of a service, such as mapping, decision analysis, data processing or storage data while the server provide the service (Mekonnen & Gorsevski, 2015). Web GIS exits benefited by providing the best agents solutions to problems that traditional GIS (Chang & Park, 2006). More recently Web-GIS has become to a cloud GIS, based in model of "Software as a Service" (SaaS) (Kerski, 2015).

Cloud computing are increasingly widespread and make possible to run cloud applications in a shared data center accessed by internet. The emergence of GIS in the cloud solved the problems associated to the increase of precision and spatial-temporal scope of information. In general, there are an accumulation of multitude data records and this dataset has varied at a daily rate (Hey, 2012) and allows network access to a set of configurable data (servers, storage, applications and services) (Yang et al., 2011). The recent emergence cloud GIS provides the ability to build a GIS service enabled for use in the cloud and can be made to scale up or down according to user needs. GIS cloud is equipped with new models of maintenance and use of geospatial data for a variety of users and to solve computing problems (Yu et al., 2014). This service provides users the ability to act in the manner of 'pay-as-you-go'. This mechanism of action has been a dream for several decades and has recently become a reality (Armbrust et al., 2010).

The cloud GIS has several characteristics namely: (i) it is not necessary to provide a software installation; (ii) not to use the computer's internal space for data storage; (iii) enable a collaborative action between different actors (flexibility); (iv) adapting services to demand and actual charge; (v) enhancing greater interoperability between various source code; (vi) decrease the time taken by decision-makers to implement processes of deliberation and (vii) the implementation of the entire system in a top-down scale (Armbrust et al., 2010; Blower, 2010; Yu et al., 2014). Cloud computing is a powerful technology that enables greater profitability in energy consumption and economic resources (Buyya et al., 2009; Lee & Chen, 2010; Marston et al., 2011). It perform massive-scale and complex computing and eliminate the need of maintain software, hardware or dedicated space (Assunção et al., 2015; Hashem et al., 2015).

Associated to this is the massive growth of generated data scale and volume which is challenging and time demanding tasks to ensure data processing, analysis and store (Hashem et al., 2015). Concerning to this Big Data concept is emerging and specified the four Vs, namely,

- Volume of data,
- Variety of data collectors,
- Velocity of data transfers, and
- Value process of discovering huge hidden values from large datasets (Gantz & Reinsel, 2011; Hashem et al., 2015).

The advancement of these technologies can enable the construction of spatial data infrastructures (SDI) and cyberinfrastructures (Schäffer et al., 2010; Yang et al., 2010). Some public cloud computing platforms are already available, including Microsoft Azure, Google App Engine or Amazon EC2. In any case a cloud can be public or private. The public cloud is available to the public while the private cloud is only used within an organization.

Cloud computing includes multiple domains, such as energy and mineral sciences, weather, traffic and simulation management systems, landscape ecology, water management, disaster management or human and environmental health (Yang et al., 2013). According to these authors the main obstacles to the success of cloud GIS are associated with policy, management, acquisition and operational requirements. Forward-looking multiple threads are identified:

- Cumulative advances for interdisciplinary approach, abreast of progress in geoscience and digital earth;
- Cloud interoperability based on standards (OGC, OGF, NIST, ISO, IEEE) and through a systemic architecture;
- Integration of innovative interaction systems for viewing and access;
- Real-time simulation for decision support;
- Security levels set to a deployed computer with a platform with distribution of certain information and the collaboration by integrating multiple platforms, and in view of the scope of science for the citizen, technology to crowdsourcing, dynamic events and challenges of education.

Another type of evolution in GIS is linked of framework of Hägerstrand (space-time model). In this context, the space-time studies the individual patterns, considering the various constraints in a particular spatial-temporal environment (Hägerstraand, 1970; Hägerstrand, 1989). There have been a number of efforts to ensure the incorporation of concepts in a GIS (e.g., Goodchild, 2013; Miller, 1991; Neutens et al., 2007; Shaw & Yu, 2009). This system presents a three dimensional orthogonal structure that consists of the union between two spatial dimensions and a temporal dimension. The spatial dimensions, structured 2D scale, represent the location of individuals, while the dimension of time represents the timing of the individual movement in a spatiotemporal system (Miller, 2004).

Several variables can represent the characteristics of the daily activity of an individual: location, time, duration, sequence and frequency of the type of activities (Ren & Kwan, 2009). It must be associated with at least one activity. Distinguished two types of activities: the movable and stationary activity. Mobile activity refers to a local motion toward another, while the stationary activity leads to a fixed location. Representation in Hägerstrand system is done in two ways: by vertical line segment when it comes to a stationary activity and a sloping straight line when there is movement toward a certain place (Chen et al., 2011).

In recent years, attempts have been incremented to store and manage the activities of an individual based on their spatial and temporal characteristics (Chen et al., 2013; Wang & Cheng, 2001). This type of analysis has been mainly used in studies that assess individual accessibility (e.g., Delafontaine et al., 2011; Kwan, 1998; Miller, 1999; Neutens, 2015; Neutens et al., 2007). These studies demonstrate that the considerations of space-time contribute to the presentation of more complex models and real human activities (Shaw & Yu, 2009).

Real-Time GIS model was assumed as a new paradigm of information science to capture the real characteristics of human undertakings (Hey, 2012) and transforming historical changed data to real-time data (Gong et al., 2015). The authors divided the model into three stages: temporal snapshots (1st stage), object-shift (2nd stage) and events and action (3rd stage) (Figure 2).

Real-Time GIS analyses have also sought to incorporate collaborative functions. These types of tools can be differentiated into several types, namely:

- The same time - same place;
- Same time - different location;
- Different time - same place;
- Different time - different location (Sun & Li, 2015).

The Real-Time Collaborative Geographic Information System (RCGIS) enables analysis of interactions in agile and flexible systems, equipped with collaborative principles.

The unpreceded growth of geographically referenced information combined with recent digital augmentation reality (AR) of places growth will become increasingly important in the future (Graham et al., 2013). AR had its recognition in 1992, when Caudell and Mizzel developed works for Boeing and designed a digital transport display in the head, so as to enable a framework of airplane schemes (Yew et al., 2016). Yet, the concept of AR is much older. This was used during the period of World War II with a project developed with the presentation of information on the windshield of the camera. In this statement, there must be a relationship between reality and the information made available in digital media. To this end, there must be a technological device (smartphone or other wireless equipment), tracking and computer software. AR is summarized in three distinct properties:

1. Combination of real and virtual objects in the real world;
2. Run interactively in real time; and
3. Registration of real and virtual objects and their connection (Azuma, 1997).

Figure 2. Three stages of real-time models
Source: (Adapted from Gong et al., 2015).

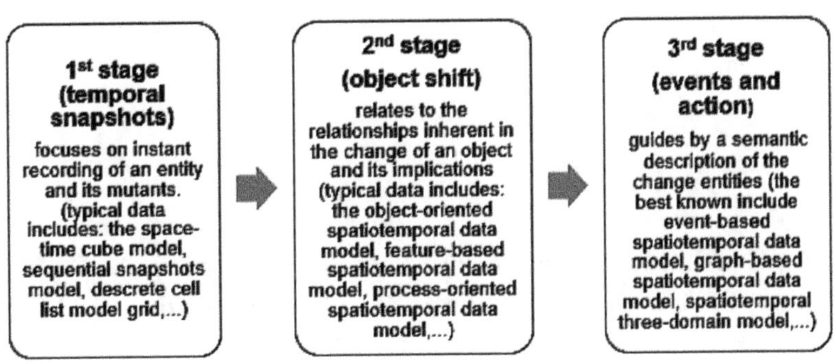

AR is receiving several applications for PC, smartphones, tablet and other devices and will be increasing in the future.

These advances in various types of GIS technologies can create significant digital divides, disadvantaging the poor, ethnic and racial minorities, residents in rural areas, the inhabitants of the global South. The own readiness for open apps coding creates exclusions face to those who can only view and contribute to these features (Elwood, 2009).

FUTURE RESEARCH DIRECTIONS

Information technology and Geographic Science are growing abreast and rapidly. Developments of GIS technology and applications must grow behind the scope of Big Data, Cloud GIS, Real-time GIS and Augmented Reality challenges. For future research, a more rigorous approach should be implemented and guided by technology interoperability, integrated multidisciplinary approach, security, how to benefit and integrate GIS database with citizen-collected data and deeply understand how to collect and analyse real time data. The advent of data needed will be exceeded by the abundance of real time data arising the challenges on how to detach different sources to canvass their quality and include them in spatial analysis. For this more deeply and integrated analysis models are required and at the same time more spread applications will be required to provide the spatial information by multiple technological devices.

CONCLUSION

The expansion and advances of GIS technologies created conditions to proliferate different approaches to work in multiple areas, such as geography, cartography, remote sensing, image processing, education or environmental sciences. This section presents the evolution of GIS from its conception to the present. Moreover, it shows the main challenges posed by multiple skills acquired by the GIS in recent decades, namely presenting a shift from traditionally confined public planning areas to multi collaborate users, from desktop to the web and from real to virtual and augmented reality. In fact, maps have always been used for the removal of political borders, but today the networking capabilities generated conditions for GIS intelligence statement. The development of the web has supported the new challenges associated with GIS, particularly in areas connected with augmented reality, the real-time GIS information, GIS space-time or Cloud GIS. Associated to this arise Big Data challenges and problems to the Geographic Science (GS). The cumulative advances in the relationship between GIS and Web can hence contribute to the expansion of information generated and treated, interoperability between servers and users and networking. Geographic information will reinforce their position in our day's life and for GS those trends will catch much more research than ever.

REFERENCES

Armbrust, M., Fox, A., Griffith, R., Joseph, A. D., Katz, R., Konwinski, A., ... Zaharia, M. (2010). A view of cloud computing. *Communications of the ACM*, *53*(4), 50–58. doi:10.1145/1721654.1721672

Assunção, M. D., Calheiros, R. N., Bianchi, S., Netto, M. A. S., & Buyya, R. (2015). Big Data computing and clouds: Trends and future directions. *Journal of Parallel and Distributed Computing*, *79–80*, 3–15. doi:10.1016/j.jpdc.2014.08.003

Azuma, R. T. (1997). A survey of augmented reality. *Presence (Cambridge, Mass.)*, *6*(4), 355–385. doi:10.1162/pres.1997.6.4.355

Bernhardsen, T. (2002). *Geographic information systems: an introduction* (3rd ed.). New York: John Wiley & Sons.

Blower, J. D. (2010). *GIS in the cloud: implementing a web map service on Google App Engine.* Paper presented at the 1st International Conference and Exhibition on Computing for Geospatial Research & Application. 10.1145/1823854.1823893

Bruno, G., & Giannikos, I. (2015). *Location and GIS. In Location Science* (pp. 509–536). Springer.

Budhathoki, N. R., & Nedovic-Budic, Z. (2008). Reconceptualizing the role of the user of spatial data infrastructure. *GeoJournal*, *72*(3-4), 149–160. doi:10.100710708-008-9189-x

Buyya, R., Yeo, C. S., Venugopal, S., Broberg, J., & Brandic, I. (2009). Cloud computing and emerging IT platforms: Vision, hype, and reality for delivering computing as the 5th utility. *Future Generation Computer Systems*, *25*(6), 599–616. doi:10.1016/j.future.2008.12.001

Chakraborty, D., Sarkar, D., Agarwal, S., Dutta, D., & Sharma, J. R. (2015). Web Based GIS Application using Open Source Software for Sharing Geospatial Data. *International Journal of Advanced Remote Sensing and GIS*, *4*(1), 1224–1228.

Chang, Y. S., & Park, H. D. (2006). XML Web Service-based development model for Internet GIS applications. *International Journal of Geographical Information Science*, *20*(4), 371–399. doi:10.1080/13658810600607857

Chen, B. Y., Li, Q., Wang, D., Shaw, S.-L., Lam, W. H., Yuan, H., & Fang, Z. (2013). Reliable space–time prisms under travel time uncertainty. *Annals of the Association of American Geographers*, *103*(6), 1502–1521. doi:10.1080/00045608.2013.834236

Chen, J., Shaw, S.-L., Yu, H., Lu, F., Chai, Y., & Jia, Q. (2011). Exploratory data analysis of activity diary data: A space–time GIS approach. *Journal of Transport Geography*, *19*(3), 394–404. doi:10.1016/j.jtrangeo.2010.11.002

Crampton, J. W. (2009). Cartography: Maps 2.0. *Progress in Human Geography*, *33*(1), 91–100. doi:10.1177/0309132508094074

Crossland, M. D. (2005). Geographical Information Systems as Decision Tools. In M. Khosrow-Pour (Ed.), *Encyclopedia of Information Science and Technology* (1st ed.; Vol. 2, pp. 1274–1277). Idea Group Reference. doi:10.4018/978-1-59140-553-5.ch224

Delafontaine, M., Neutens, T., Schwanen, T., & Van de Weghe, N. (2011). The impact of opening hours on the equity of individual space–time accessibility. *Computers, Environment and Urban Systems*, *35*(4), 276–288. doi:10.1016/j.compenvurbsys.2011.02.005

Elwood, S. (2009). Integrating participatory action research and GIS education: Negotiating methodologies, politics and technologies. *Journal of Geography in Higher Education, 33*(1), 51–65. doi:10.1080/03098260802276565

ESRI. (2013). *The language of spatial analysis*. New York: ESRI Press.

Faiz, S., & Krichen, S. (2012). *Geographical information systems and spatial optimization*. CRC Press.

Gantz, J., & Reinsel, D. (2011). Extracting value from chaos. *IDC iview, 1142*, 1-12.

Giuliani, G., Guigoz, Y., Lacroix, P., Ray, N., & Lehmann, A. (2016). Facilitating the production of ISO-compliant metadata of geospatial datasets. *International Journal of Applied Earth Observation and Geoinformation, 44*, 239–243. doi:10.1016/j.jag.2015.08.010

Gong, J., Geng, J., & Chen, Z. (2015). Real-time GIS data model and sensor web service platform for environmental data management. *International Journal of Health Geographics, 14*(1), 1. doi:10.1186/1476-072X-14-2 PMID:25572659

Goodchild, M. (2006). Geographic Information Systems. In S. Aitken & G. Valentine (Eds.), *Approaches to Human Geography* (Vol. 251-262). SAGE Publications. doi:10.4135/9781446215432.n24

Goodchild, M. (2007). Citizens as sensors: The world of volunteered geography. *GeoJournal, 69*(4), 211–221. doi:10.100710708-007-9111-y

Goodchild, M. F. (2013). Prospects for a space–time GIS: Space–time integration in geography and GIScience. *Annals of the Association of American Geographers, 103*(5), 1072–1077. doi:10.1080/00045608.2013.792175

Goodchild, M. F., Yuan, M., & Cova, T. J. (2007). Towards a general theory of geographic representation in GIS. *International Journal of Geographical Information Science, 21*(3), 239–260. doi:10.1080/13658810600965271

Graham, M., Zook, M., & Boulton, A. (2013). Augmented reality in urban places: Contested content and the duplicity of code. *Transactions of the Institute of British Geographers, 38*(3), 464–479. doi:10.1111/j.1475-5661.2012.00539.x

Grossner, K. E., Goodchild, M. F., & Clarke, K. C. (2008). Defining a digital earth system. *Transactions in GIS, 12*(1), 145–160. doi:10.1111/j.1467-9671.2008.01090.x

Hägerstraand, T. (1970). What about people in regional science? *Papers in Regional Science, 24*(1), 7–24. doi:10.1111/j.1435-5597.1970.tb01464.x

Hägerstrand, T. (1989). *Reflections on "what about people in regional science?"*. Paper presented at the Regional Science Association. 10.1111/j.1435-5597.1989.tb01166.x

Haklay, M. (2010). How Good is Volunteered Geographical Information? A Comparative Study of OpenStreetMap and Ordnance Survey Datasets. *Environment and Planning. B, Planning & Design, 37*(4), 682–703. doi:10.1068/b35097

Haklay, M., Singleton, A., & Parker, C. (2008). Web mapping 2.0: The neogeography of the GeoWeb. *Geography Compass, 2*(6), 2011–2039. doi:10.1111/j.1749-8198.2008.00167.x

Haklay, M., & Zafiri, A. (2008). Usability engineering for GIS: Learning from a screenshot. *The Cartographic Journal*, *45*(2), 87–97. doi:10.1179/174327708X305085

Hashem, I. A. T., Yaqoob, I., Anuar, N. B., Mokhtar, S., Gani, A., & Ullah Khan, S. (2015). The rise of big data on cloud computing: Review and open research issues. *Information Systems*, *47*, 98–115. doi:10.1016/j.is.2014.07.006

Hendriks, P. H. (2005). Space opera-GIS basics. In M. Khosrouw-Pour (Ed.), *Encyclopedia of Information Science and Technology* (Vol. 5, pp. 2571–2575). Idea Group Reference. doi:10.4018/978-1-59140-553-5.ch456

Hey, T. (2012). *The Fourth Paradigm–Data-Intensive Scientific Discovery. In E-Science and Information Management* (pp. 1–1). Springer.

Heywood, I., Cornelius, S., & Carver, S. (2011). *An Introduction to geographical Information systems* (4th ed.). Essex, UK: Pearson Education.

Kerski, J. J. (2015). Geo-awareness, Geo-enablement, Geotechnologies, Citizen Science, and Storytelling: Geography on the World Stage. *Geography Compass*, *9*(1), 14–26. doi:10.1111/gec3.12193

Kwan, M.-P. (1998). Space-time and integral measures of individual accessibility: A comparative analysis using a point-based framework. *Geographical Analysis*, *30*(3), 191–216. doi:10.1111/j.1538-4632.1998.tb00396.x

Lee, Y.-T., & Chen, K.-T. (2010). *Is server consolidation beneficial to MMORPG? A case study of World of Warcraft.* Paper presented at the Cloud Computing (CLOUD), 2010 IEEE 3rd International Conference. 10.1109/CLOUD.2010.57

Lin, H., Chen, M., Lu, G., Zhu, Q., Gong, J., You, X., ... Hu, M. (2013). Virtual Geographic Environments (VGEs): A New Generation of Geographic Analysis Tool. *Earth-Science Reviews*, *126*, 74–84. doi:10.1016/j.earscirev.2013.08.001

Marston, S., Li, Z., Bandyopadhyay, S., Zhang, J., & Ghalsasi, A. (2011). Cloud computing—The business perspective. *Decision Support Systems*, *51*(1), 176–189. doi:10.1016/j.dss.2010.12.006

Mekonnen, A. D., & Gorsevski, P. V. (2015). A web-based participatory GIS (PGIS) for offshore wind farm suitability within Lake Erie, Ohio. *Renewable & Sustainable Energy Reviews*, *41*, 162–177. doi:10.1016/j.rser.2014.08.030

Miller, H. J. (1991). Modelling accessibility using space-time prism concepts within geographical information systems. *International Journal of Geographical Information System*, *5*(3), 287–301. doi:10.1080/02693799108927856

Miller, H. J. (1999). Measuring space-time accessibility benefits within transportation networks: Basic theory and computational procedures. *Geographical Analysis*, *31*(1), 1–26. doi:10.1111/j.1538-4632.1999.tb00408.x

Miller, H. J. (2004). Toblers first law and spatial analysis. *Annals of the Association of American Geographers, 94*(2), 284–289. doi:10.1111/j.1467-8306.2004.09402005.x

Miller, H. J., & Goodchild, M. F. (2015). Data-driven geography. *GeoJournal, 80*(4), 449–461. doi:10.100710708-014-9602-6

Neutens, T. (2015). Accessibility, equity and health care: Review and research directions for transport geographers. *Journal of Transport Geography, 43*, 14–27. doi:10.1016/j.jtrangeo.2014.12.006

Neutens, T., Witlox, F., Van De Weghe, N., & De Maeyer, P. (2007). Space–time opportunities for multiple agents: A constraint-based approach. *International Journal of Geographical Information Science, 21*(10), 1061–1076. doi:10.1080/13658810601169873

Parker, R., & Dominguez, J. (2015). Features to Look for in a GIS Viewer. *GeoInformatics, 18*(4), 24.

Pourabbas, E. (2014). *Geographical Information Systems: Trends and Technologies*. Boca Raton, FL: CRC Press. doi:10.1201/b16871

Ren, F., & Kwan, M.-P. (2009). The impact of the Internet on human activity–travel patterns: Analysis of gender differences using multi-group structural equation models. *Journal of Transport Geography, 17*(6), 440–450. doi:10.1016/j.jtrangeo.2008.11.003

Schäffer, B., Baranski, B., & Foerster, T. (2010). *Towards spatial data infrastructures in the clouds. In Geospatial Thinking* (pp. 399–418). Springer.

Shaw, S.-L., & Yu, H. (2009). A GIS-based time-geographic approach of studying individual activities and interactions in a hybrid physical–virtual space. *Journal of Transport Geography, 17*(2), 141–149. doi:10.1016/j.jtrangeo.2008.11.012

Sun, Y., & Li, S. (2015). Real-time collaborative GIS: A technological review. *ISPRS Journal of Photogrammetry and Remote Sensing.*

Tomlinson, R. (1967). *An introduction to the geo. information system ofthe Canada Land Inventory.* Ottawa, Department ofForestry and Rural Development, ARDA.

Wang, D., & Cheng, T. (2001). A spatio-temporal data model for activity-based transport demand modelling. *International Journal of Geographical Information Science, 15*(6), 561–585. doi:10.1080/13658810110046934

Yang, C., Goodchild, M., Huang, Q., Nebert, D., Raskin, R., Xu, Y., ... Fay, D. (2011). Spatial cloud computing: How can the geospatial sciences use and help shape cloud computing? *International Journal of Digital Earth, 4*(4), 305–329. doi:10.1080/17538947.2011.587547

Yang, C., Raskin, R., Goodchild, M., & Gahegan, M. (2010). Geospatial cyberinfrastructure: Past, present and future. *Computers, Environment and Urban Systems, 34*(4), 264–277. doi:10.1016/j.compenvurbsys.2010.04.001

Yang, C., Xu, Y., & Nebert, D. (2013). Redefining the possibility of digital Earth and geosciences with spatial cloud computing. *International Journal of Digital Earth, 6*(4), 297–312. doi:10.1080/17538947.2013.769783

Yew, A., Ong, S., & Nee, A. (2016). Towards a griddable distributed manufacturing system with augmented reality interfaces. *Robotics and Computer-integrated Manufacturing*, *39*, 43–55. doi:10.1016/j.rcim.2015.12.002

Yu, G., Zhang, S., Yu, Q., Fan, Y., Zeng, Q., Wu, L., & Zhao, P. (2014). Assessing ecological security at the watershed scale based on RS/GIS: A case study from the Hanjiang River Basin. *Stochastic Environmental Research and Risk Assessment*, *28*(2), 307–318. doi:10.100700477-013-0750-x

Yu, M., Fu, P., Zhou, N., & Xia, J. (2014). ArcGIS in the cloud. In Z. Li, C. Xu, & K. Liu (Eds.), *Spatial Cloud Coputing. A practical approach*. Boca Raton, FL: CRC Press.

Yue, P., Zhou, H., Gong, J., & Hu, L. (2013). Geoprocessing in Cloud Computing platforms–a comparative analysis. *International Journal of Digital Earth*, *6*(4), 404–425. doi:10.1080/17538947.2012.748847

ADDITIONAL READING

Baker, T. R., Battersby, S., Bednarz, S. W., Bodzin, A. M., Kolvoord, B., Moore, S., ... Uttal, D. (2015). A research agenda for geospatial technologies and learning. *The Journal of Geography*, *114*(3), 118–130. doi:10.1080/00221341.2014.950684

Bhat, M. A., Shah, R. M., & Ahmad, B. (2011). Cloud Computing: A solution to Geographical Information Systems (GIS). *International Journal on Computer Science and Engineering*, *3*(2), 594–600.

Bodenhamer, D. J. (2013). Beyond GIS: Geospatial technologies and the future of history. In *History and GIS* (pp. 1–13). Netherlands: Springer. doi:10.1007/978-94-007-5009-8_1

Chakraborty, D., Sarkar, D., Agarwal, S., Dutta, D., & Sharma, J. R. (2015). Web Based GIS Application using Open Source Software for Sharing Geospatial Data. *International Journal of Advanced Remote Sensing and GIS*, *4*(1), 1224–1228.

Chen, N., Zhang, X., & Wang, C. (2015). Integrated open geospatial web service enabled cyber-physical information infrastructure for precision agriculture monitoring. *Computers and Electronics in Agriculture*, *111*, 78–91. doi:10.1016/j.compag.2014.12.009

Crampton, J. W. (2009). Cartography: Maps 2.0. *Progress in Human Geography*, *33*(1), 91–100. doi:10.1177/0309132508094074

Crampton, J. W., Graham, M., Poorthuis, A., Shelton, T., Stephens, M., Wilson, M. W., & Zook, M. (2013). Beyond the geotag: Situating 'big dataand leveraging the potential of the geoweb. *Cartography and Geographic Information Science*, *40*(2), 130–139. doi:10.1080/15230406.2013.777137

Gitis, V., Derendyaev, A., & Weinstock, A. (2016). Web-based GIS Technologies for Monitoring and Analysis of Spatio-Temporal Processes. *International Journal of Web Information Systems*, *12*(1), 102–124. doi:10.1108/IJWIS-10-2015-0032

Hedley, M. L., Templin, M. A., Czajkowski, K., & Czerniak, C. (2013). The use of geospatial technologies instruction within a student/teacher/scientist partnership: Increasing students geospatial skills and atmospheric concept knowledge. *Journal of Geoscience Education*, *61*(1), 161–169. doi:10.5408/11-237.1

Yang, C., Wong, D. W., Yang, R., Kafatos, M., & Li, Q. (2005). Performance-improving techniques in web-based GIS. *International Journal of Geographical Information Science*, *19*(3), 319–342. doi:10.1080/13658810412331280202

KEY TERMS AND DEFINITIONS

Augmented Reality: Augmented Reality (AR) into GIS assures the link between the perception of user and the relationship with the real world. The real world is represented with 2D and 3D virtual information. The computer augments the actual landscape with additional information that can be supported by inserting fields based on GIS applications.

Crowdsourcing: Crowdsourcing is an act of performing a GIS task by a user on a voluntary basis for a set of users. This type of action is based on a bottom-up approach. It is associated with the creation of data through a group dynamic. The crowdsourced data collection is carried out using portable devices (GPS, PC, mobile phones) and the data is synchronized in the central database, accessible and shareable, based on services and maps on the web.

GIS in the Cloud: GIS in the cloud or cloud GIS bases on the integrated web systems. The data generates maps to support the analysis and optimization of real time operations. The cloud integration helps the organization of complex workflows and maintaining extensive geodatabases.

GIS: GIS is a system that permits to visualize, analyze, display and understand the relationships between spatial phenomena. Nowadays, GIS is capable to transform large numbers of data, to analyze and transform momentarily alternate data and generate charts, graphs, summary and descriptive statistics. Among the main key elements to a noble GIS it is noted: computer hardware and software, operational context (people and organizations) and internet service.

Real-Time: Real-time is a term often used to describe the time of execution of a task. This tool helps the users to obtain a frequently monitoring and more efficiently. GIS technology enables the sharing of a series of real-time data. Among the main features is visualization, analysis and understanding of phenomena in reduced timescale.

Pace-Time: Space-time is suggested in Hägerstrand time-geographic framework. Presents and analyses the individual activities in time and space dimensions. In GIS environment results in the spatial representation of the dimension x and y and temporal dimensions of time in hours, minutes or seconds. The space-time patch is used for the implementation of the daily trajectory of the individual in time and space.

Web-Based GIS: Web-based GIS is based on a type of distributed information. This set of technological services is part of a communication structure between the GIS server and the client. Their relationship is expressed through URLs (created by the server) and HTTP (for the customer). Spatial data access, advanced mapping and spatial analysis are the most common type of analysis options in Web-based GIS.

This research was previously published in the Encyclopedia of Information Science and Technology, Fourth Edition edited by Mehdi Khosrow-Pour, pages 3460-3472, copyright year 2018 by Information Science Reference (an imprint of IGI Global).

Chapter 24
Geospatial Influence in Science Mapping

Carlos Granell-Canut
Universitat Jaume I of Castellón, Spain

Estefanía Aguilar-Moreno
Universitat Jaume I of Castellón, Spain

ABSTRACT

The chapter aims at drawing attention to the possibilities that geospatial technologies can bring to science mapping. The chapter briefly distinguishes the notion of mapping between the geospatial information science (GIScience) and librarianship and information science (LIS). Afterwards, an overview of recent initiatives and research work relative to (geospatial) mapping of science is presented. Based on these examples, opportunities and challenges of applying geospatial technology to science mapping are discussed. Finally, based on relevant while evolving geospatial technologies, the next steps for increasing the influence of geospatial technology in science mapping are pointed out.

INTRODUCTION

Sharing and making research data publicly available are increasingly getting attention to academia and research policy agenda. According to a 2014 European Commission guidelines for the re-use of datasets (OJEU C/240-1), geospatial data is of most importance. It raises an invaluable opportunity for libraries to play a dominant role in the not-so-distant-future for managing large collections of open (geospatial) research data. However, the geospatial dimension goes beyond data itself and embraces a wide range of spatial analysis and techniques (Smith et al., 2015). In particular, mapping and visualization techniques of geospatial data may provide endless opportunities to libraries and information science researchers in the sense of exploring the most of large open research datasets from a new perspective. With exceptions, librarians and information science professionals miss an overall perception of the possibilities that geospatial data and tools may bring them to geographically explore, analyze, and mapping research datasets and, especially, science related data. The question we pose in this chapter is whether or not geospatial technologies and mapping techniques have a role in the know–how of librarians and how these

DOI: 10.4018/978-1-5225-7359-3.ch024

technologies and techniques may influence science mapping. Furthermore the chapter aims at drawing attention to the opportunities that the geospatial dimension applied to science related data can bring to the field of science mapping.

In what follows, the chapter briefly distinguishes the notion of mapping between the Geospatial Information Science (GIScience) and Librarianship and Information Science (LIS). Afterwards, an overview about recent initiatives and research work relative to (geospatial) mapping of science is presented. Based on these examples, opportunities and challenges of applying geospatial technology to science mapping are discussed. Finally, based on relevant while evolving geospatial technologies, next steps for increasing up the influence of geospatial technology in science mapping are pointed out.

BACKGROUND

Concepts and Terminology

Science mapping, bibliographic mapping, or mapping scientific bibliography is often defined as a visual representation of how scientific disciplines and fields, authors and institutions, and scientific and technical documents and articles are related to each other (Cobo et al., 2011; Small, 1999). In order to define the aim and scope of the present chapter it is paramount to first clarify what actually science and mapping mean in the expression 'science mapping'.

Firstly, the focus on 'science' in the term science mapping refers to all the data and metadata generated during the gathering and compilation of scientific bibliographic information such as: authors, article titles, source, citations, affiliation, and related scientific data (Chen, 2013). This data will be afterwards processed, analyzed and visualized using different scientometric techniques (Garechana et al., 2012). It is worth noting that bibliography data is only one potential working area to which geospatial technology and GIScience can be applied. Other areas within LIS, such as geospatial segmentation of patrons for marketing purpose and indoor position inside library buildings for space management and for enhancing navigation of patrons, are also of interest to GIScience (Aguilar & Granell, 2013; Granell & Aguilar 2013; Scaramozzino et al., 2014; Aguilar & Granell, 2015).

The second, and most important, clarification refers to the term 'mapping', which may have different connotations from distinct fields and disciplines. As the chapter mixes ideas from two distinct fields or disciplines, namely LIS and GIScience, it is worth delimiting early the scope of 'mapping'. As introduced earlier, science mapping or bibliometric maps of science, under the lens of the Librarianship discipline, is meant to visually represent bibliographic data relative to science. Noyons (2004) provides a clear definition of the resulting science maps in which "the items are positioned in relation to each other in such a way that the ones which are cognitively related to each other are positioned in each other's vicinity, whilst the ones that are not or hardly related are distant from each other". For example, we can have citation mappings, as visualizations of citation networks from scientific documents, or author mappings, as the analysis and visualization of collaboration (joint articles) among scientists. In general, as these mappings explore datasets relative to scientific activities and results, they are altogether referred to as science mapping.

In the current literature, though, the term "mapping" often refers to "record in detail the spatial distribution of (something)" (Oxford definition's map - verb)(English Oxford Living Dictionaries, 2016), which refers to spatially arranging data over an area. For example, network diagrams using force-direct layout are common visualizations for science mapping (Boyack et al., 2015). Despite these visualizations of science maps evoke items displayed on a geographical map, such a meaning of mapping has nothing to do with a geographic map or displaying data on a map. From the GIScience perspective, though, (geospatial) mapping implies to explicitly project data on a (geographical) map. For doing so, data must be georeferenced, in other words, data must contain a clear reference to a position or location, in order to be spatially displayed in a map. Furthermore, the main difference is that science mapping in LIS usually refers to visualizations that do not necessarily include or rely on a (geographic) map. Network graphs are typical examples of visually arranged bibliographic data because they emphasize the connectivity of data, i.e. network graphs provide an easy way to quickly grasp how items of data (e.g. authors, publications, citations, institutions, etc.) are connected. Furthermore, in this chapter, the term science mapping refers to ways of visually representing bibliographic data or science–related data, but paying special attention to technologies, concepts and analysis techniques from GIScience for creating such visualizations. Indeed, as we outline later on, the geospatial influence in the science mapping literature is still anecdotal.

Before going into the literature review, it is important to briefly introduce the concepts of geospatial analysis and visualization which are used extensively later on. Geospatial analysis is concerned with statistical and analytical techniques to process geospatial data, paying particular attention to topological, geometric, or geographic properties of the data. The typology of data (geospatial) calls for slightly different types of analytics than those found in the literature (Chen et al., 2012). In short, there exist two grand categories of spatial analysis techniques according to the dichotomy of geospatial data: vector-based or raster-based data. In the case of vector-based data, typical and basic operations are map overlay (combining two or more maps or map layers), and buffering (identifying regions of a map within a specified criteria (distance, time, etc.) of one or more geographic entities or features, such as buildings, streets, or town). Raster-based data, though, is more often used in disciplines such as environmental sciences and remote sensing other than LIS. Nevertheless, analysis techniques for raster-based data still might be helpful for surface analysis. These raster-based techniques typically involve a range of actions (such as filtering and algebraic operations) applied to the grid cells (i.e. an image is often divided in regular individual units or cells that altogether form a grid) of a single raster layer (image) or multiple layers. These actions involve processing one or more raster layers according to some statistical rules leading to a new raster layer. For example a surface analysis computes each cell value by interpolating values from neighboring cells. Visualization thus complements geospatial analysis in that it centers on the use, creation and manipulation of images and maps to effectively transmit spatial information to end users (Slocum et al., 2008; Dodge et al., 2008).

Related Work

Science mapping is a sort of analytical process composed of well-defined tasks that include but are not limited to data access to sources, data preparation, computation of similarities measurement, analysis and visualization methods. Cobo et al. (2011) reviewed in detail each step of the science mapping process and provided pointers to further related work and literature for interested readers. Cobo et al. (2011) also compared various software tools that are aimed to perform mapping science analysis. The authors identified only three tools (out of nine tools) – Science of Science (Sc2) Tool, CiteSpace, and VantagePoint– that to some extent supported some basic types of geospatial analysis and visualizations. What is relevant to this chapter is whether (or not) geospatial analysis and visualization techniques are used in science mapping, which could make a big difference in yielding novel insights during the visualization and interpretation of such science maps.

Geospatial techniques commonly used are merely geocoding and geospatial thematic maps. While geocoding is the task of attaching a location and/or position to a piece of data to turn it in spatial data, geospatial thematic maps result from visualizing one or more spatial datasets over a world or base map (e.g. Figure 1). Scimago Institution Ranking web site provides data visualizations of bibliometric indicators such as international collaboration and normalization impact of worldwide research in government and higher education institutions. Figures 1 and 2 show some Scimago available visualizations, ranging from simple geospatial mapping visualizations (e.g. choropleth maps, i.e. a sort of variation of thematic maps) to more traditional, spatial representation in tabular form like distribution charts and scatterplots.

Based on this Scimago data set, researchers have explored the distinct geospatial signatures among countries and institutions attending to indicators like international collaboration and leadership (Maganote et al., 2014). The authors used scatterplots to explain how geography matters in finding unique patterns derived from bibliometric indicators taking into account distinct geographic regions to show how international collaborations behave differently from regions to regions. Similarly, Pan et al. (2012) studied the geography of worldwide citation networks using a data set other than Scimago data. The article tries to validate the Tobler's first law of geography – "Everything is related to everything else, but near

Figure 1. Choropleth map to categorize countries in function of the number of institutions from the Scimago Institution Ranking (Source: Scimago Institutions Rankings, n.d.).

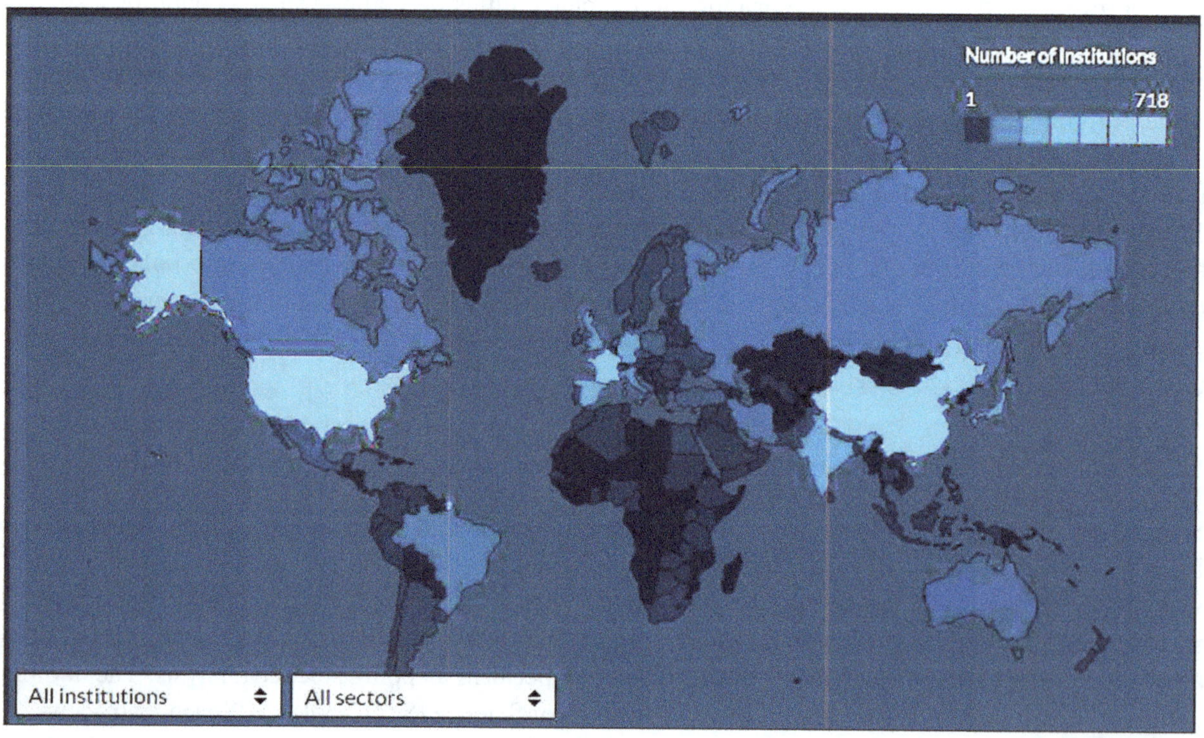

Figure 2. Map display for international collaboration indicator (Source: http://bit.ly/1RdW7lH)

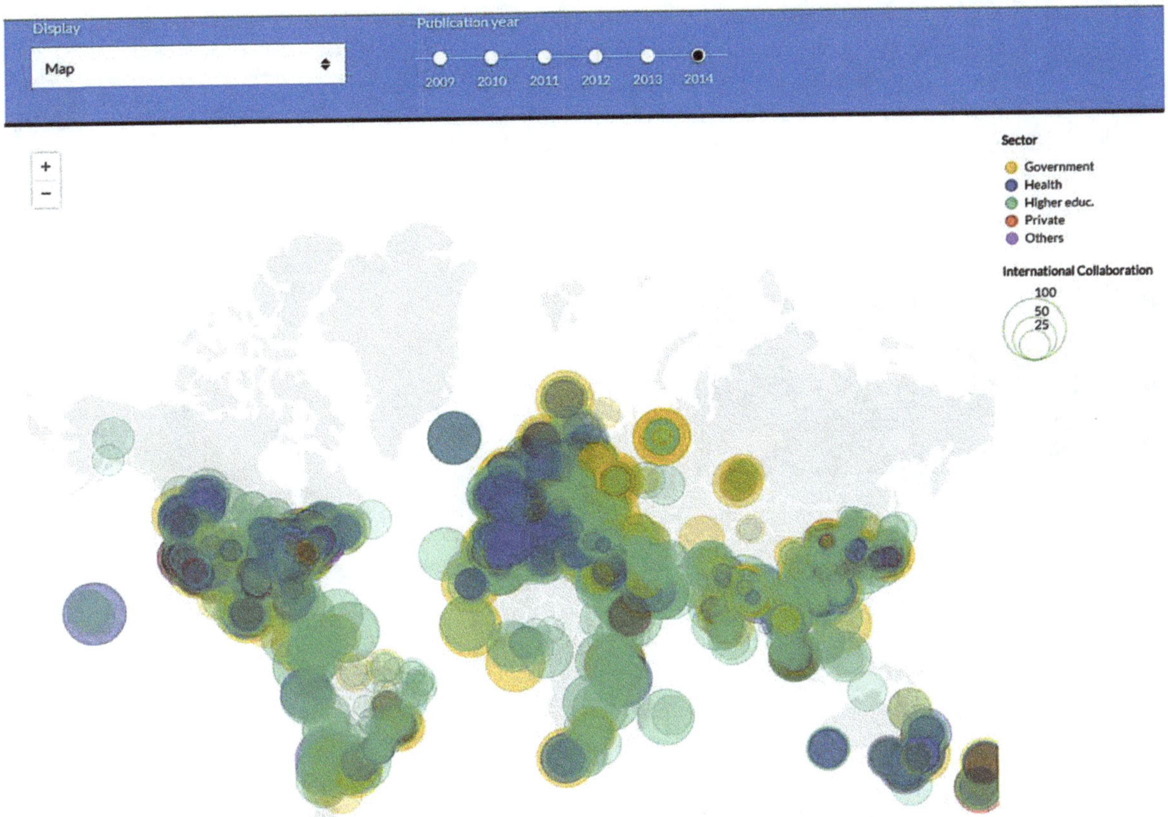

things are more related than distant things" (Tobler, 1970) – using citations networks and collaborations between scientists, without the authors explicitly mentioned it. Their hypothesis is that scientists tend to interact more frequently with other scientists who work in nearby areas forming clusters based on geographical proximity. In a similar vein, Chen (2006) described the CiteSpace toolset (http://cluster.cis.drexel.edu/~cchen/citespace/) for visualizing and analyzing trends and patterns in scientific data, especially to detect cluster and citations hotspots (Chen et al., 2010).

It is worth noting the recent work conducted by the Semantic Web journal (SWJ) by IOS Press to set up a modular, linked data-driven web portal for data relative to the journal activities, scientific papers, authors, editors, and reviewers (Hu et al., 2013). Apart from the fact that the web portal is semantically-enabled, i.e. all data is linked and encoded using semantic data formats and exposed through an SPARQL endpoint, what is relevant in the context of the chapter is the inclusion of spatio-temporal visualization to better understand how authors, co-authors, and authors citations are geographically related to each other in order to for example detect hot or densely connected regions in science (Bornmann & Waltman, 2011). Figure 3 shows some types of visualizations provided by the SWJ Web Portal. Left part of Figure 3 appears a cartogram that takes a location and resizes (distorts) the geographic area according to a value of indicator and/or variable being analyzed. Each nation's size gives a qualitative indicator of the geographical distribution of contributing authors to the SWJ journal. Cartogram are widely recognized as geographically-influenced data visualizations in information science (Olmeda-Gómez, 2014).

Figure 3. Linked data portal for the Semantic Web Journal (Source: Linked Data Portal for the Semantic Web Journal, 2010)

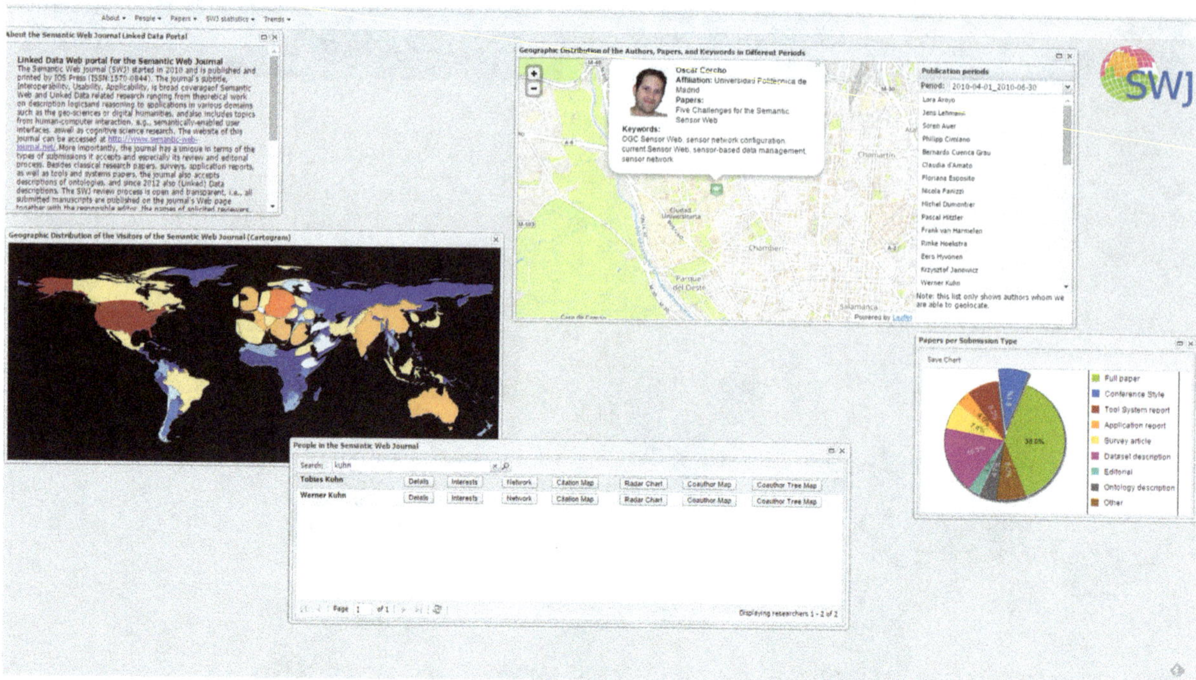

In the particular case of scientific publications, bibliographic reference management tools are useful to help users collect, categorize, and annotate their references of interest, which is a usual service offered by libraries. Also, cloud-based reference management services that incorporate social aspects (sharing, ranking or scoring references) like Mendelev or ReadCube are gaining momentum over the past years. Even traditional desktop solutions like Endnote have recently launched its counterpart cloud-based tools. A remarkable research done to let users browse, navigate and query a collection of scientific datasets through spatial and time dimensions is TimeBliography (Siabato et al, 2013; Siabato et al., 2014). It is a dynamic web tool that hosts bibliography data relative to temporal GIS topics. As depicted in Figure 4, it provides novel ways to browse an online repository through dynamic and innovative visualizations that attempt to exploit temporal and spatial attributes for each bibliography metadata record, amongst others. For example diverse types of timeliness visualizations to explore long and short periods of time are integrated into the user interface (upper part of Figure 4).

Finally, Field and O'Brien (2010) introduced the spiral spatio-temporal timeline visualization of geo-located tweets, where size of symbols larger denote recent tweets, the same color identify common words, and the location of individual markers in spiral denotes spatial association with the real origin. This experimental Twitter Map visualizations that use connected graphs may be a starting point to enable more sophisticated and informative network-based visualizations. Again space and time turn out to be key contextual information to better understand the dynamics of a connected network of people (or citations, papers, tweets, etc.)

Figure 4. TimeBliography web tool (Siabato et al., 2014) (Source: TimeBliography, n.d.)

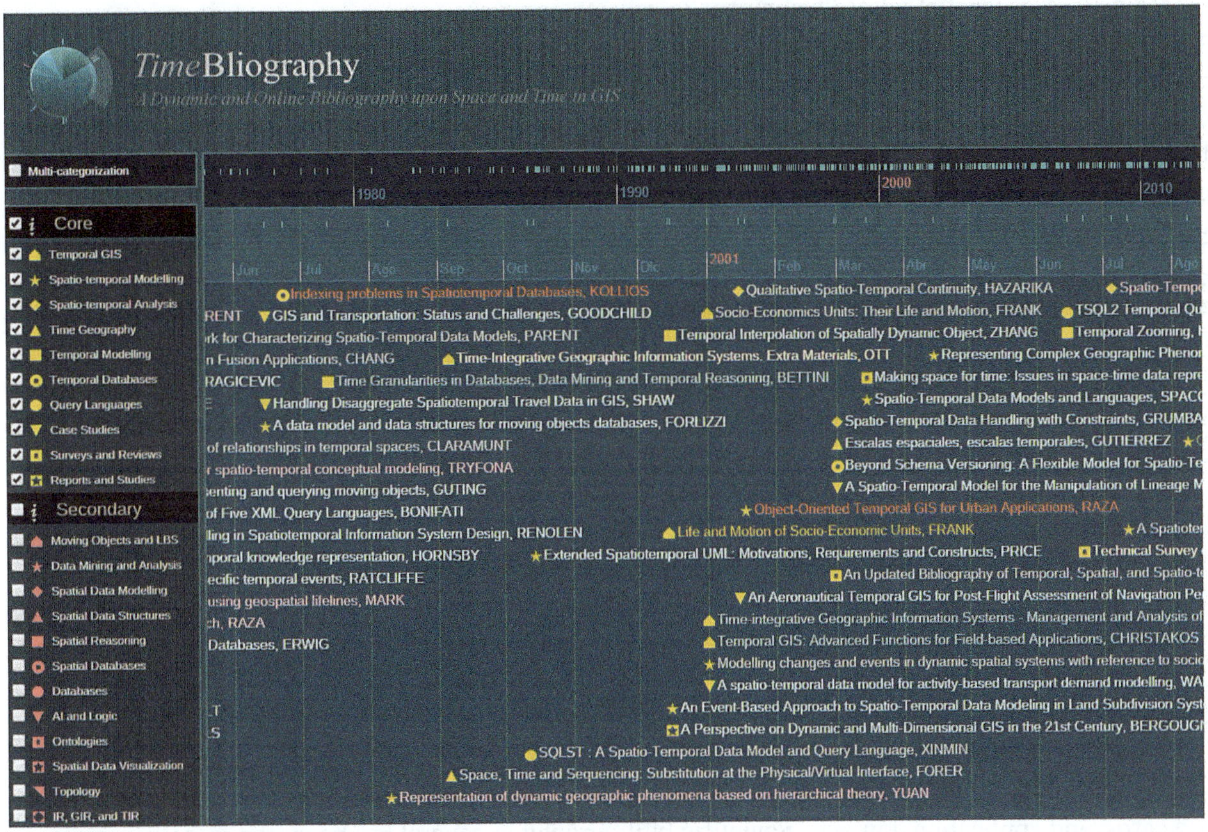

GEOSPATIAL INFLUENCE IN SCIENCE MAPPING

Despite the wide use of geographic terminology such as mapping or maps in bibliometric and scientometric, geospatial analysis and visualization methods have been largely unexplored in the literature. The brief overview described earlier presents interesting but incipient examples that attempt to include the geospatial dimension in mapping scientific bibliography data. It seems clear that there is a niche for applying geospatial aspects in science mapping, and it needs further exploration. The process of science mapping could take advantage of a wide range of geospatial analysis and visualization techniques so as to reveal new patterns and insights. Therefore, to suggest new areas where geospatial aspects can leverage science mapping to a new level, we rely on the concept of cybermetrics. Orduña-Malea and Aguilló (2015) provided a holistic view of the academic field of cybermetrics. The authors identified three main working areas within cybermetrics:

- Descriptive cybermetrics,
- Instrumental cybermetrics, and
- Applied cybermetrics.

For the sake of clarity in the exposition, this classification is taken in order to come up with unexplored uses of geospatial related techniques in science mapping.

Descriptive cybermetrics refers mainly to the definition and modelling of cybermetrics indicators. It is well beyond the scope of the chapter to discuss existing indicators, their classification and scope in the concept of cybermetrics or science mapping. What is somehow surprising is that, despite recent studies highlight the importance and relevance of geography in science mapping (Pan et al., 2012), geospatial characteristics has still poorly influence when defining bibliographic or cybermetrics indicators. Such geographically-aware indicators may give a completely new perspective of the underlying data being analyzed and visualized, and thus could serve for multiple purposes, ranging from professional career and project evaluation processes to policy making processes, as well as to even estimate the geospatial influence of an individual researcher, a scientific institution, or groups of researchers and institutions. In particular, the geospatial influence of an individual scientists has been recently explored by Zhan et al. (2014) for example. The authors devised the GeoSI index, which computes the number of countries or regions where researchers have cited a scholar's work during a given period of time, to get an estimation of the geographic coverage of the scientific influence of an individual scholar. Through the GeoSI index, the authors also examined how citations of a scholar have expanded to different countries over time, leading to the geographical influence or penetration of the research topics of one's work.

As instrumental cybermetrics is concerned with the methodological steps to obtain and process data to conveniently assess the feasibility of cybermetrics indicators, which is out of the scope of the chapter, the focus is shifted to the third and last area, applied cybermetrics, which in turn relies on both descriptive and instrumental cybermetrics.

Applied cybermetrics refers to the application of indicators (modelled by descriptive cybermetrics) combined with data analysis and visualization techniques, and altogether applied to concrete scenarios, use cases, and application domains. Scientific bibliography in general has been traditionally a recurrent domain where applied cybermetrics has been extensively used and explored. In the related literature concerning applied cybermetrics and academia, geographic network diagrams are used to represent patterns over a base map (Hu et al., 2013). Such geographic network diagrams combine network analysis methods (Boyack et al., 2005), commonly used for links analysis being for example co-authorship or citations those links, with geospatial visualizations by projecting these linked networks into a geographic map. This permits the detection of strong and weak geospatial connections among geographical areas, regions and nations. Unfortunately, geospatial visualizations in science mapping are practically constrained to geographic network diagrams, cartograms, and geospatial thematic maps. A similar deficit occurs in geospatial analysis, since geocoding is the most geospatial method used (for preprocessing data) in science mapping projects. There exist a great deal of untapped spatial analysis methods for the mapping science community (Smith et al., 2015), which for example combined with link analysis, network analysis, or data mining techniques, could most likely bring rich spatial contexts to science mapping in form of innovative geospatial data visualizations and explorations.

Apart from the lack of spatial techniques for data analysis and visualization, those few science maps that exploit spatial dimensions in bibliographic data do not consider interactive and dynamic feature in data visualizations. This prevents researchers, patrons, governors, and administrators from analyzing, understanding, and visualizing scientific maps and networks from a spatio-temporal perspective, which in the end pursues to improve the communication of information through maps.

FUTURE RESEARCH DIRECTIONS

Interactive, dynamic, and spatio-temporal visualizations are extremely useful in science mapping. Nevertheless preparing and processing geospatial data to fit the purpose of science mapping projects are time-consuming tasks. The value would not remain only in the set of references or bibliographic data, but in the availability of right tools, as spatio-temporal visualizations based on maps or other kind of charts, to enable visual analysis to inform users on the temporal, spatial, and topical variations of a given scientific topical area. In such a way, such tools could benefit researchers and scientists, as well as other collectives, as Noyons (2004) argued, such as funding agencies (to prepare research calls and evaluate research results, for example) and research policy makers (to establish topics priority and research plans in the long term).

The availability and efficiency of software tools depend strongly on the rapid changing pace of technology. Groenendyk (2013) provided an overview of emerging data geospatial visualization technologies which looked futuristic at first glance but which are partially being deployed in real projects. Holographic imaging and maps, 2-D city models, and 3-D printing were pointed by Groenendyk as key technologies for map libraries. Could existing 2-D network visualization for science mapping turn into immersive graph visualizations? Or even into physical 3-D models produced by 3D printers? Current technologies already open new opportunities for creative and innovative data visualizations for science mapping.

CONCLUSION

This chapter have presented some science mapping initiatives and research works that are being slightly influenced by geospatial technology. We argue that this trend have just started, as most of literature research works are exploratory research projects that merely grasp the surface of what geospatial technology might offer to science mapping analysis and visualization. As Field and O'Brien (2010) noted, "extracting and using the spatial and temporal dimensions reveal a contextual richness needed to truly represent and spatially organize meaningful information in a map". Location is key context, and such spatial context should be truly embedded in analytical and visualization processes for science mapping projects.

REFERENCES

Aguilar-Moreno, E., & Granell-Canut, C. (2013). Sistemas de información geográfica para unidades de información [Geographic information systems for informatio units]. *Profesional de la Información*, *22*(1), 80–86. doi:10.3145/epi.2013.ene.11

Aguilar-Moreno, E., & Granell-Canut, C. (2015). *Geobibliotecas* [Geolibraries]. Barcelona: Editorial UOC.

Bibliometrics. (2016). Retrieved from https://en.wikipedia.org/wiki/Bibliometrics

Bornmann, L., & Waltman, L. (2011). The detection of hot regions in the geography of science – A visualization approach by using density maps. *Journal of Informetrics*, *5*(4), 547–553. doi:10.1016/j.joi.2011.04.006

Boyack, K. W., Klavans, R., & Börner, K. (2005). Mapping the backbone of science. *Scientometrics, 64*(3), 351–374. doi:10.100711192-005-0255-6

Cartogram. (2016). Retrieved from https://en.wikipedia.org/wiki/Cartogram

Chen, C. (2006). CiteSpace II: Detecting and visualizing emerging trends and transient patterns in scientific literature. *Journal of the American Society for Information Science and Technology, 57*(3), 359–377. doi:10.1002/asi.20317

Chen, C. (2013). The Structure and Dynamics of Scientific Knowledge. In C. Chen (Ed.), *Mapping Scientific Frontiers: The Quest for Knowledge Visualization* (2nd ed.; pp. 163–199). London: Springer. doi:10.1007/978-1-4471-5128-9_5

Chen, C., Ibekwe-SanJuan, F., & Hou, J. (2010). The structure and dynamics of co-citation clusters: A multiple-perspective co-citation analysis. *Journal of the American Society for Information Science and Technology, 61*(7), 1386–1409. doi:10.1002/asi.21309

Chen, H., Chiang, R. H. L., & Storey, V. C. (2012). Business Intelligence and Analytics: From Big Data to Big Impact. *Management Information Systems Quarterly, 36*(4), 1165–1188.

Chloropleth Map. (2016). Retrieved from https://en.wikipedia.org/wiki/Choropleth_map

Cobo, M. J., López-Herrera, A. G., Herrera-Viedma, E., & Herrera, F. (2011). Science mapping software tools: Review, analysis, and cooperative study among tools. *Journal of the American Society for Information Science and Technology, 62*(7), 1382–1402. doi:10.1002/asi.21525

de Smith, M., Goodchild, M. F., & Longley, P. (2015). *Geospatial Analysis: a comprehensive guide to principles, techniques and software tools* (5th ed.). Leicester, UK: Troubador Publishing.

Dodge, M., McDerby, M., & Turner, T. (2008). *Geographic Visualization: Concepts, Tools and Applications*. Chichester, West Sussex, England: Wiley. doi:10.1002/9780470987643

Edmilson, J. T., Maganote, M. S. A., & Schulz, P. A. (2014). Visualization of ranking data: Geographical signatures in international collaboration, leadership and research impact. *Journal of Informetrics, 8*(3), 642–649. doi:10.1016/j.joi.2014.05.005

English Oxford Living Dictionaries. (2016). *Map*. Retrieved from https://en.oxforddictionaries.com/definition/map

Field, K., & OBrien, J. (2010). Cartoblography: Experiments in using and organizing the spatial context of micro-blogging. *Transactions in GIS, 14*(s1), 5–23. doi:10.1111/j.1467-9671.2010.01210.x

Garechana, G., Río, R., Cilleruelo, E., & Gavilanes, J. (2012). Visualizing the scientific landscape using maps of science. In S. P. Sethi & ... (Eds.), *Industrial Engineering: Innovative Networks* (pp. 103–112). London: Springer. doi:10.1007/978-1-4471-2321-7_12

Geographic information system. (2016). Retrieved from https://en.wikipedia.org/wiki/Geographic_information_system

Granell-Canut, C., & Aguilar-Moreno, E. (2013). Se busca geobibliotecario: Los datos geográficos entran en la biblioteca [Geolibrarian wanted: geographic data come to the library]. *Profesional de la Información., 22*(6), 569–575. doi:10.3145/epi.2013.nov.10

Groenendyk, M. (2013). Emerging data visualization technologies for map and geography libraries: 3-D printing, holographic imaging, 3-d city models, and 3-d model-based animations. *Journal of Map & Geography Libraries: Advances in Geospatial Information. Collections & Archives, 9*(3), 220–238.

Hu, Y., Krzysztof, J., McKenzie, G., Sengupta, K., & Hitzler, P. (2013). A Linked-data-driven and semantically-enabled journal portal for scientometrics. LNCS, 8219, 114-129. doi:10.1007/978-3-642-41338-4_8

Linked Data Portal for the Semantic Web Journal. (2010). Retrieved from http://semantic-web-journal.com/SWJPortal/

Noyons, E. C. (2004). Science maps within a science policy context. In H. F. Moed & ... (Eds.), *Handbook of Quantitative Science and Technology Research* (pp. 237–255). New York: Kluwer.

OJEU C. 240-1, 24/07/2014 (2014). *Commission notice: Guidelines on recommended standard licences, datasets and charging for the reuse of documents*. Retrieved April 10, 2015, from http://eur-lex.europa.eu/legal-content/EN/TXT/HTML/?uri=CELEX:52014XC0724(01)&from=EN

Olmeda-Gómez, C. (2014). Visualización de información [Information visualization]. *El Profesional de la Información, 23*(3), 213–219. doi:10.3145/epi.2014.may.01

Orduña-Malea, E., & Aguilló, I. F. (2014). *Cibermetría: midiendo el espacio red*. Barcelona: Editorial UOC.

Pan, R. K., Kaski, K., & Fortunato, S. (2012). World citation and collaboration networks: Uncovering the role of geography in science. *Scientific Reports, 2*, 902. doi:10.1038rep00902 PMID:23198092

Scaramozzino, J., White, R., Essic, J., Fullington, L. A., Mistry, H., Henley, A., & Olivares, M. (2014). Map room to data and GIS services: Five university libraries evolving to meet campus needs and changing technologies. *Journal of Map & Geography Libraries, 10*(6), 6–47. doi:10.1080/15420353.2014.893943

Scientometrics. (2016). Retrieved from https://en.wikipedia.org/wiki/scientometrics

Scimago Institutions Rankings. (n.d.). Retrieved from http://www.scimagoir.com/

Siabato, W., Claramunt, C., & Bernabé-Poveda, M. A. (2013). An interactive bibliography on temporal GIS. *Journal of Spatial Information Science, 7*, 99–101.

Siabato, W., Claramunt, C., Manso-Callejo, M. Á., & Bernabé-Poveda, M. Á.TimeBliography. (2014, December). A dynamic and online bibliography on temporal GIS. *Transactions in GIS, 18*(6), 799–816. doi:10.1111/tgis.12080

Slocum, T. A., McMaster, R. B., Kessler, F. C., & Howard, H. H. (2008). *Thematic Cartography and Geovisualization* (3rd ed.). Pearson Prentice Hall.

Small, H. (1999). Visualizing science by citation mapping. *Journal of the American Society for Information Science, 50*(9), 799–813. doi:10.1002/(SICI)1097-4571(1999)50:9<799::AID-ASI9>3.0.CO;2-G

Spatial Analytics. (2016). Retrieved from https://en.wikipedia.org/wiki/spatial_analytics

TimeBliography. (n.d.). Retrieved from http://spaceandtime.wsiabato.info/tGIS.html

Tobler, W. (1970). A computer movie simulating urban growth in the Detroit region. *Economic Geography, 46*(2), 234–240. doi:10.2307/143141

Zhan, F. B., Gong, X., & Xingjian, L. (2014). Mark on the globe: A quest for scientific bases of geographic information and its international influence. *International Journal of Geographical Information Science, 28*(6), 1233–1245. doi:10.1080/13658816.2013.847186

ADDITIONAL READING

Börner, K. (2010). *Atlas of Science: Visualizing What We Know*. MIT Press.

Börner, K. (2015). *Atlas of Knowledge: Anyone Can Map*. MIT Press.

Chen, C. (2013). *Mapping Scientific Frontiers: The Quest for Knowledge Visualization* (2nd ed.). London: Springer. doi:10.1007/978-1-4471-5128-9

Goodchild, M. F. (2010). Twenty years of progress: GIScience in 2010. *Journal of Spatial Information Science*, 1, http://www.josis.org/index.php/josis/article/viewArticle/32

KEY TERMS AND DEFINITIONS

Applied Cybermetrics: Applied cybermetrics refers to the application of indicators (modelled by descriptive cybermetrics) combined with data analysis and visualization techniques, and altogether applied to concrete scenarios, use cases, and application domains. Further information is available in Orduña-Malea and Aguilló (2014).

Bibliometrics: "Bibliometrics is statistical analysis of written publications, such as books or articles." Bibliometric methods are frequently used to provide quantitative analysis of academic literature or for evaluating budgetary spending. "Citation analysis is a commonly used bibliometric method" (source: Bibliometrics, 2016).

Cartograms: A cartogram is a map in which some thematic mapping variable – such as travel time, population, or socio-economic variables– is substituted for land area or distance. The geometry or space of the map is distorted in order to convey the information of this alternate variable (source: Cartogram, 2016).

Choropleth Maps: A choropleth map is a thematic map in which areas are shaded or patterned in proportion to the measurement of the statistical variable being displayed on the map, such as population density or per-capita income. The choropleth map provides an easy way to visualize how a measurement varies across a geographic area or it shows the level of variability within a region (source: Chloropleth Map, 2016).

Geospatial Mapping: Geospatial mapping is a type of spatial analysis techniques that typically employs software capable of rendering maps processing spatial data, and applying analytical methods to terrestrial or geographic datasets, including the use of geographic information systems.

GIScience: Geographic information science (GIScience) is the scientific discipline that studies data structures and computational techniques to capture, represent, process, and analyze geographic information. Goodchild (2010) summarized its core interests, including spatial analysis, visualization, and the representation of uncertainty (source. Geographical Information Systems).

Scientometric: Scientometrics is the study of measuring and analyzing science, technology and innovation. Major research issues include the measurement of impact, reference sets of articles to investigate the impact of journals and institutes, understanding of scientific citations, mapping scientific fields and the production of indicators for use in policy and management contexts (source. Scientometrics).

Spatial Analysis: Spatial analysis or geospatial analysis includes approaches to applying statistical analysis and other analytic techniques to data which has a geographical or spatial aspect, most notably in the analysis of entities using their topological, geometric, or geographic properties (source: Spatial Analysis, 2016).

This research was previously published in the Encyclopedia of Information Science and Technology, Fourth Edition edited by Mehdi Khosrow-Pour, pages 3473-3483, copyright year 2018 by Information Science Reference (an imprint of IGI Global).

Chapter 25
Parallel Development of Three Major Space Technology Systems and Human Side of Information Reference Services as an Essential Complementary Method

Joyce Gosata Maphanyane
University of Botswana, Botswana

ABSTRACT

The first human footprints dated more than 1.9 million years ago; the Homo erectus era, "upright man," marks the beginning of man's socio-economic historical development. It culminated in the Bronze Age, Iron Age, Industrial Revolution, and currently the Information Age. The current era has allowed rapid global communications and networking to shape modern society. Individuals are able to transfer information freely, with instant access to knowledge that would have been difficult or impossible previously. This chapter elaborates upon GIS/remote sensing: the highest echelon in ICT. It compares and contrasts the four globally used GIS systems of GOES, the Geosynchronous Orbiting Environmental Satellite; LANDSAT; SPOT, Satellite Pour l'Observation de la Terre; and the WorldView. Their temporal resolution, spatial resolution, spectral resolution, radiometric resolution, and their DEM characteristics are reviewed. The human side of information reference services in the form of TEK is alluded to.

INTRODUCTION

Ever since the first human footprints were made more than 1, 9 million years ago; the Homo erectus Era; 'upright man'; socio-economic-Historical Development came into being. When humans could walk; it left their other limbs, the hands free to do other things. This Era was followed by the Stone Age; the man used stone tools to defend their territories and to hunt for food. This culminated to the Bronze

DOI: 10.4018/978-1-5225-7359-3.ch025

Age and the Iron Age. These were the human development eras when man could actually work with minerals to make tools. These people could farm and rear domestic animals. The primitive smelting industries grew into the Industrial Revolution; The Industrial Revolution was a period from the 18th to the 19th century. The major changes in agriculture, manufacturing, mining, transport, technologies had a profound effect on the socio-economic and cultural conditions. This was followed by the Information Age; the Era we are in now. It is a shift from traditional industry that the previous era brought through industrialization. It is an economy based on the manipulation of information. The Information Age has allowed rapid global communications and networking to shape modern society. It is a digital world of Information Communication Technology (ICT). It is also commonly known as the Computer Age or Information Era, and it is an idea that the current age will be characterized by the ability of individuals to transfer information freely, and to have instant access to knowledge that would have been difficult or impossible to find previously. The idea is linked to the concept of a Digital Age or Digital Revolution. This chapter elaborates upon Geographical Information Systems (GIS) and Remote Sensing; the highest echelon of the ICT world. It looks at their development in studying our home, the Earth and its systems. In particular it looks at, compares and contrasts the four globally used systems; these are:

- **GOES:** The Geosynchronous Orbiting Environmental Satellite;
- LANDSAT,
- **SPOT:** Satellite Pour l'Observation de la Terre and
- The WorldView.

The first three has transacted five generation and the last system, the WorldView is the newest and most fast developing satellite system. These Earth Resources Data capture satellite systems are compared in their longevity - temporal resolution and age factor which gives results as:

- LANDSAT1 (ARTS 1) - LANDSAT 8 range July 1972 to February 1913;
- GOES 1 - GOES 15, Range October 1975 to March 2010;
- SPOT1 – SPOT 7 range February 1986 to June 2014 and
- WorldView 1 – WorldView 3, range September 2007 to August 2014.

The comparison criteria were made based on spectral resolution (Figure 5 – WorldView 3); spatial resolution (Figure 6 – WorldView 3) and radiometric resolution (Figure 7 – SPOT). Apart from that; the human side of information reference services in the form of Traditional Ecological Knowledge (TEK) is discussed as it is an essential complementary entity in GIS and remote sensing endeavours. In order to be useful information, the remote sensing data need human input in the form of referencing coordinates system, data interpretation using the visual variables of position, shape, size, texture, tone, orientation, and motion. These can then be analyzed and used for modeling the environment, disaster preparedness and decision making. The 3 Dimensional characteristics of satellite, digital elevation models (DEM) (Figure 7 – SPOT DEM) are also investigated.

The Chapter is also about the essence of data sources specific for geo-spatial science (Bossler, Jensen, McMaster, & Rizos, 2002) information for land cover mapping. The use of geospatial science techniques provides opportunities and challenges in many aspects of life including for land cover, forestry for climatic change measurements and agricultural engineering which is vital for food security (Opara, 2003)

First, it elaborated about developments of digital world in remote sensing and geographical information system (GIS) as modern day techniques for Earth surface monitoring (Mather & Koch, 2011). Satellites whose data has widespread international applications for the past four decades are discussed in details and compared. Second, it puts emphasis on the importance of the human side of information reference services that is essentially the brain behind machine based data. The human thinking is applied in the interpretation of this machine data and it acts as knowledge substitute where the remote sensing and GIS data are inadequate or unavailable. The 2012 study (Maphanyane, 2012) on the reconstruction of historical landscape for the investigation of land cover changes in the human side of information reference services based method on Ramotswa, Botswana case, had sufficiently proved that it can be applied to fill in the gap where modern technology is inadequate or unavailable (Figure 1 and Table 1). For emphasis, graphical examples are given, where the aerospace based Ramotswa digital data; its analytical techniques and their resultant information are compared to the same area data, but those that had been derived from the systems based on the human side of information reference services (Figure 2). This too has shown that the human side of information reference services was complementary.

BACKGROUND

The essence of this study hinges on technological developments of robust data collection tools. Remote sensing is one such. It shows the importance of human side of information reference services and the role it plays still, as it adds meaning and time independent data sources, even in the wake of superior tools of remote sensing.

Curiosity of Humankind: Their Cognitive Ability and Feature Naming

Humankind interest in the distribution of everything that exists on and beyond the Earth itself was triggered by what they could see (Benedict, 2002). The evidence to their cognate abilities is the byzantine names they gave to the phenomena. The Batswana were captivated by all that was around them. Some evidence in that are meaningful names they gave to celestial bodies (Kitaura & Enßlin, 2008) They also employed the skies - *legodimo* as direction tools and identification of time and its subdivisions using daily and seasonal rhythms. They put what seemed to be a scattered litter of sky furniture into distinguishable bodies with meaning. The obvious ones been the *Sun - Letsatsi*, it was used to distinguish between night and day – *Bosigo* le *Motshegare* and the stars – *Dinaledi*. *Mars* was also identified and named *Mphatalatsane*, the red Morning Star and the *Pointer*, the All-Night-Long Star – *Kgogamasigo* was observed as well. *Venus – Kopadilalelo,* the evening star - has long being observed by this people as the second brightest body after the Moon - *Kgwedi* - in the evening skies, but different from the stars as it was always in the same planetary position. The moon is the celestial body most used by the Batswana. The Motswana female-body functions, the menstruation cycle, conception, child-birth and well-being are believed to be dependent on the cadence of the moon. Also, theories about the eclipse of the Sun and the Moon, and the Comets were also devised. For example, the visit of a Comet was a bad omen, and it might signify something bad, or it might only foreshadow the death of some great chief.

Also, the humankind cognitive abilities are revealed by the work of the Egyptians who had created the precise terrestrial map of the three stars of Orion's belt in the ancient times as early as 10,500 BC; the era of Leo as depicted by the Sphinx of Giza. In this symbolic map, the Milky way is replicated

Figure 1. Botswana: South East District

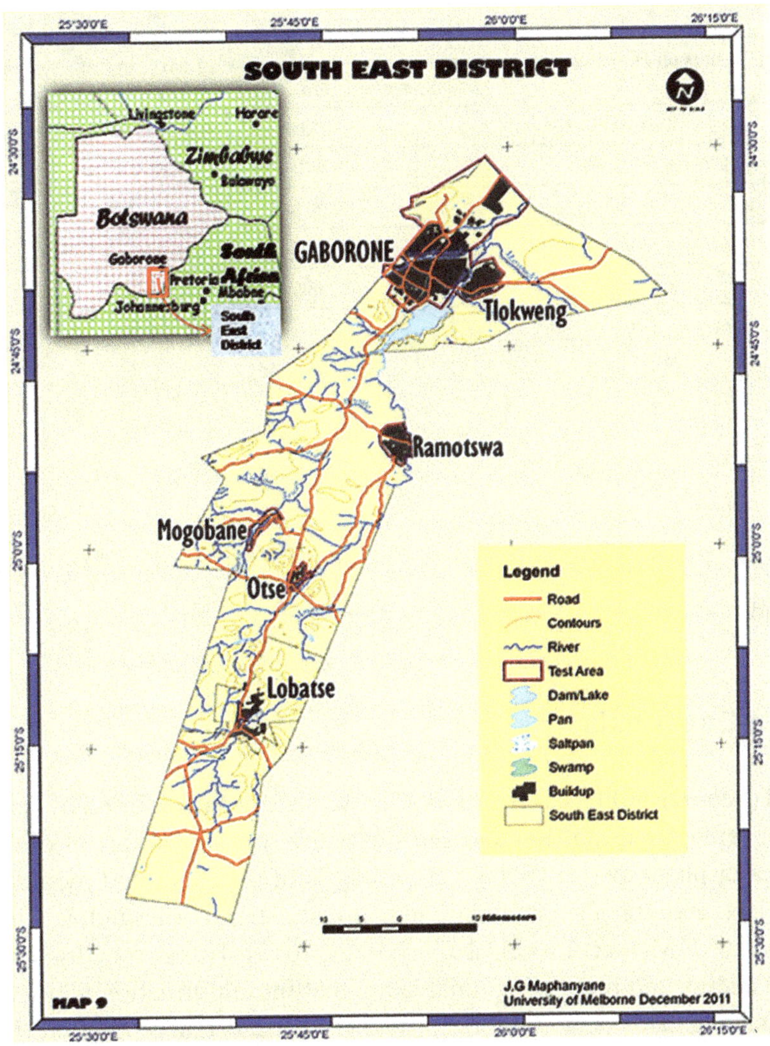

onto the ground by the position of the River Nile and the positions of the three pyramids of Giza have been plotted against the three belt stars of the Orion constellation as such the great pyramid of Khufu occupying the position of Al Nitak; the second pyramid of Khafre taking the place of Al Nilam and the third pyramid, Menkaure, representing the smallest star which is at an offset position (Hancock, 1995). Modern horoscope astrology is associated with the interaction of Indian and Hellenistic cultures in the Indo-Greek period (Pingree, 1981; Dhavale, 1984) and their eras had been named as stars thus:

- Aquarius,
- Pisces,
- Aries,
- Taurus,
- Gemini,

Table 1. Botswana: south east district data, comparison of the two methods for the investigation of land cover change.

Analysis of the Results on How Reconstruction of Historical Landscape Compares to Remote Sensing Method (Maphanyane 2012a)				
No.	**Test Area**	**1973**	**1990**	**2002**
1	Gaborone/Tlokweng	66.94%	66.13%	77.42%
2	Ramotswa	80.00%	71.79%	71.88%
3	Otse	65.27%	65.87%	67.07%
4	Mogobane	77.88%	70.72%	67.31%

Source: (Maphanyane, 2012a)

- Cancer,
- Leo,
- Virgo,
- Libra,
- Scorpio,
- Sagittarius, and
- Capricorn.

Remote Sensing Essentials

These human visual (Sensor) abilities as mentioned above have been amplified by scientific technological developments of remote sensing. Remote sensing is the science and art of obtaining information about an object, area or phenomenon through the analysis of data acquired by a device that is not in contact with the object, area, or phenomenon under investigation (Lillesand & Kiefer, 2004; Jensen, 2005; Schowengerdt, 2007; Mather & Koch, 2011; NASA, 2013). This is done by sensing and recording reflected or emitted energy and processing, analyzing, that into information which is then applied for informed decision making. This is possible because different substances like the Earth's surface phenomena absorbs, transmits and reflects electromagnetic waves energy from the source e.g. the Sun in ways that are diagnostic to the type and condition of the reflecting material. Then this reflected energy could be captured by various satellite sensors. Then, information could be extraction from the resultant images for use. One other thing that is crucial to the fundamentals of remote sensing are the distance (altitude) between the observer and the phenomena been captured and the type of the sensor.

Phenomena under study could be observed from sensors on board platforms anchored at different altitudes, determining the instantaneous field of view and the scale of the data, the spatial resolution. This means that a phenomenon can be viewed from high altitudes in its entirety at a small scale and/or only a part of it can be viewed in detail from lower altitudes at a large scale. It is always beneficial to have a closer view of phenomena; but also seeing them at small scale can be beneficial too. Also, remote sensing allows for revisits of the same place at regular intervals, this count for the temporal resolutions and makes the study of trends possible. Furthermore different types of sensors with various characteristics are possible. That is the sensor used determines the electromagnetic waves window to be captured, its spectral resolutions. And furthermore still, the level of clarity in which the phenomenon is distinguishable from its surrounding, the radiometric resolutions could be measured. That is remote sensing sees what the human eye could perceive through the visible electromagnetic waves; and far beyond that, what the eye could not see; in the gamma ray, x-ray, ultraviolet light, infrared light, microwaves and radio waves. Remote sensing gives an option of studying the scientific nature of phenomena. Even those phenomena that are hidden from normal human view like heat (thermal) and embedded moisture content in plants' leaves could be studied from this type of data as various wavelengths beyond the visible and these could later be viewed through false colour image creation by sophisticated computer software program. In the past fifty years great advances in remote sensing had been achieved by placing the sky-telescopic view, in space above the Earth's atmosphere; the Astronomical Remote Sensing (Kitaura & Enßlin, 2008). At these very high altitudes the focus is on capture gamma, x-ray, ultra-violet, visible, infra-red, microwaves and radio-waves spectrum emitted by the stars, galaxies, interstellar or intergalactic gases, and the void called "space" for their study. Space flight and exploration have led to new technologies that extend the capabilities of astronomers. For instance it is now possible to recreate and study the beginning of the Universe; the BIG BANG! And that the universe is expanding and is finite in both time and space (Jones & Lambourne, 2004). The study of detailed minute fabric of the matter – microscopic view, the Atomic Remote Sensing, which operates at particle or microscopic level, had always being possible in the field of chemistry (specific example (Linga, Kumar, & Englezos, 2007)), physics, biology, and geology and health sciences. Such devices include the microscopes, probes, electro and magnetic survey instruments, spectroscopes, scanners and x-rays. For instance, the absorption spectroscopy that uses infra-red and ultra violet and the UV light sterilisers are examples such devices (Nichol, Shaker, & Nichol, 2006). This article concentrates on the study of the Earth surface and its systems, that incorporates electromagnetic waves capture at altitudes of the ground, atmospheric and space view levels, the Earth Resources Remote Sensing. These what makes it possible for the environment to be studied in quick, frequent and more accurate ways. But, still remote sensing is not the panacea to everything; there are some problems. These normally call for human information reference services input. The four major problems related to the Earth surface and its systems remote sensing, more specifically the space view data capture and analysis issues are that:

- Although, the remote sensing data capture goes beyond the human visual capabilities, it is machine based, it lacks human mind of reason, hence it needs:
  - To be calibrated and interpreted against the real-world phenomena for it to be meaningful and useful.
  - Trained people to interpret and analyze it.
- Besides all that, it is limited by its time of first data capture. Although it began in 1939, in the form of photogrammetry, the world view dedicated coverage by satellite imagery has only been available since 1972. Consequently, for research that spans periods earlier than this period, remote sensing data in non-existent. Then, since land cover change is a slow process, it is necessary to access evidence about the land cover conditions of past eras, world coverage data from 40, 50, 60, 100 years ago and beyond!
- Furthermore, its data is not automatically available everywhere. Besides, adverse weather conditions renders satellite images useless for the remote sensing analysis of land cover.

So, although remote sensing is useful; it still requires the human side of information reference services for it to be useable and also to fill up the gap were it is inadequate or unavailable. The method for the investigation of land cover change and trends by the reconstruction of historical landscapes could be used to alleviate the problems (Maphanyane, 2012).

Human Based Reference Services in Brief

The reconstruction of historical landscapes could be used as an alternative for the investigation of land cover change. First, this method uses the elements of traditional ecological knowledge (TEK) of oral history, songs, poems, praise singing and relics. Second, it employs written records in the form of explorers' diaries (Livingstone, 1857; Rey, 1988), the government reports and development plans, the land laws and acts, verdict on major court cases on land disputes and land commissions concerning land uses. Third, it also utilizes the archival records which consist of land surveyors' field sketches, maps, photographs and aerial photographs (Shoemaker & Shoemaker, 1987). And fourth, yet again it uses the scientific endeavours like, archaeological research findings, fossil, sediments, dating such as 14C carbon, 210Pb lead, regional industrial pollutants and regionally significant biostratigraphic markers and the beta - globins, in particular, Mitochondrial deoxyribonucleic acid as research tools (Stringer, 2001).

These elements of reconstruction of historical landscape taken ensemble, enables studies that reach ages that goes far beyond dating epochs known today. Since this approach does not rely on technology in its original data capture phase, it can be applied anywhere and long time after land cover changes has taken place (Maphanyane, 2012).

The concept, the reconstruction of historical landscapes, in the context of this communiqué is defined as a method of building and mapping past landscapes by finding out how communities lived in a given historic period in a particular geographical area. This is achieved by identifying, reading and analyzing marks made at the time and left behind from certain period. This brings in a human element in the study. People's life and experiences, the knowledge that an individual has about his/her sphere of influence. It is acquired and registered in their mind through their experience of the area and it is continuously updated when new developments come up (Maphanyane, 2012). Maphanyane (2012) has further reported that each of these eight elements could be useful to the investigation of land cover change to a varying degree: That archaeological findings and pre-historic sites for instance, gave useful background knowledge of the areas studied, and at such promoted a better understanding of the areas, and later helped in the analysis of the data (Ortner, Butler, Cafarella, & Milligan, 2001). That, Written records which included informal data like travellers diaries; are unique as the authors wrote these, true to themselves as they never knew that one day their work would be used for such a purpose. And other information came from formal data like government development plans. Such documents formed sources of continuous information, which are essential for a land cover change investigation. That the elements comprising of songs, lyrics and poems (Lightfoot, 1995) were difficult to get, hence sporadically available; but, once stumbled upon could prove to be useful where upon an entire history of a group of people could be embedded in one poem or the geography of the area could be sung in a lyric which would be remembered long after the particular landscape of an area had long been eroded. The elements of place names are essential, as they gave insights into past landscapes and the history of the earlier inhabitants of an area (Atteh, 1989; Mabona, 2004). And that the element of traditional ecological knowledge sourced from oral history (with examples coming from various literature - (Brokesha, Warren, & Werner, 1980; Davis, Back, & MacLean, 1992; Leffler & Brent, 1992; Delgamuukw v. British Columbia, 3 S.C.R. 1010, 1997; Russell,

Figure 2. Graphical comparison of Ramotswa, Botswana land cover change detection by reconstruction of historical landscape (ROHL – A, C) to that by remote sensing method (RS – B, D) for two periods - 1973 to 1990 and 1990 to 2002 based on ArcGIS Software Geo-spatial Statistics Analyst Module (Maphanyane, 2012)

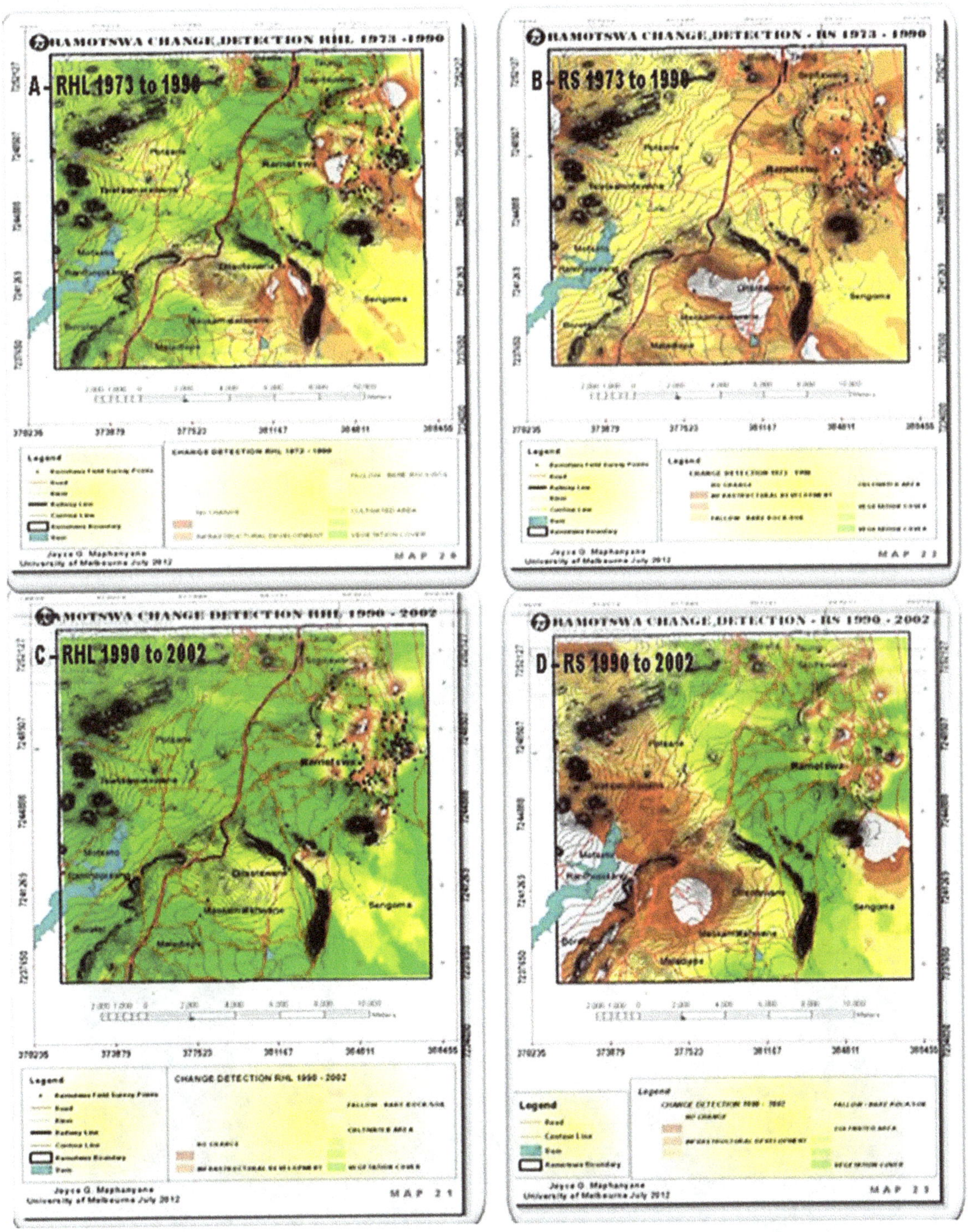

1998); the cultural norms (Berkes, Colding, & Folke, 2000; Davidson-Hunt & Berkes, 2003; Donovan & Puri, 2004); questionnaires, interviews and field (Livingstone, 1857; Rey, 1988) gave the most complete evidence data for the land cover change investigation.

The human based reference services data sources approach deepened on the understanding of current environmental issues, including global warming, desertification, loss of biodiversity, sustainable resource use and sustainability of the biosphere (Russell, 1998).

MAIN FOCUS

The focus is on two distinction methods for land cover change investigations. First, the importance and developments of remote sensing as the geo-spatial data source are discussed. To this end, analysis and comparison (Figure 3) of three satellites systems whose data had reached the most widespread global usage and that had stood the test of time that spans four decades are made. These are the Geo-synchronized Operational Environmental Satellite (GOES), for international environmental disaster monitoring and weather focusing; the Landsat Systems, which is applied for land cover mapping in green house gas (GHG) carbon loads monitoring projects for the world climate change monitoring and mitigation endevours; and the SPOT (the Satellite Pour l'Observation de la Terre), that provided the yearly images for the entire planet in Google Earth, "One World, One Year" project applications. The four trans-generational levels developments in spatial, temporal, spectral and radiometric resolutions of these systems have been

Figure 3. Landsat (LN) and SPOT satellites 1-4 generations panchromatic (GP) and multispectral (GM) resolutions

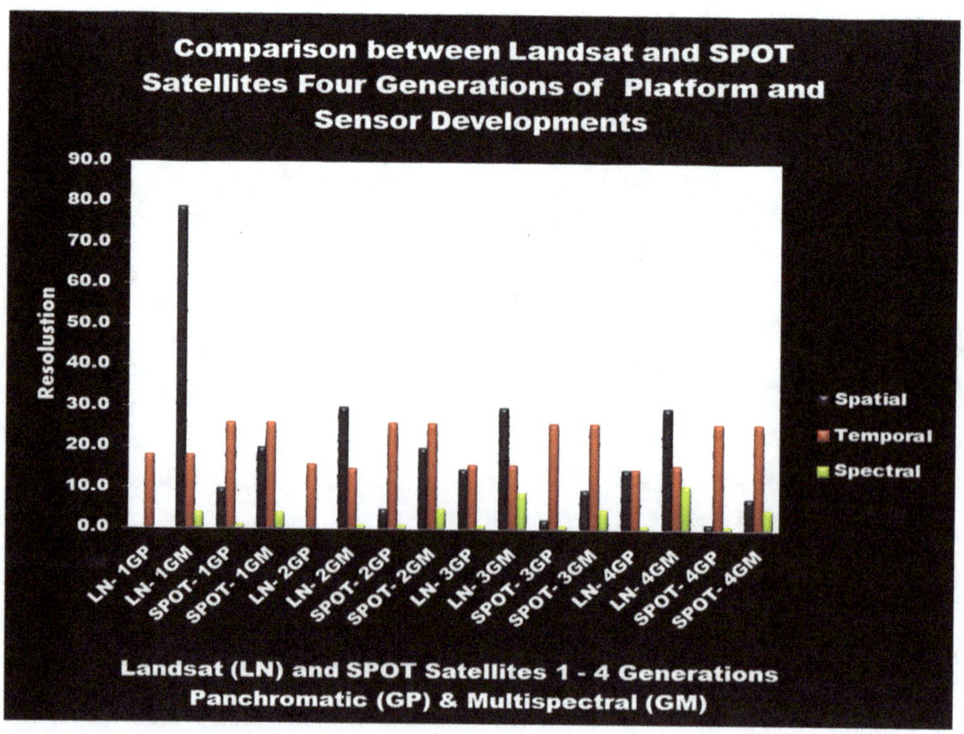

explored. Second, yet again, comparison is made between these technical sources and the human service sources (Figure 2). That is another focus that brought in the highlights from the human side of information reference services as an essential complementary method that added meaning to remote sensing and that functioned as an alternative to fill-in the gap when these digital data are inadquent or unavailable.

The Advancement in GOES System

The basic element of GOES satellite is that they are geosynchronous and geostationary. They have the ability of remaining permanently in the same area of the sky, as viewed from a particular location on Earth, and so remaining permanently within view of a given ground station. Also, they have special ability that their paths follow a circular geosynchronous orbit directly above the Earth's equator. GOES came into operation in May 1974 then the first satellite SMS 1 was launched. Its program capture data in two levels; from instruments carried by platforms orbiting the Earth at altitudes of about 35 790 km and also from ground stationed data acquisition elements such as river and rain gauges, seismometers, tide gauges, buoys, ships and automatic weather stations.

Its constellation forms part of the global network of meteorological satellites (Leese, Noar, & Pastre, 1989), spaced at approximately 70° longitude intervals around the Earth in order to provide near-global data coverage on weather forecasting, severe storm tracking, and meteorological research. The launches of these satellites are staggered to keep at least two operational satellites on orbit at all times. To date seventeen spacecraft has been put into orbit, the latest of these being GOES 15, launched in March 2010 and operates as GOES West located at 135°W over the Pacific Ocean. Other two satellites still in operational, are GOES 12 providing Latitude coverage for South America and GOES 13 operates as GOES east, and is currently located at 75°W. It provides most of the U.S. weather information. GOES R, S, T and U are scheduled for launch in 2016, 2017, 2019 and 2024 respectively. The proposed instrument package for the GOES R- series initially included: the Advanced Baseline Imager (ABI); the Hyperspectral Environmental Suite (HES); the Space Environment In-Situ Suite (SEISS), which includes two Magnetospheric Particle Sensors (MPS-HI and MPS-LO), an Energetic Heavy Ion Sensor (EHIS), and a Solar and Galactic Proton Sensor (SGPS); the Solar Imaging Suite (SIS), which includes the Solar Ultraviolet Imager (SUVI), the Solar X-Ray Sensor (XRS), and the Extreme Ultraviolet Sensor (EUVS); the Geostationary Lightning Mapper (GLM); and the Magnetometer (Hill, 2008).

GOES satellites are given an alphabetical number before launch and only designated a numerical number after launch. One launch failed, GOES G, and GOES 14 is in on-orbit storage, whereas GOES 3 and GOES 7 had been re-purposed to communications satellites duties, the former as a relay for Amundsen-Scott South Pole research station and the later by Peacesat. And among the seventeen, ten satellites have been decommissioned and boosted to graveyard orbit.

GOES is SMS 1 and 2 and GOES 1, 2 and 3 first generation make-ups, were sustained from 1974 to 1986 are characterized by latest models are the imager, a multichannel instrument that senses infrared radiant energy and visible reflected solar energy from the Earth's surface and atmosphere and the sounder that provides data for vertical atmospheric temperature and moisture profiles, surface and cloud top temperature, and ozone distribution (Gunter, 1996). They carried the visible infrared spin-scan radiometer (VISSR), the solar x-ray imager (SXI), and space environment monitoring (SEM) instruments as well. Additionally, GOES carried meteorological data collection and transmission system that relayed processed data from central weather facilities to automatic picture transmission (APT) equipped regional stations and collected and retransmitted data from remotely located ground data acquisition stations (LaViolette

& Diachok, 1974; Gunter, 1996). Five satellites, GOES 4, 5, 6, G and 7 with VISSR formed the second generation series that run from 1980 to 1995. They incorporated visible and infrared atmospheric sounder (VAS) another improvement from the earlier series that obtained atmospheric circulation patterns from frame-to-frame movement onto selected clouds at different altitudes and that captured the wind direction and its speed. The VAS added a vital third dimension to the imager; it improved the accuracy for weather prediction data (Gunter, 1996). From this system, GOES 7 had a signal transponding ability and it could locate ships and planes in distress (Doody & Stephan, 1993) The third generation includes GOES 8, 9, 10, 11 and 12 run from 1994 and is still in operational as GOES 12; the last of these series lunched in 2001 is still in service; the rest have been decommissioned. These spacecraft payloads are new-generation Imager and a Sounder. The improvements on the SEM include the magnetometer, energetic-particle sensor, high-energy proton and alpha-particle detector, and a solar X-ray sensor and are used for in-situ surveying of the near-earth space environment. They provided 24-hour monitoring and measurement of dynamic weather events in real-time, and are the first to deliver simultaneous independent imaging and sounding from geostationary orbit (Gunter, 1996; SSL, 2013). The new body-stabilized spacecraft design enabled the primary sensors to *"stare"* at Earth and thus frequently image clouds, monitor Earth's surface Atmospheric phenomena were tracked, ensuring real-time coverage of short-lived dynamic events, such as severe local storms and tropical hurricanes and cyclones. Temporal resolution of 15 minutes and the flexibility; for alteration of normal schedules that provided necessary coverage during significant weather or other events were made. The sensors have an imager with five channels/bands, so specific weather trouble spots were monitored to assist in improved short-term forecasting. The primary sensor is the *Advanced Very High Resolution Radiometer* (AVHRR). The fourth generation included GOES 13, 14, 15 lunched between 2006 and 2010. This series, like GOES 12, have a *sun-pointed Extreme Ultraviolet Sensor* (EUVS). In addition to that their spacecrafts payload includethe *Advanced Baseline Imager* (ABI); the Space Environment In-Situ Suite (SEISS), which includes two *Magnetospheric Particle Sensors* (MPS-HI and MPS-LO), an *Energetic Heavy Ion Sensor* (EHIS), and a *Solar and Galactic Proton Sensor* (SGPS); the *Solar Imaging Suite* (SIS), which includes the *Solar Ultraviolet Imager* (SUVI), the *Solar X-Ray Sensor* (XRS), and the *Geostationary Lightning Mapper* (GLM); and the *Magnetometer* (Segundo, 2010) (Figure 4).

AVHRR has lower spatial resolution (ranging from 0.9 km to 4 km) than other typical land observations sensors for example the Landsat and SPOT program, but its data is useful for study of phenomena which cover the entire globe. Advantages: Frequent coverage- very high temporal resolution, wide coverage covers whole hemisphere at one time, and that it is multi-spectral. So it is used extensively for monitoring regional, small-scale phenomena, including mapping of sea surface temperature, and natural vegetation and crop conditions (NASA - NOAA, 1992; Segundo, 2010) .

Land Resources Satellites

The weather satellites are not optimized for sensing detailed mapping of the land surface. The data they bring back is just too coarse.

In 1965, director of the U.S. Geological Survey (USGS), William Pecora, proposed the idea of a remote sensing satellite program to gather facts about the natural resources of the Earth. This was opposed by the Bureau of Budget for obvious reasons and the Department of Defense who feared that a civilian program would compromise the secrecy of their reconnaissance missions. Also, there were geo-political concerns about photographing foreign countries without permission. Eventually, after these

Figure 4. History of GOES Satellite System - GOES 1 - GOES 15, Range October 1975 to March 2010 (Source Graham, 2016).

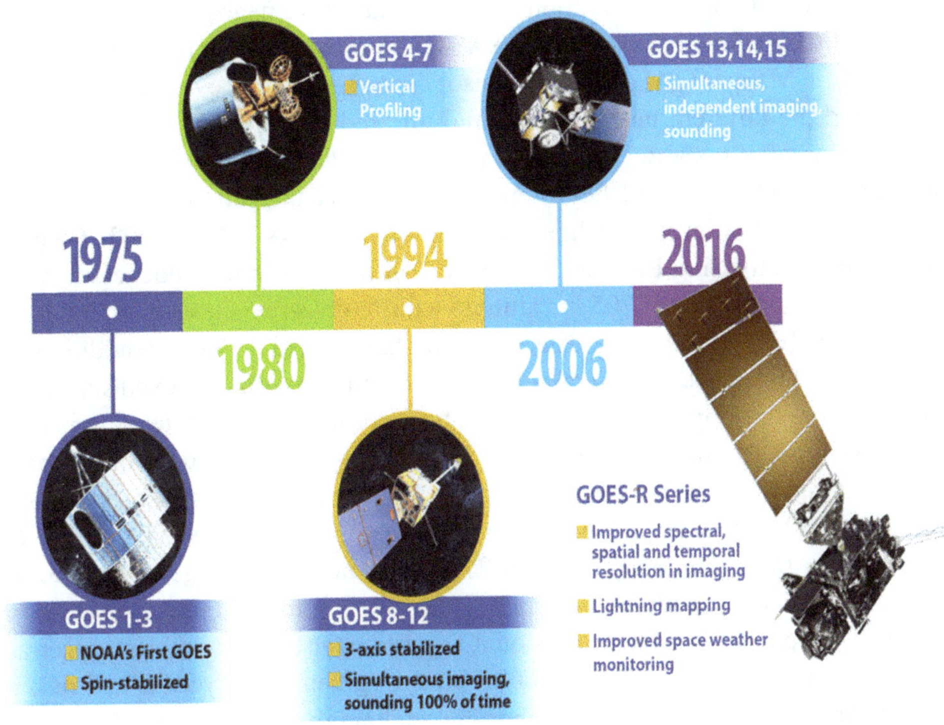

huddles had been overcome, the initial idea culminated to the lunching of Earth resources technology satellite (ERTS), which was designed to specifically to monitor the Earth's surface. In 1986, SPOT 1, a French, Landsat-like satellite was launched and it broke the United States of America spacecrafts land resource censorship monopoly.

Landsat Satellite System

ERTS-1, later renamed Landsat 1 was launched by NASA in July 1972. So far eight satellites have been launched in this system, five have been decommissioned, one, Landsat 6 was lost and two are still in service Landsat 7 and 8. The program is jointly managed by the United States Geological Survey (USGS) and NASA. The Landsat archival and latest data have been used in studies for monitoring the dynamic changes caused by both natural and manmade processes for informed decision making and resource management.

The first generation in this series is Landsat 1, 2 and 3 which were in orbit from July 1972 when Landsat 1 was launched to March 1983 when Landsat 3 was terminated. The spacecraft carried the camera system, with the Return Beam Vidicon (RBV), and the Multispectral Scanner (MSS). They orbited in near-polar, sun-synchronous at 900 km altitude (NASA - NOAA, 1992), and 18 days revisits, a much coarser temporal resolution than its predecessor, GOES. It had four multispectral bands; two bands in visible, green and red; and the other two in near infrared. The spatial resolutions were 60mx80m for

Landsat 1 and 2; and 38mx38m for Landsat 3. The System's second generation; Landsat 4 and 5 started in July 1982, when Landsat 4 was launched and ended in December 2012 when Landsat 5 was terminated (the longest-operating Earth observation satellite). They have near-polar, sun-synchronous orbits hinged at 705 km altitude with 16 days revisits (NASA - NOAA, 1992) . They carried two sensors, the MSS and the Thematic Mapper (TM). The TM sensor captured data in seven spectral bands in visible: blue, green, red' and the infrared: near-infrared, mid-infrared and thermal infrared. Bands 1-5 and 7 had a spatial resolution of 30mx30m; whereas, Band 6, was 120mx120m. Landsat 6 and 7 formed the third generation and it run from April 1999 when Landsat 7 was launched as Landsat 6 had been lost; to date. It had a relapse from the faulty scan line corrector for six weeks in May 2003. Then, the scientific community reverted back to using Landsat 5 imagery. Landsat 7 Like its predecessor was launched in near-polar, sun-synchronous orbits at 705 km altitude with 16 days revisits as well. It has the Enhanced Thematic Mapper Plus (ETM+) that replicated the TM characteristics; with an additional 15m spatial resolution panchromatic band and two thermal: Bands 6-1 and 6-2 with improved spatial resolutions of 60mx60m. Other major development is the on-board, full aperture, 5% absolute radiometric calibration, which rendered it the most accurately calibrated Earth-observing satellite and the on-board data recorder (SSL, 2013) . The future of Landsat Satellites Systems lays on the fourth generation series, Landsat 8 launched in February 2013 with orbits and revisits similar to its predecessor (NASA, 2013), (USGS, 2013). It is the Landsat Data Continuity Mission (LDCM) generation pledged to ensure the continued acquisition and availability of Landsat-like data beyond the duration of the current Landsat missions. Landsat 8 major developments are the Operational Land Imager (OLI) with nine spectral bands, one panchromatic at with 15m resolution, four visible bands in green, blue, yellow and red; four infrared bands in one in near infrared, two short wave infrared and another one cirrus infrared; all these have 30m resolution. The OLI has improved signal-to-noise ratio compared to past Landsat instruments. It also has a special Thermal Infrared Sensor (TIRS) with two bands at 100m resolution that measures land surface temperatures.

Initially, the costs of the images of this satellite series were prohibitive as they kept on rising from US$ 600, to US$ 3700 reaching a whopping price of US$ 4400 per scene. As a result of this cost recovery mandate its use dwindled. The Landsat coverage standards also languished as many observations were missed because there was no buyer. As a result of this setback, the true scientific mission based on collects as much global data as possible for future scientific study motto failed, and as data was lost forever (Landsat5, n.d.). Eventually, in January 2009 all Landsat data were made free and its use immediately increased 60-fold (Satellite Imaging Coorporation, 2001).

SPOT Satellite System

SPOT System was initiated by the French Space Agency, Centre National d'Études Spatiales (CNES), the Internationale Géographe Nationale (IGN), and Space Manufacturers; this was in collaboration with the Belgian scientific, technical and cultural services (SSTC) and the Swedish National Space Board (SNSB). SPOT is composed of both spacecrafts and a network of direct receiving stations handling images acquired by the satellites. It has been designed to improve the knowledge and management of the Earth by exploring the Earth's resources, detecting and forecasting phenomena involving climatology and oceanography, and monitoring human activities and natural phenomena (Satellite Imaging Coorporation, 2001).

Figure 5. WorldView 3; Very High Spectral Image - Madrid, Spain (Source: Satellite Imaging Cooperation, 2001).

Figure 6. WorldView 3; Very High Spatial Resolution Image – Cape Town, South Africa (Source: Satellite Imaging Cooperation, 2001).

SPOT's onboard sensors can point across the satellite track, providing a revisit capability of one to four days depending on latitude. The satellites are anchored at 832 km latitude and fly over any point on Earth within 26 days. The satellites capture both panchromatic and multispectral digital imagery from visible and infrared electromagnetic wave portion of the spectrum with a spatial resolution of 10m and 20 m respectively. SPOT system has stereo pair imagery which is vital for applications that call for 3D terrain modeling Astrium. (2013). The first generation of SPOT satellite is SPOT 1, 2 and 3 that started in February 1986 up to November 1997. They had 4 spectral bands in one panchromatic, two visible: green and red and in infra red: near infra red. The second generation is composed of SPOT 4 launched in March 1998 to July 2013. It has five spectral bands, one in panchromatic, three in visible: blue, green and red and one infra red. SPOT 5 is the third generation satellite, launched in May 2002 and had five spectral bands It gave higher resolution of 2.5 to 5 m in panchromatic mode, 10 m in multispectral mode: green and red and 20 m on infrared: near infra red and short wave And the fourth generation of these, SPOT 6 was launched in September 2012 and SPOT 7 which was 30 June 2014. This 2 satellite form a constellation at 90^0 and at 694 km altitude and has five spectral bands. It has very high resolutions of 1.5 meters in panchromatic mode, 6 m in multispectral: blue, green and red mode and 20 m on short wave infrared: near infra red. This constellation offers a wide imaging swath, which up to 60 km x 120 km and like all SPOT satellites 3D capabilities (Figure 7). It is essentially used for the creation of 3-Dimentinal data models – Digital terrain/Elevation Models – (DEM/DTM). The only three SPOT satellites which are still in orbit are SPOT 5, 6 and 7.

Figure 7. SPOT – DEM of Los Pelambres Mine, Chile (Source – Airbus Defense Space, 2013)

Satellites With Special Unique Applications

Other types of satellite systems worth mentioning here are those with unique capabilities which are indeed complementary to aero spatial data capture efforts, although, in essence, their longevity and widespread use falls far short from the three explained above. The Indian Remote Sensing Satellite (IRS) satellite series the features mimic both the Landsat sensors and those of SPOT sensors; and has two-channel Wide Field Sensor (WiFS) as well with a coarse resolution that is equivalent to GOES. The new Very High Resolution Remote Sensing Satellites (VHRRS) formats combine the spatial resolution equivalent to aerial photographs with the temporal, spectral and radiometric resolutions, which by far surpasses those possible with those three studied here. Examples are:

- WorldView (3.1m, 29 bands) (Figure 5 – Very High Spectral & Figure 6 - Spatial Resolution),
- GeoEye (0.41m, 4 Bands),
- QuickBird (0.61m, 4 bands),
- IKONOS (0.82m, 5 bands), and
- ASTER (15m, 14 bands).

The Radio Detection and Ranging (RADAR) satellites, known as the all whether, are unique, as they use the microwave energy which relatively has long wavelengths that allows these systems to "see" through clouds, smoke, and some vegetation. They generate their own energy, Active Sensors; so they can be operated day or night. All other satellite sensors, discussed above utilize the energy from the Sun; and are Passive Sensors.

Data From the Two Sources Interchangeability Problems

The main fundamental reasons for the differences between technical and human services sources are in the fabric of the origins of the data, when it happened, when and how it was preserved before it was eventually collected and used for the investigation of land cover change. The resolution of the data is paramount to its usefulness in both methods. It depends on the spatial, the thematic, temporal and the model resolutions. The spatial resolution is the smallest discernible unit in the geographical space whereas the thematic resolution addresses the level of classification aggregation used to show distributions of patterns in an area; the temporal resolution is how often data updates are made; and the model resolution gives the level of details incorporated in assessing any form of phenomena in an area.

- The basic tenet is that remote sensing and GIS and the method for the reconstruction of historical knowledge measured totally different things. For these two methods to be used interchangeably for the investigation of land cover change for a particular geographical area there is a need to reconcile what each of them measures.
- The challenge is to make them measure the same thing. So, to that end; a land cover class specification commensurate to the two methods was needed to be created.
- Besides all that, care must be taken when remote sensing cross generational data is used in the land cover change and trends investigations. This is because the remote sensing platforms and sensors have improved tremendously over the years. Consequently, land cover data from consecu-

tive time period captured by satellite from different generation could show differences pertaining to the differences in visual abilities, due to the improved tools with better clarity that might be mistakenly taken to be the change the landscape.

- So, for that reason, for efficient land cover change investigation employment of trans-generation satellite data has to be taken with caution and some form of standardization and a form of calibration of data from the different generation satellite sources is necessary. The class separability was made within the particular Landsat sensor and across different sensors.

For instance classes which had very little separability in MSS sensor were found to have huge separability in TM and ETM+. It was difficult to attribute these differences to land cover change alone as there is technology development for the better in satellite sensor sensitivity at play as well.

SOLUTIONS AND RECOMMENDATIONS

The land cover is dynamic but at the same time the technology for measuring it keeps on changing as well, there are many improvements. As a consequence, there is no uniform continuous data from which class separability alone could be determined throughout the years.

For these reasons, the percentage of how much class separability is across different land cover classes by different sensors could not be uniformly assessed because of the improved progressive difference in the quality of data itself. Maybe apparent reduction of data quality can be made by applying a statistical ratio to all the data to make it uniform so that its separability throughout the satellite capture years could be determined.

FUTURE RESEARCH DIRECTIONS

The basic tenet for future research would have to be based on the main finding of this research which is that remote sensing and GIS and the proposed new method for the reconstruction of historical knowledge measured totally different things. For these two methods to be used interchangeably for the investigation of land cover change for a particular geographical area there is a need to reconcile what each of them measures. The challenge is to make them measure the same thing, a good research topic for the future.

CONCLUSION

GOES great improvements from the earliest satellites to the present ones with more bands, so specific to weather trouble spots were monitored to assist in improved short-term forecasting. These had seen improvements on the satellite systems hardware themselves with better flexibility and more automatic ground stationed data acquisition elements such as river and rain gauges, seismometers, tide gauges, buoys, and ships weather stations.

Ever since the first satellites were launched in 1972 for Landsat System and 1986 for SPOT Satellites there has been tremendous amount of developments in platform and sensor for both system. These developments are directed towards a certain goal for each system. Landsat developments are more onto the refinement of spectral resolution variable through the improvements of the two sensors from mere 4 bands of the earliest systems to 7, 9 and 12 for the 2nd, 3rd and 4th generation. Landsat system also developments were in the improvements in the infra red more especially in thermal bands which had improved in both its spatial and spectral resolution. Also, generally these satellites have seen a slight improvement in spatial resolution and with the addition of panchromatic whereas its temporal resolutions have been maintained throughout (Figure 3). SPOT boost the very important characteristics of seeing stereo, therefore having 3-dimentions landscape creation. This also gave it increased revisits of one to four days. Although its spectral resolution are much lower 5 as compared to 12 of those of Landsat 5; it has had superior improvements in spatial resolution going from 10m, 5m, 2.5m and 1.5m for panchromatic and 20m, 20m, 10m and 8m for Multispectral from 1st, 2nd, 3rd and 4th generation satellites respectively. Generally, SPOT maintained its temporal and spectral resolutions with slight improvements on its infra red spectra but huge strides were made in the spatial resolutions improvements (Figure 3).

The results of these show that the developments of GOES, Landsat and SPOT are not in competition of one another, but are complementary as each system has maintained its niche in geo-spatial space-view (Figure 3). Other systems like the IRS, VHRRS and RADAR satellites are becoming available and these are added to increased use of digital data.

Remote sensing digital data have shown tremendous development, but on the other hand, need for the human side of information reference services for the interpretation of these more sophisticated digital data increased. It has also been shown that the accuracy of these reconstruction to historical landscape are commensurate with that of remote sensing although work need to be done to harmonize these two inherently diverse data (Figure 2 and Table 1).

REFERENCES

AIRBUS-Group. (2013). *Airbus Group Annual Review*. Toulouse: Astrium.

Atteh, O. D. (1989). *Indigenous local knowledge as key to local-level development: Possibilities, constraints and planning issues in the context of Africa*. Arusha, Tanzania: Coference Paper.

Benedict, B. M. (2002). *Curiosity: A cultural history of early modern inquiry*. Chicago: University of Chicago Press.

Berkes, F., Colding, J., & Folke, C. (2000). Rediscovery of traditional ecological knowledge as adaptive management. *Rediscovery of the Ecological Applications*, *10*, 251–262.

Bossler, J. D., Jensen, J. R., McMaster, R. B., & Rizos, C. (2002). *Manual of geospatial science and technology*. New York: Taylor and Francis.

Brokesha, D., Warren, D., & Werner, O. (1980). *Indigenous Knowledge Systems and Development*. Lanham, MD: University Press of America.

Davidson-Hunt, I., & Berkes, F. (2003). *Learning as You Journey: Anishinaabe Perception of Social-ecological Environments and Adaptive Learning*. Retrieved December 2015, from http://www.consecol.org/vol8/iss1/art5/

Davis, C., Back, K., & MacLean, K. (1992). Understanding Oral History. In K. Phyllis & J. Brent (Eds.), *Public History Readings* (pp. 264–269). Malabar, FL: Krieger.

Delgamuukw v. British Columbia, 3 S.C.R. 1010, 23799 (Supreme Court Judgments December 11, 1997).

Dhavale, D. G. (1984). The Yavanajataka of Sphujidhvaja by David Pingree. *The YavAnnals of the Bhandarkar Oriental Research Institute*, *65*(1/4), 266–267.

Donovan, D., & Puri, R. (2004). *Ecology and Society*. Retrieved December 26, 2015, from Vol 9 No. 3 Art 3: http://www.ecologyandsociety.org/vol9/iss3/art3/

Doody, D., & Stephan, G. (1993). *NASA - Jet Propulson Laboratory*. Retrieved December 26, 2015, from Basics of Space Flight: https://solarsystem.nasa.gov/basics/editorial.php

Graham, W. (2016, November 19). *Atlas V successfully launches with GOES-R advanced weather satellite*. Retrieved from https://www.nasaspaceflight.com/2016/11/atlas-v-launch-goes-r-advanced-weather-satellite/

Gunter, D. K. (1996). *Gunter's Space Page*. Retrieved August 14, 2013, from GOES: http://space.sky-rocket.de/doc_sdat/goes-d.htm

Hancock, G. (1995). The fingerprints of god: A quest for the beginning and end. London: William Heinemann Ltd.

Hill, S. (2008). *GOES-R Solar and Space Environment Data Products: Benefiting Users*. Retrieved December 26, 2015, from GOES-R Solar and Space Environment Data ProduGeostationary Operational Environmental Satellite Program (GOES): ww.weather.gov/satellite

Jensen, J. R. (2005). *Introduction to digital image processing: A remote sensing perspective*. Upper Saddle River, NJ: Prentice Hall.

Jones, M. H., & Lambourne, R. J. (2004). *An Introduction to Galaxies and Cosmology*. Cambridge, UK: Cambridge University Press.

Kitaura, F. S., & Enßlin, T. A. (2008). Bayesian reconstruction of the cosmological large-scale structure: Methodology, inverse algorithms and numerical optimization. *Monthly Notices of the Royal Astronomical Society*, *389*(2), 497–544. doi:10.1111/j.1365-2966.2008.13341.x

Landsat5. (n.d.). Retrieved from http://landsat.gsfc.nasa.gov/landsat-5/

LaViolette, P. E., & Diachok, O. I. (1974). he Use of APT satellite imagery in a subarctic airborne oceanographic survey. *Arctic*, *27*(4), 249–324. doi:10.14430/arctic2888

Leese, J. A., Noar, P. F., & Pastre, C. (1989). Operational continuity of the global meteorological satellite network. *Space Policy*, *5*(1), 12–24. doi:10.1016/0265-9646(89)90025-8

Leffler, P. K., & Brent, J. (1992). *Public history readings*. Malabar, FL: Krieger Publishing Company.

Lightfoot, K. G. (1995). Culture Contact Studies: Redifining the relationship between prehistroric and historical archaeology. *American Antiquity*, *60*(2), 199–217. doi:10.2307/282137

Lillesand, T. M., & Kiefer, R. W. (2004). *Remote sensing and image interpretation.* Chichester, UK: John Wiley & Sons.

Linga, P., Kumar, R., & Englezos, P. (2007). Gas hydrate formation from hydrogen/carbon dioxide and nitrogen/carbon dioxide gas mixtures. *Chemical Engineering Science, 62*(16), 4268–4276. doi:10.1016/j.ces.2007.04.033

Livingstone, D. (1857). *The missionary travels and researches in south africa.* London: Royal Geographical Society.

Mabona, M. (2004). *Diviners and prophets among the Xhosa (1593-1856): A study in Xhosa cultural history.* Münster: Lit Verlag.

Maphanyane, J. G. (2012, December 1). *The investigation of land cover change and trends through the reconstruction of historical landscape* (PhD Thesis). Melbourne, Victoria, Australia: The University of Melbourne.

Mather, P., & Koch, M. (2011). Computer processing of remotely-sensed images: An introduction. Wiley-Blackwell. doi:10.1002/9780470666517

NASA. (2013, July 28). *Landsat.* Retrieved August 22, 2013, from Landsat 8: http://www.nasa.gov/mission_pages/landsat/spacecraft/index.html

NASA - NOAA. (1992, August 28). *GOES - R - SERIE.* Retrieved September 06, 2013, from GOES-R Spacecraft Overview: http://www.goes-r.gov/spacesegment/overview.html

Nichol, J. E., Shaker, A., & Nichol, J. E. (2006). Application of high-resolution stereo satellite images to detailed landslide hazard assessment. *Geomorphology, 76*(1-2), 68–75. doi:10.1016/j.geomorph.2005.10.001

Opara, L. U. (2003). Traceability in agriculture and food supply chain: a review of basic concepts, technological implications, and future prospects. *Agriculture and Environment, 1*(1), 101–106.

Ortner, D. J., Butler, W., Cafarella, J., & Milligan, L. (2001). Evidence of probable scurvy in subadults from archeological sites in North America. *American Journal of Physical Anthropology, 114*(4), 343–351. doi:10.1002/ajpa.1046 PMID:11275963

Rey, C. (1988). *Monarch of all I survey: Bechunaland diaries 1929 - 1937.* Gaborone: Botswana Society: Lilian Barber Press Inc.

Russell, E. (1998). *People and the land through time: Linking ecology and history.* New Haven, CT: Yale University Press.

Satellite Imaging Coorporation. (2001). *Satellites.* Retrieved August 22, 2013, from Satallite Data: http://www.satimagingcorp.com/satellite-sensors/spot-5.html

Schowengerdt, R. A. (2007). *Remote sensing: models and methods for image processing.* Burlington, MA: Academic Press.

Segundo, E. (2010, September 01). *Boeing.* Retrieved December 26, 2015, from GOES-15 Weather Satellite: http://boeing.mediaroom.com/2010-09-01-Boeing-built-GOES-15-Weather-Satellite-Enters-Service-for-NASA-NOAA

Shoemaker, H. G., & Shoemaker, E. M. (1987). *In the Footsteps of John Wesley Powell*. Boulder, CO: Johnson Books and Powell Society.

SSL. (2013). *SPACE System Lorel*. Retrieved August 22, 2013, from GOES I: http://www.ssloral.com/html/satexp/goes.html

Stringer, C. (2001). Modern Human Origins: Progress and Prospects. *The Royal Society*, *357*, 563–579. PMID:12028792

USGS. (2013, August 29). *Landsat Mission*. Retrieved September 22, 2013, from Landsat 8: http://landsat.usgs.gov/about_ldcm.php

KEY TERMS AND DEFINITIONS

Archaeology: The importance of archaeology lies in the fact that it seeks to learn about culture from the fragmentary remains of the products of human activity (Deetz, 1992).

Biostratigraphic Markers: Biostratigraphy is a sub-discipline of sedimentary geology that relies on the physical zonation of biota, both in time and space, in order to establish the relative stratigraphic position (i.e. older, younger, same age) of sedimentary rocks between different geographic localities.

Classification: The process of classification identified and assigned each pixel of all channels of the multi-spectral images to a particular class or theme based on the statistical characteristics of the pixel brightness values known as spectral signatures.

Electromagnetic Waves Energy: The electromagnetic (EM) spectrum describes the range of wavelengths of energy that can be recorded using remote sensing and It can be broadly divided (by increasing wavelength) into Gamma rays, X-rays, Ultra-violet light, Visible light, Infrared light, Microwave and Radio waves.

Geographical Information Systems (GIS): Geographical Information Systems (GIS) is defined as a technique which people employ using computers and specific computer software and hardware to locate physical features and describe their characteristics and condition on a raster based or vector co-ordinate based digital map by interrogating attribute data and to engage in spatial analysis, and this is a sound basis for informed decision making (Burrough & McDonnell, 1998).

Geo Spatial Science (GSS): Geospatial Science is a discipline that focuses on using information technology to understand people, places, and processes of the earth. Spatial analysis of human and physically variables is fundamental to the discipline. Remote Sensing, Geographic Information Systems, and Global Positioning Systems technologies are commonly used as measurement, observation, and analysis tools for this (Radford University, 2014).

Geo-Stationary: Geostationary satellite Platforms are platforms that revolve at speeds which match the rotation of the Earth so they seem stationary, relative to the Earth's surface. This allows the satellites to observe and collect information continuously over specific areas.

Geo-Synchronized and Sun-Synchronous: The platforms are designed to follow a north-south orbit and in conjunction with the Earth's west-east rotation, they are able to capture the entire Earth's surface over a certain period of time. These satellite platforms orbits are often sun-synchronous: - They are able to cover each particular area on the Earth's surface at a constant local time of day. This ensures consistent illumination conditions for images of the same area that have been acquired at different times.

Land Cover Change: Land cover change forms a fundamental part of sustainable resource management. It is actually the main indicator of natural resource use.

Multi-Spectral: A multispectral image is one that captures image data at specific frequencies across the electromagnetic spectrum, the visible light range, and infrared. Spectral imaging sees beyond what human eye can perceive.

Panchromatic: Black and white aerial photograph or satellite images.

Platform: In order for the sensor to capture and record reflected or emitted energy from a target it must reside on a stable platform which can be on the ground (hand held camera), within the atmosphere (airplanes), and in space (the space shuttle or satellite).

Radiometric Resolution: The radiometric resolution of an imaging system describes its ability to discriminate very slight differences in energy.

Reconstruction of Historical Landscape (ROHL): *Reconstruction of historical landscapes* is defined as a method of building and mapping past landscapes by finding out how communities lived in a given historic period in a particular geographical area. This is achieved by identifying, reading and analysing marks made at the time and left behind from certain period. A historical landscape basically depicts a type of cultural landscape that contains, within a specific geographical area, both natural and human-made features that typify connected human activities, past events or patterns of physical development (Maphanyane, 2012).

Remote Sensing: Remote Sensing is the art, science, and technology of obtaining reliable information about physical objects and the environment through the processes of recording, measuring, and interpreting photographic images and patterns of electromagnetic radiant energy and other phenomena (Lillesand & Kiefer, 2004).

Sensor: Sensors are devices that functions like a human eye. They detect the reflected or emitted electromagnetic radiation from natural sources (Passive Remote Sensing) or detect responses from objects which are irradiated from artificially generated energy sources such as radar (Active Sensors).

Spatial Resolution: Spatial Resolution, Pixel Size, and Scale. With remote sensing instruments, the distance between the target being imaged and the platform, plays a large role in determining the detail of information obtained and the total area imaged by the sensor.

Spectral Resolution: Different classes of features and details in an image can often be distinguished by comparing their responses over distinct electromagnetic wavelength ranges. Spectral resolution describes the ability of a sensor to define fine wavelength intervals.

Temporal Resolution: Temporal resolution: Other than: Spatial, Spectral and radiometric resolution; there is temporal resolution. This is the time it takes to image the exact same area at the same viewing angle a second time: Revisit Time.

Traditional Ecological Knowledge (TEK): The traditional ecological knowledge as a cumulate body of knowledge and beliefs, handed down through generations by cultural transmission, about the relationship of living beings with one another and with their environment. These are gained through the teaching of family history and cultural values. They have been recorded for thousands of years by storytelling, praise-singing, and songs as oral history, and are passed on to and learned by the descendants through the recitation of the narrative at events and during ceremonies (Berkes, Colding, & Folke, 2000).

This research was previously published in the Encyclopedia of Information Science and Technology, Fourth Edition edited by Mehdi Khosrow-Pour, pages 3484-3502, copyright year 2018 by Information Science Reference (an imprint of IGI Global).

Chapter 26
Use of GIS and Remote Sensing for Landslide Susceptibility Mapping

Arzu Erener
Kocaeli University, Turkey

Gulcan Sarp
Suleyman Demirel University, Turkey

Sebnem Duzgun
Middle East Technical University, Turkey

ABSTRACT

In recent years, geographical information systems (GISs) and remote sensing (RS) have proven to be common tools adopted for different studies in different scientific disciplines. GIS is defined as a set of tools for the input, storage, retrieval, manipulation, management, modeling, analysis, and output of spatial data. RS, on the other hand, can play a role in the production of a data and in the generation of thematic maps related to spatial studies. This study focuses on use of GIS and RS data for landslide susceptibility mapping. Five factors including normalized difference vegetation index (NDVI) and topographic wetness index (TWI), slope, lineament density, and distance to roads were used for the grid-based approach for landslide susceptibility mappings. Results of this study suggest that geographic information systems can effectively be used to obtain susceptibility maps by compiling and overlaying several data layers relevant to landslide hazards.

INTRODUCTION

In recent years, geographical information systems (GISs) and Remote Sensing (RS) have proven to be common tools adopted for various studies in different scientific disciplines. GIS provides as a set of tools for the input, storage, retrieval, manipulation, management, modeling, analysis and output of spatial data. RS, on the other hand, offers earth observation data for thematic maps related to spatial studies. The use

DOI: 10.4018/978-1-5225-7359-3.ch026

of GIS and RS data for landslide susceptibility mapping are demonstrated by three different landslide susceptibility maps with five different variables (Normalized Difference Vegetation Index (NDVI), Topographic Wetness Index (TWI), slope; lineament density and distance to roads). The comparison of the generated final susceptibility maps with historical landslide locations is given with important factors affecting the accuracy of susceptibility map. The accuracy analysis of the final susceptibility maps for various weighting strategies is performed. The results indicate that assignment of weights to the slope parameter impacts the accuracy in the high susceptible zones.

BACKGROUND

Landslides are among the most common natural hazards and are the most damaging, leading to substantial economic, human, and environmental losses throughout the world. The quantitative assessment of landslide hazards for a large area is critical for mitigation of the associated risks. They are often triggered by natural phenomena and/or human activity, such as earthquakes, precipitation, erosion, deforestation etc. and are difficult to predict. One of the greatest limiting factors in predicting and mapping landslide activity is the lack of understanding of scale-dependent processes, such as erosion, weathering, and fracturing (Glenn et al., 2006). Such maps normally aim at providing a document that depicts the likelihood or possibility of new movements occurring in an area, and therefore helping to reduce future damages. To express the potential for occurrence of landslides in a quantitative manner, maps must incorporate the concept of probability, which is an assessment of the relative frequency of occurrence (Ohlmacher & Davis, 2003). Susceptibility expresses the likelihood that a landslide will occur in an area on the basis of the local terrain conditions (Soeters & Van Westen, 1996); return period or annual probability of occurrence is not considered. The main difference between susceptibility and hazard is therefore that the latter considers the temporal factor, by estimating the probability of occurrence of the phenomenon within a specified period of time (Varnes, 1984) whereas the former considers the likelihood of landslide occurrence. There are many studies in the literature about use of GIS to evaluate landslide susceptibility (e.g. Gokceoglu et al. 2005; Akgun et al. 2011; Akgun 2012; Pradhan et al. 2013; Kavzoglu et al. 2013)

MAIN FOCUS

The purpose of this study is to apply the grid based GIS techniques for landslide susceptibility mapping using five different factors including Normalized Difference Vegetation Index (NDVI), Topographic Wetness Index (TWI), slope, lineament density, and distance to roads. The scope includes the preparation of landslide susceptibility map to identify highly susceptible areas and, the accuracy assessment related to the obtained maps.

The susceptibility assessment methodology is demonstrated for More and Romsdal region in Norway (Erener & Duzgun, 2010). The study area occupies approximately 606.755 km^2 in the west part of Norway. The upper left coordinates on 112707,770408 m - 6952112,603469 m and lower right coordinates 6929466,479194 m -144909,272731 m respectively (Figure 1).

Figure 1. Study region (parts of figure adopted from NGI)

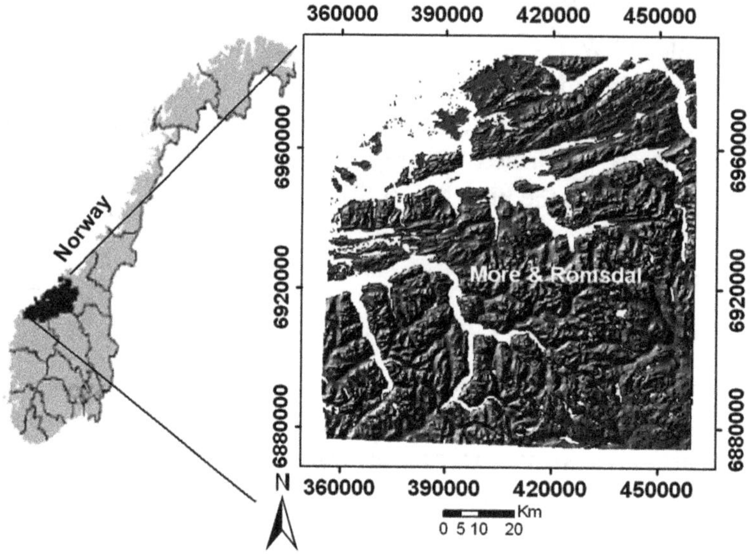

Method of Study

An empirical approach is used to map and evaluate landslide susceptibility. In this approach a grid based GIS is adopted to construct a landslide susceptibility map. Five layers of data with 30 × 30 m resolution grid were superimposed to create the landslide susceptibility map. Slope is given the most emphasis, followed by, TWI, NDVI, lineament density and distance to roads. A numerical rating system is applied and each of the five factors is grouped into three categories, and each category is assigned a value between 1 and 3, with 1 being least susceptible and 3 most susceptible to landslides. Based on their relative importance to slope instability in the study area, the five factors are assigned weights between 0.0 and 1.0 (collectively adding to 1.0). A raster-based GIS is used to overlay the five layers with 30 × 30 m resolution and calculate a Landslide Susceptibility Index (LSI) for each individual cell. The final map shows areas of low, medium, and high landslide susceptibility. The method of the study is shown in Figure 2.

Scale and properties of the data used in the study for susceptibility mapping is given in Table 1. Before beginning the processes, the water areas are masked from the layers. Vector based 50 meters

Table 1. Data scale and properties

Classification	Sub-Classification	GIS Data Type	Scale or Grid Size
Geological Hazard Map	Landslide	Point Coverage	1/500 000
Basic Map	Elevation	Line Coverage	1/100000
	Slope	GRID	30*30
Derived Maps	Topo. Wetness Index	GRID	30*30
	Lineament	Line Coverage	1/100 000
	Vegetation Index	GRID	1/100 000

Figure 2. The method of the study

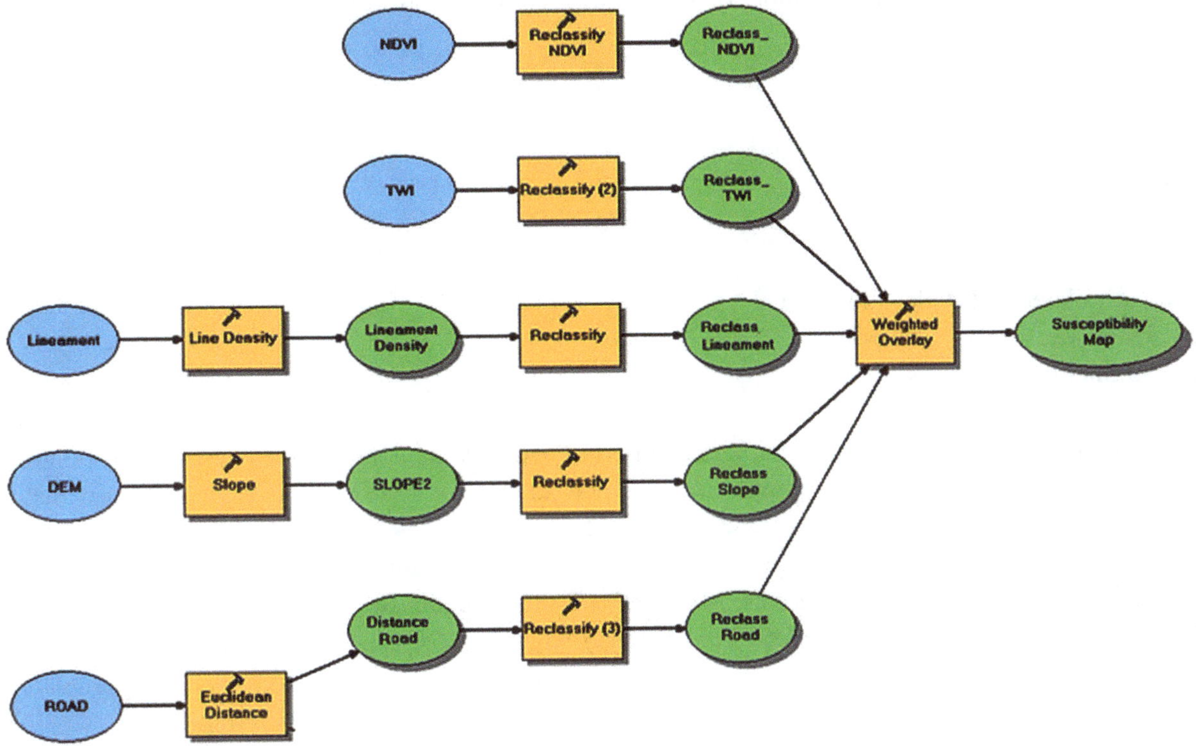

contour interval topographic maps, provided from the Geological Survey of Norway (NGU), are the main data used in the study (Figure 3a). DEM data are generated from maps with a cell size of 30 m using triangular irregular network. The result of the generated DEM includes the highest and the lowest elevation 1807.46 m-13.7282 m, respectively (Figure 3b).

The vertical accuracy of the DEM data ae tested with root-mean-square error (RMSE) statistic. For this test some principles suggested by USGS are used. According to USGS National Mapping Program

Figure 3. Contour map (a), DEM data of the area (b) categorized slope map of the area (0-25=1; 25-50=2 and 50-84=3)

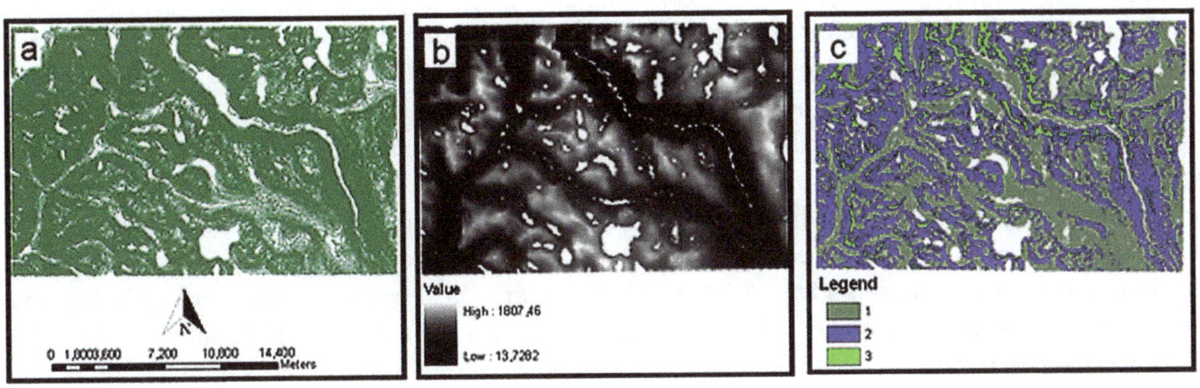

Technical Instructions, a representative sampling of test of points is used to verify the accuracy of any category of DEM. A minimum number of 28 test points per DEM is required (20 interior points and 8 edge points) (USGS).

The vertical RMSE is defined as:

$$RMSE = \sqrt{\frac{\sum \left(z_i - z_t\right)^2}{n}}$$

z_i =interpolated DEM elevation of a test point
z_t =True elevation of a test point
n = number of test points.

Accuracy is computed by a comparison of interpolated elevations in the DEM with corresponding measured elevations from the topographic map. The total vertical RMSE is found 2.30 meter which is quite admissible within the accuracy requirements of the study.

Slope, which measure of the steepness of an area on the Earth's surface, is the most important factor to affect the landslide occurrence (Lee & Min 2001). The steeper the slope, the more likely it is to slide. The slope map is derived from the DEM of the study area. The slope layer has values ranging from 0° to 84,55°, The slope image is then classified into three slope angle classes (0-25=low, 25-50= moderate and 50-84,55=high), and each class is assigned a value between 1 and 3, with 1 being least susceptible and 3 most susceptible to landslides. The result is shown in Figure 3c.

The TWI has been used extensively to describe the effect of topography on the location and size of saturated source areas of runoff generation. TWI is a function of both the slope and the upstream contributing area per unit width orthogonal to the flow direction. The calculation of the wetness index is based on logical ideas of down slope water movement and accumulation of water at the base of slopes and in depressions or swales where there is convergence of flow (Kokkila, 2002). TWI is calculated using the following expression:

$TWI = (As/\tan \alpha)$

where, *As* is the specific catchments area and α is slope angle. For TWI fill sinks operation (Figure 4a), flow direction (Figure 4b), flow accumulation (Figure 4c) operations are applied to the DEM data. Resultant map of the TWI is given in the Figure 4d.

The result of the TWI is categorized into three classes by using natural breaks algorithm. This is done by seeking to minimize each class's average deviation from the class mean, while maximizing each class's deviation from the means of the other groups. In other words, the method seeks to reduce the variance within classes and maximize the variance between classes (Jenks, 1967). TWI categorization was performed by considering the high slope areas doesn't keep rain and has less susceptibility however hollow area keeps more water has high topographic index showing high susceptibility to landslide. For this reason, low index values between 1.08-5.86 areas are given a score (1) showing low susceptibility; medium index values between 5.86 – 8.70 are given a score (2) showing medium susceptibility and, finally high index values between 8.70-20 are assigned a score (3) showing high susceptibility. The result is shown in Figure 4e.

Figure 4. Topographic wetness index generation steps: fill sinks operation 4 (a), Flow direction (b), flow accumulation (c) wetness index is (d) and reclassified topographic index (e).

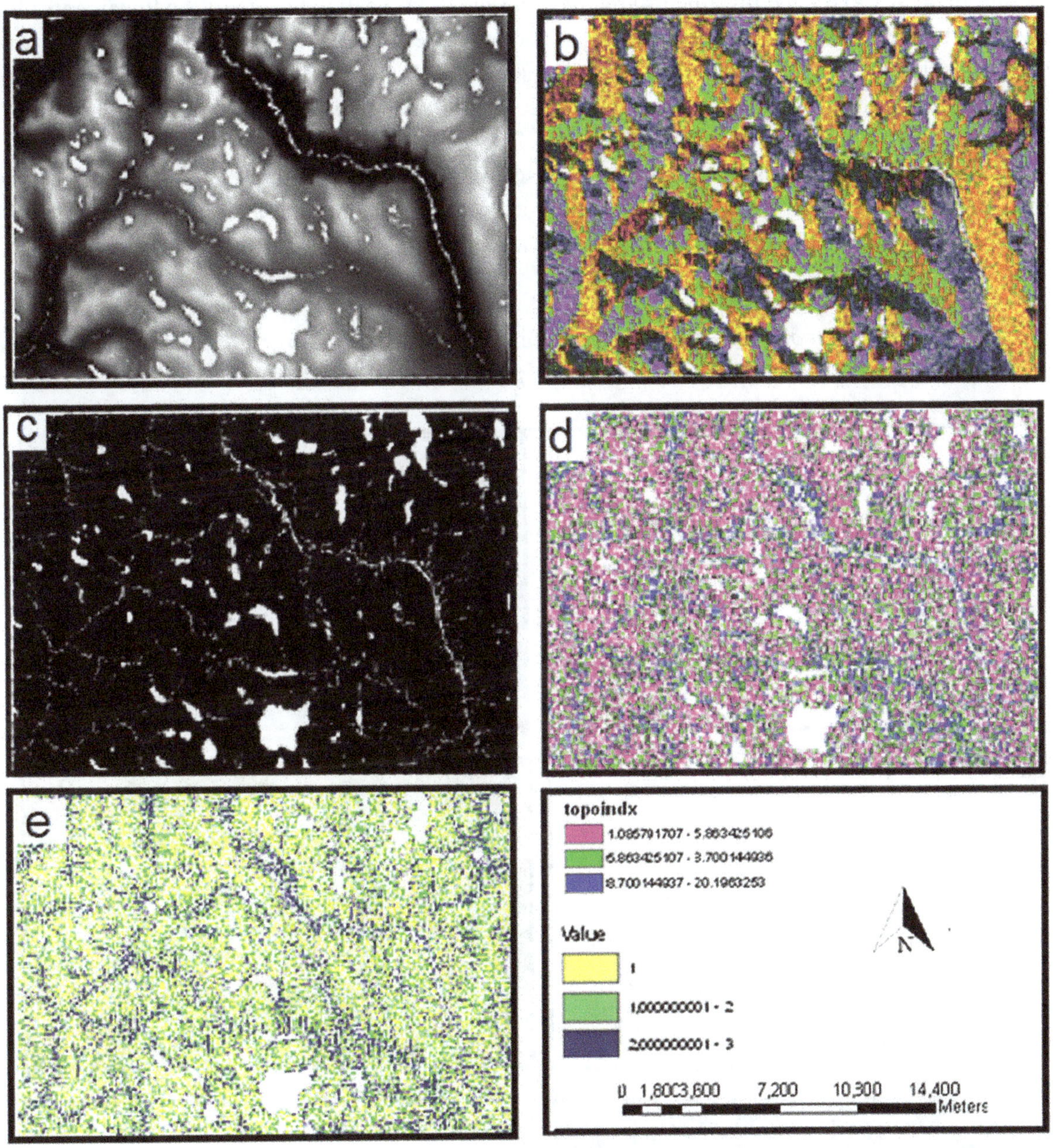

The lineament density of the area has a significant influence on the occurrence of landslides. Lineament density is also used in landslide hazard assessment (Atkinson & Massari 1998; Sarkar & Kanungo 2004; Suzen & Doyuran 2004) as it is generally considered the probability of landslides occurring is greater in highly fractured areas compared to those with a lower fracture density. In the study lineaments are extracted from Landsat ETM image by using line option of PCI Geomatica software. Band 7 with a spatial resolution 30x30 meter is selected for the extraction of the lineaments considering the purpose of this study; since this band is useful for discrimination of lineaments and other geological features such as mineral and rock types and is also sensitive to vegetation moisture content (Sabins, 1996). The results of the automatically extracted lineaments are given in Figure 5a. Spatial density of the lineaments of the area is determined and this density again categorized into three classes. Low dense lineament between 0-0.29 is considered as least susceptible areas and scored with (1); Medium dense lineament between 0-0.63 is scored with (2), density of lineament higher than 0.63 being most susceptible to landslides is scored with (3) which result is shown in Figure 5b.

The external factors such as road construction also contribute to occurrence of the landslide. For that reason, the road network of the study area also is taken into consideration as a triggering factor. The road map of the study area is provided from NGU in a vector format (Figure 5c). The highest susceptible

Figure 5. Lineament map of the study area (a); Lineament density map (b); Road network of the study area (c); Buffered and Reclassified road network of the area (d), Vegetation Indices (e)

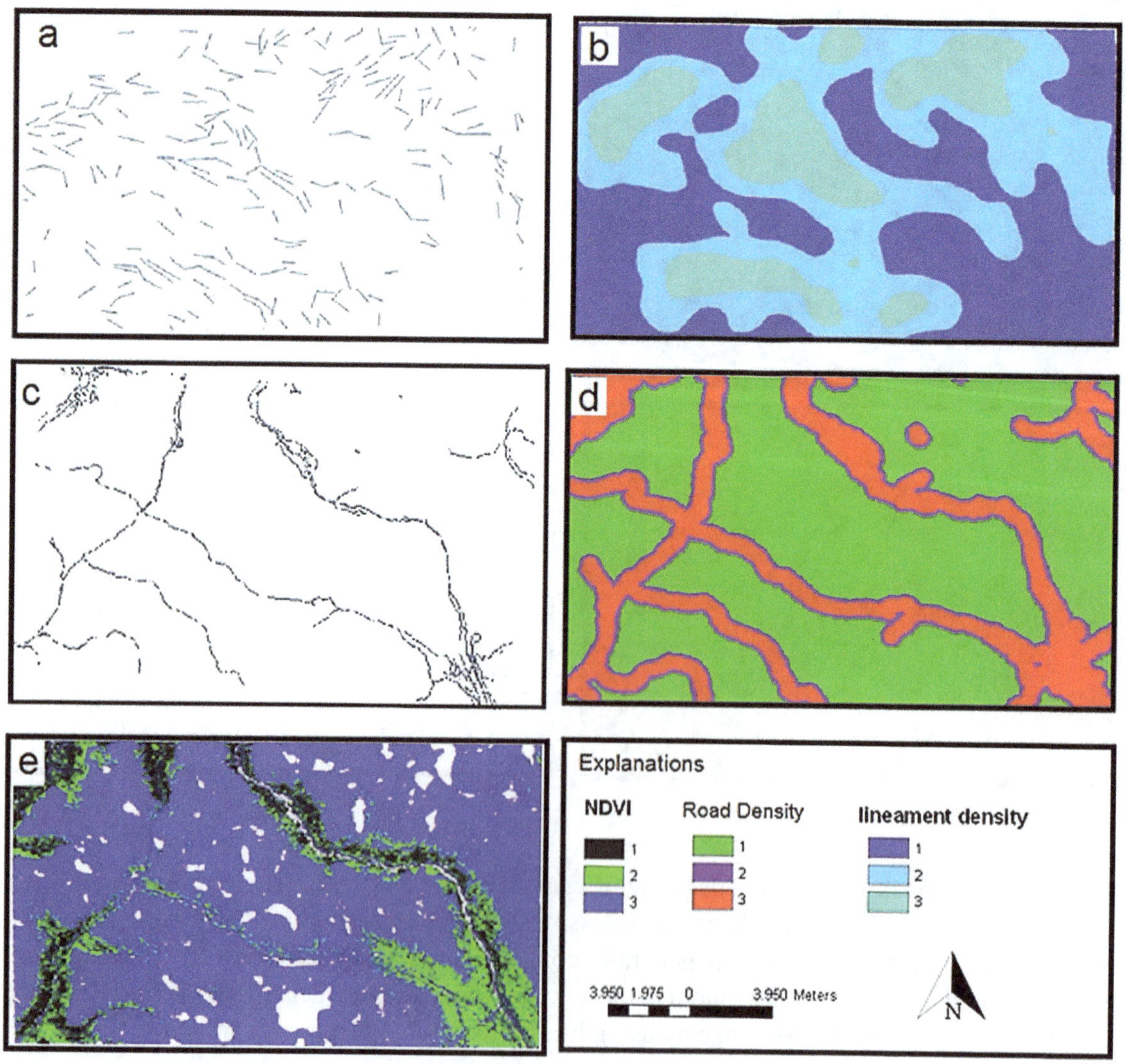

areas closest to road in 200m buffer area are given high susceptibility score (3). Road between 200-400m buffer area given medium susceptibility score (2), Road buffer>400 m is given low susceptibility score as shown in (Figure 5d).

The NDVI is produced using 4th and 3rd bands of Landsat ETM data the water areas are excluded from the map by using masking operation. The resultant map is grouped into three class and densely vegetated areas rages with NDVI values between 0.33-0.98 and given score of 1; Medium dense areas with NDVI values between 0.12-0.33 are scored as 2; Low dense areas with NDVI values less than 0.012 scored as 3. Showing 1 as least susceptible and 3 as most susceptible to landslides, the results are illustrated in Figure 5e.

SOLUTIONS AND RECOMMENDATIONS

Final Map Generation

In order to generate landslide susceptibility map; a numerical rating system is applied to five factors that contribute to landslide occurrence: slope, TWI, NDVI, lineament density and distance to roads. Each of the five factors is grouped into three categories, and each category is assigned a value between 1 and 3, with 1 being least susceptible and 3 being the most susceptible to landslides. Scoring for slope, NDVI, and lineament density is based upon previous studies, which showed that higher slopes, less vegetation, and denser lineaments increase landslide susceptibility (Sarkar et al., 1995; Pachauri et al., 1998). The landslide casual parameters are weighted by means of direct, pair-wise, and rank ordering comparison and the output is a composite index map (Castellanos and Van Westen 2007). Based on their relative importance to landslide occurrence in the study area, the five factors are assigned weights between 0.0 and 1.0 (collectively adding to 1.0). Slope is assigned the highest weight, followed by TWI, NDVI, lineament density and distance to roads. Different weights are assigned to five different layers as a result of weighing operation each weights produced different landslide susceptibility map. In order to define most suitable landslide susceptibility map of the area the accuracy of the generated map is tested with the previously defined landslide data which is shown in the accuracy assessment section in Table 3

Performance Evaluation of Final Susceptibility Maps

The accuracy of the three final landslide susceptibility maps are evaluated by using the GIS overlay analyses. For this operation a buffer zone of 30 m. is assigned to the previously defined real landslide data. 30m buffer is selected due to 1 mm shift at the location of the landslide locations on the map, which should always be considered because this amount is almost thickness of the pen used for producing landslide locations on the topographic map.

Accuracy of the results largely depends on the selection of the buffer amount. A small buffer will result in matching few landslide locations. In this study, different values for the buffer are tested. Larger buffer sizes intersect with nearly all highly susceptible zones. For this reason the amount of buffer is kept minimum, which is equal to the pixel size of the maps. Three different weighting is applied in this study in order to see the effects of the weights to the final map (Table 2). The results show that: generally, the medium susceptible areas show the highest accuracy as map 1 (Figure 6c) gives 70% (Table 3), map 2 (Figure 7c) and map 3 (Figure 8c) gives 95% accuracy (Table 3). The highest susceptibility

Figure 6. Results of final map 1 (a); High risky areas intersect with landslide points (b); Results of final map 3; Medium risky areas intersect with four landslide point s(c), and Low risky areas intersect with four landslide points (d)

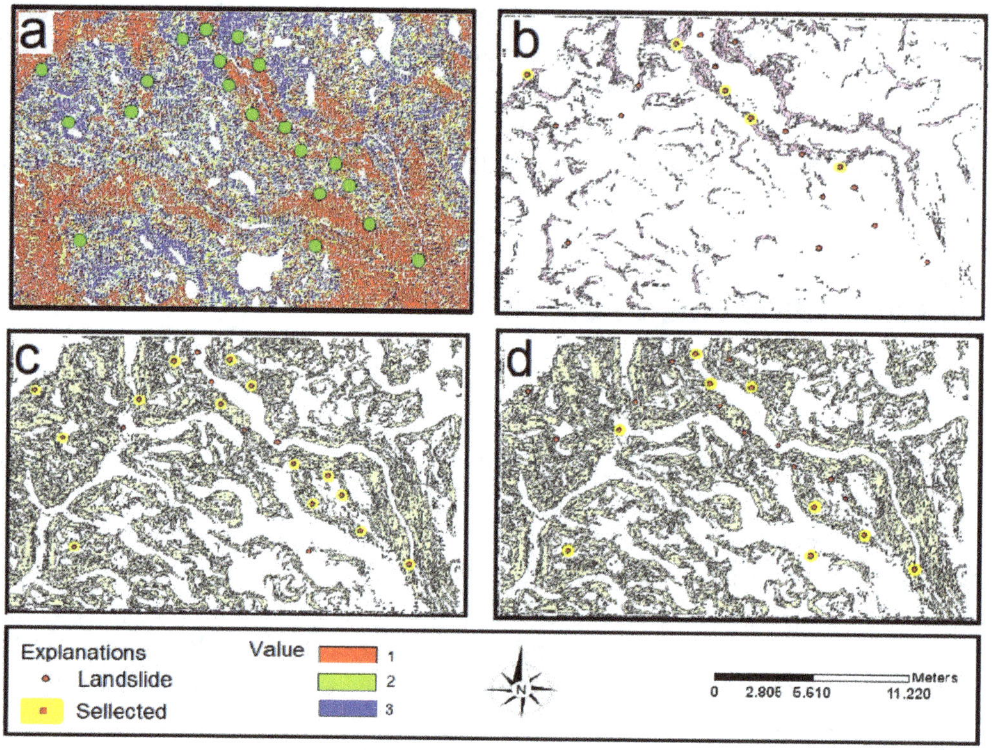

Table 2. Weighting factor used for map 1

Data Layers	Slope Gradient	Topographic Index	Vegetation Index (NDVI)	Lineament Density	Road Networks
Weighting factors of map1	0,3	0,25	0,2	0,15	0,1
Weighting factors of map2	0,4	0,3	0,15	0,1	0,05
Weighting factors of map3	0,35	0,15	0,15	0,25	0,10

Table 3. Accuracy of map 1, 2, and 3

	Total Landslide Points	Matching Points With Most Susceptible Areas	Matching Points With Medium Susceptible Areas	Matching Points With Less Susceptible Areas
Map 1	20	5	14	9
Accuracy		40%	70%	45%
Map 2	20	14	19	7
Accuracy		70%	95%	35%
Map 3	20	12	19	9
Accuracy		60%	95%	45%

Figure 7. Results of final map 2 (a); High risky areas intersect with landslide points (b); Results of final map 3; Medium risky areas intersect with four landslide point s(c), and Low risky areas intersect with four landslide points (d)

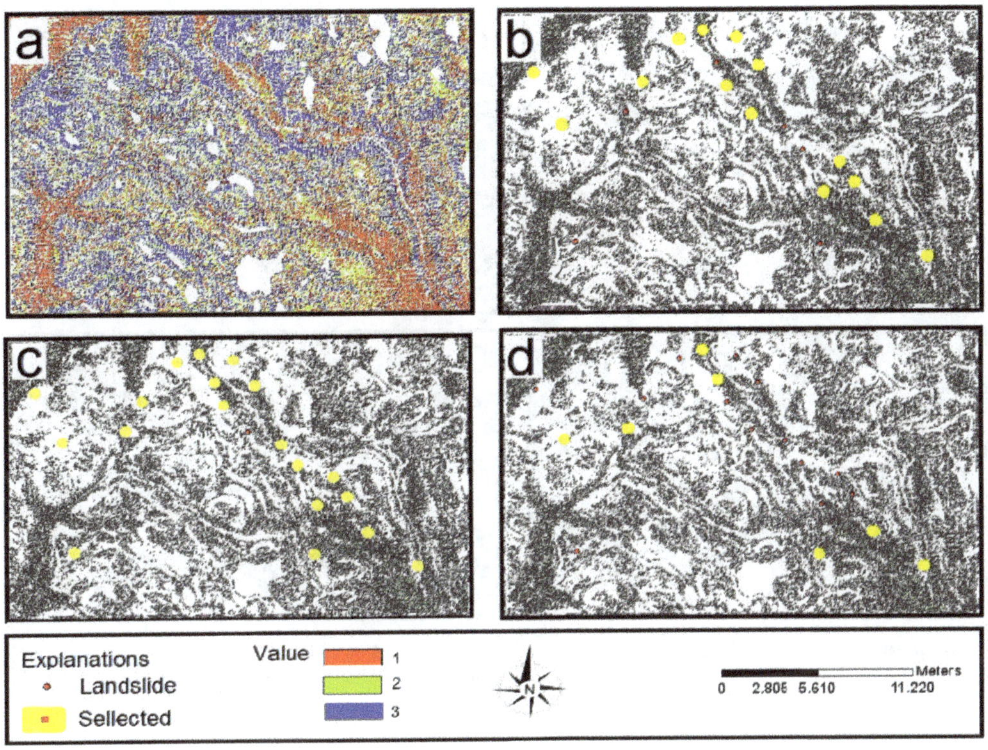

accuracy is obtined in the map 2, it may be due to a weight factor of 40%, which is assigned to slope. It can be concluded that the weighting factors assigned to map 2 gives the better estimate of susceptible regions in the study region.

The proportions of susceptibility zones of the whole region are calculated. High risk zone is reached the maximum value (97.01 km^2) in the Map 2. Minimum High risk zone value (41.26 km^2) is in the Map1 as shown in the Table 4.

Table 4. Proportions of susceptibility zones to the whole region

	Area km^2			Proportion%		
	Low	**Medium**	**High**	**Low**	**Medium**	**High**
Map1	89.88	215.67	41.26	14.81	35.55	6.80
Map2	75.16	166.39	97.01	12.39	27.42	15.99
Map3	96.85	170.57	87.46	15.96	28.11	14.415
Total Region	606.76	606.76	606.76			

Figure 8. Result of Map 3 (a); High risky areas intersect with landslide points (b); Results of final map 3; Medium risky areas intersect with four landslide point s(c), and Low risky areas intersect with four landslide points (d)

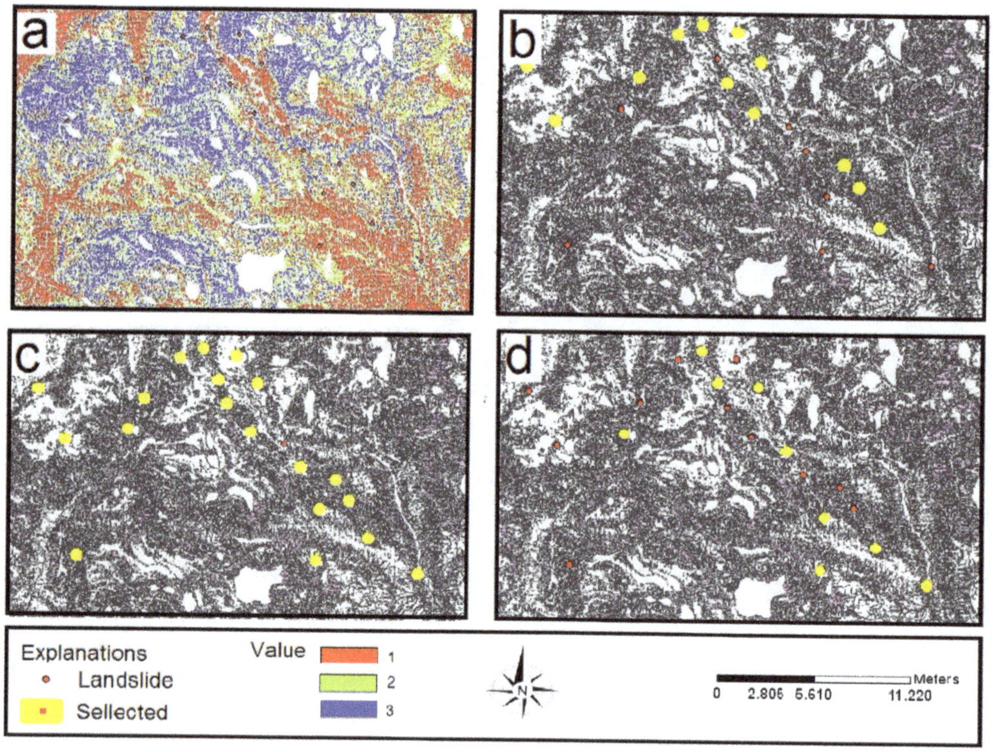

FUTURE RESEARCH DIRECTIONS

In this study it is aimed to focus on use of GIS and RS data for landslide susceptibility mapping. In order to discuss the performance of the study, three different weighting factors were evaluated. The weights were obtained from a single landslide expert. The study may be extended in the future by preparing a survey to different landslide experts in order to obtain the weights. This may provide the researcher how the expert knowledge affects the performance of the susceptibility mapping (Wang, 2008). Another issue is the factors handled in the study. In this study five different factors are used in order to obtain the susceptibility maps. It is worth to study the influence of inclusion of different factors to the susceptibility mapping in the future. One other issue may be to evaluate the quality of the factors. Factors may involve uncertainty and their propagation in susceptibility mapping should be analyzed in order to determine uncertainty in the obtained maps in a systematic way.

CONCLUSION

In the present study, landslide susceptibility mapping using combination of various factors responsible for landslide susceptibility is presented. Each factor has relative importance to probable landslide activity. A reliable and accurate susceptibility map depends on the inclusion and proper determination of the role of these parameters. The accuracy of the generated maps can change according to the buffer size applied on the previously occurred landslide test points. Large buffer size will provide higher accuracy so it is a critical value and subjective process. The result of generated three different landslide maps according to five different variables assigned with different weight factors show that different weights result in different susceptibility maps. The comparison of the generated final susceptibility maps with previously occurring landslide points shows that the most important factor affecting the accuracy is the slope of the study area. This signifies that the weight assigned to the slope generate the highest accuracy in the high risk zones. Results of this study suggest that GIS and RS can effectively be used to compile data and overlay several data layers relevant to landslide susceptibility mapping.

REFERENCES

Akgun, A., Kıncal, C., & Pradhan, B. (2012). Application of remote sensing data and GIS for landslide risk assessment as an environmental threat to Izmir city (west Turkey*). Environmental Monitoring and Assessment, 184*(9), 5453–5470. doi:10.100710661-011-2352-8 PMID:21915598

Akgun, A., Sezer, E. A., Nefeslioglu, H. A., Gokceoglu, C., & Pradhan, B. (2011). An easy-to-use MATLAB program (MamLand) for the assessment of landslide susceptibility using a Mamdani fuzzy algorithm. *Computers & Geosciences, 38*(1), 23–34. doi:10.1016/j.cageo.2011.04.012

Atkinson, P. M., & Massari, R. (1998). Generalised linear modelling of susceptibility to landsliding in the Central Apennines, Italy. *Computers & Geosciences, 24*(4), 373–385. doi:10.1016/S0098-3004(97)00117-9

Castellanos, E., & Van Westen, C. J. (2007). Generation of a landslide risk index map for Cuba using spatial multi-criteria evaluation. *Landslide, 4*(4), 311–325. doi:10.100710346-007-0087-y

Glenn, N. F., Streutker, D. R., Chadwick, D. J., Thackray, G. D., & Dorsch, S. J. (2006). Analysis of LiDAR-derived topographic information for characterizing and differentiating landslide morphology and activity. *Geomorphology, 73*(1-2), 131–148. doi:10.1016/j.geomorph.2005.07.006

Gokceoglu, C., Sonmez, H., Nefeslioglu, H. A., Duman, T. Y., & Can, T. (2005). The 17 March 2005 Kuzulu landslide (Sivas, Turkey) and landslide susceptibility map of its near vicinity. *Eng. Geol. B., 81*(1), 65–83. doi:10.1016/j.enggeo.2005.07.011

Gonzalez, A., Ramon, J., Cendrero, A., & Fabirri, A. (2003). Validation of Landslide Susceptibility Maps; Examples and Applications from a Case Study in Northern Spain. *J. Natural Hazards, 30*(3), 437–449. doi:10.1023/B:NHAZ.0000007201.80743.fc

Guzzetti, F., Paola, R., Mauro, C., Mirco, G., & Ardizzone, F. (2005). Probabilistic landslide hazard assessment at the basin scale. *Geomorphology, 72*(1-4), 272–299. doi:10.1016/j.geomorph.2005.06.002

Jenks, George, F. (1967). The Data Model Concept in Statistical Mapping. *International Yearbook of Cartography, 7*, 186–190.

Kavzoglu, T., Sahin, E. K., & Colkesen, I. (2013). Landslide susceptibility mapping using GIS- based multi-criteria decision analysis, support vector machines, and logistic regression. *Landslides*, *11*(3), 425–439. doi:10.100710346-013-0391-7

Kokkila, M. (2002). *Digital elevation models (DEM) in trafficability analysis*. ECOWOOD.

Lee, S., & Min, K. (2001). Statistical analysis of landslide susceptibility at Yongin, Korea. *Environ Geol.*, *40*(9), 1095–1113. doi:10.1007002540100310

NGI. (n.d.). Retrieved March 1, 2016, from http://www.ngi.no/en/Geohazards/Research/Stability-of-rock-slopes/

Ohlmacher, G. C., & Davis, J. C. (2003). Using multiple logistic regression and GIS technology to predict landslide hazard in northeast Kansas, USA. *Engineering Geology*, *69*(3-4), 331–343. doi:10.1016/S0013-7952(03)00069-3

Pachauri, A. K., Gupta, P. V., & Chander, R. (1998). Landslide zoning in a part of the Garhwal Himalayas. *Environ Geol.*, *36*(3-4), 325–334. doi:10.1007002540050348

Pradhan, B. (2013). A comparative study on the predictive ability of the decision tree, support vector machine and neuro-fuzzy models in landslide susceptibility mapping using GIS. *Computers & Geosciences*, *51*, 350–365. doi:10.1016/j.cageo.2012.08.023

Sabins, F. F. (1996). *Remote Sensing: Principles and Interpretation* (3rd ed.). W. H. Freeman and Company, New York Press.

Sarkar, S., & Kanungo, D. P. (2004). An integrated approach for landslide susceptibility mapping using remote sensing and GIS. *Photogrammetric Engineering and Remote Sensing*, *70*(5), 617–625. doi:10.14358/PERS.70.5.617

Sarkar, S., Kanungo, D. P., & Mehrotra, G. S. (1995). Landslide hazard zonation: A case study in Garhwal Himalaya, India. *Mountain Research and Development*, *15*(4), 301–309. doi:10.2307/3673806

Soeters, R., & Van Westen, C. J. (1996). Slope instability recognition, analysis, and zonation. In A. K. Turner & R. L. Shuster (Eds.), *Landslides Investigation and mitigation, Transp. Res. Board, Special Report 247, Natural Acad* (pp. 129–177). Washington, DC: Press.

Suzen, M. L., & Doyuran, V. (2004). Data driven bivariate landslide susceptibility assessment using geographical information systems: A method and application to Asarsuyu catchments, Turkey. *Engineering Geology*, *71*(3-4), 303–321. doi:10.1016/S0013-7952(03)00143-1

USGS. (n.d.). Retrieved March 1, 2016, from http://rockyweb.cr.usgs.gov/nmpstds/acrodocs/dem/3DEM0897.PDF

Vames, D. J. (1984). Landslide hazard zonation: A review of principles and practice, Unesco. Paris. *Natural Hazards*, *3*, 63.

Wachal, D. J., & Hudak, P. F. (2000). Mapping landslide susceptibility in Travis County, Texas, USA. *GeoJournal*, *51*(3), 245–253. doi:10.1023/A:1017524604463

Wang, R. (2008). *An expert knoeledge-based approach to landslide susceptibility mapping using GIS and fuzzy logic* (PhD Dissertation). University of Wisconsin-Madison, UMI Number: 3314311.

ADDITIONAL READING

Eker, A. M., Dikmen, M., Cambazoğlu, S., Düzgün, H. S. B., & Akgün, H. (2015). Evaluation and comparison of landslide susceptibility mapping methods: A case study for the Ulus district, Bartın, Northern Turkey. *International Journal of Geographical Information Science*, *29*(1), 132–158. doi:10.1080/13658816.2014.953164

Erener, A., & Düzgün, H. S. B. (2010). Improvement of statistical landslide susceptibility mapping by using spatial and global regression methods in the case of More and Romsdal (Norway). *Landslides*, *7*(1), 55–68. doi:10.100710346-009-0188-x

Erener, A., & Düzgün, H. S. B. (2012). Landslide susceptibility assessment: What are the effects of mapping unit and mapping method? *Environ Earth Sci.*, *66*(3), 859–877. doi:10.100712665-011-1297-0

Erener, A., & Düzgün, H. S. B. (2013). A regional scale quantitative risk assessment for landslides: Case of Kumluca watershed in Bartin, Turkey. *Landslides*, *10*(1), 55–73. doi:10.100710346-012-0317-9

Erener, A., Mutlu, A., & Düzgün, H. S. (2016). A comparative study for landslide susceptibility mapping using GIS-based multi-criteria decision analysis (MCDA), logistic regression (LR) and association rule mining (ARM). *Engineering Geology*, *203*, 45–55. doi:10.1016/j.enggeo.2015.09.007

Sassa, K., Canuti, P., & Yin, Y. (2014). Landslide Science for a Safer Geoenvironment, (Methods of Landslide Studies), New York, Springer, pp.2. Michael, E.A. and Samanta S. (2016). Landslide vulnerability mapping (LVM) using weighted linear combination (WLC) model through remote sensing and GIS techniques. Modeling Earth Systems and Environment, 2, 88.

Wang, R. (2008). *An expert knowledge-based approach to landslide susceptibility mapping using GIS and fuzzy logic*. Doctor of philosophy, at the university of Wisconsin-Madison, UMI Number: 3314311, (157)

KEY TERMS AND DEFINITIONS

DEM: Digital Elevation Model is the 3D representation of a terrain's surface.

GIS: Geographic Information Systems is a technology that is used to capture, store, manipulate, analyze, manage, and present spatial data.

Landslide Susceptibility: Landslide susceptibility is investigating the spatial likelihood of occurrence of landslides by correlating principal factors with landslide inventory data.

NDVI: Normalized Difference Vegetation Index is a most well-known index to detect vegetation and their condition in an area by using bands of remote sensing data.

RS: Remote sensing is the acquisition of information from earth without making physical contact with the object by using satellite- or aircraft-based sensor technologies.

TWI: Topographic Wetness Index describes the effect of topography on the location and size of saturated source areas of runoff generation.

Weighting Factors: Weighting Factors are estimated impact of values indicating relative importance.

This research was previously published in the Encyclopedia of Information Science and Technology, Fourth Edition edited by Mehdi Khosrow-Pour, pages 3503-3514, copyright year 2018 by Information Science Reference (an imprint of IGI Global).

Section 7

Industrial Engineering and Informatics

Chapter 27
Cuckoo Search Algorithm for Solving Real Industrial Multi-Objective Scheduling Problems

Mariappan Kadarkarainadar Marichelvam
Mepco Schlenk Enginering College, India

Mariappan Geetha
Kamaraj College of Engineering and Technology, India

ABSTRACT

A multistage hybrid flow shop scheduling problem is considered in this chapter with the objective of minimizing the makespan and mean flow time. Since the problem is NP-hard in nature, many heuristics and meta-heuristics were developed by the researchers. Cuckoo search algorithm is one of the recently developed meta-heuristic algorithms inspired by the social behavior of cuckoos. The cuckoo search algorithm is presented in this chapter to solve the hybrid flow shop scheduling problems. To illustrate the proposed algorithm, data sets from a steel furniture manufacturing company are used. The computational results show that the proposed algorithm is better than many other algorithms.

INTRODUCTION

Scheduling is an essential task in our day-to-day life that helps us to shape up our daily activities. The arrival and the departure of airplanes have to be scheduled in an airport. The class hours and the examinations are scheduled in the schools and Universities. Schedules are prepared in industries too. Today, industries need meticulous planning and scheduling to meet the customer demands. Due to globalization and liberalization the attributes of the customers have changed. Hence, industries ought to satisfy them by improving the quality, reducing the price and dispatching the goods on time. Scheduling is one of the most important decision making processes. Scheduling is defined as a process of allocating resources over time to perform the assigned tasks effectively (Baker & Trietsch 2009). The machines, equipment, facilities, computers and operators are the important resources in all organizations. Effective scheduling leads to improve the productivity, reduce the inventory, improve the production efficiency, minimize the

DOI: 10.4018/978-1-5225-7359-3.ch027

production time and cost and hence increase the efficiency of the production system. Different types of scheduling environments were addressed by Pinedo (1995). Among them the hybrid flow shop (HFS) environment plays a vital role as many industries resemble it. The HFS has also been called as a flexible flowshop, multiprocessor flow shop, flexible flow line, flow shop with multiple processors or a flow shop with parallel machines by Ribas et al. (2010). Different operations are performed in different machines in a simple flow shop. The HFS consists of a set of production stages in which each stage has multiple parallel machines. An HFS consists of both the flow shop and parallel machine environments. The parallel machine scheduling system involves the scheduling of a set of immediately available jobs, each on one of the parallel machines. The simple flow shop scheduling system is described as the sequencing of a set of immediately available jobs through each of the ordered work centers. There are two or more work centers in this system but only one machine at each work center. In the HFS, some stages may have only one machine. But, at least one stage should have two or more parallel machines. The machines may be identical, uniform or non-uniform. Some of the jobs may skip some of the stages in the HFS environment. The jobs flow in unidirectional in the HFS environment. Though many researchers have addressed the HFS scheduling problems for more than 40 years, only a few researchers have addressed the real-industrial scheduling problems with multiple objectives which are conflicting naturally with each other. Hence, in this paper a bi-objective HFS scheduling problem is considered. The objective is to minimize the makespan and mean flow time. The layout of an M–stage hybrid flow shop environment is given in figure 1.

Hoogeveen et al. (1996) proved that a two-stage hybrid flow shop scheduling problem is NP-hard in the strong sense even if there is only one machine on the first stage and two machines on the second stage. Hence, we cannot find the optimal solutions for these problems in a reasonable time. As the hybrid flow shop scheduling problems are NP-hard problems, the problems cannot be solved by exact algorithms. Researchers proposed many heuristics and meta-heuristics to solve the hybrid flow shop scheduling problems. Cuckoo search algorithm is a recently developed meta-heuristic algorithm. In this chapter, the cuckoo search algorithm is proposed to solve the multi-objective hybrid flow shop scheduling problems.

BACKGROUND

Hybrid flow shop scheduling problem was first proposed by Arthanari and Ramamurthy (1971). Researchers have developed many heuristics and meta-heuristics to solve such problems and to obtain

Figure 1. Layout of the hybrid flow shop environment

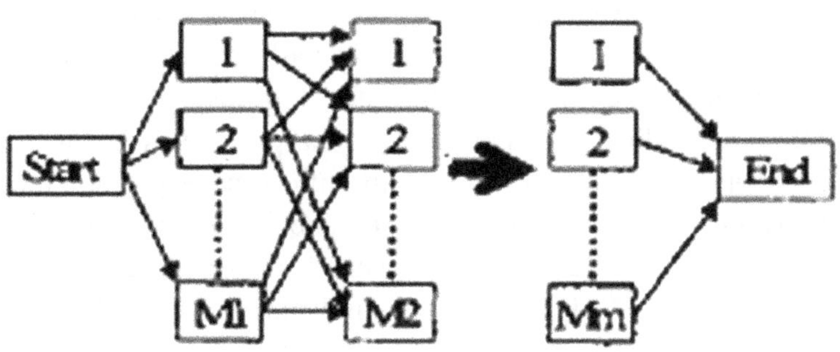

optimal or near optimal solutions with considerably less computational time. Lee and Vairaktarakis (1994) proposed heuristics to minimize makespan for hybrid flow shop scheduling problems. Riane et al. (1998) proposed efficient heuristics to minimize makespan for a three-stage flow shop problem. Brah and Loo (1999) proposed a heuristic for flow shop scheduling problems with multiple processors. Oğuz et al. (2003) also proposed some heuristics to solve multiprocessor task scheduling in a two-stage flow shop scheduling problems.

Recently, many meta-heuristics have been widely applied for solving the hybrid flow shop scheduling problems. Engin and Döyen (2004) proposed an artificial immune system algorithm for solving the hybrid flow shop scheduling problems to minimize the makespan. Yang et al. (2004) applied the tabu search simulation optimization algorithm to solve the multiprocessor flow shop scheduling problems. They applied the algorithm to solve the scheduling problem of a ceramic capacitor manufacturing company. Oğuz and Ercan (2005) have presented a genetic algorithm for the hybrid flow shop scheduling problems with multiprocessor tasks. They used a local search algorithm as a decode method to obtain the objective function value. They also proposed a new crossover operator in their work. Ruiz and Maroto (2006) have also addressed a genetic algorithm that consists of various new crossover operators to minimize makespan for the hybrid flow shop scheduling problems. The sequence dependent setup time and machine eligibility were considered by them. Ying and Lin (2006) developed an ant colony system (ACS) approach to solve the hybrid flow shop scheduling problems. Alaykýran et al. (2007) also proposed an improved ant colony optimization algorithm to minimize makespan for multistage hybrid flow shop scheduling problems. Jungwattanakit et al. (2008) formulated a 0–1 mixed integer program for hybrid flow shop scheduling problems with unrelated parallel machines. They have developed a polynomial heuristic and also a genetic algorithm to minimize makespan and total number of tardy jobs. Tseng and Liao (2008) proposed a particle swarm optimization algorithm to solve the hybrid flow shop scheduling problems to minimize makespan. Jungwattanakit et al. (2009) compared the performance of three different meta-heuristic algorithms (genetic algorithm, tabu search and simulated annealing) to minimize the convex sum of makespan and the number of tardy jobs for flexible flow shop problems with unrelated parallel machines. Jouglet et al. (2009) proposed a memetic algorithm to minimize makespan for the hybrid flow shop scheduling problems. They used a constraint programming based branch and bound algorithm as the local search in their work. Kahraman et al. (2010) solved the multiprocessor task scheduling in multistage hybrid flowshop using a parallel greedy algorithm approach. Khalouli et al. (2010) suggested an ant colony optimization algorithm for the hybrid flow shop scheduling problems to minimize the sum of the total earliness and tardiness. Rashidi et al. (2010) presented an improved multi objective parallel genetic algorithm for hybrid flow shop scheduling with unrelated parallel machines to simultaneously minimize the makespan and the maximum tardiness. Ruiz and Vázquez-Rodríguez (2010) presented a detailed literature review on exact, heuristics and meta-heuristics methods that have been proposed for the hybrid flow shop scheduling problems. They reported the different industries that resemble the hybrid flow shop scheduling environment and pointed out that only a very small percentage of woks concentrated on real industrial scheduling problems. An efficient genetic algorithm was developed by Engin et al. (2011) to solve the hybrid flow shop scheduling with multiprocessor task problems. Behnamian and Ghomi (2011) solved the hybrid flow shop scheduling problems to minimize makespan and total resource allocation costs using genetic algorithm hybridized with variable neighbourhood search. The machine and resource dependent processing time was considered by them. They have developed some random problems and compare their results. Wang et al. (2011) proposed a simulated annealing algorithm for the hybrid flow shop scheduling problems to minimize the makespan. Marichelvam and Prabaharan (2012) applied the bat algorithm for solving the realistic hybrid flowshop schedulihng problems to minimize makespan and mean flow time. Mousavi et al. (2013) also proposed a simulated annealing algorithm hybridized with some local search methods to minimize the makespan and total tardiness on a hybrid flowshop environment. Marichelvam et al. (2013) solved the hybrid flow shop scheduling problems using the bat algorithm. Marichelvam et al. (2014) also developed a discrete firefly algorithm for the multi-objective hybrid flowshop scheduling problems. Makespan and flow time are the objective functions considered by them. Marichelvam and Prabaharan (2014) developed an improved hybrid genetic scatter search (IHGSS) algorithm for multistage hybrid flow shop scheduling problems with missing operations. They hybridized the genetic algorithm and scatter search algorithm to minimize the makespan. Marichelvam and Geetha (2014) addressed the tri–objective multistage hybrid flow shop scheduling problems. They solved the problems using the discrete firefly algorithm.

Cuckoo search algorithm is one of the recently developed population-based meta-heuristic algorithms, developed by Yang and Deb (2009). Layeb (2011) applied the quantum inspired cuckoo search algorithm for solving the knapsack problems. Valian et al. (2011) developed an improved cuckoo search algorithm for solving the global optimization problems. Gherboudj et al. (2012) proposed a discrete binary version of cuckoo search algorithm for solving the 0-1 knapsack problems. Layeb and Boussalia (2012) applied a novel quantum inspired cuckoo search algorithm for bin packing problem. Marichelvam (2012) proposed an improved hybrid Cuckoo Search metaheuristics algorithm for permutation flow shop scheduling problems to minimize the makespan. Basu and Chowdhury (2013) proposed the cuckoo search algorithm for solving the economic dispatch problems. Burnwal and Deb (2013) solved the scheduling optimization problems of flexible manufacturing system using the cuckoo search-based approach. Civicioglu and Besdok (2013) presented a conceptual comparison of the cuckoo search algorithm, particle swarm optimization, differential evolution and artificial bee colony algorithms. Gandomi et al. (2013) solved the structural optimization problems using the cuckoo search algorithm. Kaveh and Bakhshpoori (2013) steel frame design optimization problems using the cuckoo search algorithm. Ouyang et al. (2013) developed a novel discrete cuckoo search algorithm for spherical traveling salesman problem. Vo et al. (2013) applied the cuckoo search algorithm for non-convex economic dispatch problems. Yang and Deb (2013) developed a multi-objective cuckoo search for design optimization problems. Marichelvam et al. (2014) proposed an improved cuckoo search algorithm for solving the hybrid flow shop scheduling problems to minimize makespan. A Discrete cuckoo search algorithm was proposed by Ouaarab et al. (2014) for solving the travelling salesman problem. Yang and Deb (2014) presented the recent advances and applications of the cuckoo search algorithm.

From the above literature review, one can easily conclude that the applications of the cuckoo search algorithm to solve the multi-objective scheduling problems require further extensive studies. Hence, in this paper the cuckoo search algorithm is proposed to minimize makespan and flow time for multistage hybrid flow shop scheduling problems.

MAIN FOCUS

This chapter considers the hybrid flow shop scheduling problems. The objective is to minimize the makespan and mean flow time. Makespan is defined as the completion time of the last job in the production system. Makespan is a performance measure used to improve the efficiency of the production system. Mean flow time is defined as the average time spent by the jobs in the production system. Reduction in mean flow time will result the reduction in work-in-process inventory. It also reduces the average response time. Hence, minimization of makespan and mean flow time are considered in this chapter. Marichelvam et al. (2014) proposed the mathematical model for the hybrid flow shop scheduling problems to minimize the weighted sum of makespan and mean flow time. The same mathematical model for the objective function of minimization of weighted sum of makespan and mean flow time is considered in this chapter. They considered several assumptions. The same assumptions are considered in this chapter also. The assumptions are:

- All n jobs are available at the beginning of scheduling.
- Each stage has infinite storage capacity.
- One machine can process only one job at a time.
- One job can be processed by only one machine at any time.
- For all the jobs, the processing times at each stage are known in advance and deterministic.
- Job set-up times are sequence-independent and are included in the job processing time of the jobs at the corresponding stage.
- Travel time between consecutive stages is negligible.
- Preemption is not allowed.

CUCKOO SEARCH ALGORITHM

Cuckoo search (CS) algorithm is a new nature-inspired metaheuristic algorithm developed by Yang and Deb (2009). Cuckoo search algorithm was inspired by the obligate brood parasitic behavior of some cuckoo species in combination with the Lévy flight behavior of some birds and fruit flies in nature. The breeding behaviour and the Lévy flights will be discussed in the following sections.

CUCKOO BREEDING BEHAVIOUR

Some of the cuckoo species lay their eggs in the nests of other host birds. The cuckoos often select the recently spawned nests instinctly. They may remove others eggs to increase the hatching probability of their own eggs. Some host birds can engage direct conflict with the intruding cuckoos. If a host bird discovers the eggs are not their owns, they will either throw these alien eggs away or simply abandon its nest and build a new nest elsewhere. Some cuckoos have evolved in such a way that female parasitic cuckoos are often very specialized in the mimicry in colour and pattern of the eggs of a few chosen host birds. This will reduce the probability of their eggs being abandoned. This also increases their reproductivity. Furthermore, the timing of egg-laying of some cuckoos is also amazing. The cuckoos often choose a nest where the host bird just laid its own eggs. In general, the cuckoo eggs hatch slightly earlier than their host eggs. Once the first cuckoo chick is hatched, the first instinct action it will take is to evict the host eggs by blindly propelling the eggs out of the nest. This will increase the cuckoo chick's share of food provided by its host bird. Moreover, a cuckoo chick can also mimic the call of host chicks to gain access to more feeding opportunity.

RULES OF CUCKOO SEARCH ALGORITHM

Based on the brooding behaviour of cuckoos, the basic steps of cuckoo search are described using the following three idealized rules/approximations:

1. Each cuckoo lays one egg (solution) at a time, and dumps its egg in a randomly chosen nest. That is, an egg represents a solution. As there is one egg in one nest, then it can be assumed that an egg is equivalent to a nest and a solution.
2. The best nests with high quality eggs/solutions will carry out to the next generation. Here is the best means the solution with the best (minimum) objective values (for minimization problems).
3. The egg laid by a cuckoo can be discovered by the host bird with a probability P_a and a nest will then be built. That is to say, a fraction P_a of the *n* nests being replaced by new nests (with new random solutions at new locations).

For minimization problems the quality or fitness function value may be the reciprocal of the objective function. Each egg in a nest represents a solution and the cuckoo egg represents a new solution. Therefore, there is no difference between an egg, a nest and solution. The pseudo code of the cuckoo search algorithm is presented below.

```
Start
Objective function f(x), x = (x1, x2, ..., xd)
Generate initial population of n host nests xi (i = 1, 2,..., n)
While (t <MaxGeneration)
Get a cuckoo randomly (say a) by Lévy flights
evaluate its quality/fitness Fa  [proportional to f(x)]
Choose a nest among n (say b) randomly
if (Fa is better than Fb),
replace solution b by the new solution;
end
A fraction (Pa) of worse nests is abandoned and new ones are built;
Keep the best solutions (or nests with quality solutions);
Rank the solutions and find the current best
end while
Post process results and visualization
End
```

When generating new solutions $x_a^{(t+1)}$ for ath cuckoo a Lévy flight is performed using the equation (1).

$$x_a^{t+1} = x_a^t + \alpha s \otimes H(p_a - \varepsilon) \otimes (x_j^t - x_k^t) \tag{1}$$

where x_j^t and x_k^t are two different solutions selected randomly by random permutation, *H*(*u*) is a Heaviside function, ε is a random number drawn from a uniform distribution, and *S* is the step size. On the other hand, the global random walk is carried out by using Lévy flights

$$x_i^{t+1} = x_i^t + \alpha L(s,\lambda), L(s,\lambda) = \frac{\lambda\Gamma(\lambda)\sin(\pi\lambda / 2)}{\pi}\frac{1}{s^{1+\lambda}}, (s \gg s_0 > 0). \tag{2}$$

Lévy flights essentially provide a random walk while their random steps are drawn from a Lévy distribution for large steps ($1<\lambda\leq3$). Here the step size vector L is drawn from the power-law distribution as given in equation (2).

LÉVY FLIGHTS

In nature, animals search for food in a random manner. In general, the foraging path of any animal is effectively a random walk because the next move is based on the current location and the transition probability to the next location. The direction of movement depends implicitly on a probability which can be modeled mathematically. The Levy flights model involve a lot of small steps, interspersed with occasional very large excursions. Yang and Deb (2013) have pointed out that the Lévy flights are a random walk whose step length is drawn from Lévy distribution. The foraging path is very important as the stopping points of a Levy flight are fractal (scale invariant is the main point here), and in complex ecosystems the distribution of food is also fractal. (i.e. there are some large areas without food). To avoid spending too much time in such unproductive areas, animals need to develop search strategies that generate a fractal distribution of stopping points. Levy flights have this property. The turning points and the trajectory of the Lévy flights are presented in Figure 2. The cuckoo search algorithm consists of three parameters P_a, α, and λ. Among them the parameters P_a and α are very important to obtain better solutions. In the literature, these parameters of the cuckoo search algorithm are kept constant. But, in this thesis work, the parameters P_a, α, and λ are not kept constant. The parameters of P_a and α are determined by the design of experiments. The values used in the simulations are based on extensive parametric studies and the parameters would be n=15 to 40, λ=1.5, and α=0.01*abs(Ub-Lb), where Ub and Lb are the upper bound and lower bound, respectively, of the problem of interested.

SOLUTIONS AND RECOMMENDATIONS

In order to validate the performance of the proposed algorithm a real industrial scheduling problem is considered in this chapter. The scheduling problem of a leading furniture manufacturing is considered in this chapter. The furniture manufacturing company is one of the leading industries in Tamilnadu, India. The layout of the collaborative company is depicted in Figure 3. The number of stages are five and the number of machines in each stage are presented in Table 1 and the processing time of jobs is presented in Table 2.

To validate the performance of the proposed algorithm the results are compared with many other algorithms addressed in the literature. Mean relative deviation index proposed by Marichelvam et al. (2014) is used in this paper as a performance measure in this chapter. MRDI is one of the most important performance measures used in the scheduling literature. Lower MRDI will be the indication of better

Figure 2. Turning points and trajectory of the Lévy flights

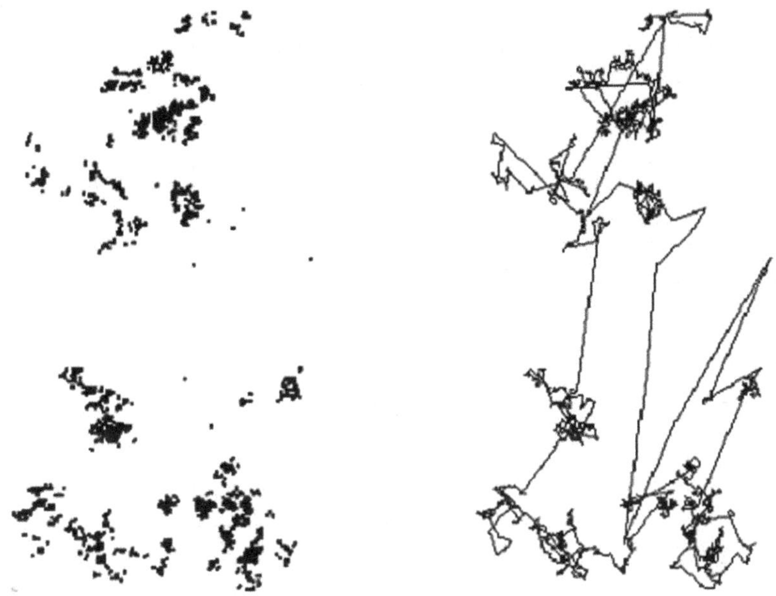

Figure 3. Layout of the case study company

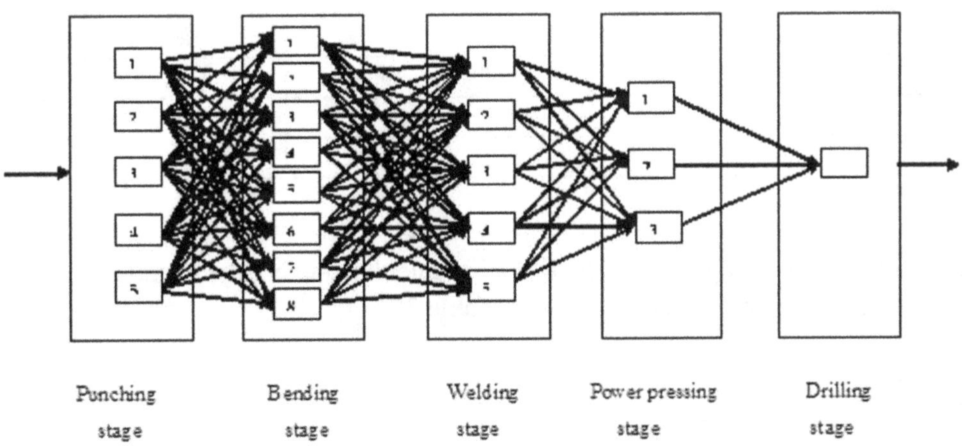

Table 1. Number of machines in each stage

Sl. No.	Stage	Number of Machines
1	Punching	5
2	Bending	8
3	Welding	5
4	Power pressing	3
5	Drilling	1

Table 2. Processing time of jobs (in seconds)

Stages → / Jobs ↓	Punching	Bending	Welding	Power Pressing	Drilling
1.	39	28	0	0	0
2.	61	66	110	0	0
3.	10	17	52	0	0
4.	22	0	0	0	0
5.	35	60	0	0	0
6.	34	66	70	0	0
7.	42	0	0	0	0
8.	42	0	0	50	0
9.	34	250	30	0	0
10.	12	18	25	0	0
11.	0	0	0	10	0
12.	0	0	0	20	0
13.	52	90	0	0	0
14.	0	0	0	20	0
15.	0	0	0	42	0
16.	12	22	0	0	32
17.	0	0	0	8	0
18.	0	0	0	6	0
19.	0	0	0	10	0
20.	0	0	0	6	0

performance of the proposed algorithm. The MRDI is calculated using the equation (3). The MRDI is comparison of different algorithms is presented in Table 3.

$$\text{MRDI} = \sum_{l=1}^{R} \frac{(Z^* - Z_{meta})}{Z^*} \times 100 / R \tag{3}$$

Table 3. MRDI Comparison of different algorithms for the test problems

Sl. No.	Algorithms	MRDI
1	Ant colony optimization algorithm	3.91
2	Cuckoo search algorithm	0
3	Genetic algorithm	2.23
4	Particle swarm optimization algorithm	1.56
5	Simulated annealing algorithm	2.77

Where,

Z^* = best objective function value

Z_{meta} = objective function value obtained by the different algorithms

R = number of runs (20)

From the result table, it is concluded easily that the performance of the proposed cuckoo search algorithm is better than many other algorithms addressed in the literature for the multi-objective real industrial scheduling problems. This indicates the effectiveness of the proposed algorithm.

FUTURE RESEARCH DIRECTIONS

In this chapter, the multi-objective hybrid flow shop scheduling problems are considered. However, many assumptions are made in this chapter. For instance, the setup time is not addressed in this chapter. Consideration of setup time and transportation time is a future research scope of this work. Consideration of due date related criteria such as earliness and tardiness would be another interesting scope of this research. The cuckoo search algorithm may also be applied for other optimization problems. The proposed algorithm might be hybridized with other algorithms.

CONCLUSION

The multi-objective hybrid flow shop scheduling problem is addressed in this chapter. A recently developed cuckoo search algorithm is proposed to minimize makespan and mean flow time for the hybrid flow shop scheduling problems. The algorithm is validated with a real industrial data. The performance of the proposed cuckoo search algorithm is compared with other algorithms such as ant colony optimization, genetic algorithm, particle swarm optimization and simulated annealing and proved to be better.

REFERENCES

Alaykýran, K., Engin, O., & Döyen, A. (2007). Using ant colony optimization to solve hybrid flow shop scheduling problems. *International Journal of Advanced Manufacturing Technology*, *35*(5-6), 541–550. doi:10.100700170-007-1048-2

Arthanari, T. S., & Ramamurthy, K. G. (1971). An extension of two machines sequencing problem. *Opsearch*, *8*(1), 10–22.

Baker, K. R., & Trietsch, D. (2009). *Principles of Sequencing and Scheduling*. John Wiley & Sons, Inc. doi:10.1002/9780470451793

Basu, M., & Chowdhury, A. (2013). Cuckoo search algorithm for economic dispatch. *Energy*, *60*, 99–108. doi:10.1016/j.energy.2013.07.011

Behnamian, J., & Ghomi, S. F. (2011). Hybrid flowshop scheduling with machine and resource-dependent processing times. *Applied Mathematical Modelling*, *35*(3), 1107–1123. doi:10.1016/j.apm.2010.07.057

Brah, S. A., & Loo, L. L. (1999). Heuristics for scheduling in a flow shop with multiple processors. *European Journal of Operational Research*, *113*(1), 113–122. doi:10.1016/S0377-2217(97)00423-2

Burnwal, S., & Deb, S. (2013). Scheduling optimization of flexible manufacturing system using cuckoo search-based approach. *International Journal of Advanced Manufacturing Technology*, *64*(5-8), 951–959. doi:10.100700170-012-4061-z

Civicioglu, P., & Besdok, E. (2013). A conceptual comparison of the Cuckoo-search, particle swarm optimization, differential evolution and artificial bee colony algorithms. *Artificial Intelligence Review*, *39*(4), 315–346. doi:10.100710462-011-9276-0

Engin, O., Ceran, G., & Yilmaz, M. K. (2011). An efficient genetic algorithm for hybrid flow shop scheduling with multiprocessor task problems. *Applied Soft Computing*, *11*(3), 3056–3065. doi:10.1016/j.asoc.2010.12.006

Engin, O., & Döyen, A. (2004). A new approach to solve hybrid flow shop scheduling problems by artificial immune system. *Future Generation Computer Systems*, *20*(6), 1083–1095. doi:10.1016/j.future.2004.03.014

Gandomi, A. H., Yang, X. S., & Alavi, A. H. (2013). Cuckoo search algorithm: A metaheuristic approach to solve structural optimization problems. *Engineering with Computers*, *29*(1), 17–35. doi:10.100700366-011-0241-y

Gherboudj, A., Layeb, A., & Chikhi, S. (2012). Solving 01 knapsack problems by a discrete binary version of cuckoo search algorithm. *International Journal of Bio-inspired Computation*, *4*(4), 229–236. doi:10.1504/IJBIC.2012.048063

Hoogeveen, J. A., Lenstra, J. K., & Veltman, B. (1996). Minimizing the makespan in a multiprocessor flow shop is strongly NP-hard. *European Journal of Operational Research*, *89*(1), 172–175. doi:10.1016/S0377-2217(96)90070-3

Jouglet, A., Oğuz, C., & Sevaux, M. (2009). Hybrid flow-shop: A memetic algorithm using constraint-based scheduling for efficient search. *Journal of Mathematical Modelling and Algorithms*, *8*(3), 271–292. doi:10.100710852-008-9101-1

Jungwattanakit, J., Reodecha, M., Chaovalitwongse, P., & Werner, F. (2008). Algorithms for flexible flow shop problems with unrelated parallel machines, setup times, and dual criteria. *International Journal of Advanced Manufacturing Technology*, *37*(3-4), 354–370. doi:10.100700170-007-0977-0

Jungwattanakit, J., Reodecha, M., Chaovalitwongse, P., & Werner, F. (2009). A comparison of scheduling algorithms for flexible flow shop problems with unrelated parallel machines, setup times, and dual criteria. *Computers & Operations Research*, *36*(2), 358–378. doi:10.1016/j.cor.2007.10.004

Kahraman, C., Engin, O., Kaya, İ., & Öztürk, R. E. (2010). Multiprocessor task scheduling in multistage hybrid flow-shops: A parallel greedy algorithm approach. *Applied Soft Computing*, *10*(4), 1293–1300. doi:10.1016/j.asoc.2010.03.008

Kaveh, A., & Bakhshpoori, T. (2013). Optimum design of steel frames using cuckoo search algorithm with lévy flights. *Structural Design of Tall and Special Buildings*, *22*(13), 1023–1036. doi:10.1002/tal.754

Khalouli, S., Ghedjati, F., & Hamzaoui, A. (2010). A meta-heuristic approach to solve a JIT scheduling problem in hybrid flow shop. *Engineering Applications of Artificial Intelligence*, *23*(5), 765–771. doi:10.1016/j.engappai.2010.01.008

Layeb, A. (2011). A novel quantum inspired cuckoo search for knapsack problems. *International Journal of Bio-inspired Computation*, *3*(5), 297–305. doi:10.1504/IJBIC.2011.042260

Layeb, A., & Boussalia, S. R. (2012). A novel quantum inspired cuckoo search algorithm for bin packing problem. *International Journal of Information Technology and Computer Science*, *4*(5), 58–67. doi:10.5815/ijitcs.2012.05.08

Lee, C. Y., & Vairaktarakis, G. L. (1994). Minimizing makespan in hybrid flowshops. *Operations Research Letters*, *16*(3), 149–158. doi:10.1016/0167-6377(94)90026-4

Marichelvam, M. K. (2012). An improved hybrid Cuckoo Search (IHCS) metaheuristics algorithm for permutation flow shop scheduling problems. *International Journal of Bio-inspired Computation*, *4*(4), 200–205. doi:10.1504/IJBIC.2012.048061

Marichelvam, M. K., & Geetha, M. (2014). Solving tri–objective multistage hybrid flow shop scheduling problems using a discrete firefly algorithm. *International Journal of Intelligent Engineering Informatics*, *2*(4), 284–303. doi:10.1504/IJIEI.2014.067190

Marichelvam, M. K., & Prabaharan, T. (2012). A bat algorithm for realistic hybrid flowshop schedulihng problems to minimize makespan and mean flow time. *ICTACT Journal on Soft Computing*, *3*(1), 428–433. doi:10.21917/ijsc.2012.0066

Marichelvam, M. K., & Prabaharan, T. (2014). Performance evaluation of an improved hybrid genetic scatter search (IHGSS) algorithm for multistage hybrid flow shop scheduling problems with missing operations. *International Journal of Industrial and Systems Engineering*, *16*(1), 120–141. doi:10.1504/IJISE.2014.057946

Marichelvam, M. K., Prabaharan, T., & Yang, X. S. (2014). Improved cuckoo search algorithm for hybrid flow shop scheduling problems to minimize makespan. *Applied Soft Computing*, *19*, 93–101. doi:10.1016/j.asoc.2014.02.005

Marichelvam, M. K., Prabaharan, T., & Yang, X. S. (2014). A discrete firefly algorithm for the multi-objective hybrid flowshop scheduling problems. *IEEE Transactions on Evolutionary Computation*, *18*(2), 301–305. doi:10.1109/TEVC.2013.2240304

Marichelvam, M. K., Prabaharan, T., Yang, X. S., & Geetha, M. (2013). Solving hybrid flow shop scheduling problems using bat algorithm. *International Journal of Logistics Economics and Globalization*, *5*(1), 15–29. doi:10.1504/IJLEG.2013.054428

Mousavi, S. M., Zandieh, M., & Yazdani, M. (2013). A simulated annealing/local search to minimize the makespan and total tardiness on a hybrid flowshop. *International Journal of Advanced Manufacturing Technology*, *64*(1-4), 369–388. doi:10.100700170-012-4031-5

Oĝuz, C., & Ercan, M. F. (2005). A genetic algorithm for hybrid flow-shop scheduling with multiprocessor tasks. *Journal of Scheduling*, *8*(4), 323–351. doi:10.100710951-005-1640-y

Oğuz, C., Ercan, M. F., Cheng, T. E., & Fung, Y. F. (2003). Heuristic algorithms for multiprocessor task scheduling in a two-stage hybrid flow-shop. *European Journal of Operational Research*, *149*(2), 390–403. doi:10.1016/S0377-2217(02)00766-X

Ouaarab, A., Ahiod, B., & Yang, X. S. (2014). Discrete cuckoo search algorithm for the travelling salesman problem. *Neural Computing & Applications*, *24*(7-8), 1659–1669. doi:10.100700521-013-1402-2

Ouyang, X., Zhou, Y., Luo, Q., & Chen, H. (2013). A novel discrete cuckoo search algorithm for spherical traveling salesman problem. *Applications of Mathematics*, *7*(2), 777–784.

Pinedo, M. (1995). *Scheduling: theory, algorithms, and systems*. Englewood Cliffs, NJ: Prentice-Hall.

Rashidi, E., Jahandar, M., & Zandieh, M. (2010). An improved hybrid multi-objective parallel genetic algorithm for hybrid flow shop scheduling with unrelated parallel machines. *International Journal of Advanced Manufacturing Technology*, *49*(9-12), 1129–1139. doi:10.100700170-009-2475-z

Riane, F., Artiba, A., & Elmaghraby, S. E. (1998). A hybrid three-stage flowshop problem: Efficient heuristics to minimize makespan. *European Journal of Operational Research*, *109*(2), 321–329. doi:10.1016/S0377-2217(98)00060-5

Ribas, I., Leisten, R., & Framiñan, J. M. (2010). Review and classification of hybrid flow shop scheduling problems from a production system and a solutions procedure perspective. *Computers & Operations Research*, *37*(8), 1439–1454. doi:10.1016/j.cor.2009.11.001

Ruiz, R., & Maroto, C. (2006). A genetic algorithm for hybrid flowshops with sequence dependent setup times and machine eligibility. *European Journal of Operational Research*, *169*(3), 781–800. doi:10.1016/j.ejor.2004.06.038

Ruiz, R., & Vázquez-Rodríguez, J. A. (2010). The hybrid flow shop scheduling problem. *European Journal of Operational Research*, *205*(1), 1–18. doi:10.1016/j.ejor.2009.09.024

Tseng, C. T., & Liao, C. J. (2008). A particle swarm optimization algorithm for hybrid flow-shop scheduling with multiprocessor tasks. *International Journal of Production Research*, *46*(17), 4655–4670. doi:10.1080/00207540701294627

Valian, E., Mohanna, S., & Tavakoli, S. (2011). Improved cuckoo search algorithm for global optimization. *International Journal of Communications and Information Technology*, *1*(1), 31–44.

Vo, D. N., Schegner, P., & Ongsakul, W. (2013). Cuckoo search algorithm for non-convex economic dispatch. Generation. *Transmission & Distribution, IET*, *7*(6), 645–654. doi:10.1049/iet-gtd.2012.0142

Wang, H. M., Chou, F. D., & Wu, F. C. (2011). A simulated annealing for hybrid flow shop scheduling with multiprocessor tasks to minimize makespan. *International Journal of Advanced Manufacturing Technology*, *53*(5-8), 761–776. doi:10.100700170-010-2868-z

Yang, T., Kuo, Y., & Chang, I. (2004). Tabu-search simulation optimization approach for flow - shop scheduling with multiple processors-a case study. *International Journal of Production Research*, *42*(19), 4015–4030. doi:10.1080/00207540410001699381

Yang, X. S., & Deb, S. (2009, December). Cuckoo search via Lévy flights. In *Nature & Biologically Inspired Computing, 2009. NaBIC 2009. World Congress on* (pp. 210-214). IEEE.

Yang, X. S., & Deb, S. (2013). Multiobjective cuckoo search for design optimization. *Computers & Operations Research*, *40*(6), 1616–1624. doi:10.1016/j.cor.2011.09.026

Yang, X. S., & Deb, S. (2014). Cuckoo search: Recent advances and applications. *Neural Computing & Applications*, *24*(1), 169–174. doi:10.100700521-013-1367-1

Ying, K. C., & Lin, S. W. (2006). Multiprocessor task scheduling in multistage hybrid flow-shops: An ant colony system approach. *International Journal of Production Research*, *44*(16), 3161–3177. doi:10.1080/00207540500536939

ADDITIONAL READING

Marichelvam, M. K., & Geetha, M. (2016). A hybrid cuckoo search metaheuristic algorithm for solving single machine total weighted tardiness scheduling problems with sequence dependent setup times. *International Journal of Computational Complexity and Intelligent Algorithms*, *1*(1), 23–34. doi:10.1504/IJCCIA.2016.077463

Marichelvam, M. K., & Geetha, M. (2016). Application of novel harmony search algorithm for solving hybrid flow shop scheduling problems to minimise makespan. *International Journal of Industrial and Systems Engineering*, *23*(4), 467–481. doi:10.1504/IJISE.2016.077698

Marichelvam, M. K., & Prabaharan, T. (2015). Solving realistic industrial scheduling problems using a multi–objective improved hybrid particle swarm optimisation algorithm. *International Journal of Operational Research*, *23*(1), 94–129. doi:10.1504/IJOR.2015.068752

Yang, X. S., & Deb, S. (2010). Engineering optimisation by cuckoo search. *International Journal of Mathematical Modelling and Numerical Optimisation*, *1*(4), 330–343. doi:10.1504/IJMMNO.2010.035430

Yildiz, A. R. (2013). Cuckoo search algorithm for the selection of optimal machining parameters in milling operations. *International Journal of Advanced Manufacturing Technology*, *64*(1-4), 55–61. doi:10.100700170-012-4013-7

KEY TERMS AND DEFINITIONS

Cuckoo Search Algorithm: A recently developed meta-heuristic algorithm to solve the optimization problems.

Makespan: Makespan is defined as the completion time of the last job to leave the system.

Mean Flow Time: Mean flow time is defined as the average time spent by the jobs in the production system.

NP-Hard Problems: Non – deterministic polynomial time hard problems.

Scheduling: Scheduling is defined as a process of allocating resources over time to perform a collection of tasks.

This research was previously published in the Encyclopedia of Information Science and Technology, Fourth Edition edited by Mehdi Khosrow-Pour, pages 4369-4381, copyright year 2018 by Information Science Reference (an imprint of IGI Global).

Chapter 28
The Trends and Challenges of 3D Printing

Edna Ho Chu Fang
University of Malaya, Malaysia

Sameer Kumar
University of Malaya, Malaysia

ABSTRACT

3D printing is a type of additive manufacturing technology where a 3D object is created by laying down subsequent layers of material at the mm scale. It is also known as rapid prototyping. 3D printing is now applied in various industries such as footwear, jewelry, architecture, engineering and construction, aerospace, dental and medical industries, education, consumer products, automotive, and industrial design. Some claim that 3D printing will put an end to traditional manufacturing, primarily since 3D printing imposes a tool-less process. Though 3D printing technology is used in weapon manufacturing, it is also being used to improve the lives of mankind. In the future, 3D printing will most probably be used to print human organs. The chapter discusses the trends and challenges faced by this exciting technology.

INTRODUCTION

3D printing is the new wave of technology advancement in the world of architecture, design and manufacturing. Also known as rapid prototyping, 3D printing is a type of additive manufacturing technology where a 3D object is created by laying down subsequent layers of material at the mm scale. 3D printers print objects by reading a CAD design file or by scanning an object (Sachs et al., 1992). Today, 3D printing is applied in various industries such as footwear, jewelry, architecture, engineering and construction, aerospace, dental and medical industries, education, consumer products, automotive and industrial design. Some claim that 3D printing will put an end to traditional manufacturing primarily since 3D printing imposes a tool-less process. Product parts can be specifically designed to avoid assembly lines, as well as ensuring maximum utilization of raw materials. In this article, the authors discuss the state-of-the-art of 3D printing its future direction.

DOI: 10.4018/978-1-5225-7359-3.ch028

BACKGROUND

In May 1980, Dr Kodama from Japan filed the very first patent application for Rapid Prototyping technology. Unfortunately, he did not file the subsequent full patent specification before the one year deadline after the application. Hence, in 1986, Charles (Chuck) Hull filed the first patent for stereolithography apparatus (SLA). He was the first to invent the SLA machine in 1983. After obtaining the patent, he went on to co-found 3D Systems Corporation, which is one of the largest organizations operating in the 3D printing world today. During the mid-nineties, the 3D printing sector started to diverge into two specific areas. First, there was the high end of 3D printing, which saw the production of complex parts. These applications include the medical, aerospace, jewelry and automotive sectors. Then there was the lower end of the market, which saw a price war among many 3D printer manufacturers, highlighting improvements in speed, accuracy and materials.

In 2007, 3D Systems came up with the first 3D printer which was priced under $10,000. The first commercial 3D printer was offered for sale in January, 2009. It was based on the RepRap concept, and came in a kit form. Makerbot Industries also developed commercial printers in April of the same year. 2013 saw Stratasys acquiring Makerbot. It was a year of significant growth and consolidation for 3D printing.

Materials for 3D printing were very limited during the early days of the technology. Today, there is an array of different types of materials available for choice.

The first step in 3D printing is to design the 3D digital model using a CAD program or scan the object with a 3D scanner. The model will then be 'sliced' into layers and converted into a printer-readable file. The printing material will be added one layer at a time.

Different materials are suited for different 3D printing technologies. Some 3D printers process powdered materials which utilize a light source to fuse layers of the powder together to make the desired shape. Others process polymer resin materials and utilize a laser to solidify the resin in ultra thin layers. Another method is the jetting of fine droplets using materials and a binder to fix the layers. One of the most commonly used 3D printing technology is the stereolithography (SLA) technology. This technology utilizes photocuring resins as raw material. New resins that are being developed will combine transparency, heat resistance and toughness.

The second commonly used technology is called 'Fused Deposition Modeling', as invented by Scott Crump, a co-founder of Stratasys Inc. The FDM technology uses thermoplastic resins as raw materials. It is the simplest 3D additive manufacturing technology, in which the thermoplastic resin softens when heat is applied. The third type of technology is called 'Selective Laser Sintering (SLS)'. This builds objects by using a laser to fuse together layers of a mixture of different powdered raw materials. The fourth type of technology is called 'Multi-jet modeling (MJM)'. Objects are built up from the layering of powder through an inkjet-like print head that also sprays a binder solution to glue the required granules together. The raw materials associated with this type of technology are sand mold or nylon resins.

THE TECHNICAL ISSUES OF 3D PRINTING

3D printing has revolutionized our society from providing medical advances; to scalable production of everything from product parts to buildings. There are, however, many issues that accompany this technology. Two main issues of 3D printing are the technical problems and the controversies. This section will be divided into two parts. Part one presents the technical problems of 3D printing, while part two presents the controversies of 3D printing.

As 3D printing is getting more widespread, the issue of its quality is always being questioned. In 3D printing, the quality of the printed object is linked to the printing speed; and the printing speed is linked to the raw materials' thermoplastic properties. Different raw materials are supposed to be printed at different speeds. To ensure a beautifully printed object, the speed of printing has to be just right – not too slow and not too fast either.

Some of the common printing problems are warping, stringing, gaps in the top layers, under-extrusion, over-extrusion, pillowing, layer-misalignment, elephant foot, etc. Furthermore, the raw materials that can be used for 3D printers are still rather limited. Presently, most of the commercial 3D printed products consist of one single material. The idea of printing electronic goods such as smart phones is already being researched but not yet foreseeable. The tough challenge here is to add different types of materials that fulfill their functionality to make up various parts of an electronic component.

Last but not least, 3D printing requires post-processing. An additive manufacturing (AM) machine cannot add finishing touches to the printed object. It requires a manually intensive process. The printed objects will still have to undergo post-printing processes such as friction-weld, paint, sand, rivet and so on.

The Controversies of 3D Printing

The controversies surrounding 3D printing include infringement of intellectual property laws, fabrication of weapons and drugs for crime purposes, compliance with FDA safety standards and ethical considerations.

Intellectual Property Laws

Online platforms such as GrabCad and Thingiverse provide users with the CAD design files needed for 3D printing. Registered users can download, upload, design and modify a 3D model on these online platforms. Infringement of copyright laws happens through the way users obtain the CAD design files. They either: (1) create an object design file; (2) modify an existing design; (3) scan the object (Mendis & Secchi, 2015). The main question here is whether a CAD file can be protected under the copyright law. Firstly, for a design to be eligible for protection, it has to qualify as a literary work. Many arguments are in favor of the notion that a CAD file is not a type of computer program (literary work), hence it is not eligible for protection under the copyright law.

Rideout (2012) asserts that CAD files cannot be categorized as copyrightable software in the USA because they are just "triangular representations of a 3D object". Unlike computer programs, the CAD files do not control the way 3D printers operate. He justifies that a 3D design file isn't a literary work and is more likely to be considered under "graphic and sculptural works". Then again, another issue will be raised if 3D designs were to be considered as sculptural works. Art works have their own set of copyrights. Apparently, a 3D model does not infringe copyright laws if the CAD file is used to create an object for use and not as an artwork. This is showcased in the example of Andrew Ainsworth, a props maker who in the film StarWars, printed the Stormtroopers' white helmets and justified that it was for use and not as an artwork. The court ruled in favor of Ainsworth.

Whilst the USA stands by the notion that a CAD file cannot be protected under the copyright law, the UK believes that a CAD file could be protected. As for now, the copyright status of CAD design files remains unclear.

Fabrication of Weapons or Drugs for Crime Purposes

3D printing provides users with countless possibilities. As much as it can be used for good purposes, it can also be misused when it falls into the hands of people with ill intentions. Anyone with sufficient modal can own a 3D printer and start printing guns or illegal drugs at the comfort of their homes. They are only limited by the costs and technological capabilities of the printers, as well as the availability of the raw materials.

Ever since Cody Wilson, a law student who became the first man to ever print a gun called 'the Liberator' from a 3D printer, firearms enthusiasts are constantly modifying gun designs to come up with more durable guns made from a combination of 3D printed materials and parts taken from traditionally manufactured guns. 3D printed guns pose threats to the society because metal detectors cannot detect them.

Generally, the United States allows its citizens to own guns though each state has its own regulatory laws regarding gun ownership. For example, in California, people with a negative track record or people with mental illness are prohibited from owning firearms. 3D printing enables almost anyone to possess a firearm through self-manufacturing.

The group of people who called for firearms to be banned in the States fear that 3D printed guns might be used for crimes or mass shootings. Even if the government does ban 3D printed guns, the technological advances in firearm printing will not stop there as it is part of the European history and culture to own guns.

While owning a firearm is legal, how about owning a gun design file? The province of New South Wales of Australia recently enacted a law that bans gun design files. The amendment made to the Firearms Act 1996 states that possessing 3D digital blueprints for manufacturing firearms is a crime punishable with a penalty of up to 14 years in jail. Individuals who possess digital gun blueprints for research purposes are legally exempted from the law.

The legal stand of manufacturing own firearms differ amongst countries all over the world. Some countries such as Australia and Japan have banned the possession of gun design files, while the US does not. Currently, making own firearms is legal in the US as long as it is intended for personal use only and not for commercial purposes.

Manufacturing of Illegal or Counterfeit Drugs

As with the printing of weapons, criminals might misuse 3D printing to manufacture illegal or counterfeit drugs. Professor Lee Cronin and a team of 45 researchers at Glasgow University are researching on additive manufacturing technology that would allow drugs to be produced through 3D printing. It was discovered that bathroom sealant which acts as the "ink" for creating reaction chambers was a suitable raw material for the 3D printer. The printer can inject "chemical inks" that would create more complex molecules through a series of reactions ("3-D Printing: The Potential Implications and Challenges for Law Enforcement", 2015).

Cronin stated that almost all drugs are made of the similar components of oxygen, carbon, and hydrogen with vegetable oils and paraffin. He foresees that in the future, people will be able to download drug recipes from online platforms and use 3D printers to produce medicines at their homes. When this happens, 3D printed drugs might not just be a possible crime threat, but a social threat as well.

Besides the issue of printing illegal drugs, individuals who want to earn easy money can also manufacture counterfeit drugs that produce similar effects, but have different chemical components from today's controlled substances. Due to the drugs' distinct chemical compositions, these counterfeit drugs may not be covered by current laws and would hence be technically legal. They might also be more dangerous than the controlled substances because less research is done to under their chemical composition and discover the side effects of these drugs.

Compliance With FDA Safety Standards

All food, toys, cosmetic and medical products made with 3D printing technologies must comply with the regulations outlined by the U.S. Food & Drug Administration (FDA). How can we ensure that 3D printed foods fulfill all the necessary safety standards? Despite its promise, 3D printing is still relatively immature when it comes to safety and quality control standards compared to conventional food manufacturing process.

3D printed food is still at its early testing phase as researchers are trying to incorporate nutrition into the printed food. 3D printed food creates a new dilemma for the FDA because the authorities are unsure whether the current GMP and food reviewing procedures are sufficient and applicable to the 3D printing industry, or should new regulations be made, said Claudia Lewis, a partner at Venable LLP.

Currently, 3D printers are capable of printing pastries, desserts and burgers or even personalized food items and can cater to people with special dietary requirements. However, a lot of stakeholders in the 3D food printing industry are not fully aware that food elements have to be regulated. The food products will not get to market unless they get some sort of approval by the FDA. How the FDA decides to regulate the food depends on which category the product belongs to. For example, FDA regulations for beverages depend on whether the beverage is classified as a conventional drink or a dietary supplement. It is required at this point for the FDA to seek to understand more about the food industry and for the industry to ask the agency for guidance.

Ethical Considerations

3D printing has made its presence known in the field of medical science and technology. Through tissue engineering, 3D printing is utilized to build biological substitutes that mimic human bodily tissues and can be used to heal wounds or restore failing organs.

Traditionally, tissue engineering starts by taking a small biopsy from a patient. The cells from the biopsy are isolated, expanded and seeded on a natural or synthetic scaffold. After the cells differentiate and proliferate in the scaffold, the resulting tissue construct is transferred to a bioreactor for it to mature before it can be used for transplantation. Through this approach, researchers have successfully printed the trachea, ear cartilage, urinary bladder and so on. However, this approach is still limited in terms of accuracy and reproductibility that can be achieved as more complex multicellular structures with vascular network integration has to be built (Lucas & Spiegel, 2015).

The innovation in bioprinting is relatively new and there aren't many similar procedures with which to compare them. There aren't many prior case studies or statistics to be used for weighing the possible contradictions and risks. Patients have to fully trust the work of the biomedical engineers and doctors. Besides, the cost of the procedure will be very high, hence raising the question of medical procedure coverage by health insurance companies. Is this procedure worth the cost and risks when it is still at its early stages of research and implementation?

Still at its early stages of research, it is only ethical for the researchers to consider the patients' safety in using bioprinted tissues or organs. The safety of the patient throughout the healing process is the utmost concern with bioprinting (Lucas & Spiegel, 2015).

Though there are biomedical engineers who have successfully printed various body parts, it will take a considerable amount of years before commercialized printing of compatible organs could take place. Currently, only microtissues can be printed (Varkey & Atala, 2015). The Food and Drug Administration will have to screen through all bioprinted tissues and organs for effectiveness and safety, and also weigh the pros and risks involved.

A number of regulatory laws will need to be cleared before the first lab-printed organ becomes commercially available. Technological advancements in terms of software and advances in medical science and technology are needed before the application of bioprinting can occur in a widespread manner (Varkey & Atala, 2015).

SOLUTIONS AND RECOMMENDATIONS

Technical problems can be easily resolved by creating a 3D printer that is of higher sophistication, is durable and cost-efficient. More raw materials can be sought after and test-runned for printing a wider range of objects in the future.

The U.S. government should implement copyright laws for CAD design files specifically on those of firearms. Though it is legal to own a firearm, the government can pass a law that prohibits the manufacturing of firearms for selling purposes. The law should also implement a strict screening of eligible firearm owners before gunsmiths start selling to them. It is reasonable that only sound-minded and good-intentioned folks are capable of handling firearms for logical reasons. The authority of printing drugs should be strictly given to registered medical institutions, practitioners and researchers to ensure quality control. Law enforcement will have to alter its investigation methods for drug crimes.

Furthermore, in terms of 3D printed food, the FDA must ensure that all food printed should adhere to the safety standards. Food printed has to be edible and contains nutrition readily absorbed by the human body.

Lastly, 3D bioprinting should be ethical by taking care of the patient's safety and concern (first and foremost). More tests has to be run in the field of organ printing to ensure maximum compatibility of cells and functioning after an organ transplant.

FUTURE RESEARCH DIRECTIONS

3D printing of hybrid objects is still an under-researched topic as most of the research done is on single-raw material printed products. The future of 3D printing would be promising if objects that require different raw materials for different functioning parts can be printed completely without the need of product assembling.

Another interesting field for research is 3D printed food. Currently, there isn't sufficient research that ensures the safety of 3D printed food in compliance with the regulations outlined by the FDA. Once 3D printed food passes the FDA regulations, they can start to be produced commercially and should garner great interest in the market.

CONCLUSION

3D technology will continue to evolve. Many stakeholders are adopting it due to the benefits it promises. A technology will prove to be useful if it falls into the right hands. Though 3D printing technology is used in weapon manufacturing, it is also being used to improve the lives of mankind. In the future, 3D printing will most probably be used to print human organs. An issue of concern is raised regarding the possibility of printing 3D drugs, through the assembly of chemical compounds at a molecular level. In the future, patients can print medicine on their own. It is only hoped that with regulations, the potential of 3D printing technology can be harvested for good and not for worse.

REFERENCES

Bose, S., Vahabzadeh, S., & Bandyopadhyay, A. (2013). Bone tissue engineering using 3D printing. *Materials Today*, *16*(12), 496–504. doi:10.1016/j.mattod.2013.11.017

Campbell, T., Williams, C., Ivanova, O., & Garrett, B. (2011). *Could 3D printing change the world. In Technologies, Potential, and Implications of Additive Manufacturing*. Washington, DC: Atlantic Council.

Connell, J. L., Ritschdorff, E. T., Whiteley, M., & Shear, J. B. (2013). 3D printing of microscopic bacterial communities. *Proceedings of the National Academy of Sciences of the United States of America*, *110*(46), 18380–18385. doi:10.1073/pnas.1309729110 PMID:24101503

Estes, A. C. (2015, January 06). *3D-Printed Guns Are Only Getting Better, and Scarier*. Retrieved April 26, 2016, from http://gizmodo.com/3d-printed-guns-are-only-getting-better-and-scarier-1677747439

Grunewald, S. J. (2015, August 19). *The Shuty Hybrid 3D Printed 9mm Pistol Raises Questions About 3D Printed Gun Control*. Retrieved April 26, 2016, from https://3dprint.com/89919/shuty-hybrid-3d-printed-pistol/

Holliman, K. (2015, April 20). *Innovations in 3D Printed Food Outpacing FDA Regulations - Food Quality & Safety*. Retrieved April 26, 2016, from http://www.foodqualityandsafety.com/article/innovations-in-3d-printed-food-outpacing-fda-regulations/

Journal, M. (2015). *The potential of 3D printing*. Retrieved from http://www.meti.go.jp/english/publications/pdf/journal2013_08a.pdf

Lucas, C., & Spiegel, M. (2015). *3D printing skin grafts for facial reconstruction*. Academic Press.

Manyika, J., Chui, M., Bughin, J., Dobbs, R., Bisson, P., & Marrs, A. (2013). *Disruptive technologies: Advances that will transform life, business, and the global economy* (Vol. 12). McKinsey Global Institute New York.

Mendis, D., & Secchi, D. (2015). *A Legal and Empirical Study of 3D Printing Online Platforms and an Analysis of User Behaviour*. Academic Press.

Peter, V., & Kinscö, I. (2013). *Developing an evaluation and progress methodology to underpin the intervention logic of the Action Plan to Boost Demand for European Innovations*. EU.

Petrick, I. J., & Simpson, T. W. (2013). 3D printing disrupts manufacturing: How economies of one create new rules of competition. *Research-Technology Management*, *56*(6), 12–16. doi:10.5437/08956308X5606193

Ratto, M., & Ree, R. (2012). Materializing information: 3D printing and social change. *First Monday*, *17*(7). doi:10.5210/fm.v17i7.3968

Rideout, B. (2012). Printing the impossible triangle: The copyright implications of three-dimensional printing. *The Journal of Business. Entrepreneurship & the Law*, *5*(1), 6.

Sachs, E., Cima, M., Williams, P., Brancazio, D., & Cornie, J. (1992). Three dimensional printing: Rapid tooling and prototypes directly from a CAD model. *Journal of Engineering for Industry*, *114*(4), 481–488. doi:10.1115/1.2900701

Schwartz, C. (2015, March). 3-D Printing: The Potential Implications and Challenges for Law Enforcement. *The Police Chief*, *82*, 64–69.

Sissons, A., & Thompson, S. (2012). *Three Dimensional Policy: Why Britain needs a policy framework for 3D printing*. London: The Big Innovation Centre.

Stahl, H. (2015). *3D Printing – Risks and Opportunities*. Retrieved from http://www.oeko.de/oekodoc/1888/2013-532-en.pdf

Stratasys. (2015). *Digital Dentistry 3D printing makes digital dentistry happen*. Retrieved from http://www.stratasys.com/industries/dental

Varkey, M., & Atala, A. (2015). Organ Bioprinting: A Closer Look at Ethics and Policies. *Wake Forest JL & Pol'y*, *5*, 275.

ADDITIONAL READING

Childers, E. P., Wang, M. O., Becker, M. L., Fisher, J. P., & Dean, D. (2015). 3D printing of resorbable poly (propylene fumarate) tissue engineering scaffolds. *MRS Bulletin*, *40*(02), 119–126. doi:10.1557/mrs.2015.2

Lee, N. (2016). Revisiting the Principle of Technological Neutrality in Patent Protection in the Age of 3D Printing Technology and Cloud Computing. InTRIPS plus 20 (pp. 361-388). Springer Berlin Heidelberg.

Lieben, L. (2016). Regenerative medicine: The future of 3D printing of human tissues is taking shape. *Nature Reviews. Rheumatology*, *12*(4), 191. doi:10.1038/nrrheum.2016.29 PMID:26935277

Loy, J., Tatham, P., Healey, R., & Tapper, C. L. (2016). 3D Printing Meets Humanitarian Design Research: Creative Technologies in Remote Regions.*Creative Technologies for Multidisciplinary Applications*, 54.

ONeill, B., Wang, D. D., Pantelic, M., Song, T., Guerrero, M., Greenbaum, A., & ONeill, W. W. (2015). Reply: the role of 3D printing in structural heart disease: all that glitters is not gold. *JACC: Cardiovascular Imaging*, *8*(8), 988–989. doi:10.1016/j.jcmg.2015.04.011 PMID:26271095

Rideout, B. (2012). Printing the impossible triangle: The copyright implications of three-dimensional printing. *The Journal of Business. Entrepreneurship & the Law*, *5*(1), 6.

Varkey, M., & Atala, A. (2015). Organ Bioprinting: A Closer Look at Ethics and Policies. *Wake Forest JL & Pol'y*, *5*, 275.

Xing, J. F., Zheng, M. L., & Duan, X. M. (2015). Two-photon polymerization microfabrication of hydrogels: An advanced 3D printing technology for tissue engineering and drug delivery. *Chemical Society Reviews*, *44*(15), 5031–5039. doi:10.1039/C5CS00278H PMID:25992492

KEY TERMS AND DEFINITIONS

Additive Manufacturing (AM): The process of making 3D objects by applying layer upon layer.

CAD Design File: Stands for computer-aided design file; used by engineers or architects to create technical illustrations.

Elephant Foot: The base of the model is slightly bulging outwards.

Friction-Weld: A welding process that generates heat through mechanical friction between workpieces to fuse materials.

Gunsmithing: The act of making, selling and repairing firearms.

Pillowing: Bumps or holes are present on the top surface of the printed model.

Post-Processing: Furbishing or decorating stage of the printed model.

Rapid Prototyping: A set of techniques used to quickly fabricate a scale model of a part using 3D CAD data.

Scalable Production: A production that can cope and perform under an expanding workload.

Stringing: Unsightly strings of plastic between parts of the printed model.

Warping: The print bends upwards at the base of the model and is no longer parallel with the print platform.

This research was previously published in the Encyclopedia of Information Science and Technology, Fourth Edition edited by Mehdi Khosrow-Pour, pages 4382-4389, copyright year 2018 by Information Science Reference (an imprint of IGI Global).

Section 8

Optical Engineering

Chapter 29
Visible Light Communication Numerous Applications

Ala' Fathi Khalifeh
German Jordan University, Jordan

Hasan Farahneh
Ryerson University, Canada

Christopher Mekhiel
Ryerson University, Canada

Xavier Fernando
Ryerson University, Canada

ABSTRACT

Visible light communication (VLC) is a promising research topic that aims at utilizing the visible light spectrum for data communication, which in turn off-loads the heavily utilized wireless radio spectrum. VLC can take advantage of the increased use of light emitting diodes (LED) for lighting purposes in different fields, such as automotive headlights, traffic signals, advertising, aviation, and general lighting. Utilizing solid-state LEDs for lighting purposes not only saves energy but also can be used for data communication since LEDs can be easily modulated by switching the light on and off in frequencies above the human eye perception. This data can be transmitted in addition to the main lighting functionality of these devices. VLC is relatively easy to implement and safe, as it does not have potential side effects like the radio signals. As such, this new technology has a great potential for adoption in several applications. This chapter will briefly introduce the technology and discuss some potential indoors and outdoors applications.

INTRODUCTION

The initial theory of Visible Light Communication (VLC) was founded in the 1880s when Alexander Graham Bell invented the photo-phone which was used to transmit a voice signal using the modulated sunlight. Since the time of Graham Bell, optical communication research has attracted the interest of scholars around the world and has evolved into a new IEEE standard namely the P802.15.7 - Standard

DOI: 10.4018/978-1-5225-7359-3.ch029

for Short-Range Wireless Optical Communication (standard, 802.15.7 (2015)). In 2003 at the Nakagawa Laboratory in Keio University, Japan, they have proposed using the Light Emitting Diodes (LEDs) for data transmission.

A major factor that contributes to the evolution of VLC technology is the existing infrastructure. Hence, previously installed facilities, such as LED traffic lights or LED sign boards are readily used. Since the transmitters for VLC are light sources, they function for lighting purposes and illuminate the surrounding environment, hence the radiation power and signal-to-noise ratio (SNR) is high; paving the way for a stable communication link (T. Yamazato, I. H. (2014)).

With respect to the emergence of green communication, VLC is highly energy efficient as it utilizes LEDs. The United States Department of Energy further corroborated the importance of LED technology, as shown in Table 1. There is superiority in terms of power consumption and operating lifetime in LED technology as compared to traditional lighting technology, such as incandescent and fluorescent lighting. This clearly shows the potential of the LED lighting technology to replace all the conventional illumination tools as well as serve as a reliable transmitter for a VLC link (Chung, Y.-Y. T.-Y. (2014)).

Radio Frequency (RF) wireless connectivity has been used for several decades as it allows for indoor and short distance links to be established without any physical connection. However, these solutions remain relatively expensive and have low to medium data rates. RF wireless links require that spectrum licensing fees are paid to federal regulatory bodies and are required to be contained within strict spectral zones. These frequency allocations are determined by local authorities and may vary from country to country, making a standard interface difficult. Since the visible light spectrum is not in the licensed band (400 to 790 THz), licensing fees can be avoided which effectively reduces system cost. In addition, the broadcast nature of the RF link is beneficial for mobile connectivity but this may result in interference between devices located within close proximity. Due to the RF wavelength, it is difficult to contain within boundaries and can impede system performance (Hranilovic, S. (2005)).

Optical radiation in the infrared or visible range is easily contained by opaque boundaries. As a result, interference between adjacent devices can be minimized easily and economically. Additionally, inexpensive LEDs and photodiodes are able to interchangeably work between baseband and transmission frequencies where as high-frequency RF circuit design techniques are required in the RF domain. Free-space optical (FSO) links with an inherent low probability of intercept and anti-jamming characteristics is among the most secure of all wide-area connectivity solutions (Hranilovic, S. (2005)).

Unlike many RF systems that radiate signals in all directions, thus making the signal available to all within the receiving range, FSO transceivers use a highly directional and cone-shaped beam with a dominant line-of-sight (LOS) propagation path. Therefore, interception is extraordinarily difficult and anyone

Table 1. Performance of the conventional and LED lighting technology

Lamp Type	Watts	Lumens	Operating Lifetime
Incandescent	60	900	1000
Compact florescent lamp	15	900	8500
LED (2011)	12.5	800	25000
LED – future(2015)	5.8	800	40000

tapping into the systems can easily be detected as the intercept equipment must be placed within the very narrow optical foot print (Ghassemlooy . Z., P. W. (2012)). Although this contributes to the security of wireless optical links and reduces interference it also greatly impacts the high mobility of such devices.

The aim of this book chapter is to introduce the concept of VLC as an emerging technology from a system and hardware design point of view, and shed the light on the rich features and potential of this technology that make it a viable substitution for other wireless technologies. The importance of this technology will be demonstrated by covering various applications and scenarios.

BACKGROUND

VLC is the process of transmitting digital data by using the visible light spectrum. This can be achieved by modulating the data using a light source such as LEDs that can be switched fast enough to avoid observable flickering or light dimming, another possible way is to change the light intensity in a way that is not observable to the human eye but can be detected using an appropriate sensor such as a photo diode (PD). Along with its prime function of lighting, LEDs can also serve to transmit data as long as there are suitable receivers.

Additionally, image sensor pixels can be used as an effective VLC receiver. The ability to spatially separate multiple sources of image sensors provide an attractive feature to VLC. Image sensor pixels can sense LED transmission sources and discards other pixels which detects ambient noise. More specifically, outdoor usage of VLC is possible by discarding pixels associated with noise sources such as the sunlight or streetlights. Hence, image sensor based VLC is an attractive solution for outdoor mobile applications (T. Yamazato, I. H. (2014)). With regards to short-range VLC applications, the SNR of a direct detection receiver is proportional to the square of the received optical power. Therefore, VLC links can tolerate only a comparatively limited amount of signal path loss.

Figure 1 shows a general block diagram for a simple VLC transceiver. As shown in the figure, the upper part represents the transmitter, where binary data is first passed to a coder such as a Reed-Solomon coder (Khalifeh, A. Y. (2010)) where error correction codes are added. After that, binary data is modu-

Figure 1. VLC transceiver block diagram

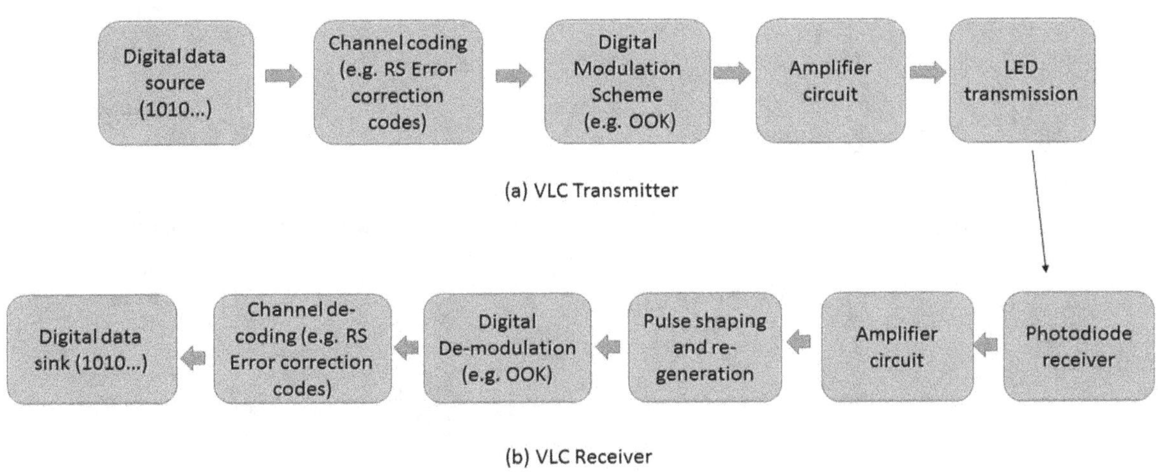

lated using a digital modulation scheme such as binary On-Off Keying (OOK) where data is modulated by passing it to an amplification circuit that drives an LED(s), which converts the data into light over the wireless communication channel.

One can notice from Figure 2 that either a single or several LEDs can be used to transmit data. The same applies to the receiving circuit, thus four different configurations can be used to increase transmission rates, as well as improve the received SNR; (a) Single-Input Single-Output (SISO), where only one LED transmitter and one photodiode receiver is used, (b) Single-Input Multiple-Output (SIMO), where one LED transmitter and multiple photodiode receivers are used, (c) Multiple-Input Single-Output(MISO) where multiple LED transmitters and one photodiode receiver is used, and finally, (d) Multiple-Input Multiple-Output (MIMO) where multiple LED transmitter and multiple photodiode receivers are used.

The data path is then transmitted to the receiver circuit, where the light pulses are detected by a photodiode(s). The output signal is then passed to an amplifier, pulse shaping and re-generation blocks, a channel decoding module to correct the introduced errors, and a demodulator circuit that will detect and convert the signal back into binary data. Without any ambient noise or distortion, the transmitted bit stream should reach the destination without any errors. However, due to the internal and external noise sources, some binary data will be falsely detected which causes bit errors. In addition to the typical additive white Gaussian noise that is locally generated by the receiver electronic circuit, VLC receivers are prone to the ambient surrounding and background lights that can cause errors in the detection process.

However, in order to mitigate the effect of these external sources, various techniques are used such as the one proposed in (Ya´ nez, V. T. (2009)) where the channel characteristic is estimated and the received signal is equalized based on the channel estimation model. Another technique presented in (Ya´ nez, V. T. (2009)) uses the wavelength filtering to reduce the unwanted inference signal.

VLC APPLICATIONS

This section discusses in details the major applications where VLC has a great potential. Applications are divided into two main categories; indoor and outdoor. VLC can be used in various indoor applications

Figure 2. Several transmitter-receiver configurations

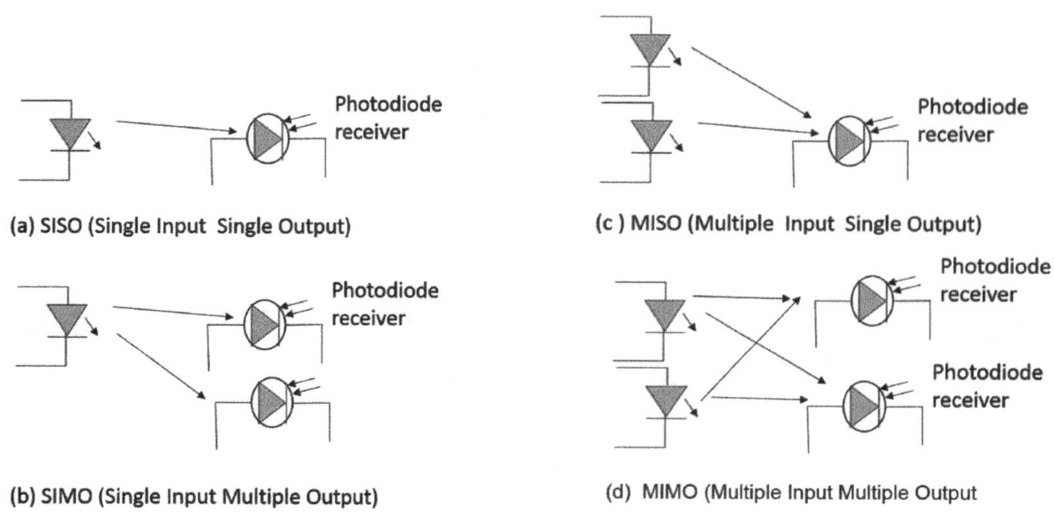

and scenarios, namely, in high-density indoor areas such as classrooms, conference halls, convention centers and other assembly spaces where light can be used to broadcast relevant information to the audience. Entertainment applications can also utilize VLC for video and audio streaming and broadcasting. Indoor navigation and localization is another important application of VLC, and finally, VLC can play an important role in data transmission in sensitive areas where radio wireless communication is not preferable, some examples are in hospitals and inside the airplanes. A closer look for these potential applications will be described next.

VLC can also be used in various environments where it is not safe or recommended to use the radio spectrum, since it may interfere with other critical electronic equipment. The other set of applications that can efficiently utilize this technology is some of the outdoor-based applications. Research has been done on the field of vehicle-to-vehicle (V2V) communications where cars can communicate with each other to exchange critical and general messages. Additionally, it can be easily used for Intelligent Transportation Systems (ITS) and vehicle-to-infrastructure and infrastructure-to-vehicle (V2I and I2V) communication.

Indoor Applications

VLC has many potential indoor applications. This is due to the fact that an indoor channel changes slowly when compared to outdoor channels, where the channel may vary more rapidly. Three main applications will be discussed: Indoor localization and positioning, medical services, and VLC in-flight communication systems.

Indoor Localization and Positioning Services

One important application of VLC is indoor localization and positioning. This is a vibrant research area that is gaining much attention recently. Positioning applications cover a wide area where the technology can be included into various consumer electronics. Indoor positioning technology can be used to guide users in large areas. Moreover, positioning systems using VLC can also detect products inside large warehouses, automating some record management processes.

As known in literature, radio signals coming from Global Positioning System (GPS) satellites cannot penetrate well through walls of large buildings, consequently, fast and accurate indoor positioning is difficult to achieve by GPS. To overcome this problem, there are two possible alternatives; RF and VLC-based techniques. Examples of RF based techniques are: wireless local area network, radio-frequency identification (RFID), cellular, ultra-wide band and Bluetooth. These technologies will have positioning accuracies from tens of centimeters to several meters. But unfortunately, this amount of accuracy is not sufficient for many indoor applications. Hence, the techniques based on VLC are gaining greater attention.

LEDs are currently being installed in many large buildings such as museums and shopping malls, because they have the advantage of much longer life and lower operating cost. For these reasons, indoor positioning techniques based on VLC and LEDs are the desirable and preferable options (LVX Minnesota lightning. (2015)).

Medical Services

Another major concern these days are regarding the compatibility of medical devices in healthcare centers with the incorporation of wireless technology (S.S. Muhlen, D. D. (2008)). RF wireless technology has always been associated with the emission of electromagnetic interference (EMI) (H. Hong, Y. R. (2008)). Intrusion of the EMI jeopardize the quality of medical monitoring, as the accuracy and efficiency of data transmission is crucial for the medical staff to provide corresponding measures or treatment, based on the real–time information received (K.S. Tan, I. H. (2001), Chung, Y.-Y. T.-Y. (2014)).

VLC can be applied to indoor medical applications by transmitting patient data as well as management data. Healthcare information such as electrocardiography (ECG), photoplethysmogram (PPG) signals and text information can be transmitted simultaneously, using a single channel VLC. This allows for a more precise and accurate monitoring and diagnosis.

VLC In-Flight Communication Systems

The use of VLC technologies for offering data networking for passengers during flight is another promising indoor application. Internet access, Wi-Fi is traditionally used to connect the passenger devices to an access point that will send their request to the internet either via a satellite connection or via plane-to-earth data link. This Wi-Fi scheme shares the radio spectrum among many passengers, which may cause congestion and reduce the speed of communication. In addition, Wi-Fi may affect the airplane navigation systems especially at take-off and landing causing EMI.

A better solution can be implemented by utilizing a hybrid radio and optical system, where passengers may use the radio spectrum for uploading their requests, which usually require less bandwidth than the downlink. The downlink can be provided to the passengers via the personal LED lights available on top of their seats. This scheme will increase the internet speed due to abundant bandwidth available in the optical domain and will reduce the interference to the airplane navigation system. Brighter overhead lighting will result in higher SNR and greater connectivity.

Outdoor Applications

In this section, potential outdoor applications for VLC will be discussed. Despite the more challenging problems such as fast time varying channels and users' high mobility, VLC, if designed properly, can be used in several outdoor applications such as automotive communication, intelligent transportation systems, and underwater communications.

Visible Light Communication in Automotive Industry

Due to VLC being optimal for short range communication, it has significant potential in the automotive industry such as car-to-car (C2C) and Intelligent transportation systems (ITS). Additionally, existing infrastructure is in place to take advantage of VLC such as headlights and streetlights. In the next few years, the implementation of LED taillights, brake lights, will all be more than 50% (Cui, K. C. (2012)). All these conditions make it desirable to consider VLC in the automotive industry. Vehicular Visible Light Communications (V2LC) offers a dependable solution to put into operation vehicle-to-vehicle (V2V) communications and requires minimal installation cost.

V2LC is challenged by dynamic ambient noise, high mobility, and moderately low transmitter heights on communicating systems such as vehicles and roadside units. These limitations make the V2LC channel modelling a challenging task. V2LC utilizes either a photodiode detector or a camera installed on the vehicle to be used as a receiving element. It was found that V2LC systems perform better in high volume vehicular traffic; this is because of the fact that visible light can mainly propagate within proximity and LOS to reduce interference and increase link scalability (CALM. (2011)).

C2C Communication System

A traditional C2C -VLC scenario is shown in Fig. 3, where the car on the left side communicates with the car on the right side using its headlamp. As shown, the projected beam pattern is single, and the received light consist of the LOS and reflected or Non-Line of Sight (NLOS) components. It can be noticed that a multi-path interference will occur due to the reflected rays from the road surface, which depends on many factors such as the pavement material, the angle of incidence as well as the weather condition (rain, snow, fog, etc.).

There are many regulations issued by Economic Commission of Europe and Federal Motor Vehicle Safety Standards in United State of America related to headlamps to insure that vehicles will provide good road illumination. At the same time, the light emitted from headlamps should not visually disturb other road users. Hence, the lamp, its reflective devices, and other associated equipment must achieve specific requirements. Therefore, the Lambertian model which has been used widely in indoor VLC LED modelling, cannot be used to model the vehicle's light pattern (Kumar, N., et al. (2012)).

In C2C VLC links, usually the LOS and the NLOS components form the received optical power. For FSO, there are additional noise sources, such as the background solar radiation, streetlights, vehicles, and secondary reflections which can be treated as artificial light (Ghassemlooy . Z., P. W. (2012)). The background solar radiation is mostly the dominant noise source as it is composed of direct and scattered radiation (Cui, K. C. (2012)). As such, good noise cancellation mechanisms are required to account for the nature of these noise sources.

In addition, appropriate signal processing algorithms are needed to equalize the received signal and improve its condition. Especially, the channel coherence time will be very short in an outdoor vehicular environment due to increased mobility of fast moving vehicles. Hence, any channel estimation and equalization has to be achieved seamlessly.

ITS Traffic Light to Car Communication

Improvements to safety, fuel consumption, and transportation time are the key objectives of an intelligent transportation system. ITS refers to the potential of adding information and smart communication technologies to the transportation infrastructure and vehicles in order to enhance transportation safety, mobility, and support productivity through the use of advanced communications technologies. VLC can

Figure 3. C2C VLC and projected LOS and NLOS beam patterns

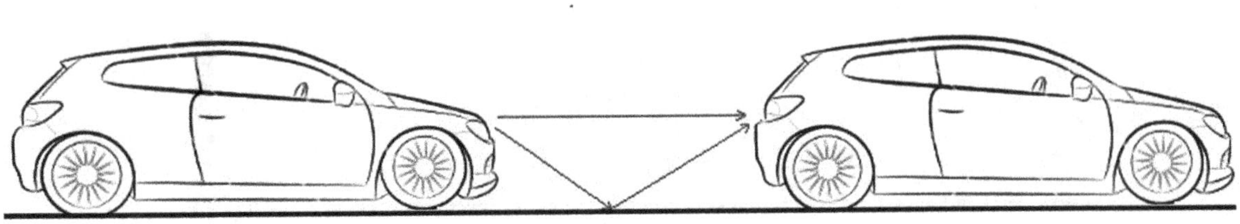

play an important role in ITS, such as broadcasting important traffic information. Here several configurations, such as V2V using LED-based rear panel lights, and bidirectional infrastructure-to-vehicle are possible. Traffic information system using LED-based traffic lights were investigated by many authors. They touched upon topics such as analyzing the communication performance and defining a service area, in which communications using a particular data rate together with intensity modulation OOK can be maintained.

VLC is suitable for both a broadcast system in I2V communication systems and effective in V2V as well. In a V2V scenario, a vehicle in front of a traffic light can receive the information sent by the traffic light, and then it can relay this information using the brake lights to the vehicle running behind. This can be extended to establish a vehicle ad-hoc network. With respect to I2V broadcasting, there is no need for extra power to broadcast traffic safety related information, which can be continuously broadcasted to support smooth traffic flow as well as reducing accidents and fatalities (CALM. (2011)).

LED-based traffic lights are also considered to be a suitable choice for traffic light Road Side Units (RSUs) in the ITS architecture without the need for additional infrastructure. One of the most possible approaches of communication between an LED-based RSU and a vehicle is presented in Fig. 4. The first stage is collecting and passing the detected information to a central control unit, which consists of many parts such as data processing, geographical distribution of the data, and information broadcasting. After the data collection stage, traffic control actions are made. Traditional ITS systems are normally based on wireless radio solutions and short range infrared systems.

In order to deliver information from the traffic control infrastructure, which is represented by traffic lights to vehicles, VLC can be effectively integrated into this scenario. This technique offers many advantages, like cost effectiveness, alleviation of additional wireless communication support, and energy efficiency. The most important benefit is that it does not require additional power to support future communication systems. Fig. 4 shows that the ITS use control technology and access networks to achieve communication facilities. In this case, it becomes important for an ITS architecture design to be flexible enough to accommodate integration of new systems (Kumar, N. L. (2012)).

Figure 4. VLC integration with ITS architecture depicting the vehicle Field-of-View (FOV) and the Variable Message System (VMS)

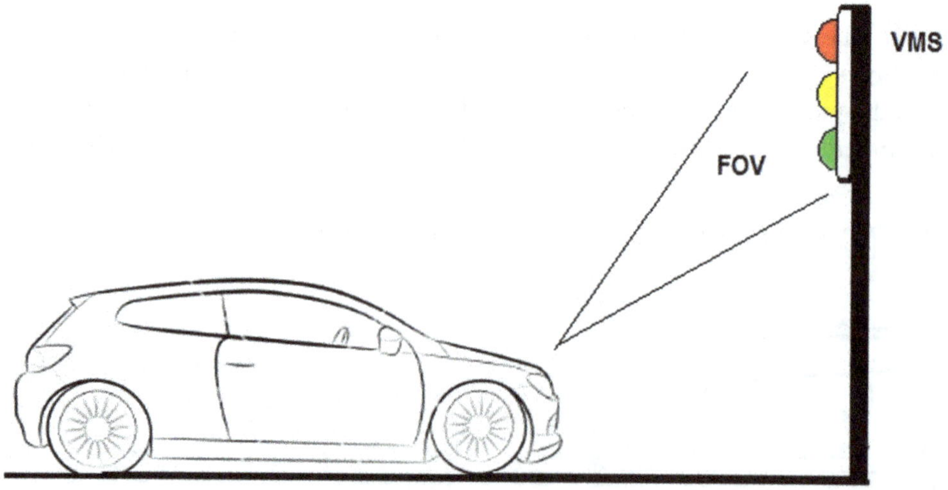

Underwater Communication

Providing an efficient and reliable data transmission for underwater communication is a challenging task, this is due to the nature of water, which absorbs the energy of the transmitted signal and refracts it. Data transmission underwater suffers from high-energy consumption, long propagation delay, and limited link bandwidth. RF-based solutions are not efficient due to the high attenuation and due to the multipath signal propagation (Gabriel, C. K. (2011)). As an alternative, acoustic waves are in use for underwater communications. Table 2 compares acoustic and optical communication methods for underwater use. As shown in the table, both acoustic and optical mediums suffer from power loss. The acoustic waves suffer from more than 0.1 dB per meter per Hz, so the higher the frequency and distance, the higher the attenuation and power loss.

For optical transmission, the amount of power loss depends on the turbidity, which is a measure of the water purines and its clarity. The higher the turbidity, the higher the power loss. In addition, the signal strength is inversely proportional with the distance (Lanbo, L. S.-H. (2008)). Although acoustic transmission has a greater range, its transmission rate is very limited. While the optical communication link has higher data rates for low distance, it becomes detrimental over a long distance. The propagation delay for optical communication is much lower when compared to an acoustic link, since light speed inside water is significantly faster than the speed of acoustic transmission. Moreover, optical transceivers are relatively cheap and less complicated when compared with acoustic transceivers. It is important to note that the performance metrics in Table 2 are based on a seawater environment, so it may vary for other underwater environments.

DISCUSSION

Ever since the growing demand and use of data, there has been a need to reliably transmit significant amounts of data in real time. Current technology, as well as a growing number of users limit the com-

Table 2. Comparison between underwater communication using acoustic and optical communication methods, assuming a seawater environment

Property	Acoustic	Optical
Require alignment	No	Yes
Receiver complexity	Higher	Lower
Recommended communication range	Up to 5 km	Up to 100m
Data rate	Up to 100 Kbps	Up to 1 Gbps
Power loss	> 0.1 dB/m/Hz	proportional to water turbidity
Propagation delay /speed	large/slow (1500-2000 m/s)	low/ high (2.255 x 10^8)
Frequency band	~ kHz	10^{15} Hz

Sources: Lanbo, L. S.-H. (2008), Yang S., K., . (2011), Lurton, X. (2010)

munication capacity. Hence, there is a great need for high bandwidth, energy efficiency, cost, and reliability for a communication link. Hence, VLC offers an excellent alternative to traditional technology.

However, there exists some challenges that were investigated to make this a viable communication technology. These challenges are dealing with the LOS requirement of VLC, inability to communicate through opaque obstacles, high mobility, ambient light interference, limited LED photon emission rate, required DC biasing, and limited modulation bandwidth as well as the non-linearity nature of LEDs (Dong-Fang Zhang, Y.-J. Z.-Y. (2013)).

As previously discussed, there are several approaches which can be taken to alleviate the main challenges with VLC. To name a few, implementing a MIMO or LED array can be used to increase the SNR and improve diversity in impeding scenarios, channel modelling and equalization at the receiver, camera image sensors used to eliminate ambient light-treated as noise, and finally advanced modulation techniques to overcome limited modulation bandwidth.

FUTURE RESEARCH DIRECTIONS

Since VLC has several challenges as previously mentioned, there has been a fair amount of lab research toward improving its performance as well as improving its distance range. In order to mitigate the channel effect on a VLC system, several techniques have been proposed, such as the usage of Orthogonal Frequency Division Multiplexing (OFDM), and the use of MIMO (Mohammed S. A. Mossaad, S. H. (2015)).

With respect to the limited LED modulation bandwidth, polarization multiplexing methods can be combined with other multiplexing methods which can further increase the communication rate. Cvijetic et al. studied Bit Error Rate (BER) performance of a polarization multiplexed optical wireless transmission (Chiharu Mukai, K. O. (2012)). Additionally, advanced modulation techniques have been carried out to achieve high data rates up to 100Tbit/s by implementing orbital angular momentum (OAM) with other multiplexing domains and present a free-space data link that uniquely combines OAM-, polarization-, and wavelength-division multiplexing by using three-dimensional multiplexing. (Hao Huang, X. (2014)).

In addition, researchers have looked at combining the Power Line Communication (PLC) technology with VLC, such that the power lines become the data source of these lights (Ding W., Y. F. (2015), Hao M., L. L. (2013)).

CONCLUSION

In this chapter, we have discussed several indoor and outdoor potential applications that VLC can utilize in the upcoming years. Despite its performance challenges, VLC showed significant potential in several indoor and outdoor applications, such as: indoor localization and positions, in-mall advertising and user behavior tracking, and in-plane multimedia broadcasting. For outdoor applications, it shows a promising potential in the automotive industry as well as underwater communications.

REFERENCES

CALM. (2015, Nov 1). *Communications in Cooperative Intelligent Transport Systems - CALM for C-ITS*. Retrieved from Introduction to Communications Access for Land Mobiles: http://calm.itsstandards. EU/

Chiharu Mukai, K. O. (2012). *BER Performance of Multiplexed CSK using Linear/Circular Polarization in Optical Wireless Communications*. ISITA.

Chung, Y.-Y. T.-Y. (2014). Mobile health-monitoring system through visible light communication. *Bio-Medical Materials and Engineering*, 608–737.

Cui, K. C., Chen, G., Xu, Z., & Roberts, R. D. (2012). Traffic light to vehicle visible light communication channel characterization. *Applied Optics*, *51*(27), 6594–6605. doi:10.1364/AO.51.006594 PMID:23033030

Ding, W. Y. F., Yang, F., Yang, H., Wang, J., Wang, X., Zhang, X., & Song, J. (2015). A hybrid power line and visible light communication system for indoor hospital applications. *Computers in Industry*, *68*, 170–178. doi:10.1016/j.compind.2015.01.006

Dong-Fang Zhang, Y.-J. Z.-Y. (2013). Multi-LED Phase-Shifted OOK Modulation Based. *IEEE Photonics Technology Letters*. Retrieved from https://www.eldoled.com

Gabriel, C. K. (2011). *Channel modeling for underwater optical communication. Proceeding in IEEE GLOBECOM Workshops*. GC Wkshps.

Ghassemlooy, Z. P. W. (2012). *Optical wireless communications: system and channel modeling with Matlab*. CRC Press.

Grantham, C. (1998). Visible Light Communication for Audio Systems. *IEEE Transactions on Consumer Electronics*, 1–11.

Hao, M. L. L. (2013). Integration of indoor visible light and power line communication systems. *17th IEEE International Symposium in Power Line Communications and Its Applications (ISPLC)*. 10.1109/ISPLC.2013.6525866

Hao Huang, X. (2014). 100 Tbit/s free-space data link enabled by three-dimensional multiplexing of orbital angular momentum, polarization, and wavelength. *Optics Letters*, *39*(2), 197–200. doi:10.1364/OL.39.000197 PMID:24562105

Hong, H., Y. R. (2008). Information illuminating system for healthcare institution. *Proceedings of International Conference on Bioinformatics and Biomedical Engineering*, 801-804.

Hranilovic, S. (2005). *Wireless Optical Communication Systems*. Springer.

Khalifeh, A. Y., & Yousefizadeh, H. (2010). Optimal Audio Transmission Over Error-Prone Wireless Links. *IEEE Transactions on Multimedia*, *12*(3), 204–214. doi:10.1109/TMM.2010.2041096

Kumar, N. L. (2012). Visible Light Communications in Intelligent Transportation Systems. *Intelligent Vehicles Symposium Alcalá de Henares*, 748-753. 10.1109/IVS.2012.6232282

Lanbo, L. S.-H. (2008). *Prospects and problems of wireless communication for underwater sensor networks*. Wirel. Commun. Mob. Comput.

Lurton, X. (2010). *An Introduction to Underwater Acoustics Principles and Applications*. Springer-Verlag Berlin Heidelberg. doi:10.1007/978-3-642-13835-5

LVX Minnesota Lightning. (2015, Nov 1). Retrieved from http://purelifi.com

Masayuki Kinoshita, T. Y. (2014). *Motion Modeling of Mobile Transmitter for Image Sensor Based I2V-VLC, V2I-VLC, and V2V-VLC*. Globecom Workshop-Optical Wireless Communications.

Mohammed, S. A., & Mossaad, S. H. (2015). Visible Light Communications Using OFDM and Multiple LEDs. *IEEE Transactions on Communications*, 4304–4313.

Muhlen, S. S. D. D. (2008). New challenges in controlling EMI in hospitals. Proceedings of International Federation for Medical & Biological Engineering, 834-837.

Pengfei, L., e. a. (2014). Fundamental Analysis of a Car to Car Visible Light Communication System. *9th International Symposium on Communication Systems, Networks & Digital Sign (CSNDSP).*

Perez-Jimenez, R. R.-H. (2011). Visible Light Communication Systems for Passenger In-Flight Data Networking. *IEEE International Conference on Consumer Electronics (ICCE)*, 445-446. 10.1109/ICCE.2011.5722675

Popoola, W. O., & Ghassemlooy, Z. (2009). BPSK subcarrier intensity modulated free-space. *Journal of Lightwave Technology*, *27*(8), 967–973. doi:10.1109/JLT.2008.2004950

Ramirez-Iniguez, R. S. M. (2008). Optical Wireless Communications: IR for Wireless Connectivity. Boca Raton, FL: CRC Press.

Standard 802.15.7. (2015, Nov 1). Retrieved from https://standards.ieee.org/develop/project/802.15.7.html

Takaya Yamazato, I. T. (2014). Image-Sensor-Based Visible Light Communication for Automotive Applications. *IEEE Communications Magazine*, 10.

Tan, K. S. (2001). Electromagnetic interference in medical devices: Health Canada's past current perspectives and activities. *IEEE International Symposium on Electromagnetic Compatibility*, 1283-1284.

U.S. Department of Energy. (2013). *Part 1: Review of Life-Cycle Energy Consumption of Incadescent, Compact Fluorescent, and LED Lamps.* Author.

Yamazato, T., Takai, I., Okada, H., Fujii, T., Yendo, T., Arai, S., ... Kawahito, S. (2014). Image sensor based visible light communication for automotive. *IEEE Communications Magazine*, *52*(7), 88–97. doi:10.1109/MCOM.2014.6852088

Ya´nez, V. T. (2009). Illumination interference reduction system for VLC Communications. *Proc. WSEAS Int. Conf. Math. Methods, Computer. Tech. Intel.*

Yang, S. K. (2011). Reduction of optical interference by wavelength filtering in RGB-LED based indoor VLC system. *Proceeding of Opto-Electronics and Communications Conference (OECC).*

Zahmati & Fernando. (2011). Emerging Wireless Applications in Aerospace: Benefits and Challenges. *Exponent, the Honeywell Technical Journal.*

ADDITIONAL READING

Gallico, P. V. (2007). *Optics research trends*. New York: Nova Science.

Haas, H. (2015). Visible Light Communication. *Optical Fiber Communication Conference*.

Haas, H., & Chen, C. (2015). What is LiFi? *2015 European Conference on Optical Communication (ECOC)*. 10.1109/ECOC.2015.7341879

Haruyama, S. (2012). Advances in Visible Light Communication Technologies. *European Conference and Exhibition on Optical Communication*. 10.1364/ECEOC.2012.We.3.B.5

Koteles, E. S., & Willner, A. E. (1995). *Emerging components and technologies for all-optical networks: Conference: Papers*. SPIE.

Whyte, B. (1999). *Networked futures: Trends for communication systems development*. Chichester, England: J. Wiley. doi:10.1002/0470841850

KEY TERMS AND DEFINITIONS

CDMA: A channel access method, which allows several users to share a band of frequencies.

Intelligent Transportation Systems (ITS): The potential of adding information and smart communications technologies to the transportation infrastructure and vehicles, in order to enhance transportation safety, mobility, and support productivity.

Light-Emitting Diode: A semiconductor device that emits visible light at a single wavelength when an electric current passes through it. The output from an LED can range from red (at a wavelength of approximately 700 nanometers) to blue-violet (about 400 nanometers).

OFDM: A method of encoding digital data on multiple carrier frequencies.

Signal to Noise Ratio: A ratio of desired signal to undesired signal (noise) in the average power level of a transmission.

Smart Grid: An electrical grid which includes a variety of operational and energy measures including smart meters, smart appliances, renewable energy resources, and energy efficiency resources.

Visible Light Communication (VLC): Data communications medium which uses visible light between 400 and 800 THz (780–375 nm).

Wireless Communication: Transfer of information between two or more points that are not connected by an electrical conductor.

This research was previously published in the Encyclopedia of Information Science and Technology, Fourth Edition edited by Mehdi Khosrow-Pour, pages 6672-6683, copyright year 2018 by Information Science Reference (an imprint of IGI Global).

Compilation of References

3XN. (2008). *Digital architecture*. Retrieved May 2, 2010, from http://www.3xn.dk

Abido, M. (2002). Optimal power flow using particle swarm optimization. *Electrical Power & Energy Systems*, *24*(7), 563–571. doi:10.1016/S0142-0615(01)00067-9

Able Software Corp. 3D-DOCTOR. (2016). Retrieved June 28, 2016, from http://ablesw.com/3d-doctor/images.html#SurfaceRendering

Abrial, J. (2009). Faultless systems: Yes we can! *Computer*, *42*(9), 30–36. doi:10.1109/MC.2009.283

Ageron, B., Gunasekaran, A., & Spalanzani, A. (2012). Sustainable supply management: An empirical study. *International Journal of Production Economics*, *140*(1), 168–182. doi:10.1016/j.ijpe.2011.04.007

Aguilar-Moreno, E., & Granell-Canut, C. (2013). Sistemas de información geográfica para unidades de información [Geographic information systems for informatio units]. *Profesional de la Información*, *22*(1), 80–86. doi:10.3145/epi.2013.ene.11

Aguilar-Moreno, E., & Granell-Canut, C. (2015). *Geobibliotecas* [Geolibraries]. Barcelona: Editorial UOC.

Ahmad, I., & Ranka, S. (2016). *Handbook of energy-aware and green computing*. CRC Press.

AIRBUS-Group. (2013). *Airbus Group Annual Review*. Toulouse: Astrium.

Akgun, A., Kıncal, C., & Pradhan, B. (2012). Application of remote sensing data and GIS for landslide risk assessment as an environmental threat to Izmir city (west Turkey). *Environmental Monitoring and Assessment*, *184*(9), 5453–5470. doi:10.100710661-011-2352-8 PMID:21915598

Akgun, A., Sezer, E. A., Nefeslioglu, H. A., Gokceoglu, C., & Pradhan, B. (2011). An easy-to-use MATLAB program (MamLand) for the assessment of landslide susceptibility using a Mamdani fuzzy algorithm. *Computers & Geosciences*, *38*(1), 23–34. doi:10.1016/j.cageo.2011.04.012

Akhgar, B., Pattinson, C., & Dastbaz, M. (2015). *Green Information Technology: A Sustainable Approach*. Waltham, MA: Elsevier, Inc.

Åklint, T., Carlström, E., Johander, P., & Stiernstedt, J. (2013, March 7). Layered manufacturing of free-form multi-material micro-components. *Patent*, *WO2013030064*, A1.

Akram, A., & Nanna, S. (2003). *Medical-based geographical information system (mGIS): diagnosis versus healing*. Global Health Network Supercourse Project. Retrieved October 2, 2015, from http://supercourse.isciii.es/lecture/lec10671/index.htm

Al Garni, H., Kassem, A., Awasthi, A., Komljenovic, D., & Al-Haddad, K. (2016). A multicriteria decision making approach for evaluating renewable power generation sources in Saudi Arabia. *Sustainable Energy Technologies and Assessments*, *16*, 137–150. doi:10.1016/j.seta.2016.05.006

AlAli, A. B., Griffin, M. F., & Butler, P. E. (2015). Three-Dimensional Printing Surgical Applications. *Eplasty*, *15*, e37. PMID:26301002

Alaykýran, K., Engin, O., & Döyen, A. (2007). Using ant colony optimization to solve hybrid flow shop scheduling problems. *International Journal of Advanced Manufacturing Technology*, *35*(5-6), 541–550. doi:10.100700170-007-1048-2

Alkan, A., & Gunay, M. (2012). Identification of EMG signals using discriminant analysis and svm classifier. *Expert Systems with Applications*, *39*(1), 44–47. doi:10.1016/j.eswa.2011.06.043

Allan, R., Borkowska, B., & Grigg, C. (1974). Probabilistic analysis of power flows. Electrical Engineers. *Proceedings of the Institution of*, *121*(12), 1551–1556.

Allan, R., Grigg, C., & Al-Shakarchi, M. (1976). Numerical techniques in probabilistic load flow problems. *International Journal for Numerical Methods in Engineering*, *10*(4), 853–860. doi:10.1002/nme.1620100412

Alwi, Hampson, & Mohamed. (2002). *Non value-adding activities: A comparative study of Indonesian and Australian construction projects*. Academic Press.

American Cancer Society. (2004). *Breast Cancer Facts and Figures 2003-2004*. Atlanta, GA: American Cancer Society. Retrieved July 28, 2005, from https://www.cancer.org/acs/groups/content/@nho/documents/document/caff2003brfpwsecuredpdf.pdf

Ampolo, C. (1980). La formazione della città nel Lazio. Le condizioni materiali della produzione. Agricoltura e paesaggio agrario. *Dialoghi di Archeologia*, *1*(2), 15–46.

Anders, G. J. (1989). *Probability concepts in electric power systems*. Academic Press.

Anonymous. (2008a). *Plant Growth, in Science*. Retrieved from http://www.steve.gb.com/images/science/wood_layers.jpg

Anonymous. (2008b). *Climate Change, Past and Future, Calspace Courses*. Retrieved from http://earthguide.ucsd.edu/virtualmuseum/climatechange1/cc1syllabus.shtml

Aoki, K., Nishikori, A., & Yokoyama, R. (1987). Constrained load flow using recursive quadratic programming. Power Systems. *IEEE Transactions on*, *2*(1), 8–16. doi:10.1109/T-ED.1987.22880

Aouad, G., Lee, A., & Wu, S. (2006). *Constructing the future: nD modelling*. Routledge.

Appolloni, A., Sun, H., Jia, F., & Li, X. (2014). Green Procurement in the private sector: A state of the art review between 1996 and 2013. *Journal of Cleaner Production*, *85*, 122–133. doi:10.1016/j.jclepro.2014.08.106

Aragonés-Beltran, P., Mendoza-Roca, J. A., Bes-Pia, A., García-Melon, M., & Parra-Ruiz, E. (2009). Application of multicriteria decision analysis to jar-test results for chemicals selection in the physical–chemical treatment of textile wastewater. *Journal of Hazardous Materials*, *164*(1), 288–295. doi:10.1016/j.jhazmat.2008.08.046 PMID:18829168

Arayici, Y., Coates, P., Koskela, L., Kagioglou, M., Usher, C., & Oreilly, K. (2011). Technology adoption in the BIM implementation for lean architectural practice. *Automation in Construction*, *20*(2), 189–195. doi:10.1016/j.autcon.2010.09.016

aric. (2008). *Global Climate Change Student Guide*. Author.

Armbrust, M., Fox, A., Griffith, R., Joseph, A. D., Katz, R., Konwinski, A., ... Zaharia, M. (2010). A view of cloud computing. *Communications of the ACM*, *53*(4), 50–58. doi:10.1145/1721654.1721672

Arslan, T., Yilmaz, V., & Aksoy, H. K. (2010). Structural Equation Model for Environmentally Conscious Purchasing Behavior. *International Journal of Environmental of Research*, *6*(1), 323–334.

Arslan, Y. Z., Adli, M. A., Akan, A., & Baslo, M. B. (2010). Prediction of externally applied forces to human hands using frequency content of surface EMG signals. *Computer Methods and Programs in Biomedicine*, *98*(1), 36–44. doi:10.1016/j.cmpb.2009.08.005 PMID:19762107

Arthanari, T. S., & Ramamurthy, K. G. (1971). An extension of two machines sequencing problem. *Opsearch*, *8*(1), 10–22.

Assunção, M. D., Calheiros, R. N., Bianchi, S., Netto, M. A. S., & Buyya, R. (2015). Big Data computing and clouds: Trends and future directions. *Journal of Parallel and Distributed Computing*, *79–80*, 3–15. doi:10.1016/j.jpdc.2014.08.003

Atkinson, P. M., & Massari, R. (1998). Generalised linear modelling of susceptibility to landsliding in the Central Apennines, Italy. *Computers & Geosciences*, *24*(4), 373–385. doi:10.1016/S0098-3004(97)00117-9

Atteh, O. D. (1989). *Indigenous local knowledge as key to local-level development: Possibilities, constraints and planning issues in the context of Africa*. Arusha, Tanzania: Coference Paper.

Ayeni, O. O., Saka, D. N., & Ikwuemesi, G. (2004). Developing a Multimedia GIS Database for Tourism Industry in Nigeria. *ISPRS 2004: 20th International Society for Photogrammetry and Remote Sensing (ISPRS) Congress: Geo-Imagery Bridging Continents*.

Ayers, S. (1999). *The cultural impact of computer technology*. Yale-New Haven Teachers Institute. Retrieved February 29, 2016 from http://www.yale.edu/ynhti/curriculum/units/1999/7/99.07.07.x.html

Azevedo, S. G., Carvalho, H., & Machado, V. C. (2011). The influence of green practices on supply chain performance: A case study approach. *Transportation Research Part E, Logistics and Transportation Review*, *47*(6), 850–871. doi:10.1016/j.tre.2011.05.017

Azuma, R. T. (1997). A survey of augmented reality. *Presence (Cambridge, Mass.)*, *6*(4), 355–385. doi:10.1162/pres.1997.6.4.355

Back, T., Hammel, U., & Schwefel, H.-P. (1997). Evolutionary computation: Comments on the history and current state. Evolutionary computation. *IEEE Transactions on*, *1*(1), 3–17.

Bag, S. (2014). *Developing a GSCM model for the Indian rubber goods manufacturing sector (Unpublished Doctoral Dissertation). University of Petroleum & Energy Studies*. Available at http://ils.ddn.upes.ac.in:8001/cgi-bin/koha/opac-authoritiesdetail.pl?authid=13645

Bag, S. (2016). Green strategy, supplier relationship building and supply chain performance: Total interpretive structural modelling approach. *International Journal of Procurement Management*, *9*(4), 398–426. doi:10.1504/IJPM.2016.077702

Bag, S. (2016b). Flexible procurement systems is key to supply chain sustainability. *Journal of Transport and Supply Chain Management*, *10*(1), 1–9. doi:10.4102/jtscm.v10i1.213

Bag, S., & Anand, N. (2014). Modeling Green Supply Chain Management framework using ISM and MICMAC analysis. *African Journal of Business Management*, *8*(22), 1053.

Bahaire, T., & Elliott-White, M. (1999). The Aplication of Geographical Information Systems GIS in Sustainable Tourism Planning: A Review. International *Journal of Sustainable Tourism*, *7*(2), 159–174.

Baker, K. R., & Trietsch, D. (2009). *Principles of Sequencing and Scheduling*. John Wiley & Sons, Inc. doi:10.1002/9780470451793

Bakirtzis, A. G., Biskas, P. N., Zoumas, C. E., & Petridis, V. (2002). Optimal power flow by enhanced genetic algorithm. Power Systems. *IEEE Transactions on, 17*(2), 229–236.

Baldé, C. P., Wang, F., Kuehr, R., & Huisman, J. (2015). The global e-waste monitor - 2014. United Nations University, IAS-SCYCLE.

Balogun, V. F., Thompson, A. F., & Sarumi, O. A. (2010). A 3D Geo-Spatial Virtual Reality System for Virtual Tourism. *The Pacific Journal of Science and Technology, 11*(2), 601–609.

Balzani, M., Santopuoli, N., Grieco, A., & Zaltron, N. (2004). *Laser scanner 3D survey in archaeological field: The Forum of Pompeii. International Conference on Remote Sensing Archaeology*, Beijing, China.

Bangemann, T., Riedl, M., Thron, M., & Diedrich, C. (2016). Integration of Classical Components Into Industrial Cyber–Physical Systems. *Proceedings of the IEEE, 104*(5), 947–959. doi:10.1109/JPROC.2015.2510981

Bartoloni, G. (2002). *La civiltà villanoviana all'inizio della storia etrusca*. Roma, Italy: Carocci.

Bartoloni, G. (2003). *Le società dell'Italia primitiva. Lo studio delle necropoli e la nascita delle aristocrazie*. Roma, Italy: Carocci.

Basel Action Network (BAN). (2011). *Toxic trade news: "Cochin Port a safe conduit for imported e-waste". "Most aspects of e-waste not regulated in U.S., Va."; "Research identifies U.S electronic wastes as likely source of toxic jewellery imports from China"; "178 countries agree to allow the ban on exports of toxic wastes to developing countries to become law"*. Retrieved March 1, 2016 from http://www.ban.org/

Basmajian, J. V., & DeLuca, C. J. (1985). *Muscles alive: Their functions revealed by electromyography*. Baltimore, MD: The Williams & Wilkins Company.

Basu, M., & Chowdhury, A. (2013). Cuckoo search algorithm for economic dispatch. *Energy, 60*, 99–108. doi:10.1016/j.energy.2013.07.011

Batlle, C., & Rodilla, P. (2009). Electricity demand response tools: Current status and outstanding issues. *European Review of Energy Markets, 3*, 1–27.

Battistella, G. (2013, October). *L'innovazione tecnologica per le attività di licensing, monitoraggio e controllo degli insediamenti IPPC soggetti ad AIA nazionale*. Paper presented at Veneto Region Educational Course 'Emissione negli impianti di processo', Mestre, Italy.

Battistella, G., & Di Marco G. (2013). *Air emissions measurements in environmental monitoring of IPPC plants*. IGI Global.

Battistella, G., & Di Marco, G. (2013, June). *Some Open Issues in Environmental Monitoring of IPPC Plants Air Emissions Measurements*. Paper presented at 4th Imeko TC19 Symposium on Environmental Instrumentation and Measurements Protecting Environment, Climate Changes and Pollution Control, Lecce, Italy.

Battistella, G., Di Marco, G., Bonaiuti, F., & Carlucci, C. (2014, March). *La Ossidazione Termica come migliore tecnica disponibile per gli Off-Gas*. Paper presented at the AIDII 20th Conference of Industrial Hygiene 'Le giornate di Corvara', Corvara, Italy.

Becker-Andre, M., Wiesenberg, I., Schaeren-Wiemers, N., Andre, E., Missbach, M., Saurat, J. H., & Carlberg, C. (1994). Pineal gland hormone melatonin binds and activates an orphan of the nuclear receptor superfamily. *The Journal of Biological Chemistry, 269*(46), 28531–28534. PMID:7961794

Bedini, E. (1997). I resti faunistici. In M. Bonghi Jovino & C. Chiaramonte Treré (Eds.), *Tarquinia. Testimonianze archeologiche e ricostruzione storica: scavi sistematici dell'abitato. Campagne 1982-1988* (Vol. 1-2, pp. 103–141). Roma, Italy: Tarchna.

Behnamian, J., & Ghomi, S. F. (2011). Hybrid flowshop scheduling with machine and resource-dependent processing times. *Applied Mathematical Modelling*, *35*(3), 1107–1123. doi:10.1016/j.apm.2010.07.057

Behzadan, A. H., & Kamat, V. R. (2012). Enabling discovery-based learning in construction using telepresent augmented reality. *Automation in Construction.*

Behzadan, A. H., Timm, B. W., & Kamat, V. R. (2008). General-purpose modular hardware and software framework for mobile outdoor augmented reality applications in engineering. *Advanced Engineering Informatics*, *22*(1), 90–105. doi:10.1016/j.aei.2007.08.005

Bellini, A. (2002). *Fused deposition of ceramics: a comprehensive experimental, analytical and computational study of material behavior, fabrication process and equipment design* (Ph.D thesis). Drexel University.

Benedict, B. M. (2002). *Curiosity: A cultural history of early modern inquiry.* Chicago: University of Chicago Press.

Benysek, G., Kazmierkowski, M., Popczyk, J., & Strzelecki, R. (2011). Power electronic systems as a crucial part of smart grid infrastructure-a survey. *Technical Sciences*, *59*(4), 455–473.

Berkes, F., Colding, J., & Folke, C. (2000). Rediscovery of traditional ecological knowledge as adaptive management. *Rediscovery of the Ecological Applications*, *10*, 251–262.

Bernhardsen, T. (2002). *Geographic information systems: an introduction* (3rd ed.). New York: John Wiley & Sons.

Berno, T. (2011). Sustainability on a Plate: Linking Agriculture and Food in the Fiji Islands Tourism Industry. In R. Torres, & J. Momsen (Eds.), Tourism and Agriculture: New Geographies of Production and Rural Restructuring, (pp. 87-103). London: Routledge.

BETA. (2013). *Beta Analytic, Radiocarbon Dating.* Retrieved from http://www.radiocarbon.com/tree-ring-calibration.htm

Bibb, R., & Brown, R. (2000). The application of computer aided product development techniques in medical modelling topic: Rehabilitation and prostheses. *Biomedical Sciences Instrumentation*, *36*, 319–324. PMID:10834252

Bibliometrics. (2016). Retrieved from https://en.wikipedia.org/wiki/Bibliometrics

Bietti Sestieri, A. M. (1996). *Protostoria. Teoria e pratica.* Roma, Italy: La Nuova Italia Scientifica.

Bietti Sestieri, A. M. (2002). L'agricoltura in Italia nell'età dei Metalli. In G. Forni & A. Marcone (Eds.), *Storia dell'agricoltura italiana. L'età antica. La preistoria* (Vol. 1-2, pp. 205–217). Firenze, Italy: Polistampa.

Bird, J. (2014). *Electrical Circuit: Theory and Technology* (5th ed.). New York: Routledge.

Birman, K. P., Chen, J., Hopkinson, E. M., Thomas, R. J., Thorp, J. S., Renesse, R. V., & Vogels, W. (2005). Overcoming communications challenges in software for monitoring and controlling power systems. *Proceedings of the IEEE*, *93*(5), 1028–1041. doi:10.1109/JPROC.2005.846339

Blower, J. D. (2010). *GIS in the cloud: implementing a web map service on Google App Engine.* Paper presented at the 1st International Conference and Exhibition on Computing for Geospatial Research & Application. 10.1145/1823854.1823893

Boerma, J. A. K. (1989). Land evaluation in prehistoric perspective: some observations. In Haex, O.C.M., Curvers, H.H., & Akkermans, P.M.M.G. (Eds). To the Euphrates and beyond (pp. 17-28). Rotterdam, The Netherlands: CRC Press.

Bohas, A., & Poussing, N. (2016). An empirical exploration of the role of strategic and responsive corporate social responsibility in the adoption of different green IT strategies. *Journal of Cleaner Production*, *122*, 240–251. doi:10.1016/j.jclepro.2016.02.029

Bon-García, J. V. (2012). *Declaración ambiental año 2011*. Consorci Hospital General Universitari Valencia.

Boran, F. E., Boran, K., & Menlik, T. (2012). The evaluation of renewable energy technologies for electricity generation in Turkey using intuitionistic fuzzy TOPSIS. *Energy Sources Part B*, *7*(1), 81–90. doi:10.1080/15567240903047483

Bornmann, L., & Waltman, L. (2011). The detection of hot regions in the geography of science – A visualization approach by using density maps. *Journal of Informetrics*, *5*(4), 547–553. doi:10.1016/j.joi.2011.04.006

Borthakur, A., & Singh, P. (2012). Electronic waste in India: Problem and Policies. *International Journal of Environment Science*, *3*, 354–362.

Borthakur, A., & Sinha, K. (2013). Generation of electronic waste in India: Current scenario, dilemmas and stakeholders. *African Journal of Environmental Science and Technology*, *7*(9), 899–910. doi:10.5897/AJEST2013.1505

Bose, A. (2005). Improved communication/computation infrastructure for better monitoring and control. In *Proceedings of the IEEE Power Engineering Society General Meeting* (pp. 2715–2716). 10.1109/PES.2005.1489755

Bose, S., Vahabzadeh, S., & Bandyopadhyay, A. (2013). Bone tissue engineering using 3D printing. *Materials Today*, *16*(12), 496–504. doi:10.1016/j.mattod.2013.11.017

Bossler, J. D., Jensen, J. R., McMaster, R. B., & Rizos, C. (2002). *Manual of geospatial science and technology*. New York: Taylor and Francis.

Botterill, D., Owen, R., Emmanuel, L. N., Foster, N. T., Gale, T., Nelson, C., & Selby, M. (2002). Perceptions from the Periphery: The Experience of Wales. In F. Brown & D. Hall (Eds.), Tourism in the Peripheral Areas, (pp. 7-38). Clevedon: Channel View Publications.

Boyack, K. W., Klavans, R., & Börner, K. (2005). Mapping the backbone of science. *Scientometrics*, *64*(3), 351–374. doi:10.100711192-005-0255-6

Bracht, U., & Masurat, T. (2006). The Digital Factory between vision and reality. *Computers in Industry*, *56*(4), 325–333. doi:10.1016/j.compind.2005.01.008

Bradley, P., Leach, M., & Torriti, J. (2013). A review of the costs and benefits of demand response for electricity in UK. *Energy Policy*, *52*, 312–327. doi:10.1016/j.enpol.2012.09.039

Bradley, R. S. (1985). *Quaternary Palaeoclimatology: Methods of Palaeoclimatic Reconstruction*. London: Unwin Hyman.

Brah, S. A., & Loo, L. L. (1999). Heuristics for scheduling in a flow shop with multiple processors. *European Journal of Operational Research*, *113*(1), 113–122. doi:10.1016/S0377-2217(97)00423-2

Brantley, S., Yamaguchi, D., Cameron, K., & Pringle, P. (1986). Tree-Ring Dating of Volcanic Deposits. *Earthquakes and Volcanoes*, *18*(5), 184–194.

Braudel, F. (2002). *Il Mediterraneo. Lo spazio la storia gli uomini le tradizioni*. Milano, Italy: Bompiani.

Breen, J., & Breen, J. (2008). Critical Observations and Strategic Perspectives at Half-time. In K. Terzidis (Ed.), *Proceedings of First International Conference on Critical Digital: What Matter(s)?* (pp. 129-136). Cambridge, MA: Harvard University Graduate School of Design.

Brijnath, B., & De Souza, D. K. (2012). Four Ways Geographic Information Systems Can Help to Enhance Health Service Planning and Delivery for Infectious Diseases in Low-Income Countries. *Journal of Health Care for the Poor and Underserved*, *23*(4), 1410–1420. doi:10.1353/hpu.2012.0146 PMID:23698658

Brokesha, D., Warren, D., & Werner, O. (1980). *Indigenous Knowledge Systems and Development*. Lanham, MD: University Press of America.

Brouwer, A. S., Van den Broek, M., Zappa, W., Turkenburg, W. C., & Faaij, A. (2016). Least-cost options for integrating intermittent renewables in low-carbon power systems. *Applied Energy*, *161*, 48–74. doi:10.1016/j.apenergy.2015.09.090

Bruner, J. S. (1966). *Towards a theory of instruction*. Harvard University Press.

Bruno, G., & Giannikos, I. (2015). *Location and GIS. In Location Science* (pp. 509–536). Springer.

Budhathoki, N. R., & Nedovic-Budic, Z. (2008). Reconceptualizing the role of the user of spatial data infrastructure. *GeoJournal*, *72*(3-4), 149–160. doi:10.100710708-008-9189-x

Buhalis, D. (1998). Strategic use of information technologies in the tourism industry. *Tourism Management*, *19*(5), 409–421. doi:10.1016/S0261-5177(98)00038-7

Buhalis, D. (1999). Tourism on the Greek Islands: Issues of Peripherality, Competitiveness and Development. *International Journal of Tourism Research*, *1*(5), 341–358. doi:10.1002/(SICI)1522-1970(199909/10)1:5<341::AID-JTR201>3.0.CO;2-0

Burnwal, S., & Deb, S. (2013). Scheduling optimization of flexible manufacturing system using cuckoo search-based approach. *International Journal of Advanced Manufacturing Technology*, *64*(5-8), 951–959. doi:10.100700170-012-4061-z

Butler, R. (1993). Alternative Tourism: The Thin Edge of the Wedge. In V. Smith & W. Eadington (Eds.), *Tourism Alternatives* (pp. 31–46). Chichester, UK: Wiley.

Buys, J., De Florio, V., & Blondia, C. (2011) Towards context-aware adaptive fault tolerance in SOA applications. In *Proceedings of the 5th ACM international conference on Distributed event-based system*, (pp. 63-74). 10.1145/2002259.2002271

Buyya, R., Yeo, C. S., Venugopal, S., Broberg, J., & Brandic, I. (2009). Cloud computing and emerging IT platforms: Vision, hype, and reality for delivering computing as the 5th utility. *Future Generation Computer Systems*, *25*(6), 599–616. doi:10.1016/j.future.2008.12.001

Cachin, C., & Tessaro, S. (2006, June). Optimal resilience for erasure-coded Byzantine distributed storage. In *Proceedings of the International Conference on Dependable Systems and Networks*, (pp. 115-124). IEEE. 10.1109/DSN.2006.56

CALM. (2015, Nov 1). *Communications in Cooperative Intelligent Transport Systems - CALM for C-ITS*. Retrieved from Introduction to Communications Access for Land Mobiles: http://calm.itsstandards. EU/

Campbell, T., Williams, C., Ivanova, O., & Garrett, B. (2011). *Could 3D printing change the world. In Technologies, Potential, and Implications of Additive Manufacturing*. Washington, DC: Atlantic Council.

Cancer (Fact sheet N 297). World Health Organization. (n.d.). Retrieved from http://www.who.int/mediacentre/factsheets/fs297/

Cao, J., & Sanders, D. B. (1996). Multivariate discriminant analysis of the electromyographic interference pattern: Statistical approach to discrimination among controls, myopathies and neuropathies. *Medical & Biological Engineering & Computing*, *34*(5), 369–374. doi:10.1007/BF02520008 PMID:8945863

Cappers, P., Mills, A., Goldman, C., Wiser, R., & Eto, J. H. (2012). An assessment of the role mass market demand response could play in contributing to the management of variable generation integration issues. *Energy Policy*, *48*, 420–429. doi:10.1016/j.enpol.2012.05.040

Carandini, A. (2003). *La nascita di Roma. Dei, Lari, eroi e uomini all'alba di una civiltà* (Vol. 1-2). Torino, Italy: Einaudi.

Cardani, E. (2006). Proposta vincente. From Lyon Confluence: Progetto, Christian de Portzamparc. *Arca*, *219*, 91.

Carnero, M. C. (2014). *Model for Sustainability in Health Care Organizations.* In J. Wang (Ed.), *Encyclopedia of Business Analytics and Optimization* (Vol. 3, pp. 437–456). Hershey, PA: IGI Global.

Carnero, M. C. (2015). Assessment of Environmental Sustainability in Health Care Organizations. *Sustainability*, *7*(7), 8270–8291. doi:10.3390u7078270

Carrer, M. (2011). Individualistic and Time-Varying Tree-Ring Growth to Climate Sensitivity. *PLoS ONE*, *6*(7), e22813. doi:10.1371/journal.pone.0022813 PMID:21829523

Cartogram. (2016). Retrieved from https://en.wikipedia.org/wiki/Cartogram

Castellanos, E., & Van Westen, C. J. (2007). Generation of a landslide risk index map for Cuba using spatial multi-criteria evaluation. *Landslide*, *4*(4), 311–325. doi:10.100710346-007-0087-y

Castillo, M. (2011, January 14). *Electronic Waste: Where does it go and what happens to it?* Retrieved March 1, 2016 from http://techland.time.com/2011/01/14/electronic-waste-where-does-it-go-and-what-happens-to-it/

Castro, M., & Liskov, B. (2002). Practical Byzantine fault tolerance and proactive recovery. *ACM Transactions on Computer Systems*, *20*(4), 398–461. doi:10.1145/571637.571640

Castro, M., Rodrigues, R., & Liskov, B. (2003). BASE: Using abstraction to improve fault tolerance. *ACM Transactions on Computer Systems*, *21*(3), 236–269. doi:10.1145/859716.859718

Caudell, T. P., & Mizell, D. W. (1992). *Augmented reality: An application of heads-up display technology to manual manufacturing processes. System Sciences, 1992. Proceedings of the Twenty-Fifth Hawaii International Conference on.* IEEE. 10.1109/HICSS.1992.183317

Cena, G., Scanzio, S., Valenzano, A., & Zunino, C. (2011). Evaluation of the real-time properties of open-source protocol stacks. In *IEEE 16th Conference on Emerging Technologies & Factory Automation (ETFA)* (pp. 1-8). 10.1109/ETFA.2011.6059003

Cena, G., Cibrario Bertolotti, I., Scanzio, S., Valenzano, A., & Zunino, C. (2012). Evaluation of EtherCAT Distributed Clock Performance. *IEEE Transactions on Industrial Informatics*, *8*(1), 20–29. doi:10.1109/TII.2011.2172434

Cena, G., Cibrario Bertolotti, I., Scanzio, S., Valenzano, A., & Zunino, C. (2013). Synchronize your watches: Part I: General-purpose solutions for distributed real-time control. *IEEE Industrial Electronics Magazine*, *7*(1), 18–29. doi:10.1109/MIE.2012.2232354

Cena, G., Scanzio, S., & Valenzano, A. (2016). Seamless Link-Level Redundancy to Improve Reliability of Industrial Wi-Fi Networks. *IEEE Transactions on Industrial Informatics*, *12*(2), 608–620. doi:10.1109/TII.2016.2522768

Cena, G., Scanzio, S., Valenzano, A., & Zunino, C. (2015). Implementation and Evaluation of the Reference Broadcast Infrastructure Synchronization Protocol. *IEEE Transactions on Industrial Informatics*, *11*(3), 801–811. doi:10.1109/TII.2015.2396003

Cereia, M., Cibrario Bertolotti, I., & Scanzio, S. (2011). Performance of a Real-Time EtherCAT Master Under Linux. *IEEE Transactions on Industrial Informatics*, *7*(4), 679–687. doi:10.1109/TII.2011.2166777

Chae, M. P., Rozen, W. M., McMenamin, P. G., Findlay, M. W., Spychal, R. T., & Hunter-Smith, D. J. (2015). Emerging Applications of Bedside 3D Printing in Plastic Surgery. *Frontiers in Surgery*, *2*, 25. doi:10.3389/fsurg.2015.00025 PMID:26137465

Chakraborty, D., Sarkar, D., Agarwal, S., Dutta, D., & Sharma, J. R. (2015). Web Based GIS Application using Open Source Software for Sharing Geospatial Data. *International Journal of Advanced Remote Sensing and GIS*, *4*(1), 1224–1228.

Chan, F. H. Y., Yang, Y. S., Lam, F. K., Zhang, Y. T., & Parker, P. A. (2000). Fuzzy EMG classification for prosthesis control. *IEEE Transactions on Rehabilitation Engineering*, *8*(3), 305–311. doi:10.1109/86.867872 PMID:11001510

Chang, D. Y. (1996). Applications of the extent analysis method on fuzzy AHP. *European Journal of Operational Research*, *95*(3), 649–655. doi:10.1016/0377-2217(95)00300-2

Chang, W. C., Kliot, M., & Sretavan, D. W. (2008). Microtechnology and nanotechnology in nerve repair. *Neurological Research*, *10*(10), 1053–1062. doi:10.1179/174313208X362532 PMID:19079980

Chang, Y. S., & Park, H. D. (2006). XML Web Service-based development model for Internet GIS applications. *International Journal of Geographical Information Science*, *20*(4), 371–399. doi:10.1080/13658810600607857

Chan, H. K., Wang, X., & Raffoni, A. (2012). An integrated approach for green design: Life-cycle, fuzzy AHP and environmental management accounting. *Procedia: Social and Behavioral Sciences*, *57*, 236–242.

Chaperon, S., & Bramwell, B. (2013). Dependency and Agency in Peripheral Tourism Development. *Annals of Tourism Research*, *40*(1), 132–154. doi:10.1016/j.annals.2012.08.003

Chapra Steven, C., & Canale Raymond, P. (2008). *Numerical methods for engineers*. McGraw-Hill Pub.

Chaudhuri, S., & Ray, N. (2015). Application of Web-Based Geographical Information System (GIS). In Tourism Development, Encyclopedia of Information Science and Technology. IGI Global.

Che, D., & Wargenau, A. (2011). Visiting Winery Tasting Rooms: Venues for Education, Differentiation and Direct Marketing. In R. Torres, & J. Momsen (Eds.), Tourism and Agriculture: New Geographies of Production and Rural Restructuring, (pp. 192-204). London: Routledge.

Chen, B. Y., Li, Q., Wang, D., Shaw, S.-L., Lam, W. H., Yuan, H., & Fang, Z. (2013). Reliable space–time prisms under travel time uncertainty. *Annals of the Association of American Geographers*, *103*(6), 1502–1521. doi:10.1080/00045608.2013.834236

Chen, C. (2006). CiteSpace II: Detecting and visualizing emerging trends and transient patterns in scientific literature. *Journal of the American Society for Information Science and Technology*, *57*(3), 359–377. doi:10.1002/asi.20317

Chen, C. (2013). The Structure and Dynamics of Scientific Knowledge. In C. Chen (Ed.), *Mapping Scientific Frontiers: The Quest for Knowledge Visualization* (2nd ed.; pp. 163–199). London: Springer. doi:10.1007/978-1-4471-5128-9_5

Chen, C. T. (2000). Extension of the TOPSIS for group decision –making under fuzzy environment. *Fuzzy Sets and Systems*, *114*(1), 1–9. doi:10.1016/S0165-0114(97)00377-1

Chen, C., Ibekwe-SanJuan, F., & Hou, J. (2010). The structure and dynamics of co-citation clusters: A multiple-perspective co-citation analysis. *Journal of the American Society for Information Science and Technology*, *61*(7), 1386–1409. doi:10.1002/asi.21309

Cheng, C., & Piccoli, G. (2002). Web-Based Training in the Hospitality Industry: A Conceptual Definition, Taxonomy and Preliminary Investigation. *International Journal of Hospitality Information Technology*, *2*(2), 19–33. doi:10.3727/153373402803617737

Chen, H., Chiang, R. H. L., & Storey, V. C. (2012). Business Intelligence and Analytics: From Big Data to Big Impact. *Management Information Systems Quarterly*, *36*(4), 1165–1188.

Chen, J., Shaw, S.-L., Yu, H., Lu, F., Chai, Y., & Jia, Q. (2011). Exploratory data analysis of activity diary data: A space–time GIS approach. *Journal of Transport Geography*, *19*(3), 394–404. doi:10.1016/j.jtrangeo.2010.11.002

Cheron, G., Draye, J. P., Bourgeios, M., & Libert, G. (1996). A dynamic neural network identification of electromyography and trajectory relationship during complex movements. *IEEE Transactions on Bio-Medical Engineering*, *43*(5), 552–558. doi:10.1109/10.488803 PMID:8849468

Chia, H. N., & Wu, B. M. (2015). Recent advances in 3D printing of biomaterials. *Journal of Biological Engineering*, *9*(1), 4. doi:10.118613036-015-0001-4 PMID:25866560

Chi, H.-L., Kang, S.-C., & Wang, X. (2013). Research trends and opportunities of augmented reality applications in architecture, engineering, and construction. *Automation in Construction*, *33*, 116–122. doi:10.1016/j.autcon.2012.12.017

Chiharu Mukai, K. O. (2012). *BER Performance of Multiplexed CSK using Linear/Circular Polarization in Optical Wireless Communications*. ISITA.

Chiou, T. Y., Chan, H. K., Lettice, F., & Chung, S. H. (2011). The influence of greening the suppliers and green innovation on environmental performance and competitive advantage in Taiwan. *Transportation Research Part E, Logistics and Transportation Review*, *47*(6), 822–836. doi:10.1016/j.tre.2011.05.016

Chloropleth Map. (2016). Retrieved from https://en.wikipedia.org/wiki/Choropleth_map

Choi, J. W., & Kim, N. (2015). Clinical application of three-dimensional printing technology in craniofacial plastic surgery. *Archives of Plastic Surgery*, *3*(3), 267–277. doi:10.5999/aps.2015.42.3.267 PMID:26015880

Choi, S. H., & Cheung, H. H. (2005). A Multi-Material Virtual Prototyping System. *Computer Aided Design*, *37*(1), 123–136. doi:10.1016/j.cad.2004.06.002

Choi, S. H., & Cheung, H. H. (2007). Multi-Material Virtual Prototyping for Rapid Product Development and Biomedical Engineering. *Computers in Industry*, *53*(5), 557–566.

Choi, S. H., & Cheung, H. H. (2009). A Topological Hierarchy-Based Approach to Layered Manufacturing of Functionally Graded Multi-Material Objects. *Computers in Industry*, *60*(5), 349–363. doi:10.1016/j.compind.2009.01.008

Choi, S. H., & Kwok, K. T. (2004). A Topological Hierarchy-Sorting Algorithm for Slice Contours. *Rapid Prototyping Journal*, *10*(2), 98–113. doi:10.1108/13552540410526971

Chu, J. U., Moon, I., Lee, Y. J., Kim, S. K., & Mun, M. S. (2007). A supervised feature-projection-based real-time EMG pattern recognition for multifunction myoelectric hand control. *IEEE/ASME Transactions on Mechatronics*, *12*(3), 282–290. doi:10.1109/TMECH.2007.897262

Chung, Y.-Y. T.-Y. (2014). Mobile health-monitoring system through visible light communication. *Bio-Medical Materials and Engineering*, 608–737.

Cibrario Bertolotti, I., & Hu, T. (Eds.). (2015). *Embedded Software Development: The Open-Source Approach*. Boca Raton, FL: CRC Press.

Cintra, Y. C., & Carter, D. (2012). Internalising sustainability: Reflections on management control in Brazil. *International Journal of Strategic Management*, *12*(2), 108–125.

Ciotti, F., & Roncaglia, G. (2000). *Il mondo digitale. Introduzione ai nuovi media*. Roma-Bari, Italy: Laterza.

Civicioglu, P., & Besdok, E. (2013). A conceptual comparison of the Cuckoo-search, particle swarm optimization, differential evolution and artificial bee colony algorithms. *Artificial Intelligence Review*, *39*(4), 315–346. doi:10.100710462-011-9276-0

CNN. (2007). *Staged cyber attack reveals vulnerability in power grid.* Retrieved from http://www.cnn.com/2007/US/09/26/power.at.risk/index.html

Cobbing, M. (2008). *Toxic Tech: Not in our backyard. Uncovering the hidden flows of e-waste.* Report from Greenpeace International, Amsterdam. Retrieved March 1, 2016 from http://www.greenpeace.org/raw/content/belgium/fr/press/reports/toxic-tech.pdf

Cobo, M. J., López-Herrera, A. G., Herrera-Viedma, E., & Herrera, F. (2011). Science mapping software tools: Review, analysis, and cooperative study among tools. *Journal of the American Society for Information Science and Technology*, *62*(7), 1382–1402. doi:10.1002/asi.21525

Coleman, M. P., & Reiter, R. J. (1992). Breast cancer, blindness and melatonin. *European Journal of Cancer*, *28*(2-3), 501–503. doi:10.1016/S0959-8049(05)80087-5 PMID:1591073

Colonna, G. (2000). I caratteri originali della civiltà etrusca. In M. Torelli (Ed.), *Gli Etruschi, Catalogo della mostra, Venezia, Palazzo Grassi* (pp. 25–41). Milano, Italy: Bompiani.

Comunidad de Madrid. (2005). *Guía para la implantación de sistemas de gestión medioambiental en centros hospitalarios.* Madrid: Dirección General de Promoción y Disciplina Ambiental.

Connell, J. (2006). Medical Tourism: Sea, Sun, Sand and Surgery. *Tourism Management*, *27*(6), 1093–1100. doi:10.1016/j.tourman.2005.11.005

Connell, J. L., Ritschdorff, E. T., Whiteley, M., & Shear, J. B. (2013). 3D printing of microscopic bacterial communities. *Proceedings of the National Academy of Sciences of the United States of America*, *110*(46), 18380–18385. doi:10.1073/pnas.1309729110 PMID:24101503

Construction, M.-H. (2008). Building Information Modeling Trends SmartMarket Report. New York: Academic Press.

Cook, G., & Pomerantz, D. (2015). *Clicking Clean: A Guide to Building the Green Internet.* Washington, DC: Greenpeace, Inc.

Coop Himmelb(l)au (2003). *Musée des Confluences.* Retrieved December 17, 2008, from http://www.arcspace.com/architects/coop_himelblau/musee_des_confluences/

Cortes, C., & Vapnik, V. (1995). Support-vector networks. *Machine Learning*, *20*(3), 273–297. doi:10.1007/BF00994018

Crampton, J. W. (2009). Cartography: Maps 2.0. *Progress in Human Geography*, *33*(1), 91–100. doi:10.1177/0309132508094074

Cremaschi, M. (1992). Insediamenti e sussistenza nella media età del Bronzo, Italia settentrionale. *Rassegna di Archeologia*, *10*, 145–188.

Cremaschi, M. (2000). *Manuale di geoarcheologia.* Roma-Bari, Italy: Laterza.

Crispim, J., Braz, J., Castro, R., & Esteves, J. (2014). Smart Grids in the EU with smart regulation: Experiences from the UK, Italy and Portugal. *Utilities Policy*, *31*, 85–93. doi:10.1016/j.jup.2014.09.006

Crossland, M. D. (2005). Geographical Information Systems as Decision Tools. In M. Khosrow-Pour (Ed.), *Encyclopedia of Information Science and Technology* (1st ed.; Vol. 2, pp. 1274–1277). Idea Group Reference. doi:10.4018/978-1-59140-553-5.ch224

Cui, K. C., Chen, G., Xu, Z., & Roberts, R. D. (2012). Traffic light to vehicle visible light communication channel characterization. *Applied Optics*, *51*(27), 6594–6605. doi:10.1364/AO.51.006594 PMID:23033030

Da Silva, A. L., & Arienti, V. (1990). Probabilistic load flow by a multilinear simulation algorithm. Generation. *Transmission and Distribution*, *137*(4), 276–282. doi:10.1049/ip-c.1990.0037

Darby, S. J., & McKenna, E. (2012). Social implication of residential demand response in cool temperate climates. *Energy Policy*, *49*, 759–769. doi:10.1016/j.enpol.2012.07.026

Das, D., & Rahaman, H. (2015). *Carbon Nanotube and Graphene Nanoribbon Interconnects*. Boca Raton, FL: CRC Press.

Davidson-Hunt, I., & Berkes, F. (2003). *Learning as You Journey: Anishinaabe Perception of Social-ecological Environments and Adaptive Learning*. Retrieved December 2015, from http://www.consecol.org/vol8/iss1/art5/

Davin, K. (1997). Effects of Computer Support on Group Decisions. *Journal of Hospitality & Tourism Research (Washington, D.C.)*, *21*(2), 44–57. doi:10.1177/109634809702100204

Davis, C., Back, K., & MacLean, K. (1992). Understanding Oral History. In K. Phyllis & J. Brent (Eds.), *Public History Readings* (pp. 264–269). Malabar, FL: Krieger.

De Santis, R., Gloria, A., Russo, T., DAmora, U., Varriale, A., Veltri, M., ... Ambrosio, L. (2014). Reverse engineering of mandible and prosthetic framework: Effect of titanium implants in conjunction with titanium milled full arch bridge prostheses on the biomechanics of the mandible. *Journal of Biomechanics*, *16*, 3825–3829. doi:10.1016/j.jbiomech.2014.10.020 PMID:25468297

de Smith, M., Goodchild, M. F., & Longley, P. (2015). *Geospatial Analysis: a comprehensive guide to principles, techniques and software tools* (5th ed.). Leicester, UK: Troubador Publishing.

Dedrick, J. (2010). Green IS: Concepts and issues for information systems research. *Communications of the Association for Information Systems*, *27*(11), 173–183.

del Valle, Y., Venayagamoorthy, G. K., Mohagheghi, S., Hernandez, J.-C., & Harley, R. G. (2008). Particle swarm optimization: Basic concepts, variants and applications in power systems. *Evolutionary Computation*, *12*(2), 171–195. doi:10.1109/TEVC.2007.896686

Delafontaine, M., Neutens, T., Schwanen, T., & Van de Weghe, N. (2011). The impact of opening hours on the equity of individual space–time accessibility. *Computers, Environment and Urban Systems*, *35*(4), 276–288. doi:10.1016/j.compenvurbsys.2011.02.005

Delagrange, P., & Guardiola-Lemaitre, B. (1997). Melatonin, its receptors, and relationships with biological rhythm disorders. *Clinical Neuropharmacology*, *20*(6), 482–510. doi:10.1097/00002826-199712000-00002 PMID:9403224

Delgamuukw v. British Columbia, 3 S.C.R. 1010, 23799 (Supreme Court Judgments December 11, 1997).

Deng, A., Xiong, R., He, W., Wei, D., & Zeng, C. (2014). Postoperative rehabilitation strategy for acetabular fracture: Application of 3D printing technique. *Nan Fang Yi Ke Da Xue Xue Bao*, *4*, 591–593. PMID:24752117

Deng, D., & Chen, Y. (2015). Origami-based self-folding structure design and fabrication using projection based stereolithography. *Journal of Mechanical Design*, *2*(2), 021701–021712. doi:10.1115/1.4029066

Desmedt, Y. (1994). Threshold cryptography. *European Transactions on Telecommunications*, *5*(4), 449–457. doi:10.1002/ett.4460050407

Devesa, S. S., Grauman, D. J., Blot, W. J., Pennello, G. A., Hoover, R. N., & Fraumeni, J. F. (1999). *Atlas of cancer mortality in the United States, 1950-94*. Bethesda, MD: National Institutes of Health, National Cancer Institute.

Devi, B. S., Shobha, S. V., & Kamble, R. K. (2004). E-Waste: The Hidden harm of Technological Revolution. *Journal IAEM*, *31*, 196–205.

Dhavale, D. G. (1984). The Yavanajataka of Sphujidhvaja by David Pingree. *The YavAnnals of the Bhandarkar Oriental Research Institute*, *65*(1/4), 266–267.

Di Gennaro, F. (1986). *Forme d'insediamento fra Tevere e Fiora dal Bronzo Finale al principio dell'età del Ferro*. Firenze, Italy: Istituto di Studi Etruschi e Italici.

Diabat, A., & Govindan, K. (2011). An analysis of the drivers affecting the implementation of green supply chain management. *Resources, Conservation and Recycling*, *55*(6), 659–667. doi:10.1016/j.resconrec.2010.12.002

DictionaryG. I. S. ESRI. (n.d.). Retrieved from http://support.esri.com/

Ding, W. Y. F., Yang, F., Yang, H., Wang, J., Wang, X., Zhang, X., & Song, J. (2015). A hybrid power line and visible light communication system for indoor hospital applications. *Computers in Industry*, *68*, 170–178. doi:10.1016/j.compind.2015.01.006

Directive 2009/72/Ec Of The European Parliament And Of The Council of 13 July 2009 concerning common rules for the internal market in electricity. (n.d.). Retrieved July 20th, 2016 at: http://eur-lex.europa.eu/legal-content/EN/TXT/PDF/?uri=CELEX:32009L0072&from=EN

Directive 2012/27/Eu Of The European Parliament And Of The Council Of 25 October 2012 On Energy Efficiency. (n.d.). Retrieved July 20th, 2016 At: http://eur-lex.europa.eu/legal-content/EN/TXT/PDF/?uri=CELEX:32012L0027&from=EN

Disabled World. (2015). *Electronic waste: Medical & health issues*. Retrieved March 3, 2016 from http://www.disabled-world.com/health/ewaste.php

Disposal of e-waste. (2009). Rajya Sabha Unstarred Question No. 1887.

Dodge, M., McDerby, M., & Turner, T. (2008). *Geographic Visualization: Concepts, Tools and Applications*. Chichester, West Sussex, England: Wiley. doi:10.1002/9780470987643

Donath, D., Beetz, J., Grether, K., Petzold, F., & Seichter, H. (2001). *Augmented reality techniques for design and revitalisation in existing built environments*. Buffalo, NY: Angenommener Beitrag zur ACADIA.

Dong-Fang Zhang, Y.-J. Z.-Y. (2013). Multi-LED Phase-Shifted OOK Modulation Based. *IEEE Photonics Technology Letters*. Retrieved from https://www.eldoled.com

Dong, S., Feng, C., & Kamat, V. R. (2012). Sensitivity analysis of augmented reality-assisted building damage reconnaissance using virtual prototyping. *Automation in Construction*.

Donovan, D., & Puri, R. (2004). *Ecology and Society*. Retrieved December 26, 2015, from Vol 9 No. 3 Art 3: http://www.ecologyandsociety.org/vol9/iss3/art3/

Doody, D., & Stephan, G. (1993). *NASA - Jet Propulson Laboratory*. Retrieved December 26, 2015, from Basics of Space Flight: https://solarsystem.nasa.gov/basics/editorial.php

Dorrendorf, L., Gutterman, Z., & Pinkas, B. (2007). Cryptanalysis of the random number generator of the windows operating system. *Proceedings of the 14th ACM Conference on Computer and Communications Security*, 476–483.

Dossick, C. S., & Neff, G. (2009). Organizational divisions in BIM-enabled commercial construction. *Journal of Construction Engineering and Management*, *136*(4), 459–467. doi:10.1061/(ASCE)CO.1943-7862.0000109

Dubey, A. K., Gupta, U., & Jain, S. (2015). Breast cancer statistics and prediction methodology: A systematic review and analysis. *Asian Pacific Journal of Cancer Prevention*, *16*(10), 4237–4245. doi:10.7314/APJCP.2015.16.10.4237 PMID:26028079

Dubey, R., & Ali, S. (2014). Identification of flexible manufacturing system dimensions and their interrelationship using total interpretive structural modelling and fuzzy MICMAC analysis. *Global Journal of Flexible Systems Management*, *15*(2), 131–143. doi:10.100740171-014-0058-9

Dubey, R., Bag, S., Ali, S., & Venkatesh, V. (2013). Green purchasing is key to superior performance: An empirical study. *International Journal of Procurement Management*, *6*(2), 187–210. doi:10.1504/IJPM.2013.052469

Du, J., Zhao, L., Zeng, Y., Zhang, L., Li, F., Liu, P., & Liu, C. (2011). Comparison of Electrical Properties between Multi-Walled Carbon Nanotube and Graphene Nanosheet/High Density Polyethylene Composites with a Segregated Network Structure. *Carbon*, *49*(4), 1094–1110. doi:10.1016/j.carbon.2010.11.013

Dupont, B., Dietrich, K., De Jonghe, C., Ramos, A., & Belmans, R. (2014). Impact of residential demand response on power system operation: A Belgian case study. *Applied Energy*, *122*, 1–10. doi:10.1016/j.apenergy.2014.02.022

DUrso, P. S., Earwaker, W. J., Barker, T. M., Redmond, M. J., Thompson, R. G., & Effeney, D. J. (2000). Custom Cranioplasty Using Stereolithography and Acrylic. *British Journal of Plastic Surgery*, *53*(3), 200–204. doi:10.1054/bjps.1999.3268 PMID:10738323

Eastman, C. M. (1995). Tourism in Kenya and the Marginalization of Swahili. *Annals of Tourism Research*, *22*(1), 172–185. doi:10.1016/0160-7383(94)00069-5

Eastman, C., Jeong, Y.-S., Sacks, R., & Kaner, I. (2009). Exchange model and exchange object concepts for implementation of national BIM standards. *Journal of Computing in Civil Engineering*, *24*(1), 25–34. doi:10.1061/(ASCE)0887-3801(2010)24:1(25)

Edmilson, J. T., Maganote, M. S. A., & Schulz, P. A. (2014). Visualization of ranking data: Geographical signatures in international collaboration, leadership and research impact. *Journal of Informetrics*, *8*(3), 642–649. doi:10.1016/j.joi.2014.05.005

Egilmez, G., Gumus, S., & Kucukvar, M. (2015). Enviromental sustainability benchmarking of the U.S. and Canada metropoles: An expert judgment-based multi-criteria decision making approach. *Cities (London, England)*, *42*, 31–41. doi:10.1016/j.cities.2014.08.006

Eid, C., Codani, P., Chen, Y., Perez, Y., & Hakvoort, R. (2015). *Aggregation of demand side flexibility in a smart grid: A review of European market design*. EEM15 – 12th International Conference on the European Energy Market, Lisboa,, Portugal.. 10.1109/EEM.2015.7216712

Eid, C., Koliou, E., Valles, M., Reneses, J., & Hakvoort, R. (2016). Time-based pricing and electricity demand response: Existing barriers and next steps. *Utilities Policy*, *40*, 15–25. doi:10.1016/j.jup.2016.04.001

Eisen, R. J., & Eisen, L. (2014). Use of Geographic Information Systems in Infectious Disease Surveillance. In N. M. M'ikanatha & J. K. Iskander (Eds.), *Concepts and Methods in Infectious Disease Surveillance*. Oxford, UK: John Wiley & Sons, Ltd. doi:10.1002/9781118928646.ch20

Eldamaty, A., Faried, S. O., & Aboreshaid, S. (2005). Damping power system oscillations using a fuzzy logic based unified power flow controller. In Electrical and Computer Engineering, (pp. 1950–1953). doi:10.1109/CCECE.2005.1557364

Elliot, S. (2007) Environmentally sustainable ICT: a critical topic for IS research? *Pacific Asia Conference on Information Systems*, *114*, 100-112.

Elwood, S. (2009). Integrating participatory action research and GIS education: Negotiating methodologies, politics and technologies. *Journal of Geography in Higher Education, 33*(1), 51–65. doi:10.1080/03098260802276565

Engeli, M. (1999). *Storie digitali. Poetiche della comunicazione*. Torino, Italy: Testo&Immagine.

Engin, O., Ceran, G., & Yilmaz, M. K. (2011). An efficient genetic algorithm for hybrid flow shop scheduling with multiprocessor task problems. *Applied Soft Computing, 11*(3), 3056–3065. doi:10.1016/j.asoc.2010.12.006

Engin, O., & Döyen, A. (2004). A new approach to solve hybrid flow shop scheduling problems by artificial immune system. *Future Generation Computer Systems, 20*(6), 1083–1095. doi:10.1016/j.future.2004.03.014

English Oxford Living Dictionaries. (2016). *Map*. Retrieved from https://en.oxforddictionaries.com/definition/map

ESRI. (2013). *The language of spatial analysis*. New York: ESRI Press.

Estes, A. C. (2015, January 06). *3D-Printed Guns Are Only Getting Better, and Scarier*. Retrieved April 26, 2016, from http://gizmodo.com/3d-printed-guns-are-only-getting-better-and-scarier-1677747439

Eufinger, H., Wehmöller, M., Harders, A., & Heuser, L. (1995). Prefabricated Prostheses for the Reconstruction of Skull Defects. *Internal Journal of Oral and Maxillofacial Surgery, 24*(1), 104-110.

European Network of Transmission System Operators for Electricity – ENTSO-E. (2015). *Market design for demand side response*. Policy paper, November 2015. Retrieved September 12th, 2016 at: https://www.entsoe.eu/Documents/Publications/Position%20papers%20and%20reports/entsoe_pp_dsr_web.pdf

European Parliament. (2003, January 27). Directive 2002/96/EC of the European Parliament and of the Council on waste electrical and electronic equipment (WEEE). *Official Journal of the European Union, L, 37*, 24–38.

European Parliament. (2012, July 4). Directive 2012/19/EU of the European Parliament and of the Council on waste electrical and electronic equipment (WEEE). *Official Journal of the European Union, L, 197*, 38–71.

Faiz, S., & Krichen, S. (2012). *Geographical information systems and spatial optimization*. CRC Press.

Fajuyigbe, O., & Balogun, V. F. (2007). Web-Enabled GIS: A Potential Tool for Sustaining Tourism in (Western) Nigeria, *Nigeria. Computers & Society*, (June): 15–22.

Fajuyigbe, O., Balogun, V. F., & Odembe, O. M. (2007). Web-Based Geographical Information System (GIS) for Tourism in Oyo State, Nigeria. *Information Technology Journal, 6*(5), 613–622. doi:10.3923/itj.2007.613.622

Fang, A., Zheng, M., & Fan, D. (2015). Application status of rapid prototyping technology in artificial bone based on reverse engineering. *Sheng Wu Yi Xue Gong Cheng Xue Za Zhi, 1*, 225–228. PMID:25997297

FAO. (2009). *Guidelines for good forestry and range practices in arid and semi-arid zones of the Near East*. Cairo: FAO Regional Office for the Near East.

Farhangi, H. (2010). The path of the smart grid. *Power and Energy Magazine, IEEE, 8*(1), 18–28. doi:10.1109/MPE.2009.934876

Farina, D., Jiang, N., Rehbaum, H., Holobar, A., Graimann, B., Dietl, H., & Aszmann, O. (2014). The extraction of neural information from the surface EMG for the control of upper limb prostheses: Emerging avenues and challenges. *IEEE Transactions on Neural Systems and Rehabilitation Engineering, 22*(4), 797–809. doi:10.1109/TNSRE.2014.2305111 PMID:24760934

Federal Energy Regulatory Commission (FERC). Retrieved from http://www.ferc.gov/industries/electric/indus-act/demand-response/dem-res-adv-metering.asp

Feiner, S. K. (2002). Augmented reality: A new way of seeing. *Scientific American, 286*(4), 48–55. doi:10.1038cientificamerican0402-48 PMID:11905108

Fesenmaier, D. R., Vogt, C. A., & Stewart, W. P. (1993). Investigating the influence of welcome center information on travel behavior. *Journal of Travel Research, 31*(3), 47–52. doi:10.1177/004728759303100308

Feuerriegel, S., & Neumann, D. (2016). Integration scenarios of Demand Response into electricity markets: Load shifting, financial savings and policy implications. *Energy Policy, 96*, 231–240. doi:10.1016/j.enpol.2016.05.050

Field, K., & OBrien, J. (2010). Cartoblography: Experiments in using and organizing the spatial context of micro-blogging. *Transactions in GIS, 14*(s1), 5–23. doi:10.1111/j.1467-9671.2010.01210.x

Forni, G. (2002). L'agricoltura: coltivazione ed allevamento. Genesi, evoluzione, contesto. In G. Forni & A. Marcone (Eds.), Storia dell'agricoltura italiana. L'età antica. La preistoria (pp. 7-157). Firenze, Italy: Polistampa.

Frank, S., Poncharal, P., Wang, Z. L., & de Heer, W. A. (1998). Carbon Nanotube Quantum Resistors. *Science, 280*(5370), 1744–1746. doi:10.1126cience.280.5370.1744 PMID:9624050

Frazzoli, C., Orisakwe, O. E., Dragone, R., & Mantovani, A. (2010). Diagnostic health risk assessment of electronic waste on the general population in developing countries scenarios. *Environmental Impact Assessment Review, 30*(6), 388–399. doi:10.1016/j.eiar.2009.12.004

Froese, T. (2005). Impact of emerging information technology on information management. *International Conference on Computing in Civil Engineering*, Cancun, Mexico. 10.1061/40794(179)162

Froese, T. M. (2010). The impact of emerging information technology on project management for construction. *Automation in Construction, 19*(5), 531–538. doi:10.1016/j.autcon.2009.11.004

Fu, K., Moreno, D., Yang, M., & Wood, K. L. (2014). Bio-inspired design: An overview investigating open questions from the broader field of design-by-analogy. *Journal of Mechanical Design, 11*(11), 111102. doi:10.1115/1.4028289

Gabriel, C. K. (2011). *Channel modeling for underwater optical communication. Proceeding in IEEE GLOBECOM Workshops*. GC Wkshps.

Gaidajis, G., Angelakoglou, K., & Aktsoglou, D. (2010). E-waste Environmental Problem and Current Management. *Journal of Engineering Science and Technology, Review, 3*(1), 193–199.

Galdakao-Usansolo Hospital. (2012). *Declaración Ambiental. Ejercicio 2012*. Available online: http://www. hospital-galdakao-usansolo.com

Galvez, D., Rakotondranaivo, A., Morel, L., Camargo, M., & Fick, M. (2015). Reverse logistics network design for a biogas plant: An approach based on MILP optimization and Analytical Hierarchical Process (AHP). *Journal of Manufacturing Systems, 37*(3), 616–623. doi:10.1016/j.jmsy.2014.12.005

Gandomi, A. H., Yang, X. S., & Alavi, A. H. (2013). Cuckoo search algorithm: A metaheuristic approach to solve structural optimization problems. *Engineering with Computers, 29*(1), 17–35. doi:10.100700366-011-0241-y

Gantz, J., & Reinsel, D. (2011). Extracting value from chaos. *IDC iview, 1142*, 1-12.

Gao, J., Xiao, Y., Liu, J., Liang, W., & Chen, C. (2012). A survey of communication/networking in smart grids. *Future Generation Computer Systems, 28*(2), 391–404. doi:10.1016/j.future.2011.04.014

Gao, W., Ramani, K., Cipra, R. J., & Siegmund, T. (2013). Kinetogami: A reconfigurable, combinatorial, and printable sheet folding. *Journal of Mechanical Design, 11*(11), 111009. doi:10.1115/1.4025506

Garcia, S., Cintra, Y., Torres, R. C. S. R., & Lima, F. G. (2016). Corporate sustainability management: A proposed multi-criteria model to support balanced decision-making. *Journal of Cleaner Production, 136*, 181–196. doi:10.1016/j.jclepro.2016.01.110

García-Suárez, A.M., Butler, C.J., & Baillie, M.G.L. (2009). Climate signal in tree-ring chronologies in a temperate climate: A multi-species approach. *Dendrochronologia, 27*, 183–198. doi:10.1016/j.dendro.2009.05.003

Gardiner, S., & Scott, N. (2014). Successful Tourism Clusters: Passion in Paradise. *Annals of Tourism Research, 46*, 163–184. doi:10.1016/j.annals.2014.01.004

Garechana, G., Río, R., Cilleruelo, E., & Gavilanes, J. (2012). Visualizing the scientific landscape using maps of science. In S. P. Sethi & ... (Eds.), *Industrial Engineering: Innovative Networks* (pp. 103–112). London: Springer. doi:10.1007/978-1-4471-2321-7_12

Garzino, G., Spallone, R., & Lo Turco, M. (2011). Strategie digitali per modelli conoscitivi / Digital strategies for knowledge based models. In G. Garzino (Ed.), Disegno (e) in_formazione. Disegno politecnico / Drawing (and) information. Polytechnic drawing (pp. 70-111). Sant'Arcangelo di Romagna, Italy: Maggioli.

Geographic information system. (2016). Retrieved from https://en.wikipedia.org/wiki/Geographic_information_system

Gerstle, T. L., Ibrahim, A. M., Kim, P. S., Lee, B. T., & Lin, S. J. (2014). A plastic surgery application in evolution: Three-dimensional printing. *Plastic and Reconstructive Surgery*, *2*(2), 446–451. doi:10.1097/01.prs.0000436844.92623.d3 PMID:24469175

Getz, D. (1992). Tourism Planning and Destination Life Cycle. *Annals of Tourism Research, 19*(4), 752–770. doi:10.1016/0160-7383(92)90065-W

Ghasem, N. M., Al-Marzouqi, M. H., & Duaidar, A. (2011). Effect of quenching temperature on the performance of poly (vinylidene fluoride) microporous hollow fiber membranes fabricated via thermally induced phase separation technique on the removal of CO_2 from CO_2 gas mixture. *International Journal of Greenhouse Gas Control, 5*(6), 1550–1558. doi:10.1016/j.ijggc.2011.08.012

Ghasem, N. M., Al-Marzouqi, M. H., & Rahim, N. A. (2013). Modeling of CO_2 absorption in a membrane contactor considering solvent evaporation. *Separation and Purification Technology, 110*, 1–10. doi:10.1016/j.seppur.2013.03.008

Ghasem, N. M., Al-Marzouqi, M., & Zhu, L. P. (2012b). Preparation and properties of polyethersulfone hollow fiber membranes with o-xylene as a additive used in membrane contactors for CO_2 absorption. *Separation and Purification Technology, 92*, 1–10. doi:10.1016/j.seppur.2012.03.005

Ghasem, N., Al-Marzouqi, M. H., & Rahim, N. A. (2012a). Effect of polymer extrusion temperature on poly (vinylidene fluoride) hollow fiber membranes: Properties and performance used as gas–liquid membrane contactor for CO_2 absorption. *Separation and Purification Technology, 99*, 91–103. doi:10.1016/j.seppur.2012.07.021

Ghasem, N., Al-Marzouqi, M. H., & Rahim, N. A. (2014). Absorption of CO_2 Form Natural Gas via Gas-liquid PVDF Hollow Fiber Membrane Contactor and Potassium Glycinate as Solvent. *Jurnal Teknologi, 69*(9), 121–126. doi:10.11113/jt.v69.3409

Ghassemlooy, Z. P. W. (2012). *Optical wireless communications: system and channel modeling with Matlab*. CRC Press.

Gherboudj, A., Layeb, A., & Chikhi, S. (2012). Solving 01 knapsack problems by a discrete binary version of cuckoo search algorithm. *International Journal of Bio-inspired Computation, 4*(4), 229–236. doi:10.1504/IJBIC.2012.048063

Gianvecchio, S., & Wang, H. (2007). Detecting covert timing channels: an entropy-based approach. *Proceedings of the 14th ACM Conference on Computer and Communications Security*, 307–316. 10.1145/1315245.1315284

Giardino, C. (1998). *I metalli nel mondo antico. Introduzione all'archeometallurgia*. Roma-Bari, Italy: Laterza.

Gils, H. C. (2014). Assessment of the theoretical demand response potential in Europe. *Energy*, *67*, 1–18. doi:10.1016/j.energy.2014.02.019

Giordano, M., Ausiello, P., & Martorelli, M. (2012). Accuracy evaluation of surgical guides in implant dentistry by non-contact reverse engineering techniques. *Dental Materials*, *9*(9), e178–e185. doi:10.1016/j.dental.2012.06.006 PMID:22796037

GIS for Marketing: Where Strategy Meets Opportunity. (2010). Retrieved on June 20, 2015, from http://www.esri.com/marketing

Giuliani, G., Guigoz, Y., Lacroix, P., Ray, N., & Lehmann, A. (2016). Facilitating the production of ISO-compliant metadata of geospatial datasets. *International Journal of Applied Earth Observation and Geoinformation*, *44*, 239–243. doi:10.1016/j.jag.2015.08.010

Glenn, N. F., Streutker, D. R., Chadwick, D. J., Thackray, G. D., & Dorsch, S. J. (2006). Analysis of LiDAR-derived topographic information for characterizing and differentiating landslide morphology and activity. *Geomorphology*, *73*(1-2), 131–148. doi:10.1016/j.geomorph.2005.07.006

Glover, J. D., Sarma, M. S., & Overbye, T. J. (2011). *Power system analysis and design*. CengageBrain.com.

Gokceoglu, C., Sonmez, H., Nefeslioglu, H. A., Duman, T. Y., & Can, T. (2005). The 17 March 2005 Kuzulu landslide (Sivas, Turkey) and landslide susceptibility map of its near vicinity. *Eng. Geol. B.*, *81*(1), 65–83. doi:10.1016/j.enggeo.2005.07.011

Goldman, D. (2012). *Google unveils 'Project Glass' virtual-reality glasses*. Money.

Gong, J., Geng, J., & Chen, Z. (2015). Real-time GIS data model and sensor web service platform for environmental data management. *International Journal of Health Geographics*, *14*(1), 1. doi:10.1186/1476-072X-14-2 PMID:25572659

Gong, S., Zhu, Z. H., & Meguid, S. A. (2015). Anisotropic Electrical Conductivity of Polymer Composites with Aligned Carbon Nanotubes. *Polymer*, *56*, 498–506. doi:10.1016/j.polymer.2014.11.038

Gonzalez, A., Ramon, J., Cendrero, A., & Fabirri, A. (2003). Validation of Landslide Susceptibility Maps; Examples and Applications from a Case Study in Northern Spain. *J. Natural Hazards*, *30*(3), 437–449. doi:10.1023/B:NHAZ.0000007201.80743.fc

Goodchild, M. (2006). Geographic Information Systems. In S. Aitken & G. Valentine (Eds.), *Approaches to Human Geography* (Vol. 251-262). SAGE Publications. doi:10.4135/9781446215432.n24

Goodchild, M. (2007). Citizens as sensors: The world of volunteered geography. *GeoJournal*, *69*(4), 211–221. doi:10.100710708-007-9111-y

Goodchild, M. F. (1984). ILACS: A Location-Allocation Model for Retail Site Selection. *Journal of Retailing*, *60*, 84–100.

Goodchild, M. F. (2013). Prospects for a space–time GIS: Space–time integration in geography and GIScience. *Annals of the Association of American Geographers*, *103*(5), 1072–1077. doi:10.1080/00045608.2013.792175

Goodchild, M. F., Yuan, M., & Cova, T. J. (2007). Towards a general theory of geographic representation in GIS. *International Journal of Geographical Information Science*, *21*(3), 239–260. doi:10.1080/13658810600965271

Grady, B. P. (2011). *Carbon Nanotube-Polymer Composites: Manufacture, Properties, and Applications*. Hoboken, NJ: John Wiley & Sons. doi:10.1002/9781118084380

Graham, W. (2016, November 19). *Atlas V successfully launches with GOES-R advanced weather satellite.* Retrieved from https://www.nasaspaceflight.com/2016/11/atlas-v-launch-goes-r-advanced-weather-satellite/

Graham, M., Zook, M., & Boulton, A. (2013). Augmented reality in urban places: Contested content and the duplicity of code. *Transactions of the Institute of British Geographers, 38*(3), 464–479. doi:10.1111/j.1475-5661.2012.00539.x

Granell-Canut, C., & Aguilar-Moreno, E. (2013). Se busca geobibliotecario: Los datos geográficos entran en la biblioteca [Geolibrarian wanted: geographic data come to the library]. *Profesional de la Información., 22*(6), 569–575. doi:10.3145/epi.2013.nov.10

Grantham, C. (1998). Visible Light Communication for Audio Systems. *IEEE Transactions on Consumer Electronics*, 1–11.

Green, K., Morton, B., & New, S. (1998). Green purchasing and supply policies: Do they improve companies environmental performance? *Supply Chain Management: An International Journal, 3*(2), 89–95. doi:10.1108/13598549810215405

Green, K., Morton, B., & New, S. (2000). Greening organizations purchasing, consumption, and innovation. *Organization & Environment, 13*(2), 206–225. doi:10.1177/1086026600132003

Greenpeace. (2008). *Take Back Blues, An assessment of e-waste takes back in India.* Retrieved February 29, 2016 from www.greenpeace.org/india/press/reports/recycling-ofelectronic-waste

Gregotti, V. (1995). Editoriale. *Rassegna*, 61.

Groenendyk, M. (2013). Emerging data visualization technologies for map and geography libraries: 3-D printing, holographic imaging, 3-d city models, and 3-d model-based animations. *Journal of Map & Geography Libraries: Advances in Geospatial Information. Collections & Archives, 9*(3), 220–238.

Grossner, K. E., Goodchild, M. F., & Clarke, K. C. (2008). Defining a digital earth system. *Transactions in GIS, 12*(1), 145–160. doi:10.1111/j.1467-9671.2008.01090.x

Grunewald, S. J. (2015, August 19). *The Shuty Hybrid 3D Printed 9mm Pistol Raises Questions About 3D Printed Gun Control.* Retrieved April 26, 2016, from https://3dprint.com/89919/shuty-hybrid-3d-printed-pistol/

Gu, P., & Li, L. (2002). Fabrication of Biomedical Prototypes with Locally Controlled Properties Using FDM. *CIRP Annals – Manufacturing Technology, 51*(1), 181-184.

Guan, X., Yang, B., Chen, C., Dai, W., & Wang, Y. (2016). A comprehensive overview of cyber-physical systems: from perspective of feedback system. *IEEE/CAA Journal of Automatica Sinica, 3*(1), 1-14.

Guidance Manual for the Implementation of the Organisation for Economic Co-operation and Development (OECD). (2007). *Recommendation C(2004)100 on Environmentally Sound Management (ESM) of Waste, Section 5.2, 12.* Retrieved March 2, 2016 from http://www.oecd.org/env/waste/39559085.pdf

Guidi, A. (1998). The emergence of the state in Central and Northern Italy. *Acta Archaeologica, 69*, 139–161.

Gumus, A. T. (2009). Evaluation of hazardous waste transportation firms by using a two step fuzzy AHP and TOPSIS methodology. *Expert Systems with Applications, 36*(2), 4067-4074.

Gunter, D. K. (1996). *Gunter's Space Page.* Retrieved August 14, 2013, from GOES: http://space.skyrocket.de/doc_sdat/goes-d.htm

Gupta, V., & Tandon, P. (2015). Heterogeneous Object Modelling with Material Convolution Surfaces. *Computer Aided Design, 62*, 236–247. doi:10.1016/j.cad.2014.12.005

Gutterman, Z., & Malkhi, D. (2005). Hold your sessions: An attack on Java session-id generation. *Lecture Notes in Computer Science*, *3376*, 44–57. doi:10.1007/978-3-540-30574-3_5

Guzzetti, F., Paola, R., Mauro, C., Mirco, G., & Ardizzone, F. (2005). Probabilistic landslide hazard assessment at the basin scale. *Geomorphology*, *72*(1-4), 272–299. doi:10.1016/j.geomorph.2005.06.002

Habibollahzadeh, H., Luo, G.-X., & Semlyen, A. (1989). Hydrothermal optimal power flow based on a combined linear and nonlinear programming methodology. *Power Systems*, *4*(2), 530–537. doi:10.1109/59.193826

Hägerstraand, T. (1970). What about people in regional science? *Papers in Regional Science*, *24*(1), 7–24. doi:10.1111/j.1435-5597.1970.tb01464.x

Hägerstrand, T. (1989). *Reflections on "what about people in regional science?"*. Paper presented at the Regional Science Association. 10.1111/j.1435-5597.1989.tb01166.x

Hakkarainen, M., Woodward, C., & Rainio, K. (2009). Software architecture for mobile mixed reality and 4D BIM interaction. *Proc. 25th CIB W78 Conference.*

Haklay, M. (2010). How Good is Volunteered Geographical Information? A Comparative Study of OpenStreetMap and Ordnance Survey Datasets. *Environment and Planning. B, Planning & Design*, *37*(4), 682–703. doi:10.1068/b35097

Haklay, M., Singleton, A., & Parker, C. (2008). Web mapping 2.0: The neogeography of the GeoWeb. *Geography Compass*, *2*(6), 2011–2039. doi:10.1111/j.1749-8198.2008.00167.x

Haklay, M., & Zafiri, A. (2008). Usability engineering for GIS: Learning from a screenshot. *The Cartographic Journal*, *45*(2), 87–97. doi:10.1179/174327708X305085

Hall, C. M. (2005). Rural Wine and Food Tourism Cluster Network Development. In D. Hall, I. Kirkpatrick & M. Mitchell (Eds.), Rural Tourism and Sustainable Business, (pp. 149-164). Clevendon: Channel View Publications.

Halpin & Woodhead. (1998). *Construction management.* Wiley.

Hammarlund, D. (2013). *Equipment and sample handling.* Retrieved from http://www.geol.lu.se/dendro/en/equipment.htm

Hancock, G. (1995). The fingerprints of god: A quest for the beginning and end. London: William Heinemann Ltd.

Hao Huang, X. (2014). 100 Tbit/s free-space data link enabled by three-dimensional multiplexing of orbital angular momentum, polarization, and wavelength. *Optics Letters*, *39*(2), 197–200. doi:10.1364/OL.39.000197 PMID:24562105

Hao, M. L. L. (2013). Integration of indoor visible light and power line communication systems. *17th IEEE International Symposium in Power Line Communications and Its Applications (ISPLC).* 10.1109/ISPLC.2013.6525866

Hargrove, L. J., Scheme, E. J., Englehart, K. B., & Hudgins, B. S. (2010). Multiple binary classifications via linear discriminant analysis for improved controllability of a powered prosthesis. *IEEE Transactions on Neural Systems and Rehabilitation Engineering*, *18*(1), 49–57. doi:10.1109/TNSRE.2009.2039590 PMID:20071277

Harrington, J. M., Aw, T. C., & Baker, E. L. (2003). Occupational and environmental health and safety. In A. W. David, M. C. Timothy, D. F. John, & J. B. Edward (Eds.), *Oxford Textbook of Medicine* (4th ed.; Vol. 1, pp. 956–960). New York: Oxford University Press.

Hashem, I. A. T., Yaqoob, I., Anuar, N. B., Mokhtar, S., Gani, A., & Ullah Khan, S. (2015). The rise of big data on cloud computing: Review and open research issues. *Information Systems*, *47*, 98–115. doi:10.1016/j.is.2014.07.006

Hassini, E., Surti, C., & Searcy, C. (2012). A literature review and a case study of sustainable supply chains with a focus on metrics. *International Journal of Production Economics*, *140*(1), 69–82. doi:10.1016/j.ijpe.2012.01.042

Haykin, S. (1994). *Neural networks: a comprehensive foundation*. Prentice Hall.

Haykin, S. (1999). *Neural Networks: A Comprehensive Foundation*. Englewood Cliffs, NJ: Prentice-Hall.

He, J. L., & Gebhardt, H. (2012). Socio-Culture Embeddedness of Spatial Agglomeration of Creative Industries in Shanghai. *Sociology Study*, *2*(8), 605–614.

Hencken, H. (1968). *Tarquinia, Villanovans and Early Etruscans. American School of Prehistoric Research Bulletin, 23.*

Hendriks, P. H. (2005). Space opera-GIS basics. In M. Khosrouw-Pour (Ed.), *Encyclopedia of Information Science and Technology* (Vol. 5, pp. 2571–2575). Idea Group Reference. doi:10.4018/978-1-59140-553-5.ch456

Heo, E., Kim, J., & Boo, K. J. (2010). Analysis of the assessment factors for renewable energy dissemination program evaluation using fuzzy AHP. *Renewable & Sustainable Energy Reviews*, *14*(8), 2214–2220. doi:10.1016/j.rser.2010.01.020

Hermann, B. G., Kroeze, C., & Jawjit, W. (2007). Assessing environmental performance by combining life cycle assessment, multi-criteria analysis and environmental performance indicators. *Journal of Cleaner Production*, *15*(18), 1787–1796. doi:10.1016/j.jclepro.2006.04.004

Hervani, A. A., Helms, M. M., & Sarkis, J. (2005). Performance measurement for green supply chain management. *Benchmarking. International Journal (Toronto, Ont.)*, *12*(4), 330–353.

He, Y., Xue, G. H., & Fu, J. Z. (2015). Fabrication of low cost soft tissue prostheses with the desktop 3D printer. *Scientific Reports*, *4*, 6973. doi:10.1038rep06973 PMID:25427880

Hey, T. (2012). *The Fourth Paradigm–Data-Intensive Scientific Discovery. In E-Science and Information Management* (pp. 1–1). Springer.

Heywood, I., Cornelius, S., & Carver, S. (2011). *An Introduction to geographical Information systems* (4th ed.). Essex, UK: Pearson Education.

Higgs, G., Berry, R., Kidner, D., & Langford, M. (2008). Using IT approaches to promote public participation in renewable energy planning: Prospects and challenges. *Land Use Policy*, *25*(4), 596–607. doi:10.1016/j.landusepol.2007.12.001

Hill, S. (2008). *GOES-R Solar and Space Environment Data Products: Benefiting Users*. Retrieved December 26, 2015, from GOES-R Solar and Space Environment Data ProduGeostationary Operational Environmental Satellite Program (GOES): ww.weather.gov/satellite

Himmelblau, C. (2003). Musée des Confluences. *GA Document*, *73*, 12–17.

Hledik, R. (2009). How green is the smart grid? *The Electricity Journal*, *22*(3), 29–41. doi:10.1016/j.tej.2009.03.001

Holden, W. (2005). *Juniper research*. White Paper.

Holliman, K. (2015, April 20). *Innovations in 3D Printed Food Outpacing FDA Regulations - Food Quality & Safety*. Retrieved April 26, 2016, from http://www.foodqualityandsafety.com/article/innovations-in-3d-printed-food-outpacing-fda-regulations/

Hong, H., Y. R. (2008). Information illuminating system for healthcare institution. *Proceedings of International Conference on Bioinformatics and Biomedical Engineering*, 801-804.

Hoogeveen, J. A., Lenstra, J. K., & Veltman, B. (1996). Minimizing the makespan in a multiprocessor flow shop is strongly NP-hard. *European Journal of Operational Research*, *89*(1), 172–175. doi:10.1016/S0377-2217(96)90070-3

Hossenlopp, L. (2007). Engineering perspectives on iec 61850. *IEEE Power and Energy Magazine*, *5*(3), 45–50. doi:10.1109/MPAE.2007.365819

Hou, L., & Wang, X. (2011). Experimental Framework for Evaluating Cognitive Workload of Using AR System for General Assembly Task. *Proceedings of the 28th International Symposium on Automation and Robotics in Construction.*

Hranilovic, S. (2005). *Wireless Optical Communication Systems.* Springer.

Hsu, C. W., & Hu, A. H. (2008). Green supply chain management in the electronic industry. *International Journal of Environmental Science and Technology*, *5*(2), 205–216. doi:10.1007/BF03326014

Hsueh, Y. H. (2013b). Mushroom huts diffusion analysis based on GPS data. In Y. H. Hsueh (Ed.), Global positioning systems: Signal structure, applications and sources of error and biases, (pp. 115-124). New York: Nova Science Publishers, Inc.

Hsueh, Y. H. (2012). The Influence of Transportation Revolution to Local Tourism Development-Example for Puli Area of Taiwan. *Sociology Study*, *2*(8), 615–625.

Hsueh, Y. H. (2013a). Exploring the Clustering Location of Accommodation Units through the Tourism Development in the Cing-Jing Area of Taiwan. *International Journal of Basic & Applied Sciences*, *13*(4), 34–43.

Hsueh, Y. H., & Chen, C. H. (2015). Clustering of Tea Processing Industry Based on GIS Analysis. *International Journal of Agricultural Sciences and Natural Resources*, *2*(3), 58–66.

Hu, H., & Speizer, F. E. (2001). Specific environmental and occupational hazards. In Harrison's Principles of Internal Medicine (15th ed.; vol. 2, pp. 2591-2592). McGraw-Hill Inc.

Hu, Y., Krzysztof, J., McKenzie, G., Sengupta, K., & Hitzler, P. (2013). A Linked-data-driven and semantically-enabled journal portal for scientometrics. LNCS, 8219, 114-129. doi:10.1007/978-3-642-41338-4_8

Huang, I. B., Keisler, J., & Linkov, I. (2011). Multi-criteria decision analysis in environmental sciences: Ten years of applications and trends. *The Science of the Total Environment*, *409*(19), 3578–3594. doi:10.1016/j.scitotenv.2011.06.022 PMID:21764422

Huber, M., & Kipman, U. (2011). The Mental Health of Deaf Adolescents with Cochlear Implants Compared to Their Hearing Peers. *International Journal of Audiology*, *53*(3), 146–154. doi:10.3109/14992027.2010.533704 PMID:21309643

Hudgins, B., Parker, P., & Scott, R. N. (1993). A new strategy for multifunction myoelectric control. *IEEE Transactions on Bio-Medical Engineering*, *40*(1), 82–94. doi:10.1109/10.204774 PMID:8468080

Hughes, M., & Touchan, R. (1997). *Preparation for Collaborative Work on Past Climate in Jordan.* Science Storm, A world of wonder. Retrieved from http://www.sciencestorm.com/

Hu, K., Kulkarni, D. D., Choi, I., & Tsukruk, V. V. (2014). Graphene-Polymer Nanocomposites for Structural and Functional Applications. *Progress in Polymer Science*, *39*(11), 1934–1972. doi:10.1016/j.progpolymsci.2014.03.001

Hunt, C., Malone, C., Sevink, J., & Stoddart, S. (1990). Environment, soils and early agriculture in Apennine Central Italy. *World Archaeology*, *22*(1), 34–44. doi:10.1080/00438243.1990.9980127

Hu, Z., Kim, J., Wang, J., & Byrne, J. (2015). Review of dynamic pricing programs in the U.S. and Europe: Status quo and policy recommendations. *Renewable & Sustainable Energy Reviews*, *42*, 743–751. doi:10.1016/j.rser.2014.10.078

Hu, Z., & Wang, X. (2006). A probabilistic load flow method considering branch outages. *Power Systems*, *21*(2), 507–514. doi:10.1109/TPWRS.2006.873118

Hwang, B.-G., Thomas, S. R., Haas, C. T., & Caldas, C. H. (2009). Measuring the impact of rework on construction cost performance. *Journal of Construction Engineering and Management*, *135*(3), 187–198. doi:10.1061/(ASCE)0733-9364(2009)135:3(187)

Hyers, K. (2006). *US mobile broadcast video market: Five predictions*. ABI Research.

Iaia, C. (1999). *Simbolismo funerario e ideologia alle origini di una civiltà urbana. Forme rituali nelle sepolture "villanoviane" a Tarquinia e Vulci, e nel loro entroterra. Grandi contesti e problemi della protostoria italiana*. Firenze, Italy: All'Insegna del Giglio.

ICT & Tourism USAID from the American People. (n.d.). *International Journal of Knowledge and Learning, 3*(6), 628-648. doi:10.1504/IJKL.2007.016837

Indiana State University. (2013). *Dendrochronology Lab, Indiana State University*. Retrieved from http://dendrolab.indstate.edu/equip_fac.htm

International Agency for Research on Cancer (IARC). (1990). Chromium. IACR Monograph, 49, 677.

International Agency for Research on Cancer (IARC). (1993). Beryllium and Beryllium Compounds. International Agency for Research on Cancer (IARC) Monograph, 58, 444.

Ipakchi, A., & Albuyeh, F. (2009). Grid of the future. *Power and Energy Magazine, IEEE, 7*(2), 52–62. doi:10.1109/MPE.2008.931384

Irizarry, J., Gheisari, M., Williams, G., & Walker, B. N. (2012). InfoSPOT: A mobile Augmented Reality method for accessing building information through a situation awareness approach. *Automation in Construction*.

Isaai, M. T., Kanani, A., Tootoonchi, M., & Afzali, H. R. (2011). Intelligent timetable evaluation using fuzzy AHP. *Expert Systems with Applications, 38*(4), 3718–3723. doi:10.1016/j.eswa.2010.09.030

Ishida, S., Nojima, T., & Hagiwara, I. (2014). Mathematical approach to model foldable conical structures using conformal mapping. *Journal of Mechanical Design, 136*(9), 091007. doi:10.1115/1.4027848

Iwamoto, S., & Tamura, Y. (1981). A load flow calculation method for ill-conditioned power systems. *Power Apparatus and Systems*, (4), 1736–1743.

Jafar Jafari. (2003). Encyclopedia of Tourism (p. 308). Academic Press.

Jafari, M. & Khorankeh S. (2013). Impact of climate and environmental changes on forest ecosystem's productivity (case study: Galugah). *Iranian Journal of Forest and Poplar Research, 21*(1), 166-183.

Jafari, M. (2007). *Review on needfulness for plant ecophysiological study and investigation on climate change's effects on forest, rangeland and desert ecosystems*. Presented at Workshop, Climate Change in South-Eastern European Countries: Causes, Impacts, Solutions, Orangerie, Burggarten, Graz, Austria. Retrieved from http://www.intecol.net/pages/002_personal.php?mode=list&table=blog&tb_kind=Research&id=mjafari&vvt=aa&bidx=27531

Jafari, M. (2010). Climate Change Impacts on Iranian Ecosystems with review on Climate Change Study Methods (Pub. No. 421). RIFR.

Jafari, M. (2013a). *New method on dendrochronology study in forest ecosystems; two new approaches: Dendro-productivity and Dendro-genetic*. Paper ID: 1076-ADA2013. Presented in The 3rd International Conference of Asian Dendrochronology Association (ADA2013), Climate Change and tree Rings, Tehran, Iran. Retrieved from http://ada2013.ut.ac.ir/paper?manu=7148

Jafari, M. (2014a). *Dendro-Medical as New Approach on Dendrochronological Study method*. Presented in the 9th International Conference on Dendrochronology, as Poster No. PO086, Ref. 1590, ThemeZ01, Melbourne, Australia.

Jafari, M. (2014b). Change and vulnerability of Net Primary Production (NPP) in Iranian forest, rangeland and desert ecosystems impacted by climate change. *Iranian Journal of Range and Desert Research, 21*(1), 139-153.

Jafari, M. (2012a). A new approach to dendroecological studies: Climate change impact on forest' wood production in Astara (Gilan). *Iranian Journal of Wood and Paper Science Research, 27*(4), 690–706.

Jafari, M. (2013b). Integration of forestry and rangeland management institutions in the Islamic Republic of Iran. In *Challenges in adopting an integrated approach to managing forest and rangelands in the Near East Region, Part II, Country studies* (pp. 37–67). Cairo: FAO.

Jafari, M. (2013c, July-December). Challenges in Climate Change and Environmental Crisis: Impacts of Aviation Industry on Human. *Urban and Natural Environments in International Journal of Space Technology Management and Innovation, 3*(2), 24–46.

Jafari, M. (2015). Dendrochronology and Climate Change. In *Encyclopedia of Information Science and Technology* (3rd ed., pp. 2917–2930). IGI Global.

Jafari, M. A., Han, W., Mohammadi, F., Safari, A., Danforth, S. C., & Langrana, N. (2000). A Novel System for Fused Deposition of Advanced Multiple Ceramics. *Rapid Prototyping Journal, 6*(3), 161–175. doi:10.1108/13552540010337047

Jafari, M., Maghsudloo, M. K., Khanjani Shiraz, B., Karimidoost, A., & Hemmati, A. (2012b). New approach in dendrogenetic method for identification of beech tree variability. *Iranian Journal of Rangelands and Forests Plant Breeding and Genetic Research, 20*(2), 284–294.

Jain, A. (Ed.). (2008). *E-waste: Implications, regulations, and management in India and current global best practices.* New Delhi: Teri Press.

Jardim-Goncalves, R., & Grilo, A. (2010). SOA4BIM: Putting the building and construction industry in the Single European Information Space. *Automation in Construction, 19*(4), 388–397. doi:10.1016/j.autcon.2009.11.009

Jenks, George, F. (1967). The Data Model Concept in Statistical Mapping. *International Yearbook of Cartography, 7*, 186–190.

Jensen, J. R. (2005). *Introduction to digital image processing: A remote sensing perspective.* Upper Saddle River, NJ: Prentice Hall.

Jiao, Y., Zhang, S., Li, Y., Wang, Y., & Yang, B. (2012). Towards cloud Augmented Reality for construction application by BIM and SNS integration. *Automation in Construction.*

Ji, P., Ma, X., & Li, G. (2014). Developing green purchasing relationships for the manufacturing industry: An evolutionary game theory perspective. *International Journal of Production Economics.*

Johnston, R. (2005). *Obtaining high performance phasor measurements in a geographically distributed status dissemination network* (Master's thesis). Washington State University.

Jomaa, M. H., Masenelli-Varlot, K., Seveyrat, L., Lebrun, L., Dib Jawhar Dib, M. C., Beyou, E., & Cavaillé, J. Y. (2015). Investigation of Elastic, Electrical and Electromechanical Properties of Polyurethane/Grafted Carbon Nanotubes Nanocomposites. *Composites Science and Technology, 121*, 1–8. doi:10.1016/j.compscitech.2015.10.019

Jones, D. B., Sung, R., Weinberg, C., Korelitz, T., & Andrews, R. (2016). Three-Dimensional Modeling May Improve Surgical Education and Clinical Practice. *Surgical Innovation, 2*(2), 189–195. doi:10.1177/1553350615607641 PMID:26423911

Jones, M. H., & Lambourne, R. J. (2004). *An Introduction to Galaxies and Cosmology.* Cambridge, UK: Cambridge University Press.

Joseph, K. (2007). Electronic waste management in India - Issues and Strategies. *Proceedings of the Eleventh International Waste Management and Landfill Symposium.*

Joseph, S. T., DeChello, L. M., Kulldorff, M., Gregorio, D. I., Gershman, S., & Mroszczyk, M. (2004). The geographic distribution of breast cancer incidence in Massachusetts 1988 to 1997, adjusted for covariates. *International Journal of Health Geographics, 3*(1), 17. doi:10.1186/1476-072X-3-17 PMID:15291960

Jouglet, A., Oğuz, C., & Sevaux, M. (2009). Hybrid flow-shop: A memetic algorithm using constraint-based scheduling for efficient search. *Journal of Mathematical Modelling and Algorithms, 8*(3), 271–292. doi:10.100710852-008-9101-1

Journal, M. (2015). *The potential of 3D printing*. Retrieved from http://www.meti.go.jp/english/publications/pdf/journal2013_08a.pdf

Juckett, D. A., & Rosenberg, B. (1997). Time series analysis supporting the hypothesis that enhanced cosmic radiation during germ cell formation can increase breast cancer mortality in germ cell cohorts. *International Journal of Biometeorology, 40*(4), 206–221. doi:10.1007004840050043 PMID:9225597

Jug, K., Hilbert, P. C., & Shail, S. (2001). σ-π Energy Separation in Modern Electronic Theory for Ground States of Conjugated Systems. *Chemical Reviews, 101*(5), 1477–1500. doi:10.1021/cr990328e PMID:11710230

Jungwattanakit, J., Reodecha, M., Chaovalitwongse, P., & Werner, F. (2008). Algorithms for flexible flow shop problems with unrelated parallel machines, setup times, and dual criteria. *International Journal of Advanced Manufacturing Technology, 37*(3-4), 354–370. doi:10.100700170-007-0977-0

Jungwattanakit, J., Reodecha, M., Chaovalitwongse, P., & Werner, F. (2009). A comparison of scheduling algorithms for flexible flow shop problems with unrelated parallel machines, setup times, and dual criteria. *Computers & Operations Research, 36*(2), 358–378. doi:10.1016/j.cor.2007.10.004

Kahraman, C., Engin, O., Kaya, İ., & Öztürk, R. E. (2010). Multiprocessor task scheduling in multistage hybrid flow-shops: A parallel greedy algorithm approach. *Applied Soft Computing, 10*(4), 1293–1300. doi:10.1016/j.asoc.2010.03.008

Kallenberg, L. A. C., Schulte, E., Disselhorst-Klug, C., & Hermensa, H. J. (2007). Myoelectric manifestations of fatigue at low contraction levels in subjects with and without chronic pain. *Journal of Electromyography and Kinesiology, 17*(3), 264–274. doi:10.1016/j.jelekin.2006.04.004 PMID:16790358

Kamat, V. R., Martinez, J. C., Fischer, M., Golparvar-Fard, M., Peña-Mora, F., & Savarese, S. (2010). Research in visualization techniques for field construction. *Journal of Construction Engineering and Management, 137*(10), 853–862. doi:10.1061/(ASCE)CO.1943-7862.0000262

Kamermans, H., Loving, S. H., & Voorrips, A. (1985). Changing patterns of prehistoric landuse in the Agro Pontino. In C. Malone & S. Stoddart (Eds.), Papers in Italian Archaeology, IV, Part I, The Human Landscape (pp. 53-68). Oxford, UK: BAR International Series 243.

Kandil, Hastak, & Dunston. (2014). Creating 3D web-based game environment using BIM models for virtual on-site visiting of building HVAC systems. *Bridges, 10.*

Kang, H. Y., Ogunseitan, O., Shapiro, A. A., & Schoenung, J. M. (2007). A comparative hierarchical decision framework on toxics use reduction effectiveness for electronic and electrical industries. *Environmental Science & Technology, 41*(2), 373–379. doi:10.1021/es060739u PMID:17310694

Kang, J. H., Anderson, S. D., & Clayton, M. J. (2007). Empirical study on the merit of web-based 4D visualization in collaborative construction planning and scheduling. *Journal of Construction Engineering and Management, 133*(6), 447–461. doi:10.1061/(ASCE)0733-9364(2007)133:6(447)

Kaufmann, A., & Gupta, M. M. (1988). *Fuzzy Mathematical Models in Engineering and Management Science*. Amsterdam: North Holland.

Kaushik, A., & Vidyarthi, D. P. (2016). A green energy model for resource allocation in computational grid using dynamic threshold and GA. *Sustainable Computing: Informatics and Systems*, *9*, 42–56.

Kaveh, A., & Bakhshpoori, T. (2013). Optimum design of steel frames using cuckoo search algorithm with lévy flights. *Structural Design of Tall and Special Buildings*, *22*(13), 1023–1036. doi:10.1002/tal.754

Kavzoglu, T., Sahin, E. K., & Colkesen, I. (2013). Landslide susceptibility mapping using GIS- based multi-criteria decision analysis, support vector machines, and logistic regression. *Landslides*, *11*(3), 425–439. doi:10.100710346-013-0391-7

Kaya, T., & Kahraman, C. (2011). An integrated fuzzy AHP–ELECTRE methodology for environmental impact assessment. *Expert Systems with Applications*, *38*(7), 8553–8562. doi:10.1016/j.eswa.2011.01.057

Kerski, J. J. (2015). Geo-awareness, Geo-enablement, Geotechnologies, Citizen Science, and Storytelling: Geography on the World Stage. *Geography Compass*, *9*(1), 14–26. doi:10.1111/gec3.12193

Khalifeh, A. Y., & Yousefizadeh, H. (2010). Optimal Audio Transmission Over Error-Prone Wireless Links. *IEEE Transactions on Multimedia*, *12*(3), 204–214. doi:10.1109/TMM.2010.2041096

Khalili, N. R., & Duecker, S. (2013). Application of multi-criteria decision analysis in design of sustainable environmental management system framework. *Journal of Cleaner Production*, *47*, 188–198. doi:10.1016/j.jclepro.2012.10.044

Khalouli, S., Ghedjati, F., & Hamzaoui, A. (2010). A meta-heuristic approach to solve a JIT scheduling problem in hybrid flow shop. *Engineering Applications of Artificial Intelligence*, *23*(5), 765–771. doi:10.1016/j.engappai.2010.01.008

Khan, M., & Dickens, P. (2014). Selective Laser Melting (SLM) of Pure Gold for Manufacturing Dental Crowns. *Rapid Prototyping Journal*, *20*(6), 471–479. doi:10.1108/RPJ-03-2013-0034

Khan, R. A. J., Junaid, M., & Asgher, M. M. (2009). Analyses and monitoring of 132 kv grid using etap software. In *Electrical and Electronics Engineering* (p. I–113). IEEE.

Khatwani, G., Singh, S. P., Trivedi, A., & Chauhan, A. (2015). Fuzzy-TISM: A fuzzy extension of TISM for group decision making. *Global Journal of Flexible Systems Management*, *16*(1), 97–112. doi:10.100740171-014-0087-4

Khidir ElTayeb, T., Zailani, S., & Jayaraman, K. (2010). The examination on the drivers for green purchasing adoption among EMS 14001 certified companies in Malaysia. *Journal of Manufacturing Technology Management*, *21*(2), 206–225. doi:10.1108/17410381011014378

Khurana, H., Hadley, M., Lu, N., & Frincke, D. A. (2010). Smart-grid security issues. *Security & Privacy, IEEE*, *8*(1), 81–85. doi:10.1109/MSP.2010.49

Kiaee, M., & Jafari, M. (2014). Investigation and consideration of forest tree reaction to climate and environmental changes (Case study: Lavizan forest park). *Journal of Plant Research*, *27*(1), 130–141.

Kiguchi, K., Tanaka, T., Watanabe, K., & Fukuda, T. (2003). Exoskeleton for human upper-limb motion support, *IEEE International Conference on Robotics and Automation*, 2206 – 2211.

Kim, H., Lee, J. K., Park, J. H., Park, B. J., & Jang, D. S. (2002). Applying digital manufacturing technology to ship production and the maritime environment. *Integrated Manufacturing Systems*, *13*(5), 295–305. doi:10.1108/09576060210429748

Kim, K. S., Choi, H. H., Moon, C. S., & Mun, C. W. (2011). Comparison of k-nearest neighbor, quadratic discriminant and linear discriminant analysis in classification of electromyogram signals based on the wrist-motion directions. *Current Applied Physics*, *11*(3), 740–745. doi:10.1016/j.cap.2010.11.051

Kitaura, F. S., & Enßlin, T. A. (2008). Bayesian reconstruction of the cosmological large-scale structure: Methodology, inverse algorithms and numerical optimization. *Monthly Notices of the Royal Astronomical Society, 389*(2), 497–544. doi:10.1111/j.1365-2966.2008.13341.x

Klein, G. T., Lu, Y., & Wang, M. Y. (2013). 3D printing and neurosurgery - ready for prime time? *World Neurosurgery, 3-4*(3-4), 233–235. doi:10.1016/j.wneu.2013.07.009 PMID:23871811

Kliukiene, J., Tynes, T., & Andersen, A. (2001). Risk of breast cancer among Norwegian women with visual impairment. *British Journal of Cancer, 84*(3), 397–399. doi:10.1054/bjoc.2000.1617 PMID:11161406

Kohler, T. A., Gumerman, G. J., & Reynolds, R. G. (2005). Simulating Ancient Societies. *Scientific American, 293*(1), 77–84. doi:10.1038cientificamerican0705-76 PMID:16008305

Kokkila, M. (2002). *Digital elevation models (DEM) in trafficability analysis.* ECOWOOD.

Kotla, R., Alvisi, L., Dahlin, M., Clement, A., & Wong, E. (2007). Zyzzyva: Speculative byzantine fault tolerance. *Proceedings of 21st ACM Symposium on Operating Systems Principles*, 45-58.

Krishna, B. V., Xue, W. C., Bose, S., & Bandyopadhyay, A. (2008). Functionally Graded Co-Cr-Mo Coating on Ti-6Al-4V Alloy Structures. *Acta Biomaterialia, 4*(3), 697–706. doi:10.1016/j.actbio.2007.10.005 PMID:18054298

Kulldorff, M., Feuer, E. J., Miller, B. A., & Freedman, L. S. (1997). Breast cancer clusters in the northeast United States: A geographic analysis. *American Journal of Epidemiology, 146*(2), 161–170. doi:10.1093/oxfordjournals.aje.a009247 PMID:9230778

Kumar, R., & Shah, J. D. (2014). Current Status of Recycling Waste Printed Circuit Boards in India. *Journal of Environmental Protection*, 9-16.

Kumar, N. L. (2012). Visible Light Communications in Intelligent Transportation Systems. *Intelligent Vehicles Symposium Alcalá de Henares*, 748-753. 10.1109/IVS.2012.6232282

Kuniholm, P. I. (2001). *Dendrochronology and Other Applications of Tree-ring Studies in Archaeology. In The Handbook of Archaeological Sciences.* London: John Wiley & Sons, Ltd.

Kwan, M.-P. (1998). Space-time and integral measures of individual accessibility: A comparative analysis using a point-based framework. *Geographical Analysis, 30*(3), 191–216. doi:10.1111/j.1538-4632.1998.tb00396.x

Lacy, J., Mitchell, D., & Schell, W. (1993). Cryptolib: Cryptography in software. *Proceedings of the 4th USENIX Security Symposium*, 1–17.

Laden, F., Spiegelman, D., Neas, L. M., Colditz, G. A., Hankinson, S. E., Manson, J. E., ... Hunter, D. J. (1997). Geographic variation in breast cancer incidence rates in a cohort of U.S. women. *Journal of the National Cancer Institute, 89*(18), 1373–1378. doi:10.1093/jnci/89.18.1373 PMID:9308708

Lahdelma, R., Salminen, P., & Hokkanen, J. (2002). Locating a waste treatment facility by using stochastic multicriteria acceptability analysis with ordinal criteria. *European Journal of Operational Research, 142*(2), 345–356. doi:10.1016/S0377-2217(01)00303-4

Lalchandani, N. (2010, April 24). E-scare. *The Times of India.*

Lamelas, M. T., Marinoni, O., Hoppe, A., & de la Riva, J. (2008). Suitability analysis for sand and gravel extraction site location in the context of a sustainable development in the surroundings of Zaragoza (Spain). *Environmental Geology, 55*(8), 1673–1686. doi:10.100700254-007-1116-9

Lamport, L., Shostak, R., & Pease, M. (1982). The Byzantine generals problem. *ACM Transactions on Programming Languages and Systems*, *4*(3), 382–401. doi:10.1145/357172.357176

Lanbo, L. S.-H. (2008). *Prospects and problems of wireless communication for underwater sensor networks*. Wirel. Commun. Mob. Comput.

Landsat5. (n.d.). Retrieved from http://landsat.gsfc.nasa.gov/landsat-5/

Larimian, T., Zarabadi, Z. S. S., & Sadeghi, A. (2013). Developing a fuzzy AHP model to evaluate environmental sustainability from the perspective of Secured by Design scheme-A case study. *Sustainable Cities and Society*, *7*, 25–36. doi:10.1016/j.scs.2012.10.001

LaViolette, P. E., & Diachok, O. I. (1974). he Use of APT satellite imagery in a subarctic airborne oceanographic survey. *Arctic*, *27*(4), 249–324. doi:10.14430/arctic2888

Layeb, A. (2011). A novel quantum inspired cuckoo search for knapsack problems. *International Journal of Bio-inspired Computation*, *3*(5), 297–305. doi:10.1504/IJBIC.2011.042260

Layeb, A., & Boussalia, S. R. (2012). A novel quantum inspired cuckoo search algorithm for bin packing problem. *International Journal of Information Technology and Computer Science*, *4*(5), 58–67. doi:10.5815/ijitcs.2012.05.08

Le Nguyen, H. (1997). Newton-raphson method in complex form power system load flow analysis. *Power Systems*, *12*(3), 1355–1359. doi:10.1109/59.630481

Lee, Y.-T., & Chen, K.-T. (2010). *Is server consolidation beneficial to MMORPG? A case study of World of Warcraft.* Paper presented at the Cloud Computing (CLOUD), 2010 IEEE 3rd International Conference. 10.1109/CLOUD.2010.57

Lee, C. Y., & Vairaktarakis, G. L. (1994). Minimizing makespan in hybrid flowshops. *Operations Research Letters*, *16*(3), 149–158. doi:10.1016/0167-6377(94)90026-4

Lee, K.-S., & Moon, K.-J. (2014). Hospital distribution in a metropolitan city: Assessment by a geographical information system grid modelling approach. *Geospatial Health*, *8*(2), 537–544. doi:10.4081/gh.2014.43 PMID:24893031

Lee, S. J., Jung, I. Y., Lee, C. Y., Choi, S. Y., & Kum, K. Y. (2001). Clinical Application of Computer-Aided Rapid Prototyping for Tooth Transplantation. *Dental Traumatology. Munksgaard*, *17*(3), 114–119. doi:10.1034/j.1600-9657.2001.017003114.x PMID:11499760

Lee, S. K., Mogi, G., Lee, S. K., & Kim, J. W. (2011). Prioritizing the weights of hydrogen energy technologies in the sector of the hydrogen economy by using a fuzzy AHP approach. *International Journal of Hydrogen Energy*, *36*(2), 1897–1902. doi:10.1016/j.ijhydene.2010.01.035

Lee, S., & Min, K. (2001). Statistical analysis of landslide susceptibility at Yongin, Korea. *Environ Geol.*, *40*(9), 1095–1113. doi:10.1007002540100310

Leese, J. A., Noar, P. F., & Pastre, C. (1989). Operational continuity of the global meteorological satellite network. *Space Policy*, *5*(1), 12–24. doi:10.1016/0265-9646(89)90025-8

Leffler, P. K., & Brent, J. (1992). *Public history readings*. Malabar, FL: Krieger Publishing Company.

Le, G. M., Gomez, S. L., Clarke, C. A., Glaser, S. L., & West, D. W. (2002). Cancer incidence patterns among Vietnamese in the United States and Ha Noi, Vietnam. *International Journal of Cancer*, *102*(4), 412–417. doi:10.1002/ijc.10725 PMID:12402312

Leijnse, J. N., & Spoor, C. W. (2012). Reverse engineering finger extensor apparatus morphology from measured coupled interphalangeal joint angle trajectories - a generic 2D kinematic model. *Journal of Biomechanics*, *3*(3), 569–578. doi:10.1016/j.jbiomech.2011.11.002 PMID:22134183

Lei, S., Frank, M. C., Anderson, D. D., & Brown, T. D. (2014). A Method to Represent Heterogeneous Materials for Rapid Prototyping: The Matryoshka Approach. *Rapid Prototyping Journal*, *5*, 390–402. doi:10.1108/RPJ-10-2012-0095 PMID:26120277

Lekawa-Raus, A., Patmore, J., Kurzepa, L., Bulmer, J., & Koziol, K. (2014). Electrical Properties of Carbon Nanotube based Fibers and Their Use in Electrical Wiring. *Advanced Functional Materials*, *24*(24), 3661–3682. doi:10.1002/adfm.201303716

Leon, M., Gutierrez, J. M., Leija, L., & Munoz, R. (2011). EMG pattern recognition using support vector machines classifier for myoelectric control purposes. *Pan American Health Care Exchanges (PAHCE)*, 175–178.

Lepawsky, J., & McNabb, C. (2010). Mapping international flows of electronic waste. *Canadian Geographer*, *54*(2), 177–195. doi:10.1111/j.1541-0064.2009.00279.x

Li, G., & Zhang, X.-P. (2009). Comparison between two probabilistic load flow methods for reliability assessment. In Power & Energy Society General Meeting, (pp. 1-7). IEEE. doi:10.1109/PES.2009.5275534

Liang, Z. H., Yang, K., Sun, Y. W., Yuan, J. H., Zhang, H. W., & Zhang, Z. Z. (2006). Decision support for choice optimal power generation projects: Fuzzy comprehensive evaluation model based on the electricity market. *Energy Policy*, *34*(17), 3359–3364. doi:10.1016/j.enpol.2005.06.021

Liberatore, V. (2006). Integrated playback, sensing and network control. *Proceedings of the 25th IEEE International Conference on Computer Communications*, 1–12.

Li, F., Qiao, W., Sun, H., Wan, H., Wang, J., Xia, Y., ... Zhang, P. (2010). Smart transmission grid: Vision and framework. *Smart Grid*, *1*(2), 168–177. doi:10.1109/TSG.2010.2053726

Lightfoot, K. G. (1995). Culture Contact Studies: Redifining the relationship between prehistroric and historical archaeology. *American Antiquity*, *60*(2), 199–217. doi:10.2307/282137

Li, L., Saedan, M., Feng, W., Fuh, J., Wong, Y., Loh, H., ... Lu, L. (2009). Development of a multi-nozzle drop-on-demand system for multi-material dispensing. *Journal of Materials Processing Technology*, *209*(9), 4444–4448. doi:10.1016/j.jmatprotec.2008.10.040

Lillesand, T. M., & Kiefer, R. W. (2004). *Remote sensing and image interpretation*. Chichester, UK: John Wiley & Sons.

Lin, D., Li, Q., Li, W., Zhou, W. L., & Swain, M. V. (2009). Design Optimization of Functionally Graded Dental Implant for Bone Remodeling. *Composites. Part B, Engineering*, *40*(7), 668–675. doi:10.1016/j.compositesb.2009.04.015

Linga, P., Kumar, R., & Englezos, P. (2007). Gas hydrate formation from hydrogen/carbon dioxide and nitrogen/carbon dioxide gas mixtures. *Chemical Engineering Science*, *62*(16), 4268–4276. doi:10.1016/j.ces.2007.04.033

Lin, H., Chen, M., Lu, G., Zhu, Q., Gong, J., You, X., ... Hu, M. (2013). Virtual Geographic Environments (VGEs): A New Generation of Geographic Analysis Tool. *Earth-Science Reviews*, *126*, 74–84. doi:10.1016/j.earscirev.2013.08.001

Linington, R. E. (1982). Il villaggio protostorico nella località Calvario sui Monterozzi a Tarquinia. In Studi in onore di Ferrante Rittatore Vonwiller (pp. 245-256). Como, Italy: Società archeologica comense.

Linked Data Portal for the Semantic Web Journal. (2010). Retrieved from http://semantic-web-journal.com/SWJPortal/

Li, Q. W., Li, Y., Zhang, X. F., Chikkannanavar, S. B., Zhao, Y. H., Dangelewicz, A. M., ... Zhu, Y. T. (2007). Structure-Dependent Electrical Properties of Carbon Nanotube Fibers. *Advanced Materials*, *19*(20), 3358–3363. doi:10.1002/adma.200602966

Li, S., & Gong, J. (2008). MASHUP: A New Way of Providing Web Mapping/GIS Services, *The International Archives of the Photogrammetry. Remote Sensing and Spatial Information Sciences*, *37*(B4), 639–648.

Lissoni, P., Rovelli, F., Malugani, F., Bucovec, R., Conti, A., & Maestroni, G. J. (2001). Anti-angiogenic activity of melatonin in advanced cancer patients. *Neuroendocrinology Letters*, *22*(1), 45–47. PMID:11335879

Liverani, M. (1988). *Antico Oriente. Storia società economia*. Roma-Bari, Italy: Laterza.

Livingstone, D. (1857). *The missionary travels and researches in south africa*. London: Royal Geographical Society.

Longmatey, D., Samuel, A., & Prah, B. K. (2002). *Management and Promotion of Tourism in Ghana: A GIS Approach from African Tourism Websites*. Retrieved on May 20, 2015. http://www.africantourism.com/ghana.htm

Lorrain, T., Jiang, N., & Farina, D. (2011). Influence of the training set on the accuracy of surface EMG classification in dynamic contractions for the control of multifunction prostheses. *Journal of Neuroengineering and Rehabilitation*, *8*(1), 25. doi:10.1186/1743-0003-8-25 PMID:21554700

Love, P. E., Edwards, D. J., Han, S., & Goh, Y. M. (2011). Design error reduction: Toward the effective utilization of building information modeling. *Research in Engineering Design*, *22*(3), 173–187. doi:10.100700163-011-0105-x

Lu, C., Saifullah, A., Li, B., Sha, M., Gonzalez, H., Gunatilaka, D., ... Chen, Y. (2016). Real-Time Wireless Sensor-Actuator Networks for Industrial Cyber-Physical Systems. *Proceedings of the IEEE*, *104*(5), 1013–1024. doi:10.1109/JPROC.2015.2497161

Lucas, C., & Spiegel, M. (2015). *3D printing skin grafts for facial reconstruction*. Academic Press.

Lucas, M. F., Gaufriau, A., Pascual, S., Doncarli, C., & Farina, D. (2008). Multi-channel surface EMG classification using support vector machines and signal-based wavelet optimization. *Biomedical Signal Processing and Control*, *3*(2), 169–174. doi:10.1016/j.bspc.2007.09.002

Luh, J. J., Chang, G. C., Cheng, C. K., Lai, J. S., & Kuo, T. S. (1999). Isokinetic elbow joint torques estimation from surface EMG and joint kinematic data: Using an artificial neural network model. *Journal of Electromyography and Kinesiology*, *9*(3), 173–183. doi:10.1016/S1050-6411(98)00030-3 PMID:10328412

Lundgren, K. (2012a). The Global Impact of e-waste: Addressing the Challenge. International Labour Office, Programme on Safety and Health at Work and the Environment (Safe Work), Sectoral Activities Department (Sector). Geneva: ILO.

Lundgren, K. (2012b). The Global Impact of e-waste: Addressing the Challenge. International Labour Office, Programme on Safety and Health at Work and the Environment (Safe Work), and Sectoral Activities Department (SECTOR). Geneva: ILO, ISBN 978-92-2-126897-0 (print)

Lurton, X. (2010). *An Introduction to Underwater Acoustics Principles and Applications*. Springer-Verlag Berlin Heidelberg. doi:10.1007/978-3-642-13835-5

LVX Minnesota Lightning. (2015, Nov 1). Retrieved from http://purelifi.com

Lyseen, A. K., Nohr, C., Sorensen, E. M., Gudes, O., Geraghty, E. M., Shaw, N. T., & Bivona-Tellez, C. (2014). A Review and framework for categorizing current research and development in health related geographical information systems (GIS) studies. *Yearbook of Medical Informatics*, *9*(1), 110–124. doi:10.15265/IY-2014-0008 PMID:25123730

Mabona, M. (2004). *Diviners and prophets among the Xhosa (1593-1856): A study in Xhosa cultural history*. Münster: Lit Verlag.

Macko, M., Mikołajewska, E., Szczepański, Z., Augustyńska, B., & Mikołajewski, D. (2016). Repository of images for reverse engineering and medical simulation purposes. *Medical and Biological Sciences*, *3*, 23–29.

Madu, C. N., Kuei, C., & Madu, I. E. (2002). A hierarchic metric approach for integration of green issues in manufacturing: A paper recycling application. *Journal of Environmental Management*, *64*(3), 261–272. doi:10.1006/jema.2001.0498 PMID:12040959

Mahmood, A., Exel, R., & Sauter, T. (2014). Delay and Jitter Characterization for Software-Based Clock Synchronization Over WLAN Using PTP. *IEEE Transactions on Industrial Informatics*, *10*(2), 1198–1206. doi:10.1109/TII.2014.2304413

Maji, P. K., Banerjee, A. F., Banerjee, P. S., & Karmakar, S. (2014). Additive Manufacturing in Prosthesis Development – A Case Study. *Rapid Prototyping Journal*, *20*(6), 480–489. doi:10.1108/RPJ-07-2012-0066

Maldonado, T. (2005). *Reale e virtuale*. Milano, Italy: Feltrinelli.

Mallegni, F., Rottoli, M., & Bruni, S. (2001). I dati delle discipline consociate. In A.M. Moretti Sgubini (Ed.), Tarquinia etrusca. Una nuova storia, Catalogo della mostra. Tarquinia, Museo Archeologico Nazionale (pp. 59-61). Roma, Italy: L'Erma di Bretschneider.

Mallick, A. K., & Hota, B. S. (2015). *Load flow study in power system*. Retrieved from http://ethesis.nitrkl.ac.in/2464/1/load flow in power system.pdf

Mandolesi, A. (1999). *La "prima" Tarquinia. L'insediamento protostorico sulla Civita e nel territorio circostante. Grandi contesti e problemi della Protostoria italiana*. Firenze, Italy: All'Insegna del Giglio.

Mann, R. W., & Reimers, S. D. (1970). Kinesthetic sensing for the EMG controlled Boston Arm. *IEEE Transactions on Man-Machine Systems*, *11*(1), 110–115. doi:10.1109/TMMS.1970.299971

Manovich, L. (2001). *The Language of New Media*. Cambridge, MA: MIT Press.

Manyika, J., Chui, M., Bughin, J., Dobbs, R., Bisson, P., & Marrs, A. (2013). *Disruptive technologies: Advances that will transform life, business, and the global economy* (Vol. 12). McKinsey Global Institute New York.

Maphanyane, J. G. (2012, December 1). *The investigation of land cover change and trends through the reconstruction of historical landscape* (PhD Thesis). Melbourne, Victoria, Australia: The University of Melbourne.

Marichelvam, M. K. (2012). An improved hybrid Cuckoo Search (IHCS) metaheuristics algorithm for permutation flow shop scheduling problems. *International Journal of Bio-inspired Computation*, *4*(4), 200–205. doi:10.1504/IJBIC.2012.048061

Marichelvam, M. K., & Geetha, M. (2014). Solving tri–objective multistage hybrid flow shop scheduling problems using a discrete firefly algorithm. *International Journal of Intelligent Engineering Informatics*, *2*(4), 284–303. doi:10.1504/IJIEI.2014.067190

Marichelvam, M. K., & Prabaharan, T. (2012). A bat algorithm for realistic hybrid flowshop schedulihng problems to minimize makespan and mean flow time. *ICTACT Journal on Soft Computing*, *3*(1), 428–433. doi:10.21917/ijsc.2012.0066

Marichelvam, M. K., & Prabaharan, T. (2014). Performance evaluation of an improved hybrid genetic scatter search (IHGSS) algorithm for multistage hybrid flow shop scheduling problems with missing operations. *International Journal of Industrial and Systems Engineering*, *16*(1), 120–141. doi:10.1504/IJISE.2014.057946

Marichelvam, M. K., Prabaharan, T., & Yang, X. S. (2014). A discrete firefly algorithm for the multi-objective hybrid flowshop scheduling problems. *IEEE Transactions on Evolutionary Computation, 18*(2), 301–305. doi:10.1109/TEVC.2013.2240304

Marichelvam, M. K., Prabaharan, T., & Yang, X. S. (2014). Improved cuckoo search algorithm for hybrid flow shop scheduling problems to minimize makespan. *Applied Soft Computing, 19*, 93–101. doi:10.1016/j.asoc.2014.02.005

Marichelvam, M. K., Prabaharan, T., Yang, X. S., & Geetha, M. (2013). Solving hybrid flow shop scheduling problems using bat algorithm. *International Journal of Logistics Economics and Globalization, 5*(1), 15–29. doi:10.1504/IJLEG.2013.054428

Marinho, B., Ghislandi, M., Tkalya, E., Koning, C. E., & de With, G. (2012). Electrical Conductivity of Compacts of Graphene, Multi-Wall Carbon Nanotubes, Carbon Black, and Graphite Powder. *Powder Technology, 221*, 351–358. doi:10.1016/j.powtec.2012.01.024

Markram, H. (2012). The human brain project. *Scientific American, 6*(6), 50–55. doi:10.1038cientificamerican0612-50 PMID:22649994

Marshall, J. (2001). Women and Strangers: Issues of Marginalization in Seasonal Tourism. *Tourism Geographies, 3*(2), 165–186. doi:10.1080/14616680010030266

Marston, S., Li, Z., Bandyopadhyay, S., Zhang, J., & Ghalsasi, A. (2011). Cloud computing—The business perspective. *Decision Support Systems, 51*(1), 176–189. doi:10.1016/j.dss.2010.12.006

Maruthappu, M., & Keogh, B. (2014). How might 3D printing affect clinical practice? *BMJ (Clinical Research Ed.), 349*, g7709. PMID:25550064

Masayuki Kinoshita, T. Y. (2014). *Motion Modeling of Mobile Transmitter for Image Sensor Based I2V-VLC, V2I-VLC, and V2V-VLC*. Globecom Workshop-Optical Wireless Communications.

Mata, F., García, J. A., Sierra, S., Giordani, V., Gutiérrez, F. J., Márquez, M., ... Cerrato, R. J. (2011). *EMAS 2009. Declaración ambiental EP Hospital Costa del Sol*. Consejería de Salud de la Junta de Andalucía.

Mather, P., & Koch, M. (2011). Computer processing of remotely-sensed images: An introduction. Wiley-Blackwell. doi:10.1002/9780470666517

Mavi, R. K., Kazemi, S., Najafabadi, A. F., & Mousaabadi, H. B. (2013). Identification and assessment of logistical factors to evaluate a green supplier using the fuzzy logic DEMATEL method. *Polish Journal of Environmental Studies, 22*(2), 445–455.

McCarthy, B. C. (2008). *Introduction to Dendrochronology, Lab procedures*. Retrieved from http://www.plantbio.ohiou.edu/dendro/

McCullagh, M. C. (2011). Effects of a Low Intensity Intervention to Increase Hearing Protector Use among Noise-exposed Workers. *American Journal of Industrial Medicine, 54*(3), 210–215. doi:10.1002/ajim.20884 PMID:20721900

McKibben, D., & Furlonger, D. (2009). *Magic quadrant for operational risk management software for financial services*. Gartner Research. Available: http://www. gartner. com/technology/home. jsp

McLaughlin, E. (2013). *Definition of green IT (green information technology)*. Retrieved March 30, 2016 from http://searchcio.techtarget.com/definition/green-IT-green-information-technology

Mekonnen, A. D., & Gorsevski, P. V. (2015). A web-based participatory GIS (PGIS) for offshore wind farm suitability within Lake Erie, Ohio. *Renewable & Sustainable Energy Reviews, 41*, 162–177. doi:10.1016/j.rser.2014.08.030

Melliopoulos, A. P. S. (2007). Substation automation: Are we there yet? *IEEE Power and Energy Magazine*, *5*(3), 28–30. doi:10.1109/MPAE.2007.365815

Mendis, D., & Secchi, D. (2015). *A Legal and Empirical Study of 3D Printing Online Platforms and an Analysis of User Behaviour*. Academic Press.

Merideth, M., Iyengar, A., Mikalsen, T., Tai, S., Rouvellou, I., & Narasimhan, P. (2005). Thema: Byzantine-fault-tolerant middleware for web services applications. *Proceedings of the IEEE Symposium on Reliable Distributed Systems*, 131–142. 10.1109/RELDIS.2005.28

Metke, A. R., & Ekl, R. L. (2010). Security technology for smart grid networks. *Smart Grid*, *1*(1), 99–107. doi:10.1109/TSG.2010.2046347

Michael, E. J. (2007). Micro-Clusters: Antiques, Retailing and Business Practice. In E. J. Michael (Ed.), Micro-Clusters and Networks: The Growth of Tourism, (pp. 63-78). London: Elsevier Publications.

Michael, E. (2002). Antiques and Tourism in Australia. *Tourism Management*, *23*(2), 117–125. doi:10.1016/S0261-5177(01)00053-X

Milano, F. (2005). An open source power system analysis toolbox. *Power Systems*, *20*(3), 1199–1206. doi:10.1109/TPWRS.2005.851911

Miller, H. J. (1991). Modelling accessibility using space-time prism concepts within geographical information systems. *International Journal of Geographical Information System*, *5*(3), 287–301. doi:10.1080/02693799108927856

Miller, H. J. (1999). Measuring space-time accessibility benefits within transportation networks: Basic theory and computational procedures. *Geographical Analysis*, *31*(1), 1–26. doi:10.1111/j.1538-4632.1999.tb00408.x

Miller, H. J. (2004). Toblers first law and spatial analysis. *Annals of the Association of American Geographers*, *94*(2), 284–289. doi:10.1111/j.1467-8306.2004.09402005.x

Miller, H. J., & Goodchild, M. F. (2015). Data-driven geography. *GeoJournal*, *80*(4), 449–461. doi:10.100710708-014-9602-6

Min, C., Shen, X., Shi, Z., Chen, L., & Xu, Z. (2010). The Electrical Properties and Conducting Mechanisms of Carbon Nanotube/Polymer Nanocomposites: A Review. *Polymer-Plastics Technology and Engineering*, *49*(12), 1172–1181. doi:10.1080/03602559.2010.496405

Min, H., & Galle, W. P. (1997). Green purchasing strategies: Trends and implications. *International Journal of Purchasing and Materials Management*, *33*(2), 10–17. doi:10.1111/j.1745-493X.1997.tb00026.x

Min, H., & Galle, W. P. (2001). Green purchasing practices of US firms. *International Journal of Operations & Production Management*, *21*(9), 1222–1238. doi:10.1108/EUM0000000005923

Miranda, V., & Saraiva, J. (1991). Fuzzy modelling of power system optimal load flow. In *Power Industry Computer Application* (pp. 386–392). IEEE; doi:10.1109/PICA.1991.160606

Mitchell, R. (2013). IT gets its green on. *Computerworld*, 26–29.

Mittal, G., Dhand, V., Rhee, K. Y., Park, S.-J., & Lee, W. R. (2015). A Review on Carbon Nanotubes and Graphene as Fillers in Reinforced Polymer Nanocomposites. *Journal of Industrial and Engineering Chemistry*, *21*, 11–25. doi:10.1016/j.jiec.2014.03.022

Mohammed, S. A., & Mossaad, S. H. (2015). Visible Light Communications Using OFDM and Multiple LEDs. *IEEE Transactions on Communications*, 4304–4313.

Molla, A. (2009). Organizational motivations for Green IT: exploring Green IT matrix and motivation models. *Pacific Asia Conference on Information Systems*.

Molla, A., Cooper, V., Corbitt, B., Deng, H., Peszynski, K., Pittayachawan, S., & Teoh, S. Y. (2008). E-readiness to G-readiness: developing a green information technology readiness framework. *Australasian Conference on Information Systems*, *35*, 669-678.

Momoh, J. (2012). *Smart grid: fundamentals of design and analysis* (Vol. 63). Wiley; doi:10.1002/9781118156117

Momoh, J. A. (2009). Smart grid design for efficient and flexible power networks operation and control. In *Power Systems Conference and Exposition* (pp. 1–8). IEEE; doi:10.1109/PSCE.2009.4840074

Momsen, J. (1998). Caribbean Tourism and Agriculture: New Linkages in the Global Era? In T. Klak (Ed.), *Globalization and Neoliberalism: The Caribbean Context*. Oxford, UK: Rowman & Littleman Publishers, Inc.

Mongelli, M., & Scanzio, S. (2016). A neural approach to synchronization in wireless networks with heterogeneous sources of noise. *Ad Hoc Networks*, *49*, 1–16. doi:10.1016/j.adhoc.2016.06.002

Moodie, A. (2015). *Is the Obama Administration lowering the bar on green electronics?* Retrieved March 20, 2016 from http://www.theguardian.com/sustainable-business/2015/apr/23/obama-epeat-epa-technology-green-electronics-council

Moriconi, C. (2001). Il modello virtuale. In A. Soletti, P. Belardi, & F. Cataliotti (Eds.), *Claudio Moriconi professione infografico* (pp. 20–30). Perugia, Italy: Università degli Studi di Perugia.

Moscati, S. (2001). *Civiltà del mare. I fondamenti della storia mediterranea*. Napoli, Italy: Liguori.

Moslehi, K., & Kumar, R. (2010). A reliability perspective of the smart grid. *Smart Grid*, *1*(1), 57–64. doi:10.1109/TSG.2010.2046346

Mousavi, S. M., Zandieh, M., & Yazdani, M. (2013). A simulated annealing/local search to minimize the makespan and total tardiness on a hybrid flowshop. *International Journal of Advanced Manufacturing Technology*, *64*(1-4), 369–388. doi:10.100700170-012-4031-5

Muhlen, S. S. D. D. (2008). New challenges in controlling EMI in hospitals. Proceedings of International Federation for Medical & Biological Engineering, 834-837.

Mukherjee, A. B. (1998). Chromium in the environment of Finland. *The Science of the Total Environment*, *217*(1-2), 9–19. doi:10.1016/S0048-9697(98)00163-6 PMID:9695169

Munda, G. (2005). Measuring sustainability: A multi-criterion framework. *Environment, Development and Sustainability*, *2*(1), 117–134. doi:10.100710668-003-4713-0

Murphy, S. V., & Atala, A. (2014). 3D bioprinting of tissues and organs. *Nature Biotechnology*, *32*(8), 773–785. doi:10.1038/nbt.2958 PMID:25093879

Murugesan, S. (2008). Harnessing Green IT: Principles and practices. *IT Professional*, *10*(1), 24–33. doi:10.1109/MITP.2008.10

Murugesan, S., & Gangadharan, G. (2012). *Harnessing green IT: principles and practices*. Chichester, UK: John Wiley and Sons. doi:10.1002/9781118305393

Naim, R., & Ismail, A. F. (2013). Effect of polymer concentration on the structure and performance of PEI hollow fiber membrane contactor for CO2 stripping. *Journal of Hazardous Materials*, *250*, 354–361. doi:10.1016/j.jhazmat.2013.01.083 PMID:23474409

NASA - NOAA. (1992, August 28). *GOES - R - SERIE.* Retrieved September 06, 2013, from GOES-R Spacecraft Overview: http://www.goes-r.gov/spacesegment/overview.html

NASA. (2013, July 28). *Landsat.* Retrieved August 22, 2013, from Landsat 8: http://www.nasa.gov/mission_pages/landsat/spacecraft/index.html

National Cancer Institute. (n.d.). *SEER*Stat version 5.3.1.* Bethesda, MD: National Cancer Institute. Retrieved from http://www.seer.cancer.gov/seerstat

Negroponte, N. (1995). *Being digital.* New York: Alfred A. Knopf.

Neste, J., & Karjalainen T. P. (2013). *A literature review – The use of multi-criteria decision analysis in Environmental Impact Assessment, Annex 7.2.5.* Inception report IMPERIA EU LIFE11 ENV/FI/905 Action: B1B Report on the use of MCDA in EIA and SEA, 25.4.2013

Neutens, T. (2015). Accessibility, equity and health care: Review and research directions for transport geographers. *Journal of Transport Geography, 43*, 14–27. doi:10.1016/j.jtrangeo.2014.12.006

Neutens, T., Witlox, F., Van De Weghe, N., & De Maeyer, P. (2007). Space–time opportunities for multiple agents: A constraint-based approach. *International Journal of Geographical Information Science, 21*(10), 1061–1076. doi:10.1080/13658810601169873

NGI. (n.d.). Retrieved March 1, 2016, from http://www.ngi.no/en/Geohazards/Research/Stability-of-rock-slopes/

Nguyen, K. K., Cheriet, M., Lemay, M., Reijs, V., Mackarel, A., & Pastrama, A. (2012). Environmental-aware virtual data center network. *Computer Networks, 56*(10), 2538–2550. doi:10.1016/j.comnet.2012.03.008

Nichol, J. E., Shaker, A., & Nichol, J. E. (2006). Application of high-resolution stereo satellite images to detailed landslide hazard assessment. *Geomorphology, 76*(1-2), 68–75. doi:10.1016/j.geomorph.2005.10.001

Noyons, E. C. (2004). Science maps within a science policy context. In H. F. Moed & ... (Eds.), *Handbook of Quantitative Science and Technology Research* (pp. 237–255). New York: Kluwer.

Nykiforuk, C., & Flaman, L. (2011). Geographic information systems (GIS) for health promotion and public health: A review. *Health Promotion Practice, 12*(1), 63–73. doi:10.1177/1524839909334624 PMID:19546198

Oddo, F. (2004). *Squint/Opera. Nuove professionalità per raccontare l'architettura contemporanea.* Retrieved October 3, 2015, from http://architettura.it/files/20040123/index.htm

Ogden, J. (2013). Blog: Green IT – a key to long-term savings. *CIO Asia.* Retrieved June 20, 2016 from http://www.cio-asia.com/tech/data-center/blog-green-it--a-key-for-long-term-savings/?page=1

Oĝuz, C., & Ercan, M. F. (2005). A genetic algorithm for hybrid flow-shop scheduling with multiprocessor tasks. *Journal of Scheduling, 8*(4), 323–351. doi:10.100710951-005-1640-y

Oğuz, C., Ercan, M. F., Cheng, T. E., & Fung, Y. F. (2003). Heuristic algorithms for multiprocessor task scheduling in a two-stage hybrid flow-shop. *European Journal of Operational Research, 149*(2), 390–403. doi:10.1016/S0377-2217(02)00766-X

Ohlmacher, G. C., & Davis, J. C. (2003). Using multiple logistic regression and GIS technology to predict landslide hazard in northeast Kansas, USA. *Engineering Geology, 69*(3-4), 331–343. doi:10.1016/S0013-7952(03)00069-3

OJEU C. 240-1, 24/07/2014 (2014). *Commission notice: Guidelines on recommended standard licences, datasets and charging for the reuse of documents.* Retrieved April 10, 2015, from http://eur-lex.europa.eu/legal-content/EN/TXT/HTML/?uri=CELEX:52014XC0724(01)&from=EN

Olmeda-Gómez, C. (2014). Visualización de información [Information visualization]. *El Profesional de la Información*, *23*(3), 213–219. doi:10.3145/epi.2014.may.01

Opara, L. U. (2003). Traceability in agriculture and food supply chain: a review of basic concepts, technological implications, and future prospects. *Agriculture and Environment*, *1*(1), 101–106.

Orduña-Malea, E., & Aguilló, I. F. (2014). *Cibermetría: midiendo el espacio red.* Barcelona: Editorial UOC.

Ortner, D. J., Butler, W., Cafarella, J., & Milligan, L. (2001). Evidence of probable scurvy in subadults from archeological sites in North America. *American Journal of Physical Anthropology*, *114*(4), 343–351. doi:10.1002/ajpa.1046 PMID:11275963

Oskoei, M. A., & Hu, H. (2007). Myoelectric control systems-a survey. *Biomedical Signal Processing and Control*, *2*(4), 275–294. doi:10.1016/j.bspc.2007.07.009

Oskoei, M. A., & Hu, H. (2008). Support vector machine-based classification scheme for myoelectric control applied to upper limb. *IEEE Transactions on Bio-Medical Engineering*, *55*(8), 1956–1965. doi:10.1109/TBME.2008.919734 PMID:18632358

Otto Bock Gmb. (2012). Retrieved from http://www.ottobock.com/

Ouaarab, A., Ahiod, B., & Yang, X. S. (2014). Discrete cuckoo search algorithm for the travelling salesman problem. *Neural Computing & Applications*, *24*(7-8), 1659–1669. doi:10.100700521-013-1402-2

Ouma, Y. O., Opudo, J., & Nyambenya, S. (2015). Comparison of Fuzzy AHP and Fuzzy TOPSIS for Road Pavement Maintenance Prioritization: Methodological Exposition and Case Study. *Advances in Civil Engineering*, Article ID 140189. .10.1155/2015/140189

Ouyang, X., Zhou, Y., Luo, Q., & Chen, H. (2013). A novel discrete cuckoo search algorithm for spherical traveling salesman problem. *Applications of Mathematics*, *7*(2), 777–784.

Pacciarelli, M. (2000). *Dal villaggio alla città. La svolta protourbana del 1000 a.C. nell'Italia tirrenica. Grandi contesti e problemi della Protostoria Italiana*. Firenze, Italy: All'Insegna del Giglio.

Pachauri, A. K., Gupta, P. V., & Chander, R. (1998). Landslide zoning in a part of the Garhwal Himalayas. *Environ Geol.*, *36*(3-4), 325–334. doi:10.1007002540050348

Pallottino, M. (1937). Tarquinia. *Monumenti Antichi della R. Accademia Nazionale dei Lincei*. 36.

Pandey, O. K., & Chakraborty, R. (2007). *GIS: Information technology for planning management and development of GOA tourism master plan in next millennium*. Retrieved on May 20, 2015. http://www.gisdevelopment.net/application

Pandi, K. M., & Somasundaram, K. (2016). Energy efficient in virtual infrastructure and green cloud computing: A review. *Indian Journal of Science & Technology*, *9*(11), 1–8. doi:10.17485/ijst/2016/v9i11/89399

Pan, R. K., Kaski, K., & Fortunato, S. (2012). World citation and collaboration networks: Uncovering the role of geography in science. *Scientific Reports*, *2*, 902. doi:10.1038rep00902 PMID:23198092

Papagiannakis, G., Singh, G., & Magnenat-Thalmann, N. (2008). A survey of mobile and wireless technologies for augmented reality systems. *Computer Animation and Virtual Worlds*, *19*(1), 3–22. doi:10.1002/cav.221

Parisi, D. (2001). *Simulazioni. La realtà rifatta nel computer*. Bologna, Italy: Il Mulino.

Park, C.-S., & Kim, H.-J. (2012). A framework for construction safety management and visualization system. *Automation in Construction.*

Park, C.-S., Lee, D.-Y., Kwon, O.-S., & Wang, X. (2012). A Framework for Proactive Construction Defect Management using BIM, Augmented Reality and Ontology-based Data Collection Template. *Automation in Construction.*

Parker, R., & Dominguez, J. (2015). Features to Look for in a GIS Viewer. *GeoInformatics, 18*(4), 24.

Parsons, J. R., Le, C. K., Langrana, N. A., Clemow, A. J., & Chen, E. H. (1992). *Functional and Biocompatible Intervertebral Disc Spacer Containing Elastomeric Material of Varying Hardness.* United States Patent.

Pengfei, L., e. a. (2014). Fundamental Analysis of a Car to Car Visible Light Communication System. *9th International Symposium on Communication Systems, Networks & Digital Sign (CSNDSP).*

Penttilä, H. (2006). *Describing the changes in architectural information technology to understand design complexity and free-form architectural expression.* ITcon.

Perez-Jimenez, R. R.-H. (2011). Visible Light Communication Systems for Passenger In-Flight Data Networking. *IEEE International Conference on Consumer Electronics (ICCE)*, 445-446. 10.1109/ICCE.2011.5722675

Peroni, R. (1969). Per uno studio dell'economia di scambio in Italia nel quadro dell'ambiente culturale dei secoli intorno al 1000 a.C. *La Parola del Passato, 125*, 134–160.

Peroni, R. (1994). *Introduzione alla protostoria italiana.* Roma-Bari, Italy: Laterza.

Peroni, R. (1996). *L'Italia alle soglie della storia.* Roma-Bari, Italy: Laterza.

Peter, V., & Kinscö, I. (2013). *Developing an evaluation and progress methodology to underpin the intervention logic of the Action Plan to Boost Demand for European Innovations.* EU.

Petrick, I. J., & Simpson, T. W. (2013). 3D printing disrupts manufacturing: How economies of one create new rules of competition. *Research-Technology Management, 56*(6), 12–16. doi:10.5437/08956308X5606193

Phinyomark, A., Phukpattaranont, P., & Limsakul, C. (2012). Feature reduction and selection for EMG signal classification. *Expert Systems with Applications, 39*(8), 7420–7431. doi:10.1016/j.eswa.2012.01.102

Phinyomark, A., Quaine, F., Charbonnier, S., Serviere, C., Tarpin-Bernard, F., & Laurillau, Y. (2013). EMG feature evaluation for improving myoelectric pattern recognition robustness. *Expert Systems with Applications, 40*(12), 4832–4840. doi:10.1016/j.eswa.2013.02.023

Phys.org. (2014). *German Supercomputer is a world champion in saving energy.* Retrieved March 10, 2016 from http://phys.org/news/2014-11-german-supercomputer-world-champion-engergy.html

Pierce, T., Dobson, G., & Phillips, M. (2009). GIS Mashups on the Web: Using Google, ESRI, and Open Source to Develop Weather and Climate Applications for the Web. *Geospatial Infrastructure Solutions Conference*, Tampa FL

Pilavachi, P. A., Chatzipanagi, A. I., & Spyropoulou, A. I. (2009). Evaluation of hydrogen production methods using the analytic hierarchy process. *International Journal of Hydrogen Energy, 34*(13), 5294–5303. doi:10.1016/j.ijhydene.2009.04.026

Pillay, M., & Rogerson, C. M. (2013). Agriculture-Tourism Linkages and Pro-Poor Impacts: The Accommodation Sector of Urban Coastal KwaZulu-Natal, South Africa. *Applied Geography (Sevenoaks, England), 36*, 49–58. doi:10.1016/j.apgeog.2012.06.005

Pinde, F., & Sun, J. (2011). *Web GIS, Principles and Applications.* ESRI Press.

Pinedo, M. (1995). *Scheduling: theory, algorithms, and systems.* Englewood Cliffs, NJ: Prentice-Hall.

Pinnock, C. B., Meier, E. M., Joshi, N. J., Wu, B., & Lam, M. T. (2016). Customizable Engineered Blood Vessels using 3D Printed Inserts. *Methods (San Diego, Calif.)*, *99*, 20–27. doi:10.1016/j.ymeth.2015.12.015 PMID:26732049

Pompe, W., Worch, H., Epple, M., Friess, W., Gelinsky, M., Greil, P., ... Schulte, K. (2003). Functionally Graded Materials for Biomedical Applications. *Materials Science and Engineering A*, *362*(1-2), 40–60. doi:10.1016/S0921-5093(03)00580-X

Popoola, W. O., & Ghassemlooy, Z. (2009). BPSK subcarrier intensity modulated free-space. *Journal of Lightwave Technology*, *27*(8), 967–973. doi:10.1109/JLT.2008.2004950

Poulsen, K. (2004a). *Software bug contributed to blackout*. Retrieved from http://www.securityfocus.com/news/8016

Poulsen, K. (2004b). *Tracking the blackout bug*. Retrieved from http://www.securityfocus.com/news/8412

Pourabbas, E. (2014). *Geographical Information Systems: Trends and Technologies*. Boca Raton, FL: CRC Press. doi:10.1201/b16871

Pourebrahim, S., Hadipour, M., Mokhtar, M. B., & Taghavi, S. (2014). Application of VIKOR and fuzzy AHP for conservation priority assessment in coastal areas: Case of Khuzestan district, Iran. *Ocean and Coastal Management*, *98*, 20–26. doi:10.1016/j.ocecoaman.2014.05.009

Powell, D. (Ed.). (1991). *Delta-4: A Generic Architecture for Dependable Distributed Computing*. Springer-Verlag. doi:10.1007/978-3-642-84696-0

Pradhan, B. (2013). A comparative study on the predictive ability of the decision tree, support vector machine and neuro-fuzzy models in landslide susceptibility mapping using GIS. *Computers & Geosciences*, *51*, 350–365. doi:10.1016/j.cageo.2012.08.023

Prideaux, B. (2002). Building Visitor Attractions in Peripheral Areas: Can Uniqueness Overcome Isolation to Produce Viability? *International Journal of Tourism Research*, *4*(5), 379–389. doi:10.1002/jtr.387

Pruss-Ustun, A., & Corvalan, C. (2006). *Preventing disease through healthy environments: Towards an estimate of environmental burden of disease*. WHO Publication.

Puckett, J., Byster, L., Westervelt, S., Gutierrez, R., Davis, S., Hussain, A., & Dutta, M. (2002). *Exporting Harm - The High-Tech Trashing of Asia; The Basel Action Network (BAN) Silicon Valley Toxics Coalition*. Seattle, WA: SVTC.

Rabin, T. (1998). A simplified approach to threshold and proactive RSA. *Proceedings of the 18th Annual International Cryptology Conference*, 89-104. 10.1007/BFb0055722

Rahaman, M. (2010). Application of GIS in EcoTourism Developemnt: A Case Study in Sundarbans, Bangladesh. *International Journal of Entrepreneurial Behaviour And Research*, *10*(4),247-59.

Rahim, N. A., Ghasem, N., & Al-Marzouqi, M. H. (2014). Stripping of CO_2 from different aqueous solvents using PVDF hollow fiber membrane contacting process. *Journal of Natural Gas Science and Engineering*, *21*, 886–893. doi:10.1016/j.jngse.2014.10.016

Rahim, N. A., Ghasem, N., & Al-Marzouqi, M. H. (2015). Absorption of CO_2 from natural gas using different amino acid salt solutions and regeneration using hollow fiber membrane contactors. *Journal of Natural Gas Science and Engineering*, *26*, 108–117. doi:10.1016/j.jngse.2015.06.010

Ramachandra, T. V., & Saira, V. K. (2004, March). Environmentally sound options for waste management, *Envis Journal of Human Settlements*. Retrieved March 2, 2016 from http://www.ces.iisc.ernet.in/energy/paper/ewaste/ewaste.html

Ramesh, L., Chowdhury, S., Chowdhury, S., Natarajan, A., & Gaunt, C. (2009). Minimization of power loss in distribution networks by different techniques. *Energy and Power Engineering*, *2*(1).

Ramirez-Iniguez, R. S. M. (2008). Optical Wireless Communications: IR for Wireless Connectivity. Boca Raton, FL: CRC Press.

Rashidi, E., Jahandar, M., & Zandieh, M. (2010). An improved hybrid multi-objective parallel genetic algorithm for hybrid flow shop scheduling with unrelated parallel machines. *International Journal of Advanced Manufacturing Technology, 49*(9-12), 1129–1139. doi:10.100700170-009-2475-z

Ratto, M., & Ree, R. (2012). Materializing information: 3D printing and social change. *First Monday, 17*(7). doi:10.5210/fm.v17i7.3968

Ray, N., Das, D. K., & Chaudhuri, S. (2014). Tourism Promotion Through Web. In Strategic Marketing in Fragile Economic Condition. IGI Global.

Ray, N., Das, D. K., Ghosh, S., & Banerjee, A. (2011). A Study on Application of Information Technology on Tourism Development, IEEE Proceeding Catalog No: CFP1141H-ART ISBN978-1-4244-6579- China

Reaz, M. B. I., Hussain, M. S., & Mohd-Yasin, F. (2006). Techniques of EMG signal analysis: Detection, processing, classification and applications. *Biological Procedures Online, 8*(1), 11–35. doi:10.1251/bpo115 PMID:19565309

Reiter, R. J. (1991). Pineal melatonin: Cell biology of its synthesis and its physiological interactions. *Endocrine Reviews, 12*(2), 151–180. doi:10.1210/edrv-12-2-151 PMID:1649044

Ren, F., & Kwan, M.-P. (2009). The impact of the Internet on human activity–travel patterns: Analysis of gender differences using multi-group structural equation models. *Journal of Transport Geography, 17*(6), 440–450. doi:10.1016/j.jtrangeo.2008.11.003

Renfrew, C., & Bahn, P. (1995). *Archeologia. Teorie, Metodi, Pratica.* Bologna, Italy: Zanichelli.

Rentschler, M., & Laukemann, P. (2012). Towards a reliable parallel redundant WLAN black channel. In *9th IEEE International Workshop on Factory Communication Systems (WFCS)* (pp. 255-264). Lemgo, Germany. 10.1109/WFCS.2012.6242573

Report, W. (2013). *State of the industry, Annual worldwide progress report, Wohlers Associates Inc.* United States of America.

Resnik, L. (2010). Research update: VA study to optimize DEKA arm. *Journal of Rehabilitation Research and Development, 47*(3), ix–x. doi:10.1682/JRRD.2010.03.0034 PMID:20665342

Rey, C. (1988). *Monarch of all I survey: Bechunaland diaries 1929 - 1937.* Gaborone: Botswana Society: Lilian Barber Press Inc.

Reza, B., Sadiq, R., & Hewage, K. (2011). Sustainability assessment of flooring systems in the city of Tehran: An AHP-based life cycle analysis. *Construction & Building Materials, 25*(4), 2053–2066. doi:10.1016/j.conbuildmat.2010.11.041

Rhea, S., Eaton, P., Geels, D., Weatherspoon, H., Zhao, B., & Kubiatowicz, J. (2003). Pond: the oceanstore prototype. *Proceedings of the 2nd USENIX Conference on File and Storage Technologies* (Vol. 3, pp. 1-14).

Riane, F., Artiba, A., & Elmaghraby, S. E. (1998). A hybrid three-stage flowshop problem: Efficient heuristics to minimize makespan. *European Journal of Operational Research, 109*(2), 321–329. doi:10.1016/S0377-2217(98)00060-5

Ribas, I., Leisten, R., & Framiñan, J. M. (2010). Review and classification of hybrid flow shop scheduling problems from a production system and a solutions procedure perspective. *Computers & Operations Research, 37*(8), 1439–1454. doi:10.1016/j.cor.2009.11.001

Rideout, B. (2012). Printing the impossible triangle: The copyright implications of three-dimensional printing. *The Journal of Business. Entrepreneurship & the Law*, *5*(1), 6.

Ripley, B., Kelil, T., Cheezum, M. K., Goncalves, A., Carli, M. F. D., Rybicki, F. J., ... Blankstein, R. (2016). 3D Printing based on Cardiac CT Assists Anatomic Visualization prior to Transcatheter Aortic Valve Replacement. *Journal of Cardiovascular Computed Tomography*, *10*(1), 28–36. doi:10.1016/j.jcct.2015.12.004 PMID:26732862

RMTRR. (2008). *Rocky Mountain Tree-Ring Research (RMTRR)*. Retrieved from http://www.rmtrr.org/gallery.html

Robert, N. P. (1996). *Cancer wars: How politics shapes what we know and don't know about cancer*. New York: Basic Books.

Robinson, W. J., & Evans, R. (1980). A microcomputer-based tree-ring measuring system. *Tree-Ring Bulletin*, *40*, 59–64.

Roch, A., Greifzu, M., Talens, E. R., Stepien, L., Roch, T., Hege, J., ... Leson, A. (2015). Ambient Effects on the Electrical Conductivity of Carbon Nanotubes. *Carbon*, *95*, 347–353. doi:10.1016/j.carbon.2015.08.045

Roche, L. M. (1998). *Breast cancer in New Jersey, 1979-1995*. New Jersey Dept. of Health and Senior Services, Cancer Surveillance Program, Office of Cancer Epidemiology.

Rodrigues, F., Osiro, L., & Ribeiro, L. C. (2014). A comparison between Fuzzy AHP and Fuzzy TOPSIS methods to supplier selection. *Applied Soft Computing*, *21*, 194–209. doi:10.1016/j.asoc.2014.03.014

Rodríguez, E., Carmona, M., & Soto, J. (2005). *Memoria sostenibilidad del hospital clínico San Carlos*. Madrid, Spain: Hospital Clínico San Carlos.

Rogers, E. N. (1990). *Gli elementi del fenomeno architettonico*. Napoli, Italy: Guida Editore.

Rose, N. (2014). The Human Brain Project: Social and ethical challenges. *Neuron*, *82*(6), 1212–1215. doi:10.1016/j.neuron.2014.06.001 PMID:24945767

Rossiter, D. G. (1994). *Land Evaluation*. Cornell University, College of Agriculture & Life Sciences, Department of Soil, Crop & Atmosferic Sciences. Retrieved August 1994, from http://www.css.cornell.edu/faculty/dgr2/docs/Scas494/s494ch1.pdf

Rottoli, M. (1997). I resti vegetali. In M. Bonghi Jovino & C. Chiaramonte Treré (Eds.), *Tarquinia. Testimonianze archeologiche e ricostruzione storica: scavi sistematici dell'abitato. Campagne 1982-1988* (Vol. 1-2, pp. 92–96). Roma, Italy: Tarchna.

Ruiz, R., & Maroto, C. (2006). A genetic algorithm for hybrid flowshops with sequence dependent setup times and machine eligibility. *European Journal of Operational Research*, *169*(3), 781–800. doi:10.1016/j.ejor.2004.06.038

Ruiz, R., & Vázquez-Rodríguez, J. A. (2010). The hybrid flow shop scheduling problem. *European Journal of Operational Research*, *205*(1), 1–18. doi:10.1016/j.ejor.2009.09.024

Russell, E. (1998). *People and the land through time: Linking ecology and history*. New Haven, CT: Yale University Press.

Rutten, J., Kwiatkowska, M. G. N., & Parker, D. (2004). *Mathematical techniques for analyzing concurrent and probabilistic systems*. American Mathematical Society.

Sabins, F. F. (1996). *Remote Sensing: Principles and Interpretation* (3rd ed.). W. H. Freeman and Company, New York Press.

Sachs, E., Cima, M., Williams, P., Brancazio, D., & Cornie, J. (1992). Three dimensional printing: Rapid tooling and prototypes directly from a CAD model. *Journal of Engineering for Industry*, *114*(4), 481–488. doi:10.1115/1.2900701

Salomons, W., & Frostner, U. (1984). Metals in hydrocycle. Springer-Verlag. doi:10.1007/978-3-642-69325-0

Salvado, M. F., Azevedo, S. G., Matias, J. C. O., & Ferreira, L. M. (2015). Proposal of a Sustainability Index for the Automotive Industry. *Sustainability*, *7*(2), 2113–2144. doi:10.3390u7022113

Samel, A., & Wegmann, H. M. (1997). Bright light: A countermeasure for jet lag? *Chronobiology International*, *14*(2), 173–183. doi:10.3109/07420529709001154 PMID:9095377

Sanghera, B., Naique, S., Papaharilaou, Y., & Amis, A. (2001). Preliminary Study of Rapid Prototype Medical Models. *Rapid Prototyping Journal*, *7*(5), 275–284. doi:10.1108/13552540110410486

Sannomiya, E. K., Brito, A. A., & Agelieri, F. (2008). Surgical Planning for Resection of an Ameloblastoma and Reconstruction of the Mandible Using a Selective Laser Sintering 3D Biomodel. *Oral Surgery, Oral Medicine, Oral Pathology, Oral Radiology, and Endodontics*, *106*(1), 36–40. doi:10.1016/j.tripleo.2008.01.014 PMID:18585610

Santoyo-Castelazo, E., & Azapagic, A. (2014). Sustainability assessment of energy systems: Integrating environmental, economic and social aspects. *Journal of Cleaner Production*, *80*, 1–20. doi:10.1016/j.jclepro.2014.05.061

Sarkar, S., & Kanungo, D. P. (2004). An integrated approach for landslide susceptibility mapping using remote sensing and GIS. *Photogrammetric Engineering and Remote Sensing*, *70*(5), 617–625. doi:10.14358/PERS.70.5.617

Sarkar, S., Kanungo, D. P., & Mehrotra, G. S. (1995). Landslide hazard zonation: A case study in Garhwal Himalaya, India. *Mountain Research and Development*, *15*(4), 301–309. doi:10.2307/3673806

Sassatelli, G. (2009). Gli Etruschi e la loro economia. In V. Castronovo (Ed.), *Storia dell'economia mondiale. Dall'antichità al Medioevo. Dal Neolitico agli albori del I millennio* (Vol. 1-2, pp. 99–116). Roma-Bari, Italy: Laterza.

Sasson, A. M. (1969). Nonlinear programming solutions for load-flow, minimum-loss, and economic dispatching problems. *Power Apparatus and Systems*, (4), 399–409.

Satellite Imaging Coorporation. (2001). *Satellites*. Retrieved August 22, 2013, from Satallite Data: http://www.satimagingcorp.com/satellite-sensors/spot-5.html

Saxena, D., Singh, S., & Verma, K. (2010). Application of computational intelligence in emerging power systems. Engineering. *Science and Technology*, *2*(3), 1–7.

Scanzio, S. (2012). SoftPLC-Based Control: A Comparison between Commercial and Open-Source EtherCAT Technologies. In M. Khan & A. Ansari (Eds.), *Handbook of Research on Industrial Informatics and Manufacturing Intelligence: Innovations and Solutions* (pp. 440–463). Hershey, PA: Information Science Reference. doi:10.4018/978-1-4666-0294-6.ch019

Scaramozzino, J., White, R., Essic, J., Fullington, L. A., Mistry, H., Henley, A., & Olivares, M. (2014). Map room to data and GIS services: Five university libraries evolving to meet campus needs and changing technologies. *Journal of Map & Geography Libraries*, *10*(6), 6–47. doi:10.1080/15420353.2014.893943

Schäffer, B., Baranski, B., & Foerster, T. (2010). *Towards spatial data infrastructures in the clouds. In Geospatial Thinking* (pp. 399–418). Springer.

Schleich, J., Klobasa, M., Gölz, S., & Brunner, M. (2013). Effects of feedback on residential electricity demand – Findings from a field trial in Austria. *Energy Policy*, *61*, 1097–1106. doi:10.1016/j.enpol.2013.05.012

Schluepa, M., Hageluekenb, C., Kuehrc, R., Magalinic, F., Maurerc, C., Meskersb, C., . . . Wang, F. (2009). *Recycling: From e-waste to resources, Sustainable Innovation and Technology Transfer Industrial Sector Studies (Nairobi and Bonn, UNEP and STeP)*. Retrieved March 2, 2016 from http://www.unep.org/pdf/Recycling_From_e-waste_to_resources.pdf

Schneider, K. P., Chassin, D., Chen, Y., & Fuller, J. C. (2009). Distribution power flow for smart grid technologies. In Power Systems Conference and Exposition, 2009. PSCE'09. IEEE/PES, (pp. 1-7). IEEE. doi:10.1109/PSCE.2009.4840078

Schneider, F. (1990). Implementing fault-tolerant services using the state machine approach: A tutorial. *ACM Computing Surveys*, *22*(4), 299–319. doi:10.1145/98163.98167

Schowengerdt, R. A. (2007). *Remote sensing: models and methods for image processing*. Burlington, MA: Academic Press.

Schumacher, P. (2004). *Hadid digitale. Paesaggi in movimento*. Torino, Italy: Testo & Immagine.

Schutz, A., Olsson, M., Jensen, A., Gerhardsson, L., Borjesson, J., Mattsson, S., & Skerfving, S. (2005). Lead in finger bone, whole blood, plasma and urine in lead-smelter workers: Extended exposure range. *International Archives of Occupational and Environmental Health*, *78*(1), 35–43. doi:10.100700420-004-0559-5 PMID:15750821

Schwartz, C. (2015, March). 3-D Printing: The Potential Implications and Challenges for Law Enforcement. *The Police Chief*, *82*, 64–69.

Schwarzer, S., Giuliani, A., De Bono, G., Kluser, S., & Peduzzi, P. (2005). *"E-waste: the hidden side of IT equipment's manufacturing and use". Early Warnings on Emerging Environmental Threats, No. 5, United Nations Environment Programme (UNEP)*. GRID Europe.

Schweingruber, F. H. (1988). *Tree Rings: Basics and Applications of Dendrochronology*. Dordrecht: Reidel; doi:10.1007/978-94-009-1273-1

Scientometrics. (2016). Retrieved from https://en.wikipedia.org/wiki/scientometrics

Scimago Institutions Rankings. (n.d.). Retrieved from http://www.scimagoir.com/

Scott, St. G., & Nielsen, E. (2002). Dendrochronological analysis of a coffin board recovered from Pukatawagan, Bay, Manitoba. Terrain Sciences Division, Geological Survey of Canada and Manitoba Geological Survey, Manitoba Industry, Trade and Mines.

Seaton, A. V. (1999). Book Towns as Tourism Developments in Peripheral Areas. *International Journal of Tourism Research*, *1*(5), 389–399. doi:10.1002/(SICI)1522-1970(199909/10)1:5<389::AID-JTR204>3.0.CO;2-0

Sebelius, F., Eriksson, L., Holmberg, H., Levinsson, A., Lundborg, G., Danielsen, N., ... Montelius, L. (2005). Classification of motor commands using a modified self-organising feature map. *Medical Engineering & Physics*, *27*(5), 403–413. doi:10.1016/j.medengphy.2004.09.008 PMID:15863349

Sedighizadeh, M., & Rezazadeh, A. (2008). Using genetic algorithm for distributed generation allocation to reduce losses and improve voltage profile. World Academy of Science. *Engineering and Technology*, *37*, 251–256.

Segundo, E. (2010, September 01). *Boeing*. Retrieved December 26, 2015, from GOES-15 Weather Satellite: http://boeing.mediaroom.com/2010-09-01-Boeing-built-GOES-15-Weather-Satellite-Enters-Service-for-NASA-NOAA

Seo, D.-J., & Hong, W.-H. (1996). Solubilities of carbon dioxide in aqueous mixtures of diethanolamine and 2-amino-2-methyl-1-propanol. *Journal of Chemical & Engineering Data*, *41*(2), 258–260. doi:10.1021/je950197o

Seol, Y. J., Kang, H. W., Lee, S., Atala, A., & Yoo, J. J. (2014). Bioprinting technology and its applications. *European Journal of Cardio-Thoracic Surgery*, *46*(3), 342–348. doi:10.1093/ejcts/ezu148 PMID:25061217

Sevani, V., Raman, B., & Joshi, P. (2014). Implementation-Based Evaluation of a Full-Fledged Multihop TDMA-MAC for WiFi Mesh Networks. *IEEE Transactions on Mobile Computing*, *13*(2), 392–406. doi:10.1109/TMC.2012.251

Shah, G., Molina, A., & Blaze, M. (2006). Keyboards and covert channels. *Proceedings of the 15th USENIX Security Symposium*, 59–75.

Shahmiri, R., Das, R., Aarts, J. M., & Bennani, V. (2014). Finite element analysis of an implant-assisted removable partial denture during bilateral loading: Occlusal rests position. *The Journal of Prosthetic Dentistry*, *5*(5), 1126–1133. doi:10.1016/j.prosdent.2014.04.023 PMID:24951387

Shah, R. (2012). Eco-friendly IT: Greener approach to IT. *Journal of Management and Business Research*, *2*(2), 4–25.

Shanghai Kesheng Prosthese Corporation Ltd. (2012). Retrieved from http://www.keshen.com/index-en.asp

Shanker, S. (2008). *ICT and Tourism: Challenges and Opportunities, Conference on Tourism in India – Challenges Ahead.* IIMK.

Shanzhen, Y., Lizhu, Z., Chunxiao, X., Qilun, L., & Yong, Z. (2001). Semantic and Interoperable WebGIS. *Proceedings of the Second International Conference on Web Information Systems Engineering*, 2, 42-47. 10.1109/WISE.2001.996702

Sharkey, D. A., & Momsen, J. (1995). Tourism in Dominica, West Indies: Problems and Prospects. *Caribbean Geography*, *6*(1), 40–51.

Shaw, S.-L., & Yu, H. (2009). A GIS-based time-geographic approach of studying individual activities and interactions in a hybrid physical–virtual space. *Journal of Transport Geography*, *17*(2), 141–149. doi:10.1016/j.jtrangeo.2008.11.012

Shen, B., Ghatikar, G., Lei, Z., Li, J., Wikler, G., & Martin, P. (2014). The role of regulatory reforms, market changes, and technology development to make demand response a viable resource in meeting energy challenges. *Applied Energy*, *130*, 814–823. doi:10.1016/j.apenergy.2013.12.069

Sheppard, P. R. (2013). *Crossdating Tree Rings.* Laboratory of Tree-Ring Research, The University of Arizona. Retrieved from http://www.ltrr.arizona.edu/skeletonplot/applications.htm

Shin, D. H., Park, J., Woo, S., & Jang, W.-S. (2012). Representations for imagining the scene of non-existing buildings in an existing environment. *Automation in Construction.*

Shoemaker, H. G., & Shoemaker, E. M. (1987). *In the Footsteps of John Wesley Powell.* Boulder, CO: Johnson Books and Powell Society.

Shoup, V. (2000). Practical threshold signatures. *Lecture Notes in Computer Science*, *1097*, 207–220. doi:10.1007/3-540-45539-6_15

Shuai, Z., Wang, L., & Song, C. (2012). Theory of Charge Transport in Carbon Electronic Materials. Springer. doi:10.1007/978-3-642-25076-7

Siabato, W., Claramunt, C., & Bernabé-Poveda, M. A. (2013). An interactive bibliography on temporal GIS. *Journal of Spatial Information Science*, *7*, 99–101.

Siabato, W., Claramunt, C., Manso-Callejo, M. Á., & Bernabé-Poveda, M. Á.TimeBliography. (2014, December). A dynamic and online bibliography on temporal GIS. *Transactions in GIS*, *18*(6), 799–816. doi:10.1111/tgis.12080

Siano, P. (2014). Demand response and smart grids – A survey. *Renewable & Sustainable Energy Reviews*, *30*, 461–478. doi:10.1016/j.rser.2013.10.022

Siegel, R. L., Miller, K. D., & Jemal, A. (2015). Cancer statistics, 2015. *CA: a Cancer Journal for Clinicians*, *65*(1), 5–29. doi:10.3322/caac.21254 PMID:25559415

Sigala, M. (2008). *Web 2.0, social marketing strategies and distribution channels for city destinations.* Academic Press.

Sikdar, M., & Vaniya, S. (2014). The New Millennium and Emerging Concerns. *International Journal of Scientific and Research Publications, 4*(2).

Sims, R. (2009). Food, Place and Authenticity: Local Food and the Sustainable Tourism Experience. *Journal of Sustainable Tourism, 17*(3), 321–336. doi:10.1080/09669580802359293

Sindiga, I. (1996). International Tourism in Kenya and the Marginalization of the Swahili. *Tourism Management, 17*(6), 425–432. doi:10.1016/0261-5177(96)00051-9

Singare, S., Lian, Q., Wang, W. P., Wang, F., Liu, Y. X., Li, D. C., & Lu, B. H. (2009). Rapid Prototyping Assisted Surgery Planning and Custom Implant Design. *Rapid Prototyping Journal, 15*(1), 19–23. doi:10.1108/13552540910925027

Sissons, A., & Thompson, S. (2012). *Three Dimensional Policy: Why Britain needs a policy framework for 3D printing.* London: The Big Innovation Centre.

Skene, D. J. (2003). Optimization of light and melatonin to phase-shift human circadian rhythms. *Journal of Neuroendocrinology, 15*(4), 438–441. doi:10.1046/j.1365-2826.2003.01006.x PMID:12622847

Slember, J., & Narasimhan, P. (2006). Living with nondeterminism in replicated middleware applications. *Proceedings of the ACM/IFIP/USENIX 7th International Middleware Conference*, 81–100. 10.1007/11925071_5

Slocum, T. A., McMaster, R. B., Kessler, F. C., & Howard, H. H. (2008). *Thematic Cartography and Geovisualization* (3rd ed.). Pearson Prentice Hall.

Small, H. (1999). Visualizing science by citation mapping. *Journal of the American Society for Information Science, 50*(9), 799–813. doi:10.1002/(SICI)1097-4571(1999)50:9<799::AID-ASI9>3.0.CO;2-G

Smart Energy Demand Coalition. (2015). *Mapping Demand Response in Europe Today 2015*. Retrieved July 20th, 2016 at: http://www.smartenergydemand.eu/wp-content/uploads/2015/09/Mapping-Demand-Response-in-Europe-Today-2015.pdf

Smith, T., Sonnenfeld, D. A., & Naguib Pellow, D. (Eds.). (2006). Challenging the chip: Labor rights and environmental justice in the global electronics industry. Philadelphia, PA: Temple University Press.

Smith, P., Bustamante, M., Ahammad, H., Clark, H., Dong, H., Elsiddig, E. A., ... Tubiello, F. (2014). Agriculture, Forestry and Other Land Use (AFOLU). In *Climate Change 2014: Mitigation of Climate Change. Contribution of Working Group III to the Fifth Assessment Report of the Intergovernmental Panel on Climate Change.* Cambridge, UK: Cambridge University Press.

Soeters, R., & Van Westen, C. J. (1996). Slope instability recognition, analysis, and zonation. In A. K. Turner & R. L. Shuster (Eds.), *Landslides Investigation and mitigation, Transp. Res. Board, Special Report 247, Natural Acad* (pp. 129–177). Washington, DC: Press.

Spallone, R., Lo Turco, M., & Sanna, M. (2010). From the place of research to the research of place: new digital technologies for simulating future architectures in building environments. *Proceedings of ARCC 2010. International Conference on Architectural Research.*

Spallone, R. (2014). Vero, falso, finto. Rappresentazioni digitali di paesaggi urbani in trasformazione fra comunicazione e seduzione [True, false, fake. Digital representations of urban landscapes subject to transformation between communication and seduction]. *DisegnareCon, 13*, 1–8.

Spallone, R., Lo Turco, M., & Sanna, M. (2009a). Representing ideas by animated digital models in architectural competitions. *Proceedings of ARCC 2009. Leadership in Architectural Research, between academia and the profession.*

Spallone, R., Lo Turco, M., & Sanna, M. (2009b). Dimension "time" to represent contemporary architectural design. *Proceedings of IDCT. Innovative Design & Construction Technologies. Building complex shapes and beyond*, 243-258.

Spatial Analytics. (2016). Retrieved from https://en.wikipedia.org/wiki/spatial_analytics

Sreeprasad, T. S., & Berry, V. (2013). How Do the Electrical Properties of Graphene change with Its Functionalization? *Small*, *9*(3), 341–350. doi:10.1002mll.201202196 PMID:23169614

SSL. (2013). *SPACE System Lorel*. Retrieved August 22, 2013, from GOES I: http://www.ssloral.com/html/satexp/goes.html

Stahl, H. (2015). *3D Printing – Risks and Opportunities*. Retrieved from http://www.oeko.de/oekodoc/1888/2013-532-en.pdf

Stanco, F., Battiato, S., & Gallo, G. (2011). *Digital Imaging for Cultural Heritage Preservation: Analysis, Restoration, and Reconstruction of Ancient Artworks*. CRC Press.

Standard 802.15.7. (2015, Nov 1). Retrieved from https://standards.ieee.org/develop/project/802.15.7.html

Stewart, B. W., & Kliehues, P. (2003). *World Cancer Report*. Lyon: IARC Press.

Stott, B. (1974). Review of load-flow calculation methods. *Proceedings of the IEEE*, *62*(7), 916–929. doi:10.1109/PROC.1974.9544

Stout, R. W., & Crawford, V. (1991). Seasonal variations in fibrinogen concentration among elderly people. *Lancet*, *338*(8758), 9–13. doi:10.1016/0140-6736(91)90004-9 PMID:1676131

Stratasys. (2015). *Digital Dentistry 3D printing makes digital dentistry happen*. Retrieved from http://www.stratasys.com/industries/dental

Strickland, P. T., & Kensler, T. W. (1995). Chemical and physical agents in our environment. In Clinical Oncology. Churchill Livingston Inc.

Stringer, C. (2001). Modern Human Origins: Progress and Prospects. *The Royal Society*, *357*, 563–579. PMID:12028792

Subasi, A. (2012). Classification of EMG signals using combined features and soft computing techniques. *Applied Soft Computing*, *12*(8), 2188–2198. doi:10.1016/j.asoc.2012.03.035

Subburaj, S., Kulkarni, S., & Jia, L. (2014). Green IT: Sustainability by aligning business requirements with IT resource utilization. *International Journal of Communication Networks and Distributed Systems*, *12*(1), 30–46. doi:10.1504/IJCNDS.2014.057986

Su, C.-L. (2005). Probabilistic load-flow computation using point estimate method. *Power Systems*, *20*(4), 1843–1851. doi:10.1109/TPWRS.2005.857921

Sum, E. Y. L. (1991). The recovery of metals from electronic scrap. *Journal of Metallurgy*, *43*, 53–61.

Sun, D. I., Ashley, B., Brewer, B., Hughes, A., & Tinney, W. F. (1984). Optimal power flow by newton approach. *Power Apparatus and Systems*, (10), 2864–2880.

Sun, W., Starly, B., Nam, J., & Darling, A. (2005). Bio-CAD Modeling and its Applications in Computer-Aided Tissue Engineering. *Computer Aided Design*, *37*(11), 1097–1114. doi:10.1016/j.cad.2005.02.002

Sun, Y., & Li, S. (2015). Real-time collaborative GIS: A technological review. *ISPRS Journal of Photogrammetry and Remote Sensing*.

Suzen, M. L., & Doyuran, V. (2004). Data driven bivariate landslide susceptibility assessment using geographical information systems: A method and application to Asarsuyu catchments, Turkey. *Engineering Geology*, *71*(3-4), 303–321. doi:10.1016/S0013-7952(03)00143-1

Swaine, J. (2009). *Two Google Searches 'Produce Same CO_2 as Boiling a Kettle*. Retrieved March 30, 2016 from http://www.telegraph.co.uk/technology/google/4217055/Two-Google-searches-produce-same-CO2-as-boiling-a-kettle.html

Swetnam Thomas, W., & Christopher, H. B. (2002). Fire Regimes in Sierran Mixed-Conifer Forests. In Sierra Nature Notes, (vol. 2). Laboratory of Tree-Ring Research, The University of Arizona.

Takagi, T., & Sugeno, M. (1985). Fuzzy identification of systems and its applications to modeling and control. *Systems, Man and Cybernetics*, (1), 116–132.

Takaya Yamazato, I. T. (2014). Image-Sensor-Based Visible Light Communication for Automotive Applications. *IEEE Communications Magazine*, 10.

Taksali, S. M. D., Jonathan, N., Grauer, M. D., Alexander, R., & Vaccaro, M. D. (2004). Material Considerations for Intervertebral Disc Replacement Implants. *The Spine Journal*, *4*(11), 231–238. doi:10.1016/j.spinee.2004.07.012 PMID:15541671

Tanaka, K., & Iijima, S. (Eds.). (2014). *Carbon Nanotubes and Graphene* (2nd ed.). Kidlington: Oxford, Elsevier.

Tan, K. S. (2001). Electromagnetic interference in medical devices: Health Canada's past current perspectives and activities. *IEEE International Symposium on Electromagnetic Compatibility*, 1283-1284.

Taylor, J. E., & Bernstein, P. G. (2009). Paradigm trajectories of building information modeling practice in project networks. *Journal of Management Engineering*, *25*(2), 69–76. doi:10.1061/(ASCE)0742-597X(2009)25:2(69)

Taylor, T. P., Ding, M., Ehler, D. D., Foreman, T. M., Kaszuba, J. P., & Sauer, N. N. (2003). Beryllium in the environment: A Review. *Journal of Environmental Science and Health. Part A, Toxic/Hazardous Substances & Environmental Engineering*, *38*(2), 439–469. doi:10.1081/ESE-120016906 PMID:12638707

Tazzite, A., Jouhadi, H., Saiss, K., Benider, A., & Nadifi, S. (2013). Relationship between family history of breast cancer and clinicopathological features in Moroccan patients. *Ethiopian Journal of Health Sciences*, *23*(2), 150–157. PMID:23950631

Tejedor, J. (n.d.). *Declaración Ambiental Enero de 2011-Diciembre de 2011, hm Hospitales*. Available online: http://www.hmhospitales.com/grupohm/Paginas/HMHospitales.aspx

Telfer, D. J. (2001). Strategic Alliances along the Niagara Wine Route. *Tourism Management*, *22*(1), 21–30. doi:10.1016/S0261-5177(00)00033-9

Telfer, D. J., & Wall, G. (1996). Linkages between Tourism and Food Production. *Annals of Tourism Research*, *23*(3), 635–653. doi:10.1016/0160-7383(95)00087-9

Telfer, D. J., & Wall, G. (2000). Strengthening Backward Economic Linkages: Local Food Purchasing by Three Indonesian Hotels. *Tourism Geographies*, *2*(4), 421–447. doi:10.1080/146166800750035521

Telfer, S., Abbott, M., Steultjens, M., Rafferty, D., & Woodburn, J. (2013). Dose-response effects of customised foot orthoses on lower limb muscle activity and plantar pressures in pronated foot type. *Gait & Posture*, *3*(3), 443–449. doi:10.1016/j.gaitpost.2013.01.012 PMID:23391752

Teng, J.-H. (2002). A modified Gauss–Seidel algorithm of three-phase power flow analysis in distribution networks. *Electrical Power & Energy Systems*, *24*(2), 97–102.

Text of the Basel Convention. (2014). *The Control of Transboundary Movements of Hazardous Wastes and Their Disposal*. United Nations Environment Programme (UNEP). Retrieved March 1, 2016 from http://www.basel.int/portals/4/basel%20convention/docs/text/baselconventiontext-e.pdf

TGTC. (2008). *Tree growth and the tree-site complex*. TGTC.

The Ohio State University. (2013). *Woody plants*. Department of Horticulture & Crop Science, The Ohio State University. Retrieved from http://www.hcs.ohio-state.edu/hcs300/anat2.htm, http://hcs.osu.edu/

The University of Arizona. (2013). *Paleo Slide Set: Tree Rings: Ancient Chronicles of Environmental Change*. Laboratory of Tree-Ring Research, The University of Arizona.

The University of North Carolina. (2013). *Tree-Ring Dating, Lesson 2.4*. Research Laboratories of Archeology. The University of North Carolina at Chapel Hill. Retrieved from http://rla.unc.edu/lessons/Lesson/L204/L204.htm

Third World Network. (1991). Toxic Terror: Dumping of Hazardous Wastes in the Third World, Penang, Malaysia. Author.

Thomson, S., & van Belle, J. (2015). Antecedents of Green IT adoption in South African higher education institutions. *The Electronic Journal Information Systems Evaluation*, *18*(2), 172–186.

TimeBliography. (n.d.). Retrieved from http://spaceandtime.wsiabato.info/tGIS.html

Tobler, W. (1970). A computer movie simulating urban growth in the Detroit region. *Economic Geography*, *46*(2), 234–240. doi:10.2307/143141

Tomlinson, R. (1967). *An introduction to the geo. information system ofthe Canada Land Inventory*. Ottawa, Department ofForestry and Rural Development, ARDA.

Torres, R. (2003). Linkages between Tourism and Agriculture in Mexico. *Annals of Tourism Research*, *3*(3), 546–566. doi:10.1016/S0160-7383(02)00103-2

Torres, R., & Momsen, J. H. (2004). Challenges and Potential for Linking Tourism and Agriculture to Achieve Pro-Poor Tourism Objectives. *Progress in Development Studies*, *4*(4), 294–318. doi:10.1191/1464993404ps092oa

Torriti, J. (2012). Price-based demand side management: Assessing the impacts of time-of-use tariffs on residential electricity demand and peak shifting in Northern Italy. *Energy*, *44*(1), 576–583. doi:10.1016/j.energy.2012.05.043

Trias, A. (2012). The holomorphic embedding load flow method. In *Power and Energy Society General Meeting*, (pp. 1–8). IEEE.

Trigger, B. G., Kemp, B. J., O'Connor, D., & Lloyd, A. B. (2000). *Storia sociale dell'antico Egitto*. Roma-Bari, Italy: Laterza.

Tseng, C. T., & Liao, C. J. (2008). A particle swarm optimization algorithm for hybrid flow-shop scheduling with multiprocessor tasks. *International Journal of Production Research*, *46*(17), 4655–4670. doi:10.1080/00207540701294627

Tseng, M. L., Lin, Y. H., & Chiu, A. S. F. (2009). Fuzzy AHP-based study of cleaner production implementation in Taiwan PWB manufacturer. *Journal of Cleaner Production*, *17*(14), 1249–1256. doi:10.1016/j.jclepro.2009.03.022

Tsydenova, O., & Bengtsson, M. (2011a). Chemical hazards associated with treatment of waste electrical and electronic equipment. *Waste Management (New York, N.Y.)*, *31*(1), 45–58. doi:10.1016/j.wasman.2010.08.014 PMID:20869229

Tzeng, G. H., Lin, C. W., & Opricovic, S. (2005). Multi-criteria analysis of alternative-fuel buses for public transportation. *Energy Policy*, *33*(11), 1373–1383. doi:10.1016/j.enpol.2003.12.014

U.S. Department of Energy. (2013). *Part 1: Review of Life-Cycle Energy Consumption of Incadescent, Compact Fluorescent, and LED Lamps*. Author.

Underwood, H., & Calaban, M. (1987). Pineal melatonin rhythms in the lizard Anolis carolinensis: I. Response to light and temperature cycles. *Journal of Biological Rhythms*, *2*(3), 179–193. doi:10.1177/074873048700200302 PMID:2979659

United States Geography. (n.d.). Retrieved from http://www.netstate.com

USGS. (2013, August 29). *Landsat Mission*. Retrieved September 22, 2013, from Landsat 8: http://landsat.usgs.gov/about_ldcm.php

USGS. (n.d.). Retrieved March 1, 2016, from http://rockyweb.cr.usgs.gov/nmpstds/acrodocs/dem/3DEM0897.PDF

Valian, E., Mohanna, S., & Tavakoli, S. (2011). Improved cuckoo search algorithm for global optimization. *International Journal of Communications and Information Technology*, *1*(1), 31–44.

Vames, D. J. (1984). Landslide hazard zonation: A review of principles and practice, Unesco. Paris. *Natural Hazards*, *3*, 63.

Van Benthem, J., & Doets, K. (2001). Higher-order-logic. In *Handbook of Philosophical Logic* (pp. 189–243). Springer.

Van Bussel, G., Smitter, N., & Vandepas, J. (2015). Digital archiving, Green IT and environment. Deleting data to manage critical effects of the data deluge. *The Electronic Journal Information Systems Evaluation*, *18*(2), 187–198.

Van Calker, K. J., Berentsen, P. B. M., Romero, C., Giesen, G. W. J., & Huirne, R. B. M. (2006). Development and application of a multi-attribute sustainability function for Dutch dairy farming systems. *Ecological Economics*, *57*(4), 640–658. doi:10.1016/j.ecolecon.2005.05.016

Van Joolen, E. (2003). *Archaeological land evaluation. A reconstruction of the suitability of ancient landscapes for various land uses in Italy focused on the first millennium B.C.* (Doctoral dissertation). Retrieved from http://dissertations.ub.rug.nl/faculties/arts/2003/e.van.joolen/

Van Krevelen, D., & Poelman, R. (2010). A survey of augmented reality technologies, applications and limitations. *International Journal of Virtual Reality*, *9*(2), 1.

Van Leusen, M. (2002). *Pattern to process. Methodological investigations into the formation and interpretation of spatial patterns in archaeological landscapes* (Doctoral dissertation). Retrieved from http://dissertations.ub.rug.nl/faculties/arts/2002/p.m.van.leusen

Vardakas, J. S., Zorba, N., & Verikoukis, C. V. (2015). A survey on demand response programs in smart grids: Pricing methods and optimization algorithms. *IEEE Communications Surveys and Tutorials*, *17*(1), 152–178. doi:10.1109/COMST.2014.2341586

Vargas-Bernal, R., & Herrera-Pérez, G. (2012). Carbon Nanotube- and Graphene based Devices, Circuits and Sensors for VLSI Design. In VLSI Design, (pp. 41-66). Rijeka, Croatia: Intech.

Vargas-Bernal, R. (2015a). Performance Analysis of Interconnects based on Carbon Nanotubes for AMS/RF IC Design. In M. Fakhfakh, E. Tlelo-Cuautle, & M. H. Fino (Eds.), *Performance Optimization Techniques in Analog, Mixed-Signal, and Radio-Frequency Circuit Design* (pp. 336–363). Hershey, PA: IGI Global. doi:10.4018/978-1-4666-6627-6.ch014

Vargas-Bernal, R. (2015b). Advances in Computational Modeling of Electronic Devices based on Graphene. *IEEE Journal on Emerging and Selected Topics in Circuits and Systems*, *5*(1), 109–116. doi:10.1109/JETCAS.2015.2398219

Vargas-Bernal, R. (2015c). Performance Analysis Electromagnetic Interference Shielding based on Carbon Nanomaterials used in AMS/RF IC Design. In M. Fakhfakh, E. Tlelo-Cuautle, & M. H. Fino (Eds.), *Performance Optimization Techniques in Analog, Mixed-Signal, and Radio-Frequency Circuit Design* (pp. 268–294). Hershey, PA: IGI Global. doi:10.4018/978-1-4666-6627-6.ch011

Varkey, M., & Atala, A. (2015). Organ Bioprinting: A Closer Look at Ethics and Policies. *Wake Forest JL & Pol'y*, *5*, 275.

Viaggi, D., Finn, J. A., Kurz, I., & Bartolini, F. (2011). Multicriteria analysis for environmental assessment of agri-environment schemes: How to use partial information from Mid-Term Evaluations? *Agricultural Economics Research*, *12*(1), 5–21.

Viega, J., & McGraw, G. (2002). *Building Secure Software*. Addison-Wesley.

Vinodh, S., Prasanna, M., & Prakash, N. H. (2014). Integrated fuzzy AHP–TOPSIS for selecting the best plastic recycling method: A case study. *Applied Mathematical Modelling*, *38*(19-20), 4662–4672. doi:10.1016/j.apm.2014.03.007

Vlachogiannis, J. G., Hatziargyriou, N. D., & Lee, K. Y. (2005). Ant colony system-based algorithm for constrained load flow problem. *Power Systems*, *20*(3), 1241–1249. doi:10.1109/TPWRS.2005.851969

Vo, D. N., Schegner, P., & Ongsakul, W. (2013). Cuckoo search algorithm for non-convex economic dispatch. Generation. *Transmission & Distribution, IET*, *7*(6), 645–654. doi:10.1049/iet-gtd.2012.0142

von Tscharner, V. (2000). Intensity analysis in time-frequency space of surface myoelectric signals by wavelets of specified resolution. *Journal of Electromyography and Kinesiology*, *10*(6), 433–445. doi:10.1016/S1050-6411(00)00030-4 PMID:11102846

Vorsic, J., Muzek, V., & Skerbinek, G. (1991). Stochastic load flow analysis. In *Electrotechnical Conference*, (pp. 1445–1448). IEEE.

Wachal, D. J., & Hudak, P. F. (2000). Mapping landslide susceptibility in Travis County, Texas, USA. *GeoJournal*, *51*(3), 245–253. doi:10.1023/A:1017524604463

Wachsmuth, J. (2008). *Multiple Independent Extrusion Heads for Fused Deposition Modeling. Master of Science*. Virginia Polytechnic Institute and State University.

Wang & Dunston (2007). *Design, strategies, and issues towards an augmented reality-based construction training platform*. ITcon.

Wang, L. X. (1994). Adaptive fuzzy systems and control. *Design and Stability Analysis*, 29–31.

Wang, R. (2008). *An expert knoeledge-based approach to landslide susceptibility mapping using GIS and fuzzy logic* (PhD Dissertation). University of Wisconsin-Madison, UMI Number: 3314311.

Wang, X., & Love, P. (2012). BIM+ AR: Onsite information sharing and communication via advanced visualization. *Computer Supported Cooperative Work in Design (CSCWD), 2012 IEEE 16th International Conference on*. IEEE.

Wang, C. S., Wang, W. H. A., & Lin, M. C. (2010). STL Rapid Prototyping Bio-CAD Model for CT Medical Image Segmentation. *Computers in Industry*, *61*(3), 187–197. doi:10.1016/j.compind.2009.09.005

Wang, D., & Cheng, T. (2001). A spatio-temporal data model for activity-based transport demand modelling. *International Journal of Geographical Information Science*, *15*(6), 561–585. doi:10.1080/13658810110046934

Wang, H. M., Chou, F. D., & Wu, F. C. (2011). A simulated annealing for hybrid flow shop scheduling with multiprocessor tasks to minimize makespan. *International Journal of Advanced Manufacturing Technology*, *53*(5-8), 761–776. doi:10.100700170-010-2868-z

Wang, J. W., & Shaw, L. L. (2006). Fabrication of functionally graded materials via inkjet color printing. *Journal of the American Ceramic Society*, *89*(10), 3285–3289. doi:10.1111/j.1551-2916.2006.01206.x

Wang, Q. J., & Li, J. R. (2006). Interactive visualization of complex dynamic virtual environments for industrial assemblies. *Computers in Industry*, *57*(4), 366–377. doi:10.1016/j.compind.2005.11.002

Wang, W., Xu, Y., & Khanna, M. (2011). A survey on the communication architectures in smart grid. *Computer Networks*, *55*(15), 3604–3629. doi:10.1016/j.comnet.2011.07.010

Wang, X., Chan, H. K., Yee, R. W. Y., & Diaz-Rainey, I. (2012). A two-stage fuzzy AHP model for risk assessment of implementing green initiatives in the fashion supply chain. *International Journal of Production Economics*, *135*(2), 595–606. doi:10.1016/j.ijpe.2011.03.021

Wang, X., Love, P. E., Kim, M. J., Park, C.-S., Sing, C.-P., & Hou, L. (2013). A conceptual framework for integrating building information modeling with augmented reality. *Automation in Construction*, *34*, 37–44. doi:10.1016/j.autcon.2012.10.012

Wang, Z., & Alvarado, F. L. (1992). Interval arithmetic in power flow analysis. *Power Systems*, *7*(3), 1341–1349. doi:10.1109/59.207353

Warman, M. (2009). *Green I.T.: How many Google searches does it take to boil a kettle?* Retrieved March 30, 2016 from http://www.telegraph.co.uk/technology/google/4241791/Green-I.T.-how-many-Google-searches-does-it-take-to-boil-a-kettle.html

Watanabe, T., Hatakeyama, T., & Tomiita, M. (2015). Improving Assistive Technology Service by Using 3D Printing: Three Case Studies. *Studies in Health Technology and Informatics*, *217*, 1047–1052. PMID:26294609

Watari, F., Yokoyama, A., Omori, M., Hirai, T., Kondo, H., Uo, M., & Kawasaki, T. (2004). Biocompatibility of Materials and Development to Functionally Graded Implant for Bio-medical Application. *Composites Science and Technology*, *64*(6), 893–908. doi:10.1016/j.compscitech.2003.09.005

WDC. (2008). *The World Data Center for Paleoclimatology Mirror Site (WDC)*. Retrieved from http://wdc.obs-mip.fr/treering.html

Weaver, D. B. (1998). Peripheries of the Periphery Tourism in Tobago and Barbuda. *Annals of Tourism Research*, *25*(2), 292–313. doi:10.1016/S0160-7383(97)00094-7

Wehner, M. (2012). Man to machine, applications in electromyography. In M. Schwartz (Ed.), *EMG Methods for Evaluating Muscle and Nerve Function* (pp. 427–454). InTech. doi:10.5772/26495

Weisman, J. A., Nicholson, J. C., Tappa, K., Jammalamadaka, U., Wilson, C. G., & Mills, D. K. (2015). Antibiotic and chemotherapeutic enhanced three-dimensional printer filaments and constructs for biomedica l applications. *International Journal of Nanomedicine*, *10*, 357–370. PMID:25624758

Weiss, H. R. (2010). Brace technology thematic series - the Gensingen brace in the treatment of scoliosis. *Scoliosis*, *5*(1), 22. doi:10.1186/1748-7161-5-22 PMID:20942970

Wei, X., & Dong, F. (2011). Development of computer aided forming techniques in manufacturing scaffolds for bone tissue engineering. *Zhongguo Xiu Fu Chong Jian Wai Ke Za Zhi*, *12*, 1508–1512. PMID:22242356

Wen, M., Sun, X., Su, L., Shen, J., Li, J., & Guo, S. (2012). The Electrical Conductivity of Carbon Nanotube/Carbon Black/Polypropylene Composites prepared through Multistage Stretching Extrusion. *Polymer*, *53*(7), 1602–1610. doi:10.1016/j.polymer.2012.02.003

Wheatley, D. (1995). Cumulative viewshed analysis: a GIS-based method for investigating intervisibility, and its archaeological application. In G. Lock & Z. Stancic (Eds.), *Archaeology and Geographical Information Systems: A European Perspective* (pp. 171–186). London: Taylor & Francis.

White, C. T., & Todorov, T. N. (1998). Carbon Nanotubes as Long Ballistic Conductors. *Nature*, *393*(6682), 240–242. doi:10.1038/30420

WHO Cancer Mortality Database. (n.d.). World Health Organization. Retrieved from http://www-dep.iarc.fr/WHOdb/WHOdb.htm

Widmer, R., Oswald-Krapf, H., Sinha-Khetriwal, D., Schnellmann, M., & Bo¨ni, H. (2005). Global perspectives on e-waste. *Environmental Impact Assessment Review*, *25*(5), 436–458. doi:10.1016/j.eiar.2005.04.001

Winder, J., Cooke, R. S., Gray, J., Fannin, T., & Fegan, T. (1999). Medical Rapid Prototyping and 3D CT in the Manufacture of Custom Made Cranial Titanium Plates. *Journal of Medical Engineering & Technology*, *23*(1), 26–28. doi:10.1080/030919099294401 PMID:10202700

Wöhlke, G., & Schiller, E. (2005). Digital Planning Validation in automotive industry. *Computers in Industry*, *56*(4), 393–405. doi:10.1016/j.compind.2005.01.010

Wong, J. K., Li, H., Wang, H., Huang, T., Luo, E., & Li, V. (2012). Toward low-carbon construction processes: The visualisation of predicted emission via virtual prototyping technology. *Automation in Construction.*

Wong, S. H., Swartz, S. L., & Sarkar, D. (2002, April). A Middleware Architecture for Open and Interoperable GISs. *IEEE MultiMedia*, *9*(2), 62–76. doi:10.1109/93.998065

World Health Organization (WHO). (1989). Mercury, Environmental Health Criteria, 86. WHO.

Wu, A. H., Ziegler, R. G., Horn-Ross, P. L., Nomura, A. M., West, D. W., Kolonel, L. N., ... Pike, M. C. (1996). Tofu and risk of breast cancer in Asian-Americans. *Cancer Epidemiology, Biomarkers & Prevention*, *5*(11), 901–906. PMID:8922298

Wu, G. H., & Hsu, S. H. (2015). Review: Polymeric-Based 3D Printing for Tissue Engineering. *Journal of Medical and Biological Engineering*, *3*(3), 285–292. doi:10.100740846-015-0038-3 PMID:26167139

Xu, Z., Liu, Z., Sun, H., & Gao, C. (2013). Highly Electrically Conductive Ag-doped Graphene Fibers as Stretchable Conductors. *Advanced Materials*, *25*(23), 3249–3253. doi:10.1002/adma.201300774 PMID:23640822

Ya´nez, V. T. (2009). Illumination interference reduction system for VLC Communications. *Proc. WSEAS Int. Conf. Math. Methods, Computer. Tech. Intel.*

Yamazato, T., Takai, I., Okada, H., Fujii, T., Yendo, T., Arai, S., ... Kawahito, S. (2014). Image sensor based visible light communication for automotive. *IEEE Communications Magazine*, *52*(7), 88–97. doi:10.1109/MCOM.2014.6852088

Yang, T.-J., Tsai, Y.-M., & Chen, L.-G. (2011). Smart display: A mobile self-adaptive projector-camera system. *Multimedia and Expo (ICME), 2011 IEEE International Conference on*. IEEE.

Yang, X. S., & Deb, S. (2009, December). Cuckoo search via Lévy flights. In *Nature & Biologically Inspired Computing, 2009. NaBIC 2009. World Congress on* (pp. 210-214). IEEE.

Yang, C., Goodchild, M., Huang, Q., Nebert, D., Raskin, R., Xu, Y., ... Fay, D. (2011). Spatial cloud computing: How can the geospatial sciences use and help shape cloud computing? *International Journal of Digital Earth*, *4*(4), 305–329. doi:10.1080/17538947.2011.587547

Yang, C., Raskin, R., Goodchild, M., & Gahegan, M. (2010). Geospatial cyberinfrastructure: Past, present and future. *Computers, Environment and Urban Systems*, *34*(4), 264–277. doi:10.1016/j.compenvurbsys.2010.04.001

Yang, C., Xu, Y., & Nebert, D. (2013). Redefining the possibility of digital Earth and geosciences with spatial cloud computing. *International Journal of Digital Earth*, *6*(4), 297–312. doi:10.1080/17538947.2013.769783

Yang, M.-D., Chao, C.-F., Huang, K.-S., Lu, L.-Y., & Chen, Y.-P. (2012). Image-based 3D scene reconstruction and exploration in augmented reality. *Automation in Construction.*

Yang, S. K. (2011). Reduction of optical interference by wavelength filtering in RGB-LED based indoor VLC system. *Proceeding of Opto-Electronics and Communications Conference (OECC).*

Yang, T., Kuo, Y., & Chang, I. (2004). Tabu-search simulation optimization approach for flow - shop scheduling with multiple processors-a case study. *International Journal of Production Research*, *42*(19), 4015–4030. doi:10.1080/00207540410001699381

Yanguas, G. (2012). *Perfil ambiental de España. Informe basado en indicadores.* Ministerio de Agricultura, Alimentación y Medio Ambiental Español. Retrieved may 6, 2013 from http://www.magrama.gob.es/es/

Yang, X. S., & Deb, S. (2013). Multiobjective cuckoo search for design optimization. *Computers & Operations Research*, *40*(6), 1616–1624. doi:10.1016/j.cor.2011.09.026

Yang, X. S., & Deb, S. (2014). Cuckoo search: Recent advances and applications. *Neural Computing & Applications*, *24*(1), 169–174. doi:10.100700521-013-1367-1

Yeh, K.-C., Tsai, M.-H., & Kang, S.-C. (2012). On-site building information retrieval by using projection-based augmented reality. *Journal of Computing in Civil Engineering*, *26*(3), 342–355. doi:10.1061/(ASCE)CP.1943-5487.0000156

Yew, A., Ong, S., & Nee, A. (2016). Towards a griddable distributed manufacturing system with augmented reality interfaces. *Robotics and Computer-integrated Manufacturing*, *39*, 43–55. doi:10.1016/j.rcim.2015.12.002

Ying, K. C., & Lin, S. W. (2006). Multiprocessor task scheduling in multistage hybrid flow-shops: An ant colony system approach. *International Journal of Production Research*, *44*(16), 3161–3177. doi:10.1080/00207540500536939

Yin, J., Martin, J.-P., Venkataramani, A., Alvisi, L., & Dahlin, M. (2003). Separating agreement from execution for byzantine fault tolerant services. *Proceedings of the ACM Symposium on Operating Systems Principles*, 253–267. 10.1145/945445.945470

Yokokohji, Y., Sugawara, Y., & Yoshikawa, T. (2000). Accurate image overlay on video see-through HMDs using vision and accelerometers. *Virtual Reality (Waltham Cross).*

Young, A., & Yung, M. (2004). *Malicious Cryptography: Exposing Cryptovirology.* Wiley.

Young, M., Kate, A., Goldberg, I., & Karsten, M. (2010, June). Practical robust communication in dhts tolerating a byzantine adversary. In *Proceedings of the IEEE 30th International Conference on Distributed Computing Systems* (pp. 263-272). IEEE. 10.1109/ICDCS.2010.31

Yousefi, N., Sun, X., Lin, X., Shen, X., Jia, J., Zhang, B., ... Kim, J. K. (2014). Highly Aligned Graphene/Polymer Nanocomposites with Excellent Dielectric Properties for High-Performance Electromagnetic Interference Shielding. *Advanced Materials*, *26*(31), 5480–5487. doi:10.1002/adma.201305293 PMID:24715671

Youssef, R. F., Spradling, K., Yoon, R., Dolan, B., Chamberlin, J., Okhunov, Z., ... Landman, J. (2015). Applications of three-dimensional printing technology in urological practice. *BJU International*, *5*(5), 697–702. doi:10.1111/bju.13183 PMID:26010346

Yuan, F. S., Sun, Y. C., Wang, Y., & Lü, P. J. (2013). Accuracy evaluation of a new three-dimensional reproduction method of edentulous dental casts, and wax occlusion rims with jaw relation. *International Journal of Oral Science, 3*(3), 155–161. doi:10.1038/ijos.2013.49 PMID:23907676

Yue, P., Zhou, H., Gong, J., & Hu, L. (2013). Geoprocessing in Cloud Computing platforms–a comparative analysis. *International Journal of Digital Earth, 6*(4), 404–425. doi:10.1080/17538947.2012.748847

Yu, G., Zhang, S., Yu, Q., Fan, Y., Zeng, Q., Wu, L., & Zhao, P. (2014). Assessing ecological security at the watershed scale based on RS/GIS: A case study from the Hanjiang River Basin. *Stochastic Environmental Research and Risk Assessment, 28*(2), 307–318. doi:10.100700477-013-0750-x

Yu, J., Welford, R., & Hills, P. (2006 a). Industry responses to EU WEEE and ROHS Directives: Perspectives from China. *Corporate Social Responsibility and Environmental Management, 13*(5), 286–299. doi:10.1002/csr.131

Yu, M., Fu, P., Zhou, N., & Xia, J. (2014). ArcGIS in the cloud. In Z. Li, C. Xu, & K. Liu (Eds.), *Spatial Cloud Coputing. A practical approach.* Boca Raton, FL: CRC Press.

Yusoff, A. R. M. (Ed.). (2015). Graphene-based Energy Devices. Wiley-VCH.

Yu, X., & Xue, Y. (2016). Smart Grids: A Cyber–Physical Systems Perspective. *Proceedings of the IEEE, 104*(5), 1058–1070. doi:10.1109/JPROC.2015.2503119

Zadeh, L. A. (1973). Outline of a new approach to the analysis of complex systems and decision process. *IEEE Transactions on Systems, Man, and Cybernetics, SMC-3*(1), 28–44. doi:10.1109/TSMC.1973.5408575

Zahmati & Fernando. (2011). Emerging Wireless Applications in Aerospace: Benefits and Challenges. *Exponent, the Honeywell Technical Journal.*

Zaman, N., Ragab, K., & Abdullah, A. (2012). *Wireless Sensor Networks and Energy Efficiency: Protocols, Routing and Management.* Hershey, PA: IGI Global. doi:10.4018/978-1-4666-0101-7

Zecca, M., Micera, S., Carrozza, M. C., & Dario, P. (2002). Control of multifunctional prosthetic hands by processing the electromyographic signal. *Critical Reviews in Biomedical Engineering, 30*(4-6), 459–485. doi:10.1615/CritRevBiomedEng.v30.i456.80 PMID:12739757

Ze, Z., & Sx, J. (2014). Hollow fiber membrane contactor absorption of CO_2 from the flue gas: Review and perspective. *Global NEST Journal, 16*, 355–374.

Zhan, F. B., Gong, X., & Xingjian, L. (2014). Mark on the globe: A quest for scientific bases of geographic information and its international influence. *International Journal of Geographical Information Science, 28*(6), 1233–1245. doi:10.1080/13658816.2013.847186

Zhang, Y., Wang, L., & Xiang, Y. (2015). Power System Reliability Analysis With Intrusion Tolerance in SCADA Systems. *IEEE Transactions on Smart Grid.*

Zhang, C., & Hammad, A. (2012). Improving lifting motion planning and re-planning of cranes with consideration for safety and efficiency. *Advanced Engineering Informatics, 26*(2), 396–410. doi:10.1016/j.aei.2012.01.003

Zhang, H., Zhao, W., Moser, L. E., & Melliar-Smith, P. M. (2011). Design and Implementation of a Byzantine Fault Tolerance Framework for Non-Deterministic Applications. *IET Software, 5*(3), 342–356. doi:10.1049/iet-sen.2010.0013

Zhang, L., Xu, Y., Yeh, C. H., Liu, Y., & Zhou, D. (2016). City sustainability evaluation using multi-criteria decision making with objective weights of interdependent criteria. *Journal of Cleaner Production, 131*(10), 491–499. doi:10.1016/j.jclepro.2016.04.153

Zhang, P., & Lee, S. T. (2004). Probabilistic load flow computation using the method of combined cumulants and gram-charlier expansion. *Power Systems*, *19*(1), 676–682. doi:10.1109/TPWRS.2003.818743

Zhao, J., Xie, Z., Jiang, L., Cai, H., Lio, H., & Hirzinger, G. (2006). EMG control for a five-fingered interactuated prosthetic hand based on wavelet transform and sample entropy. *Proceedings of the IEEE/RSJ International Conference on Intelligent Robots and Systems.*

Zhao, W. (2009). Design and implementation of a Byzantine fault tolerance framework for Web services. *Journal of Systems and Software*, *82*(6), 1004–1015. doi:10.1016/j.jss.2008.12.037

Zhao, W. (2014). *Building dependable distributed systems.* John Wiley & Sons. doi:10.1002/9781118912744

Zheng, Q., & Kim, J.-K. (2015). *Graphene for Transparent Conductors: Synthesis, Properties and Applications.* New York, NY: Springer. doi:10.1007/978-1-4939-2769-2

Zhu, Q., Sarkis, J., Cordeiro, J. J., & Lai, K. H. (2008). Firm-level correlates of emergent green supply chain management practices in the Chinese context. *Omega*, *36*(4), 577–591. doi:10.1016/j.omega.2006.11.009

Zhu, W., OBrien, C., OBrien, J. R., & Zhang, L. G. (2014). 3D nano/microfabrication techniques and nanobiomaterials for neural tissue regeneration. *Nanomedicine (London)*, *6*(6), 859–875. doi:10.2217/nnm.14.36 PMID:24981651

Zimmerman, R. D., & Chiang, H.-D. (1995). Fast decoupled power flow for unbalanced radial distribution systems. *Power Systems*, *10*(4), 2045–2052. doi:10.1109/59.476074

About the Contributors

Mehdi Khosrow-Pour, D.B.A., received his Doctorate in Business Administration from the Nova Southeastern University (Florida, USA). Dr. Khosrow-Pour taught undergraduate and graduate information system courses at the Pennsylvania State University – Harrisburg for almost 20 years. He is currently Executive Editor at IGI Global (www.igi-global.com). He also serves as Executive Director of the Information Resources Management Association (IRMA) (www.irma-international.org), and Executive Director of the World Forgotten Children's Foundation (www.world-forgotten-children.org). He is the author/editor of more than 100 books in information technology management. He is also the Editor-in-Chief of the *International Journal of E-Politics, International Journal of Responsible Management Education, International Journal of Social and Organizational Dynamics in IT, International Journal of Handheld Computing Research,* and *International Journal of Information Security and Privacy,* and is also the founding Editor-in-Chief of the *Information Resources Management Journal, Journal of Electronic Commerce in Organizations, Journal of Cases on Information Technology,* and the *Journal of Information Technology Research,* and has authored more than 50 articles published in various conference proceedings and scholarly journals.

* * *

Mohamed Al-Marzouqi obtained his BSc and MSc from Washington State University and PhD from Oregon State University, all in chemical engineering. His specialization is in the area of gas processing and membrane gas separation. Prof. Al Marzouqi joined Chemical & Petroleum Engineering Department at the UAE University in 1996. Prof. Al Marzouqi has several years of experience in the area of transport of gaseous specious through different types of membranes. Prof. Al Marzouqi has 3 patents, 51 international journal publications, two book chapters, and more than 50 local and international conference presentations.

Ali Alharbi is an assistant professor of computer science at Qassim University, Saudi Arabia. He holds a PhD degree in computer science. His research interests include health informatics, e-learning, and computer science education.

Yunus Ziya Arslan received the B.S. degree in 2002, the M.S. degree in 2005, and the Ph.D. degree in 2009 all from the Department of Mechanical Engineering, Faculty of Engineering, Istanbul University, Turkey. He spent six months at the University of Calgary as a Visitor Researcher. He serves currently as an Associate Professor at the Department of Mechanical Engineering, Faculty of Engineering, Istanbul University. His research areas include musculoskeletal biomechanics, dental biomechanics, tissue biomechanics, computational biomechanics, and modeling and control of robotic manipulators.

Surajit Bag has completed his PhD in Logistics and Supply Chain Management from College of Management and Economic Studies, University of Petroleum & Energy Studies, Dehradun. He has attended Management Development Program in Multivariate data analysis from Indian Institute of Technology, Kharagpur. He has also attended Management Development Program in Research Methodology from Banaras Hindu University. He has trained in Case Study Teaching from Indian Institute of Management, Calcutta. He has also trained in Total Qual¬ity Management from Indian Institute of Technology, Kanpur. Presently he is working in the position of Head of Procurement with Tega Industries South Africa Pty Ltd, South Africa. He is having more than eight years of industry experience. His areas of research interest are Buyer-Supplier Relationship, Sustainable Supply Chain Management, Supply Chain Risk Management, Humanitarian logistics strategy and Business excellence. He has attended several National and International conferences and has published more than 50 papers in reputed journals. He is the recipient of AIMS-IRMA Young Management Researcher Award 2016 for his significant contribution towards management research. He is an editorial board member of International Journal of Applied Logistics, IRJ Science and Amity Journal of Operations Management. He is a regular reviewer of several prestigious journals such as ANOR, JHLSCM, VISION, IJIS, IJAL and IJPM. He is a life member of professional bodies such as Indian Rubber Institute (IRI), Operational Research Society of India (ORSI), Society of Operations Management (SOM), AIMS International and Asian Council of Logistics Management (ACLM).

Federica Bonaiuti, Registered Professional Engineer, has an education in Chemical Engineering since 2003, specialized in Environmental Engineering, Safety Engineering, Integrated Management Systems (OHSAS18001, ISO14001, ISO9001). Independent environmental, health and safety advisor, with working experience in industrial plants engineering and environmental and safety field since 2004, Lead Environmental Auditor officially certified, RSPP officially certified for all the ATECO codes. Collaboration with ISPRA for technical support to the Italian Ministry of Environment Commission for IPPC Permits releases' for Refineries, Chemical Industries, Thermoelectric Power Plants and Iron and Steel Production Industries.

Clementina Bruno holds master degrees in Business (University of Piemonte Orientale, Novara, Italy) and Economics (Univsersité Catholique de Louvain, Louvain-la-Neuve, Belgium) and a PhD in Economics and Management of the Technology (University of Bergamo, Italy). She currently works as a research fellow at University of Piemonte Orientale. Her research interests involve efficiency analysis, regulation, management, management of network services. This research is supported by the Associazione per lo sviluppo dell'Università nel territorio novarese.

Carlo Carlucci graduated in July 2009, in Chemical Engineering and Materials Engineering at the University of Calabria, has developed the knowledge in chemical plants. He has provided service at the Laboratory of Transport Phenomena and Biotechnology, University of Calabria, dealing with optimization of bioreactors for anaerobic digestion. Currently he lends his activities at the Institute for Environmental Protection and Research (ISPRA) as technical support to the IPPC (Integrated Pollution Prevent and Control) Commission for release of national IPPC Environmental Permits (AIA). In particular, its activity has dealt with the assessment procedures for the IPPC Environmental Permits of chemical plants and refineries, the comparison of plants subject to authorization with the documents adopted by the European Union for the implementation of Directives 96/61 / EC and 2010/75 / EU bearing the criteria for the identification and use of Best Available Techniques (BAT), analysis of technical solutions for the abatement and / or minimization of emissions of pollutants. It also carried out an activity of support to the IMELS (Italian Ministry for the Environment, Land and Sea) and IPPC Bureau in Seville, for reprocessing and re-adaptation to the requirements of Directive 2010/75 / EU of some European Reference Documents (BRefs) currently under revision.

María Carmen Carnero has a Doctorate in Industrial Engineering from the University of Castilla-La Mancha (2001) and is a lecturer at the Technical School of Industrial Engineering at the University of Castilla-La Mancha. She has published two books, seven chapters in science books, and research articles in Omega, Decision Support System, European Journal of Operational Research, Reliability Engineering and System Safety, Mechanical Systems and Signal Processing, Journal of Manufacturing Systems, BMC Medical Informatics and Decision Making, Sustainability, European Journal of Industrial Engineering, Production Planning and Control, etc. She has 60 contributions to congress and conferences. She was the lead researcher in three projects, and has been a part of 16 other European, national, regional and local projects.

Somnath Chaudhuri has a M. Tech. in Computer Science and Engineering, B. Tech. in Computer Science and Engineering and B. Sc. in Physics, all from the University of Calcutta, India. He is working as an International faculty, in the Department of IT, Faculty of Science, Maldives National University. Before this he was working as Lecturer and Database Advisor for Royal Thimphu College, Bhutan for 4 years. His research interest lies in the domain of GIS, Web GIS, Map Mashups, Cloud Computing, Automated System, and Application of Information Technology in Business and Industry Sectors. He has some National and International publications in these domains. He is the editor of few books published by IGI Global, USA. Mr. Chaudhuri has worked as the project guide for some real-time projects in the government and private sectors of Bhutan and Maldives. Mr. Chaudhuri has received certificate of appreciation from the Minister of Agriculture, Royal Government of Bhutan and from the Secretary of the Department of Livestock, Royal Government of Bhutan.

Edward T. Chen is a professor of Management Information Systems of Operations and Information Systems Department in the Manning School of Business at University of Massachusetts Lowell. Dr. Chen has published numerous refereed research articles in scholarly journals and refereed chapters in reference books. Dr. Chen has served as vice-president, board director, track chair, and session chair of many IS/IT professional associations and conferences. Professor Chen has also served as journal editor, editorial reviewer, and ad hoc reviewer for various academic journals. Dr. Chen has received the Irwin Distinguished Paper Award at the Southwestern Federation of Administrative Disciplines conference and the Best Paper Award at the International Conference on Accounting and Information Technology. His main research interests are in the areas of Social Media, Project Management, Knowledge Management, and Software Development.

H. H. Cheung is a lecturer in the Department of Industrial and Manufacturing Systems Engineering at the University of Hong Kong. He obtained his B.Eng., M.Phil., and Ph.D. degrees at the University of Hong Kong. Prior to joining the University of Hong Kong, he has worked in manufacturing and IT industries as a consultant for a number of years. He has gained substantial experiences in developing and managing RFID-based solutions, and the provision of consultancy services to implement RFID applications to various industries/enterprises. Dr. Cheung has also participated in a number of industrial application projects of RFID and IoT technologies, including RFID-based systems for product anti-counterfeiting, food processing management, manufacturing and logistics management of a global printing enterprise, governmental project for inventory management, and baggage management of a major international airport. His research interests include CAD/CAM, Layered Manufacturing (3D Printing), Virtual Prototyping and Virtual Manufacturing, Product Development, E-commerce, RFID, NFC and IoT applications.

S. H. Choi is associate professor in the Department of Industrial and Manufacturing Systems Engineering at the University of Hong Kong. He gained both his B.Sc. and Ph.D degrees in Mechanical Engineering at the University of Birmingham in the UK. Dr. Choi worked in computer industry as CADCAM consultant before joining the University of Hong Kong. He has undertaken a number of industrial projects, and his current research interests focus mainly on computer applications in various areas, including e-commerce, CADCAM, advanced manufacturing systems, virtual prototyping, and RFID technology. Dr. Choi has extensive teaching experience at both undergraduate and postgraduate levels, as well as a wide range of publications on his research work.

H. Sebnem Düzgün is professor of Mining Engineering Department at Middle East Technical University (METU), Ankara Turkey. Prof. Düzgün is also affiliated to Geodetic and Geographic Information Technologies Program, Centers of Earthquake Engineering (EEC), Modeling and Simulation (MOSIMMER), Image Analysis Centre (OGAM), Institutes of Applied Mathematics and Informatics at METU. She has been performing research on earthquake and landslide hazard and risk assessment as well as integration of spatial and spatio-temporal data analysis, GIS and remote sensing in earthquake and landslide hazard and risk assessment. She is the author of the book entitled remote sensing of the mine environment. She has ten book chapters and more than 180 publications on the related topics.

Arzu Erener is a Remote Sensing and Geographic Information System specialist with the research interests in risk management, building detection and spatial statistics. She obtained a B.S. in Geomatics Engineering from Istanbul Technical University in 1999, M.S. in Geomatics Engineering from Black sea Technical University in 2002 and Ph.D. in Geodesy and Geographical Information Technologies from Middle East Technical University in 2009. She is currently working at Kocaeli University as a Assoc. Prof. Dr.

Edna Fang is a Masters student at Asia-Europe Institute, University of Malaya.

Xavier N. Fernando (http://www.ee.ryerson.ca/~fernando) is a Professor and Director of Ryerson Communications Lab. He was an IEEE Distinguished Lecturer and delivered over 40 invited lectures worldwide. He has (co-) authored over 150 research articles and holds three patents (one pending). He has monographed a widely selling book on Radio over Fiber systems, which is translated to Chinese. He is also a co-author of IEEE COMSOC WEBOK. He was a member in the IEEE COMSOC Education Board Working Group on Wireless Communications, 2010-2012. He was a member of Ryerson Board of Governors during 2010-2011. He is a program evaluator for ABET and has visited American Universities for accreditation. He was a finalist for the Top 25 Immigrant Award of Canada in 2012. He is the General Chair of IEEE International Humanitarian Technology Conference, 2017. He was the General Chair for IEEE Canadian Conference on Electrical and Computer Engineering (CCECE) 2014 and General Chair of IEEE Toronto International Conference – Science and Technology for Humanity (TIC-STH) 2009. He was the Chair of IEEE Toronto Section and Member of IEEE Canada Board during 2012-13. His lab is well funded. He has received over $650,000 in research grants received just in 2013-2015 from industry and government (including two CRD grants and on MITACS cluster grant). He and his students have won several awards and prizes including; the First prize at the Humanitarian Initiatives Workshop of CCECE 2014; Second Prize at the 2014 IHTC Conference in Montreal; Best Paper Award at the International Conference of Smart Grid Engineering (SEGE 2014), UOIT, Canada, 2014; IEEE Microwave Theory and Techniques Society Prize in 2010, Sarnoff Symposium prize in 2009, Opto-Canada best poster prize in 2003 and CCECE best paper prize in 2001. Xavier Fernando earned his PhD from the University of Calgary, Alberta in 2001 in affiliation with TRLabs. His PhD work won the best Canadian paper award and a US patent. He joined Ryerson University in 2001, received early tenure and established Ryerson Communications Lab.

Akram Gasmelseed earned his B.Sc. Degree in Electrical Engineering and Informatics – major in Computer Engineering – and M.Sc. Degree in Electrical Engineering and Informatics – major in Biomedical Engineering – (all with honors) from Budapest University of Technology and Economics (BME), Budapest, Hungary, in 1993 and 1999, respectively. He received the Ph.D. Degree in Electrical Engineering – major in Biomedical Engineering – from Universiti Teknologi Malasysia (UTM), Malaysia, in 2009. He received the Best Student Award and Pro-Chancellor Award in 2010. His research is in the areas of electromagnetic biological effects, biophotonics, and computer signal / image processing application to biomedical engineering. Dr. Akram is a member of The Scientific Research Society (Sigma Xi), the IEEE (Institute of Electrical and Electronics Engineers), European Bioelectromagnetics Association (EBEA), Bioelectromagnetics Society (BMES), The Rosacea Research & Development Institute (RRDi), an Associate Editor of International Journal of Advances in Engineering & Technology, and is on the reviewer boards of of various technical journals, including IEEE Transaction on Cybernetics, the International Journal of Biomedical Engineering Technology (IJBET), IEEE Transactions on Instrumentation & Measurement (TIM), Wireless Networks, Journal of Electromagnetic Waves & Applications (JEMWA), Progress in Electromagnetic Research (PIER, PIER B, C, M, PIER letters), IEEE Microwave & Wireless Components Letter, IEEE Journal of Photovoltaics, and IEEE Transaction in Education. He is the Founder and Director of the Biomilitary Science Research Group (BSrG), and the Malaysian Arab Scientific Association (MASA). He also the Chair of the Asia-Pacific Chapter in Biomedical Wireless Technology (APC-BWT) since 2008. Dr. Akram is presently an associate professor at the Faculty of Engineering, University of Science & Technology (UST), Omdurman, Sudan. He is the Founder and chair of IEEE EMBS Malaysia section since 2008. His biography has listed in the 25th Silver Anniversary Edition of Who's Who in the World 2009 and 2013.

M. Geetha is working as an Assistant Professor in the Department of Mathematics, Kamaraj College of Engineering and Technology, Virudhunagar, Tamilnadu, India. She received her BSc, MSc, and MPhil in Mathematics from the Madurai Kamaraj University, Madurai, India. Her area of interest is operations research. She has published three papers in international journals.

Nayef Ghasem is a professor of Chemical Engineering, United Arab Emirates University, UAE Before joining UAE University he was a faculty member in the University Malaya, Malaysia for five years. He taught various chemical engineering courses at both undergraduate and graduate levels. He supervised few master and PhD students. He has more than 100 articles published in international journals and local and international conferences. He published four books, three of which are published by CRC Press, USA; "Principles of Chemical Engineering Processes", "Principles of Chemical Engineering Processes, Material and Energy Balances" and "Computer methods in Chemical Engineering". His area of research is modeling and simulation, polymer reaction engineering and fabrication of polymeric hollow fiber membrane and utilizing these membranes in the CO2 capture from flue gas and natural gas.

Carlos Granell received his PhD in Computer Science at the Universitat Jaume I of Castellón (2006). Before re-joining GEOTEC in March 2014, he worked for 3 years as a post-doc in the Digital Earth and Reference Data Unit of the European Commission's Joint Research Centre (JRC). His research interests lie in multi-disciplinary GIS, model web, and spatial analysis & visualization.

Osman Hasan received the B.Eng. (Hons.) degree from the N-W.F.P University of Engineering and Technology, Pakistan, in 1997, and the M.Eng. and Ph.D. degrees from Concordia University, Montreal, Canada, in 2001 and 2008, respectively. He served as an ASIC design Engineer from 2001 to 2003 at LSI Logic Corporation in Ottawa, Canada and as a Research Associate at Concordia University, Montreal, Canada, for 18 months after his doctoral degree. Currently, he is an Assistant Professor at the School of Electrical Engineering and Computer Science, National University of Science and Technology (NUST), Islamabad, Pakistan. He is the founder and director of System Analysis and Verification (SAVe) Lab at NUST, which mainly focuses on the design and formal verification of energy, embedded and e-health related systems. He has received several awards and distinctions, including the Pakistan's Higher Education Commission's Best University Teacher (2010) and Best Young Researcher Award (2011) and the President's gold medal for the best teacher of the University from NUST in 2015. Dr. Hasan is a Senior member of IEEE, member of Association for Automated Reasoning (AAR) and member of the Pakistan Engineering Council.

Syed Rafay Hasan received the B.Eng degree from NED University of Engineering and Technology, Pakistan, the M.Eng and the Ph.D. degrees from Concordia University, Montreal, QC, Canada, all in electrical engineering. He was an adjunct faculty in Concordia University from 2006 to 2009. From 2009 to 2011, he was a research associate at École Polytechnique de Montréal. Since September 2011, he is an assistant professor in the Electrical and Computer Engineering Department of Tennessee Tech University, Cookeville, TN, USA. Dr. Hasan won the Post-Doctoral Fellowship award from Scholarship Regroupment Stratégigue en Microsystèmes du Québec in 2011, Sigma Xi Outstanding Research Award in 2012, Faculty Research Award from Tennessee Tech University in 2013 – 2014, and Kinslow outstanding research paper award from College of Engineering of Tennessee Tech University (2015). He has published more than 36 peer reviewed journal, and conference papers. His research interests include hardware design security, 3-D IC, and design reliability of VLSI circuits. Dr. Hasan is a member of IEEE, full member of Sigma Xi, and life member of Pakistan engineering council.

Mostafa Jafari was born in year 1956 in Tehran, Iran. He started in school in year 1962 and obtained Natural Diploma in 1974 (Tehran, after 12 years). He finished his first degree (B.Sc.) in Forest and Range, graduated in 1978 (Iran), and finished his Ph.D. in Plant Science (Ecology) in 1990 (UK). His Post doctorate research was in Plant Ecophysiology Methodology in 1997 (Japan). He is member of scientific board since 1990. His interest is in plant ecology, forestry, and climate change, and giving lectures in Universities on Ecophysiology, Ecology, Range Rehabilitation, Plant Geography. He is an advisor of several post graduate students in different universities. He enjoy from Membership to the different Professional Organizations. He was Director (President) of Research Institute of Forest and Rangelands, I.R. Iran, from 1992 to Nov. 1997 (With 600 staff including 250 scientific researcher and 900 research projects, publishing about 200 books in this period). He was Director of First Vice President Office of WMO, from August 2004 to 2009. He is Head of TP Secretariat of Low Forest Cover Countries (LFCCs, International Intergovernmental Organization). Since, March 2003, he is also International Affair Advisor to the Deputy Minister and Head of IRIMO since August 2004. He published 82 articles, 12 books, and 2 university textbooks. He has done and reported two national research projects on climate changes as Managing Director of National Research Project on Climate Change in RIFR with seven sub-projects entitled: "Investigation on Climate Change Effects on Forest Ecosystems in Hyrcanian forests with Emphasize of Wood Dendrology Studies."

Derya Karabulut received the B.S. degree in 2014 and the M.S. degree in 2016 all from the Department of Mechanical Engineering, Faculty of Engineering, Istanbul University, Turkey. She is currently a Ph.D. student under the supervision of Dr. Yunus Ziya Arslan at Istanbul University since September 2016. She serves currently as a Lecturer at the Department of Mechanical Engineering, Faculty of Engineering, Halic University. Her research areas include modeling of musculoskeletal systems, computational biomechanics, sport biomechanics, dental biomechanics, and robotics.

Ala' Khalifeh received the PhD degree in Electrical and Computer Engineering from the University of California, Irvine -USA in 2010. He is currently an Assistant Professor in the Communication Engineering department at the German Jordanian University and serving as the department head. Dr. Khalifeh is the recipient of the Fulbright scholarship (2005-2007) sponsored by the Bureau of Educational and Cultural Affairs of the United States Department of State. He is currently the IEEE Chair for the communication society in Jordan. His research is in communications technology, signal processing, multimedia, and wired and wireless networking with particular emphasis on E-health and smart grids applications. He has published numerous research papers at national and international journals, conference proceedings as well as chapters of books.

Sameer Kumar has a PhD in Social Networking and is presently working as Senior Lecturer at Asia-Europe Institute, University of Malaya. He has extensive prior work experience as IT Project Manager and also holds a PMP® (Project Management Professional) certification from PMI, USA. At AEI, Dr. Sameer has taught courses on History, Society and Culture of Southeast Asia, K-Economy and the Social and economic dimensions of Digital Economy at the Masters level. He has been a Program Co-ordinator for International masters in Information Management and International Masters in ASEAN Studies programs at AEI. Dr. Sameer is a Project leader of 2 grant-based research projects. He has several papers published in ISI indexed journals and supervises candidates both at Masters and PhD levels. His current research interests are in the field of Social Networks, Social Media and Academic networks.

Hélder Tiago da Silva Lopes was born in Felgueiras on December 7, 1993. Graduated in Geography and Planning in 2014 from the Institute of Social Sciences at the University of Minho. He is MsC in Geography - Planning and Territorial Management in the same year at the same institution. In the last years, he joined the Students Association of Geography and Planning (GeoPlanUM), having served as Vice President in 2014 and as President of the General Assembly since 2015. He is also the Class Delegate of the Masters class in Geography since 2015. My main research interests are GIS, Health and Planning, Planning and Management of the Tourism, Transportation planning, Regional and Local Development and climatology. In addition, I am a collaborator research scientist in Lab2PT (R&D unit of the U. Minho focused on the Social Sciences and Humanities).

Marek Macko, MSc Eng, PhD, DSc, is engineer and scientist, specialist in mechanics and automatics. He heads Department of Mechatronics in Institute of Mechanics and Applied Computer Science, Kazimierz Wielki University in Bydgoszcz, Poland. He was Vice-Dean of the Faculty of Math, Physics, and Technology (2012-2016), and now is the Vice-Rector of the Kazimierz Wielki University (2016-2020). He is an author and co-author of many scientific publications and inventions, well-known organizer and popularizer of science and novel technologies.

Awais Mahmood received his B.Eng. degree from National University of Science and Technology (NUST), Pakistan in 2014. His research interests include smart grids, power distribution systems and model checking.

M. K. Marichelvam is currently working as Assistant Professor in the department of Mechanical Engineering at Mepco Schlenk Engineering College, Sivakasi, Tamilnadu, India. He received his BE in Mechanical Engineering from Madurai Kamaraj University in 2000 and his ME in Industrial Engineering from Thiagarajar College of Engineering, Madurai, India in 2002. His area of interest is manufacturing scheduling, multi-objective optimization, heuristics and hybrid meta-heuristics. Currently, he is pursuing his PhD at Anna University, Chennai, Tamilnadu, India. He has published more than ten papers in referred journals.

Emilia Mikolajewska, MSc, PhD, PT, is physiotherapist, assistant professor in Department of Physiotherapy, Ludwik Rydygier Collegium Medicum of the Nicolaus Copernicus University in Bydgoszcz, Poland, senior assistant in Rehabilitation Clinic, Military Clinical Hospital No. 10 with Polyclinic in Bydgoszcz, Poland, and researcher in Neurocognitive Laboratory, Centre for Modern Interdisciplinary Technologies, Nicolaus Copernicus University in Torun, Poland. She is also member of the InteRDoCTor research team. Author of 12 books, 50 book chapters, 250 articles in the area of rehabilitation and physiotherapy, especially neurorehabilitation and neurological physiotherapy. Reviewer in 102 scientific journals, several book chapters, and grant applications. Member of 16 Editorial Boards. She also teaches courses on neurorehabilitation and assistive technology and acts as supervisor of specialization candidates. Emilia's research interests cover areas: neurorehabilitation of adult patients, including post-stroke, neurorehabilitation of children, increasing of patients' quality of life using: assistive devices, telemedicine, telerehabilitation, cybertherapy, rehabilitative robotics, applications of IT, AI and integrated systems in medicine, co-operation within therapeutic multidisciplinary team, evidence based medicine in clinical practice, biomedical engineering in rehabilitation. Emilia completed her PhD Thesis entitled: 'Application of NDT-Bobath method in rehabilitation of patients after ischemic stroke' in 2007 at the Medical University in Poznan under the supervision of prof. Wanda Stryla. She has also completed four-years postgraduate studies for governmental title 'specialist in physiotherapy', postgraduate studies in R&D projects management, and many additional courses in neurorehabilitation of children and adults: NDT-Bobath basic for children (EBTA recognized), Bobath-Baby for newborn and premature children (EBTA recognize), basic and advanced courses Bobath for adults (IBITA recognized), PNF course, Kinesiology Taping course, 7-module Cyriax course, geriatric rehabilitation, etc.

Dariusz Mikolajewski, PhD Eng., is engineer and scientist, specialist in IT, applications of IT in medical sciences, biocybernetics, and medical robotics. He works for three research centers: 1) Institute of Mechanics and Applied Computer Science, Kazimierz Wielki University in Bydgoszcz, 2) Department of Informatics, Nicolaus Copernicus University in Torun, 3) Neurocognitive Laboratory, Centre for Modern Interdisciplinary Technologies, Nicolaus Copernicus University in Torun. He is also member of the InteRDoCTor research team. Author of 125 articles and chapters in the area of medical IT, biocybernetics, and biomedical engineering. His research interests cover areas medical IT, Assistive Technology and intelligent environments of disabled people, including brain computer interfaces, biocybernetics and neurocybernetics, including computational models of brain stem, generation of consciousness and its disorders, and simulation of cognitive aspects of various diseases, biomedical engineering. He graduated from Military University of Technology in Warsaw and four postgraduate studies: in digital telecommunication, computer networks, project management (various methodologies), and management of R&D projects. His PhD Thesis covers topic of computational simulation of brain stem function (Thesis advisor: Prof. Wlodzislaw Duch).

Estefanía Aguilar Moreno works at GEOTEC as project assistant for EC lifelong learning programme projects. In 2013-2014 she worked at the EC – JRC focused on emerging ethical threats triggered by ICT, but her professional background was mainly related to Competitive Intelligence and Information Management. Graduated in Information Science and Master in Information Knowledge and Society by Universitat Oberta de Catalunya (Spain), broadly speaking her interests have to do with how ICT has impacted on Society at all levels, looking for confluence points, e.g. geospatial bridges, between humanities and sciences.

Faruk Ortes is a PhD candidate in biomechanics at the Istanbul University, Mechanical Engineering Department under the advisement of Dr. Yunus Ziya ARSLAN. He received the B.S. degree from Karadeniz Technical University, Department of Mechanical Engineering in 2010 and the M.S. degree from Marmara University, Department of Mechanical Engineering in 2014. He also serves currently as a Research Assistant at the Department of Mechanical Engineering, Faculty of Engineering, Istanbul University. His research work spans a variety of areas including musculoskeletal biomechanics,computational biomechanics, applied mechanics, finite element analysis, design and production of surgical instruments.

Giuliano Pelfer obtained PhD in Ancient Topography and Landscape Archaeology at the University of Bologna (2004). His research activity was on computational applications to archaeology. In his degree at University of Florence (1998) he analysed, through image processing techniques, the Etruscan and Roman roads in the Tarquinia territory. In PhD thesis he built an archaeological GIS for analysing the settlement evolution and the city origin in the territory of the protohistorical Tarquinia. From 2001 to 2003 he collaborated in research activities with Universitat Autonoma of Barcelona and Dublin Institute of Technology. He is researcher associated to Center for Study of Complex Dynamics at the University of Florence. First proposer of the ArchaeoGRID Project (2005 - 2006) for deploying complex archaeological applications on the Grid within EGEE, he participated to many Italian and international Conferences and he published scientific papers in conference proceedings and in archaeological journals.

Nihmiya Abdul Rahim is a postdoc fellow at the Department of Chemical Engineering, UAE University.

Nilanjan Ray is from Kolkata, India. He has obtained his PhD(Mkt); M.Com (Mkt); MBA (Mkt), STC FMRM (IIT-Kgp). He has 8 years teaching experience in BBA, MBA, BCom and 6 years Research experience and guided around 56 Post Graduate students' project. Dr. Ray has contributed over 57 research papers in reputed National and International Referred, Peer Reviewed Journals, Proceedings and 6 Edited Research Hand Books from Springer, IGI-Global USA and Apple Academic Publisher CRC Press A Taylor & Francis Group, USA . He has also associated himself as a reviewer of Tourism Management Elsevier, Journal of Service Marketing Emerald Group Publishing Limited, Journal of Business and Economics, Research Journal of Business and Management Accounting and as an Editorial Board Member of several referred Journals. He has also chaired in a technical session at IJAS Conference 2012, at Harvard University, Boston, USA. Dr. Ray is a life-member of the International Business Studies Academia, Fellow Member of Institute of Research Engineers and Doctors Universal Association of Arts and Management Professionals (UAAMP) New York, USA.

Paula Cristina Almeida Cadima Remoaldo is associate professor with habilitation of Human Geography in the Department of Geography of the Social Sciences Institute, University of Minho, Portugal. She got her PhD in Human Geography, in 1999, at the University of Minho with collaboration of Louvain-la-Neuve University (Belgium). She is at present Head of the PhD in Geography at University of Minho and Co-coordinator of the

Investigation Group "Space and Representations (SpaceR)" of Lab2pt (Laboratory of Landscape, Heritage and Territory) at University of Minho. She was Head of the Geography and Planning Research Unit, University of Minho (2000-2004; 2006-2007; 2008-2015). Co-ordinator of the research field called "Organizations, Development and Territory" of the Social Sciences Research Centre, University of Minho (2010-2014). Director of the Master Program in Geography (Planning and Management of Territory) of the University of Minho, Braga, Portugal (2013-2015). She has published in the last years in the leading journals of Tourism such as Annals of Tourism Research and Current Issues in Tourism. Her main research fields are Cultural Tourism, Mega Events; Urban Tourism and Regional and Local Development.

Vitor Ribeiro is a Geography professor with strong interest in Geographic Information Systems applications. He holds a PhD in Urban and Regional Planning and is conducting a postdoctoral research on New Geographic Information Technologies applications to enhance teaching and learning process. He is a Lecturer in Geography Education at Geography Department, Minho University and at Basic Department Education, Higher School of Education of Paula Frassinetti. He is dealing with GIS, since 2000, in several research fields, e.g. flood hazard analysis, water stress management, health planning, transportation, urban transport planning, accessibility and location analysis, tourism management, crime geography, web mapping and geographic education.

Gulcan Sarp is a Geographic Information System and Remote Sensing specialist with the research interests in active tectonics and geomorphology. She obtained a B.S. in geology from Ankara University in 1998, M.S. and Ph.D. in Geodesy and Geographical Information Technologies from Middle East Technical University in 2005 and 2012 respectively. She is currently working at Süleyman Demirel University as a Assist Prof. Dr.

Stefano Scanzio (S'06-M'12) received the Laurea and Ph.D. degrees in Computer Science from Politecnico di Torino, Torino, Italy, in 2004 and 2008, respectively. He was with the Department of Computer Engineering, Politecnico di Torino, from 2004 to 2009, where he was involved in research on speech recognition and, in particular, he has been active in classification methods and algorithms. Since 2009, he has been with the National Research Council of Italy (CNR), where he is a tenured Researcher with the Institute of Electronics, Computer and Telecommunication Engineering (IEIIT), Turin. Dr. Scanzio teaches several courses on Computer Science at Politecnico di Torino. He has authored and co-authored several papers in international journals and conferences in the area of industrial communication systems, real-time networks, wireless networks and clock synchronization protocols.

Sara Catarina Gomes da Silva is doctoral student in Human Geography in the Department of Geography, University of Minho, Guimarães. She is member of the Research Group Investigation Group "Space and Representations (SpaceR)" of Lab2pt (Laboratory of Landscape, Heritage and Territory) at University of Minho. Her main research field is Geography of Crime.

Roberta Spallone, Architect and Ph.D. in Drawing and Survey of Architectural Heritage, from 1999 was Assistant Professor, since 2005 is Associate Professor of Drawing at the Politecnico di Torino, Interuniversity Department of Regional and Urban Studies and Planning (DIST). Member of the School of Architecture, she teaches in the "Laboratory of Drawing and Survey" and "Digital techniques of representation". Member of the Board of the Doctorate of Architectural and Landscape Heritage, she is responsible of the Ph.D. course "Drawing, survey and modelling for representing architectural and urban heritage". Coordinator of the first year of the bachelor course in Architecture. Awarded with Silver plaque U.I.D. (Unione Italiana per il Disegno) 1997 1st prize for her PhD Thesis. She carries on researches, both theoretical and applied, in the fields of history and criticism of architectural drawing, digital techniques of representation and urban survey. She is author of a hundred publications including monographs and essays, and selected speaker at many International Congresses.

Zbigniew Szczepanski, MSc Eng, PhD candidate, is engineer and scientist, specialist in mechanics, including its applications in medical sciences. He works in Institute of Mechanics and Applied Computer Science, Kazimierz Wielki University in Bydgoszcz. He authored and co-authored many scientific publications.

Aydin Tabrizi has a PhD in architectural and civil engineering. His work focuses on energy management and sustainable construction with the consideration of building information modeling.

Rafael Vargas-Bernal was born in Irapuato, Guanajuato, México in September 30th, 1972. He received the B. E. degree in Electronics in October 6th, 1995 from the University of Guanajuato. After, the M. S. degree in Electronics from INAOE in September 4th, 1997. Lastly, the Sc. D. degree in Electronics from the same institute in July 19th, 2000. During October 2000 - September 2001, he worked as professor of the University of Guanajuato in the Department of Communications and Electronics Engineering. Since January 2002 to December 2009, he has been with the Department of Electronics Engineering of ITESI as Professor-Researcher. From August 2005 until today, he is also working in the Department of Materials Science in the area of Advanced Materials. He is technical reviewer of standards on microtechnology and MEMS in SEMI from November 2006 to the date. Moreover, he is reviewer of the magazine IEEE Latin America Transactions of IEEE from August 2003, Sensors and Actuators B: Chemical of Elsevier from February 2010, IEEE Sensors Journal from October 2012, Nanoscale of RSC November 2012, and RSC Advances published by RSC from April 2013 until today.

Wenbing Zhao received the Ph.D. degree in Electrical and Computer Engineering from the University of California, Santa Barbara, in 2002. Currently, he is an associate professor in the Department of Electrical and Computer Engineering at Cleveland State University. His current research interests include distributed systems, computer networks, fault tolerance and security, and smart and connected health. Dr. Zhao has more than 120 academic publications.

Wenkai Zhu obtained his BEng degree at South China University of Technology. He received his MPhil and PhD degrees at The University of Hong Kong, focusing in 3D Printing and Robotics, respectively. Now he is a Software Engineer in the United Kingdom. His research interests include Robotics, 3D Printing, CAD/CAM, and Embedded Systems.

Index

www.ingramcontent.com/pod-product-compliance
Lightning Source LLC
LaVergne TN
LVHW060255011026
849161LV00011B/2715